Lecture Notes on Multidisciplinary Industrial Engineering

Series editor

J. Paulo Davim, Department of Mechanical Engineering, University of Aveiro, Aveiro, Portugal

More information about this series at http://www.springer.com/series/15734

Sanjay S. Pande · Uday S. Dixit
Editors

Precision Product-Process Design and Optimization

Select Papers from AIMTDR 2016

Springer

Editors
Sanjay S. Pande
Department of Mechanical Engineering
Indian Institute of Technology Bombay
Mumbai, Maharashtra
India

Uday S. Dixit
Department of Mechanical Engineering
Indian Institute of Technology Guwahati
Guwahati, Assam
India

ISSN 2522-5022 ISSN 2522-5030 (electronic)
Lecture Notes on Multidisciplinary Industrial Engineering
ISBN 978-981-10-8766-0 ISBN 978-981-10-8767-7 (eBook)
https://doi.org/10.1007/978-981-10-8767-7

Library of Congress Control Number: 2018934457

Printed on acid-free paper

This Springer imprint is published by the registered company Springer Nature Singapore Pte Ltd.
part of Springer Nature
The registered company address is: 152 Beach Road, #21-01/04 Gateway East, Singapore 189721, Singapore

Preface

Product development activities in the twenty-first century are characterized by precision and miniaturization of parts and assemblies. Researchers are attempting to design products which are precise, robust and reliable. This, in effect, entails critical design, analysis and optimization of the constituent parts, assemblies and processes. Integrated efforts on these fronts are being taken worldwide by researchers from academia and industry.

The focus of this book is on providing exposure to the latest research in the area of Precision Product-Process Design and Optimization. The objective is to introduce various tools and techniques for the design of precision, miniature products, assemblies and the associated manufacturing processes. In particular, precision mechanisms, robotic devices and their mechatronic control strategies are presented with case studies. On the process front, process modeling and optimization of machining, forming and 3-D printing manufacturing processes are presented. Most of these processes use non-conventional sources of energy, e.g., laser beam, ion beam and plasma. The physics-based modeling of these processes is carried out using tools like finite element method (FEM), and the process parameter optimization is carried out using optimization techniques such as evolutionary genetic algorithm (GA).

This book comprises 16 chapters. The first three focus on Precision Product Design and Control, while the rest are related to Process Modeling and Optimization. These chapters are the extended versions of peer-reviewed manuscripts presented at the 6th International and 27th All Indian Manufacturing Technology, Design and Research (AIMTDR) Conference held during December 16–18, 2016, at College of Engineering, Pune, India.

Shastri et al. reported the conceptual design and optimization of a parallel kinematic Stewart–Gough platform for multi-directional (6-axis) 3-D printing FDM process. Mishra et al. presented a comprehensive work of developing a robot-based flexible assembly system with machine vision guidance and dexterous multi-finger gripper. Various design aspects of the system such as assembly sequence planning, optimization using genetic algorithm, expert systems, control algorithm and testing on an industrial robot have been reported. Jain et al. reported the design of a novel miniature wireless mobile micromanipulation system (WMMS) for robotic assembly in which a three piezoelectric actuators fingers based compact gripper is developed for handling small objects. Detailed kinematic analysis of the system was carried out, and a prototype was developed.

Joshi and Bollar carried out a 3-D thermo-mechanical FEM numerical modeling of low-rigidity thin-wall milling process for an aerospace grade aluminum alloy. The developed model was validated with experimental results, suggesting guidelines for improving machining accuracy. Mishra and Kumar modeled laser-based (SLM) manufacturing process and used it to study the additive manufacturing of AZ91D magnesium alloy. Vinodh and Shinde carried out experimental investigations on FDM additive manufacturing process and suggested optimum process conditions using a multi-criteria decision-making technique. Kuppuswamy and Airey developed an intelligent polycrystalline diamond (PCD) tool testing system that establishes the failure characteristics of the PCD cutting inserts as well as predicts the product performance. The testing system comprises a feature extraction engine with dashboard and a predictive tool-life model using neural networks. Garg et al. investigated focused ion beam process for the fabrication of micro/nano 3-D features in optical components and presented algorithm and optimization strategy for improving the process accuracy.

Rathod et al. carried out an extensive experimental investigation on electrochemical micromachining process and reported guidelines for fabricating microgrooves of different cross sections. A brief section on mathematical analysis of the process is also included in their chapter. Tripathi and Tripathi reported a detailed experimental investigation on graphite powder mixed electrical discharge machining process and its optimization following a design of experiments (DoE) procedure. Mohapatra et al. investigated the single pitch error and material removal rate in a WEDM gear cutting process and suggested optimum process conditions for improving the accuracy of precision miniature gears. Dev et al. presented the development of a low-power medium-pressure cold plasma atomistic finishing process to achieve isotropic polishing of complex 3-D surfaces. A finite element-based COMSOL® software package was used to assist the modeling of the process.

Sharma et al. investigated ultrasonic-assisted turning process from surface integrity considerations using conventional and textured tools. The focus is on the surface roughness and residual stresses produced on parts. A FEM model for predicting the residual stresses is also reported. Kibria et al. presented an experimental study on pulsed Nd:YAG laser-based surface texturing of pure titanium material. The response surface modeling and multi-objective optimization have been carried out for achieving minimum values of surface roughness and contact angle. Nirala and Agrawal reported extensive numerical and experimental investigations to predict sheet thinning in a single-stage incremental sheet-forming process for various product shapes. Numerical results correlate well with the experiments. Ashraf et al. presented the optimization of a reconfigurable manufacturing system. They carried out optimum machine allocation in a serial production line using NSGA-II and TOPSIS techniques.

The chapters in this book have attempted to present state-of- the-art research for Precision Product Design and associated manufacturing technology. They clearly demonstrate the application of scientific principles, use of mathematical tools and the growing importance of optimization for the design of critical products. The chapters provide directions for future research. We expect that this book will be useful for postgraduate students, researchers and industry professionals and will, in general, be welcomed by the manufacturing community. Any feedback on the book is welcomed.

Mumbai, India Sanjay S. Pande
Guwahati, India Uday S. Dixit

About the AIMTDR Conference

AIMTDR Conference is a highly prestigious, biennial event organized in the field of mechanical and production engineering in India. The conference has a glorious history of organization since its inception. The first conference entitled 'All India Machine Tool Design and Research Conference' was held at Jadavpur University, Kolkata, in 1967. In early 1990s, it was thought appropriate to widen the scope of the conference to encompass areas related to different manufacturing process technologies and systems. Accordingly, it was renamed as 'All India Manufacturing Technology, Design and Research' Conference and the 16th in the series was organized at Central Machine Tool Institute, Bangalore, in 1994. It became an international event with the first international conference being held at Indian Institute of Technology Roorkee in 2006. The international conference aimed to bring together academicians, researchers and industry professionals working worldwide in the field of manufacturing to exchange and disseminate ideas.

The subsequent international AIMTDR Conferences were held at IIT Madras and Andhra University, Visakhapatnam, respectively. Jadavpur University hosted silver jubilee of the conference and organized 4th International and 25th AIMTDR Conference at Kolkata. The fifth international conference was organized at IIT Guwahati in December 2014, while the 6th one (AIMTDR 2016) was organized by the Department of Production Engineering and Industrial Management, College of Engineering, Pune (COEP), during December 16–18, 2016.

Mission, Vision, Challenges and Direction of AIMTDR Conference

(Excerpt from the address of Prof. Amitabha Ghosh, the Chief Guest of 26th AIMTDR Conference, held at IIT Guwahati during December 12–14, 2014)

Personally, I have a close association with AIMTDR Conference from the very first one for which I happened to be a humble and young member of the organizing team. That event was organized jointly by Late Prof. Amitabha Bhattacharyya and Prof. A. K. De at Jadavpur University in the year 1967. Keeping in view the growing industrialization of India, the AIMTDR Conference was planned along the line of MTDR Conference that used to be organized by Profs. Tobias and Koenigsberger at Cambridge and Birmingham alternately.

India being an emerging economy, the importance of 'manufacturing' was well recognized and one of the primary goals of AIMTDR Conference has been to bring the academicians, researchers and the engineers from the industry to a common platform for exchanging ideas and developing a deeper mutual understanding among all concerned. The organizers of the AIMTDR Conference in the past were eminently successful in this regard. With time, this event has gained maturity and has emerged as one of the most important international and national conferences held in India for all who are associated with the field of manufacturing.

Since the economic development of any country is very critically linked to the manufacturing sector, it is only very natural that the current political leadership of India has taken up 'Manufacturing in India' as one of its key objectives. In fact, 'manufacturing' should be a common objective for all the South Asian and the Southeast Asian countries to develop a good mutual understanding and cooperation to enhance the overall manufacturing capabilities of this region. Then only, this region of the world can become a powerhouse for economic growth and play a center-stage role in the world economy. This is essential if we have to eliminate the poverty that has plagued this region for a very long time.

From my long association with the evolution of AIMTDR Conference and my involvement in teaching and R&D in the field of manufacturing for almost half a century, placing a few observations before this august gathering may not be out of place.

Traditionally, the 'primary' manufacturing processes have remained in the domain of mechanical engineering and metallurgy. At the same time, the secondary and finishing processes along with the machine tools and systems involved in manufacturing have remained exclusively as part and parcel of mechanical engineering. This, in my personal opinion, has not always helped the manufacturing activities in India to take advantage of the progress made in physical and applied sciences. Barring some isolated cases, this has rendered the manufacturing activities in India to be largely devoid of major fundamental innovations. As a result, the manufacturing activities in India have remained mostly confined to the traditional lines without giving much attention to 'value addition.' Thus, for example, our earnings from the export of a couple of hundred 'made in India' cars to Europe can be offset by that through the sale of a single focused ion beam machine, measuring 1.5 m × 1 m × 1 m, by USA to India. This scenario must be changed in the coming days.

Over the years, advanced manufacturing has gradually developed into a multidisciplinary activity and real 'value addition' through 'manufacturing' can be achieved only when the advancement in physical, chemical and other sciences is used in innovating newer processes and possibilities of 'manufacturing.' Establishing a close link between the manufacturing and the related sciences has to be recognized as a necessary task. I believe that an event like AIMTDR can play a very significant role in this regard. Besides, revamping of the old-fashioned curricula for training engineers in manufacturing is essential to render them capable of facing the challenges from futuristic manufacturing; AIMTDR Conference can take a leading role in that direction by providing a separate session to discuss the issues involved in the matter. In the coming years, I am very hopeful that AIMTDR Conference will attract not only manufacturing engineers but also researchers from basic and applied sciences whose works are closely related to and important for innovations in manufacturing.

From my half-a-century-old teaching experience, I find that the current young generation is gradually becoming somewhat disinclined to take up careers in R&D related to manufacturing. This is particularly so for the brighter section of the student community; there is a feeling among them that there is not much intellectual challenge in the subjects related to manufacturing. There can be nothing farther from the truth. Perhaps, to a large extent, this is so as the curricula and syllabi have remained archaic in many universities and institutions. In fact, application of many advances in scientific principles to manufacturing is the key requirement to open the gate for the impending next Industrial Revolution. ‘Manufacturing’ needs the brilliant young minds to take up R&D careers in academia and industry with equal eagerness and enthusiasm. However, this can be possible only on receiving adequate and aggressive support from the industry houses—both financially and administratively.

Although the world has reaped the benefit of the 2nd Industrial Revolution which was triggered by the R&D in the Silicon Valley, California, except for some software-related activities, India (and many other countries in this region) had really nothing to do with the actual developments and related manufacturing. Remaining a good follower cannot take India to any leadership position though it may provide some financial relief. India and the countries in this region cannot afford to miss the opportunity to take up important position when the next Industrial Revolution comes.

Dear colleagues and friends, the silicon-based 2nd Industrial Revolution has reached a plateau and there are indications that ‘carbon’ may play a more important role and carbon-based devices will play the key role in ushering the world into the reign of the 3rd Industrial Revolution. Already enough indications are coming in that direction. If that be the case, India should play a major role and initiate well-planned pioneering activities so that manufacturing engineering becomes a multidisciplinary area involving relevant basic science and engineering subjects for the emergence and growth of ‘carbon-based technology’ in this region of the world. I am happy to notice that some of the leading world authorities on carbon devices and advanced fabrication are present in this conference. This event can be a great opportunity for the manufacturing community to take advantage of their presence and plan an appropriate course of action, to initiate planned activities, and to innovate carbon devices.

Many areas of manufacturing in the not-too-distant future will be very different from what we recognize as ‘manufacturing’ today; ‘self-assembly,’ ‘self-regulation,’ ‘self-correction’ and ‘self-replication’ will become the keywords in futuristic manufacturing. Obviously, it will be too drastic to think of redirecting all the R&D on manufacturing in this direction, but, at the same time, India should be well prepared to take active role in such areas of futuristic manufacturing (a name for that was coined a few years ago in a workshop at IIT Kanpur—‘Fabrionics’) as that will help the country to gain expertise for incorporating significant ‘value addition’ in our manufacturing activities.

AIMTDR Conference is one of the very few events that draw researchers and practitioners from the academia and the industry with equal enthusiasm. I am very hopeful that using this grand platform, we all can take India and the neighboring countries along the path of growing technological excellence and engineering marvel.

AIMTDR 2016 Conference: Objectives and Organization

Manufacturing has revolutionized itself from its contemporary form to its current digital access, more so in the era of Industry 4.0. With every Industrial Revolution, we have seen labor and asset productivity multiply and structural shifts emerge in the manufacturing world order. Several core technologies are driving Industry 4.0, be it Simulation, Autonomous Robots, Big Data and Analytics, Augmented Reality or Additive Manufacturing. If our economy needs to grow multifold to achieve sustainable development, the existing 15% share of manufacturing sector in India's GDP needs to scale to 25% in the immediate near future.

Manufacturing today seeks innovation to be ubiquitous by inventing ways to produce more with less inputs. In an era of integration where technologies complement one another, design and manufacturing face a daunting task with regard to the quality and cost-effectiveness of products. Concentrated efforts focusing on quality research need to be endorsed for improving the manufacturing processes, technology and systems to adopt world-class manufacturing technologies. The manufacturing education should also emphasize its importance to attract the talented young to this area and equip them with skills that embrace knowledge, information and techniques.

With this broad focus, College of Engineering, Pune (COEP), presented as a sequel to the AIMTDR Conferences of the past, the 6th International and 27th AIMTDR Conference in December 2016. The theme of the conference was 'Recontouring Manufacturing.' Several invited lectures and keynote addresses on cutting-edge technologies were presented in the conference by leading researchers from USA, Singapore and India. In all, 380 papers were selected for oral presentations and 80 for poster presentations. The organizing team brought out the proceedings on a CD covering all the papers presented. These papers will help to provide insights into the realistic exposure of current research and development trends in the field of manufacturing technology, design and research. During the exhibition in the conference, 40 leading companies participated by displaying new technology equipment, products and measuring equipment. The manufacturing community in India received this conference with appreciation and applause.

Through the deliberations in the conference, I hope to see the culmination of great thoughts and ideas that would introduce to develop technological solutions in the domain of manufacturing and design by the fusion of technologies straddling physical, digital and biological worlds.

B. B. Ahuja
Organizing Secretary, AIMTDR 2016
Professor of Production Engineering and Director
College of Engineering, Pune

Editorial Acknowledgements

We sincerely thank all the authors for contributing their research work as extended chapters for this book. Special thanks are due to the organizers and members of the National Advisory Committee of AIMTDR 2016 for their guidance and support. We are indeed grateful to Prof. Amitabha Ghosh, former Director of IIT Kharagpur, for his guidance and timely suggestions. Prof. J. Paulo Davim of the University of Aveiro, the Series Editor of this book, deserves a special mention for his encouragement and guidance. Finally, we sincerely acknowledge the dedicated support provided by the editorial staff of Springer Ms. Swati Meherishi, Ms. Aparajita Singh, Dr. Mayra Castro and Ms. Bhavana while bringing out this book.

Contents

Optimal Design of a Stewart–Gough Platform for Multidirectional 3-D Printing 1
Shricharana Shastry, Ritwik Avaneesh, K. A. Desai and S. V. Shah

Development of a Flexible Assembly System Using Industrial Robot with Machine Vision Guidance and Dexterous Multi-finger Gripper 31
Atul Mishra, I. A. Sainul, Sudipta Bhuyan, Sankha Deb, Debashis Sen and A. K. Deb

A Piezoelectric Actuator Based Compact Micro-manipulation System for Robotic Assembly 73
R. K. Jain, Surajit Saha and Bhaskar Ghosh

Numerical Modeling and Experimental Validation of Machining of Low-Rigidity Thin-Wall Parts 99
Gururaj Bolar and Shrikrishna N. Joshi

Modelling of SLM Additive Manufacturing for Magnesium Alloy 123
Ashish Kumar Mishra and Arvind Kumar

Parametric Optimization of 3D Printing Process Using MCDM Method 141
S. Vinodh and Priyanka Shinde

Intelligent PCD Tool Testing and Prediction of Performance 161
Ramesh Kuppuswamy and Kerry Anne Airey

Focused Ion Beam Fabrication: Process Development and Optimization Strategy for Optical Applications 189
Vivek Garg, Rakesh G. Mote and Jing Fu

Machining Guidelines for Fabricating Microgrooves of Varied Cross Sections by Electrochemical Micromachining 211
V. Rathod, B. Doloi and B. Bhattacharyya

Optimization of Process Parameters in Powder-Mixed EDM 239
S. Tripathy and D. K. Tripathy

Optimization of Single Pitch Error and MRR in a WEDM Gear Cutting Process 285
Kasinath Das Mohapatra and Susanta Kumar Sahoo

Novel Finishing Process Development for Precision Complex-Shaped Hemispherical Shell by Bulk Plasma Processing 313
D. Sam Dayala Dev, Enni Krishna and Manas Das

Ultrasonic Assisted Turning: A Comparative Study of Surface Integrity 337
Varun Sharma, Pulak M. Pandey, Uday S. Dixit, Anish Roy and Vadim V. Silberschmidt

Pulsed Nd:YAG Laser Surface Texturing of Pure Titanium Material 361
G. Kibria, A. Sen, H. M. Tariq Aziz, B. Doloi and B. Bhattacharyya

Sheet Thinning Prediction and Calculation in Incremental Sheet Forming 391
Harish K. Nirala and Anupam Agrawal

Optimum Machines Allocation in a Serial Production Line Using NSGA-II and TOPSIS 411
Masood Ashraf, Faisal Hasan and Qasim Murtaza

About the Editors

Dr. Sanjay S. Pande is a Professor of Mechanical Engineering at the Indian Institute of Technology Bombay, Mumbai, India. He was a Visiting Professor at the University of Cincinnati, USA. His research interests include Computer-Aided Design and Computer-Aided Manufacturing (CAD/CAM) with a focus on Manufacturing Process Modeling, Simulation and Optimization, Algorithms for Additive Manufacturing, Computer Numerical Control (CNC) Machining and Computational Metrology, Artificial Intelligence Techniques, and Internet-based Collaborative Product Development. He has published 140 research papers in international journals and has authored a book titled 'Computer Graphics and Product Modelling for CAD/CAM' published by Alpha Science (UK)/Narosa (India). He is the Associate Editor of the International Journal of Precision Technology, Inderscience, UK.

Dr. Uday S. Dixit received his BE in Mechanical Engineering from the University of Roorkee (now the Indian Institute of Technology Roorkee) in 1987, his MTech in Mechanical Engineering from the Indian Institute of Technology Kanpur (IIT Kanpur) in 1993, and his PhD in Mechanical Engineering from the IIT Kanpur in 1998. He has worked in two industries, HMT, Pinjore; and Indomag Steel Technology, New Delhi, where his main responsibility was designing various machines. He joined the Department of Mechanical Engineering, Indian Institute of Technology Guwahati, where he is currently a Professor, in 1998. He was also the Officiating Director of the Central Institute of Technology, Kokrajhar, from February 2014 to May 2015. He has been actively engaged in research in various areas of design and manufacturing for the last 25 years. He has authored/co-authored 75 journal papers, 91 conference papers, 20 chapters, and 6 books on mechanical engineering. He has also co-edited 3 books on manufacturing. Presently, he is an Associate Editor of the Journal of The Institution of Engineers (India), Series C, and Vice-President of the AIMTDR Conference.

Optimal Design of a Stewart–Gough Platform for Multidirectional 3-D Printing

Shricharana Shastry, Ritwik Avaneesh, K. A. Desai and S. V. Shah

Abstract The existing 3-D printing techniques have several disadvantages such as aliasing and difficulty in building around inserts due to limited motions associated with the equipment. The limitation of build direction results in poor surface finish due to aliasing (or layer stair-stepping) and adverse material properties in certain directions which limits use of 3-D printing for many industrial applications. The present study investigates the application of Parallel Kinematic Machines (PKMs) in achieving multidirectional 3-D printing. The proposed architecture addresses some of the limitations of existing Fused Deposition Modelling (FDM)-based 3-D printer by allowing six-axis motions between extruder and platform while building the component. The study explores the application of Stewart–Gough Platform (SGP) further for 3-D printing and illustrates its capability as a viable solution for multi-axis FDM. The design of SGP for multidirectional FDM is realized for optimal dexterity using bulk dexterity index. The study discusses details of the optimization formulation and consequent results associated with the same. A conceptual design of the SGP is subsequently proposed based on the results of the optimization. The proposed SGP-based machine architecture is expected to offer advantages such as improved surface finish and control of directional properties, which signifies push towards freeform fabrication using multidirectional 3-D printing.

Keywords Parallel kinematic machines (PKMs) · Stewart–Gough platform (SGP) 3-D printing · Freeform fabrication

---

S. Shastry · R. Avaneesh · K. A. Desai (✉) · S. V. Shah
Department of Mechanical Engineering, Indian Institute of Technology Jodhpur, Jodhpur 342037, Rajasthan, India
e-mail: kadesai@iitj.ac.in

S. Shastry
e-mail: charanshastry@gmail.com

R. Avaneesh
e-mail: avaneeshritwik@gmail.com

S. V. Shah
e-mail: surilshah@iitj.ac.in

S. S. Pande and U. S. Dixit (eds.), *Precision Product-Process Design and Optimization*, Lecture Notes on Multidisciplinary Industrial Engineering,
https://doi.org/10.1007/978-981-10-8767-7_1

Nomenclature

Acronyms

PKM	Parallel Kinematic Machine
FDM	Fused Deposition Modelling
SGP	Stewart–Gough Platform
AM	Additive Manufacturing
CAD	Computer-Aided Design
W/F Ratio	Workspace-to-Footprint Ratio
DOF	Degrees of Freedom
GD	Group Decoupling
SA-PM	Selectively Actuated Parallel Machine
RPY	Roll–Pitch–Yaw
KPI	Kinetostatic Performance Index
GCI	Global Conditioning Index
GMI	Global Manipulability Index
ME	Manipulability Ellipsoid
SVD	Singular Value Decomposition
SRSGP	Semi-Regular Stewart–Gough Platform
GA	Genetic Algorithm

Symbols

$v_{a=x,y,z}$	Translational velocity of end effector of robot with subscript indicating axis
$\omega_{a=x,y,z}$	Rotational velocity of end effector of robot with subscript indicating axis
θ_n	Displacement of actuator with subscript indicating actuator number
$\{B\}$	Base frame of reference of Stewart–Gough platform
$\{P\}$	Platform frame of reference of Stewart–Gough platform
O_B	Origin of base frame
O_P	Origin of platform frame
B_i	Actuator connecting points with subscript i indicating connection point on base
P_i	Actuator connecting points with subscript i indicating connection point on platform
$\mathbf{b}_i$	Vector from centre of base to actuator connecting point i
$\mathbf{p}_i$	Vector from centre of platform to actuator connecting point i
${}^B\mathbf{t}$	Vector indicating tool point represented in frame B
${}^B\mathbf{R}_P$	Standard rotation matrix for rotating vector in frame B to frame P
A	Roll angle of platform
B	Pitch angle of platform
Γ	Yaw angle of platform
$c\alpha$	Cosine of angle α
$s\alpha$	Sine of angle α

l_i	Length of strut/leg i
${}^{B}\mathbf{\Omega}$	Angular velocity matrix in frame B
${}^{B}\mathbf{s}_i$	Vector along leg i represented in frame B
$\mathbf{J}$	Jacobian matrix transforming tool point velocities to actuator velocities
$\mathbf{J}^{-1}$	Inverse of Jacobian matrix transforming actuator velocities to tool point velocities
Q	Vector representing current pose of platform
$\dot{q}$	Velocity vector of tool points
$\dot{l}$	Velocity vector of actuators
$\sigma_i(\mathbf{J})$	ith singular values of Jacobian matrix
$\lambda_i(\mathbf{J}\mathbf{J}^{\mathrm{T}})$	ith eigenvalue of $\mathbf{J}\mathbf{J}^{\mathrm{T}}$
$\mu(\mathbf{J})$	Manipulability measure of Stewart–Gough platform
$\kappa(\mathbf{J})$	Condition number of Jacobian
w	Workspace of robot
$\mathrm{d}w$	Infinitesimal volume element of workspace
r_B	Radius of the base circle
r_P	Radius of the platform circle
ϕ_B	Spacing angle between a set of base passive joints B_1–B_2, B_3–B_4, B_5–B_6
ϕ_P	Spacing angle between a set of base passive joints P_1–P_2, P_3–P_4, P_5–P_6
h	Distance between the origins of the base and platform frames when the platform is at neutral position
$\mathbf{x}$	Design space vector
d_{b1}	Distance between two passive joints in base
d_{p1}	Distance between two passive joints in platform
n_{CW}	Number of divisions in Cartesian workspace
n_{EW}	Number of divisions in Euclidean workspace

1 Introduction

Three-Dimensional (3-D) printing brought innovative ways to design, manufacture and distribute products to the end users in majority of the industries such as automotive, aerospace, engineering, medicine, biological systems and food supply chains. This technology has gained significant academic as well as industrial interest due to its ability to generate complex geometries with customizable material properties (Gao et al. 2015). 3-D printing can fabricate components directly from Computer-Aided Design (CAD) models by selective accumulation of materials. During preprocessing stage, the 3-D model of the component is sliced into a series of two-Dimensional (2-D) layers. The printing equipment builds a physical part by stacking these 2-D layers together to approximate the given CAD model. This approach of fabricating the components has several advantages such as simplified

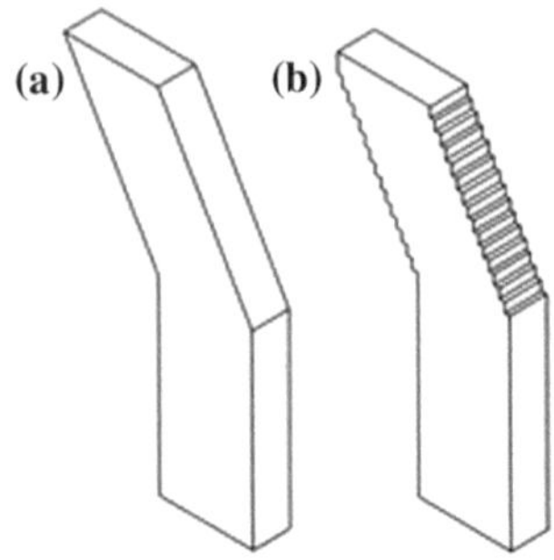

Fig. 1 Stair-casing effect on build part: **a** CAD model of actual part; **b** component fabricated using 3-D printing

tool path planning and hardware design as well as the capability to manufacture complex shapes with relative ease. However, the process suffers from a few drawbacks, which limit its applications compared to other manufacturing processes. Some of the major drawbacks of the process are as follows:

(1) *Stair-casing (aliasing) effect*: Surfaces with non-parallel orientation to the build plate exhibit jagged edges resulting in the reduced accuracy and surface finish of the built components. This is primarily due to approximation of 3-D geometry with layers of uniform thickness (Oropallo and Piegl 2016). The stair-casing effect is demonstrated in Fig. 1.
(2) *Non-uniform material properties*: The build orientation also affects properties of the component fabricated using 3-D printing. The decision related to optimal part orientation for uniform material properties is challenging and still an unsolved process-planning problem (Oropallo and Piegl 2016). As the component is built along a single axis, the properties normal to the build plate are controllable, which may be undesirable from a functional viewpoint. It has been highlighted by many researchers that the biggest challenge for implementation of 3-D printing stems from the fact that the properties of built component can be controlled in single direction only, i.e. build direction. This limitation reduces the effectiveness of the process for components requiring control of material properties in multiple directions.
(3) *Inability of building around inserts*: The tool motions are limited in most of commercial 3-D printing systems that do not allow building of material around inserts, e.g. embedded electronic or optical systems. The use of inserts presents a challenge to the machine that cannot build a component around the inserts (Oropallo and Piegl 2016).
(4) *Support structure requirements*: The use of support structures is mandatory for components with bulky over-hanging parts requiring building along oblique direction. These increases build time and cost for the component with necessity of additional post-process removal operations.

The detailed investigations are carried out by various researchers to address these drawbacks and number of solutions are suggested in the literature. Some of the solutions suggested in the literature to improve surface finish are controlled cure depth (Sager 2006; Sager and Rosen 2008), post-processing techniques (Pandey et al. 2003; Mason 2006) and meniscus methods (Narahara and Saito 1995; Pan et al. 2012). In order to enhance fabrication capability of the process and address building around inserts, techniques such as model shape modification (Kataria and Rosen 2001) and hybrid process development (Ruan et al. 2005) have also been employed. Song et al. (2015) observed that these approaches suggested in the literature improve one or at most a few drawbacks only and a generic solution is not attempted for these issues. It has been highlighted that the primary reason for above issues with the process is the use of single build direction and uniform layer thickness in fabrication of components.

Zhang et al. (2015) enhanced the capability of a 3-D printer by building at oblique angles and showed that the parts can be fabricated along curved surfaces. It was also highlighted that the component can be built according to physical property requirements such as higher mechanical strength, specific thermal, electrical and biomedical characteristics using proposed solution. This design innovation in the form of multi-axis motion capability of a 3-D printer represented a big step towards near-net shape freeform fabrication using 3-D printing. The need for multi-axis motion necessitated investigation of various robotic architectures to meet the requirements. Primarily, two types of robotic architectures are used in various manufacturing applications to achieve multi-axis motions: Serial architectures and parallel architectures. In order to determine suitability of the robotic architecture for 3-D printing application, a comparison is made on the basis of following four different criteria:

(1) *Workspace-to-Footprint (W/F) Ratio*: The workspace of a robot refers to the total reachable volume of its end effector in the space. The nominal length of its base determines footprint of the architecture. It has been referred in literature that higher workspace-to-footprint ratio of a robotic configuration ensures compactness of the design and enables portability. It is important to have a compact design for 3-D printing application; therefore, the configuration with higher *W*/*F* ratio is preferable.
(2) *Nature of Workspace*: The workspace of a robotic manipulator is the total volume swept by end effector when all possible motions are executed. A simple and regular workspace is preferred for 3-D printing application as it can be segmented into independent Cartesian and orientation spaces. This also simplifies the problem of optimizing the workspace, planning of tool path and control of motions.
(3) *Accuracy and Error Propagation*: The closeness of a manipulator to a given point within the workspace is called accuracy of robot. The presence of open-loop structures in the configuration results in error in an individual joint propagating to other joints. This leads to magnification of error in the workspace necessitating modelling of additional feedback and error compensation system for accuracy improvement. On the other hand, closed-loop structures allow motions with high accuracy and repeatability with simplified error compensation techniques.

(4) *Rigidity*: The rigidity or stiffness of a robot plays crucial role in its control as it conditions its bandwidth. Although 3-D printing does not generate higher joint forces during the process, rigidity of individual joint plays a vital role in maintaining the position of an extruder. The rigidity, and hence, bandwidth, of serial manipulators is low, only reaching a few Hz, whereas it is much better for parallel structures.

The multiple Degrees of Freedom (DOF) to the accumulative tool with respect to build platform can be provided using either serial kinematic chain structure (Cajal et al. 2013) or PKMs (Huang et al. 2005; Bi and Jin 2011). Industrial serial robots such as ABB or KUKA are used for variety of applications in different fields of engineering, e.g. building of parts for architectural artwork (Gausemeier et al. 2011). There are several problems associated with the use of serial robots for different manufacturing applications. Serial robots typically have lower rigidity, higher mass and suffer from poor accuracy in open-loop control. The error in single joint propagates into subsequent joints resulting in large position/orientation error for the end effector in the workspace. Although reinforced joints and servo control have been used for industrial applications to eliminate these issues, it is expensive for use in small-scale industries and retail applications. The other robotic configuration in the form of Parallel Kinematic Machines (PKMs) is extremely modular, reconfigurable and precise structure. It has been demonstrated in past research that PKMs offer higher stiffness, lower moving mass, higher acceleration and accuracy, reduced installation requirements, and mechanical simplicity relative to conventional machine tools for various manufacturing applications (Warnecke et al. 1998; Boer et al. 1999; Rehsteiner et al. 1999). The other potential advantages of PKMs are higher dexterity, smaller footprint and multi-mode manufacturing capability. It has been reported in the literature that PKM can be about seven times precise, five times rigid and four times faster than conventional machine tools for certain applications (Zhang 2010). This motivates the present study of investigating alternate machine architectures in the form of PKMs and in particular Stewart–Gough Platform (SGP) for FDM-based 3-D printers.

The rest of the chapter is organized as follows. Section 2 presents investigation of 6-DOF parallel and hybrid architectures developed in the previous literature. It also highlights the strength of SGP in comparison to other PKM architectures and summarizes its applications in various domains. Section 3 presents the kinematic model of the SGP and details the requirements for its use in AM. Section 4 presents an optimal design methodology used in the present study for deriving dimensions of SGP. Section 5 discusses the prototype machine configuration and compares results of the present study with conventional techniques. The chapter ends with summary of findings and further directions of the work in Sect. 6.

2 Parallel and Hybrid 6-DOF Machine Architectures

This section presents discussion about various 6-DOF architectures with potential application in multi-axis 3-D printing. The concept of group decoupling has been used in this study to appreciate the capabilities of various robotic architectures for 3-D printing application. The concept of group decoupling was proposed by Jin et al. (2009) and the same has been presented subsequently. For a 6-DOF machine, $\left(v_x v_y v_z\right)$ represent the output translational velocities of the end effector in Cartesian coordinates, and $\left(\omega_x \omega_y \omega_z\right)$ represent the output rotary velocities of the end effector about the axes (xyz), respectively, and $(\theta_1, \theta_2, \theta_3, \theta_4, \theta_5, \theta_6)$ are the six input displacements of actuators. The classification can be defined as follows:

Complete coupling: Complete coupling means that each of the output motions has contribution from all input axes. The kinematic relationships between input and output variables can be written as

$$\left(v_x v_y v_z \omega_x \omega_y \omega_z\right) = f(\theta_1, \theta_2, \theta_3, \theta_4, \theta_5, \theta_6) \tag{1}$$

Group decoupling (GD): Group decoupling means that the 6-DOF output motion of the end effector can be divided into two or more motion groups, and the different motion groups are controlled by different actuators following a certain order. Taking a 3–3 GD with partially decoupled translation and rotation, for example, the kinematic relationships between outputs and inputs can be written as

$$\begin{aligned} \left(\omega_x \omega_y \omega_z\right) &= f(\theta_4, \theta_5, \theta_6) \\ \left(v_x v_y v_z\right) &= f(\theta_1, \theta_2, \theta_3) \end{aligned} \tag{2}$$

Complete decoupling: Complete decoupling means that each independent output motion is controlled by only one input axis. The kinematic relationships of 6-DOF robots with complete decoupling between outputs and inputs can be written as

$$\begin{aligned} &v_x = f_1(\theta_1), v_y = f_2(\theta_2), v_z = f_3(\theta_3), \\ &\omega_x = f_4(\theta_4), \omega_y = f_5(\theta_5), \omega_z = f_6(\theta_6) \end{aligned} \tag{3}$$

Typical 6-DOF decoupled manipulators have the following advantages (Jin et al. 2009):

1. Decoupled motion behaviour and can be selectively actuated for different tasks allowing for specialized workspaces.
2. Closed-form inverse and forward kinematic solutions.
3. Simplified kinematic optimization, motion planning and control scheme.
4. Ease of calibration and error compensation.

Serial manipulators are usually completely decoupled or translation–orientation group decoupled by nature of their open-loop design. However, parallel manipulators have closed-loop mechanisms and hence need to be specially designed for decoupling. Hybrid architectures are usually parallel mechanisms in series and are effectively group decoupled. The definition as discussed above forms the motivation for investigating a 6-DOF parallel and hybrid mechanisms further for the proposed work.

2.1 Coupled Parallel Kinematic Machines

The conventional coupled 6-DOF manipulator is commonly referred as Stewart–Gough Platform (SGP) (Merlet 2003). The SGP has applications in machine tool technology, flight simulators, crane technology, satellite antenna positioning, telescopes, surgery, etc. The SGP consists of six prismatic actuators (called legs) connected in pairs to two different bodies (which are usually plate-like structures) via passive joints on both ends. These two bodies are generally named as the base and the platform, where the base is fixed and the platform is mobile with six degrees of freedom (three translational and three rotational). The passive joints used are typically two spherical joints or a combination of spherical and universal joints. The use of two spherical joints per leg introduces an additional passive torsional DOF for each leg, but torque cannot be transmitted through the same. The different designs of the SGP depend solely on the location of the passive joints with respect to each other on the base and platform, respectively.

Figure 2 illustrates the standard structure of the SGP commonly investigated in literature with well-established design (Husty 1996; Dasgupta and Mruthyunjaya 2000; Bandyopadhyay and Ghosal 2008). The presence of closed-loop structures introduces nonlinearity into the system which result in difficulty in computation of forward kinematics for the system (Dietmaier 1998). For the general platform, the

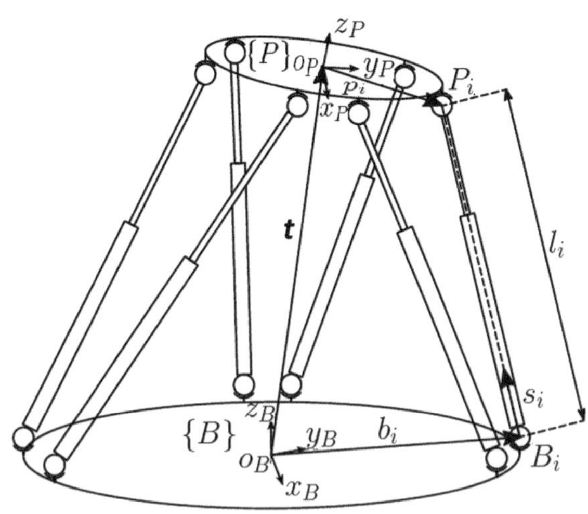

Fig. 2 Illustration of an SGP with base and platform frames

forward kinematics lacks an explicit analytical expression and usually a Jacobian-based numerical approach is used to compute the same. The coupled motion of the links results into robot with complex six-dimensional workspace which is difficult to visualize and analyse and has discontinuities (e.g. singular configurations). This leads to inability of geometric techniques in determining the workspace. Furthermore, *W/F* ratio of the SGP is also smaller compared to its serial counterparts. The coupled motion also results in higher complexity of motion planning and control schemes. However, SGP has excellent stiffness, higher positional and orientation accuracy in its limited workspace (Son et al. 2003). Also, it has comparatively lower backlash errors in joints although a vision-based feedback system can be used to correct these errors. The SGP is also able to achieve much higher speeds than its serial counterparts still maintaining its rigidity which gives it a decisive advantage for multi-axis AM. There are no shared joints in the SGP; therefore, it is simple to manufacture and easily scalable for larger workspaces as link sizes can be increased without a corresponding trade-off in the stiffness.

2.2 Decoupled Parallel Kinematic Machines

As discussed in the previous section, closed-loop kinematic structure of the SGP resulted in difficult forward kinematics, coupled motion characteristics, limited and complex workspaces, thereby making motion planning and control a complex task in most applications. A 6-DOF decoupled PKM is best suited to replace conventional PKMs without a trade-off. Decoupled parallel machines have an architecture allowing specific motion axes to be driven in a certain order using independently designated actuators. From the perspective of multi-axis 3D printing, 6-DOF machines with 3-3 GD (translational–orientation group decoupling) are of great interest. Jin et al. (2009) discussed the syntheses of decoupled mechanisms in detail. Yang et al. (2004) developed prototype of a mechanism with decoupled motion architecture. The developed mechanism has larger cylindrical workspace, simple kinematics and high stiffness in the direction of machining. The prototype of the mechanism is shown in Fig. 3.

With the introduction of 2-DOF actuators (actuators with selectively actuated independent rotation and translation), entirely new class of decoupled manipulators termed as selectively actuated manipulators (SA-PM) has been conceptualized (Law 2002). The end effector of these machines can work in multiple modes such as 3-DOF spherical motion, 3-DOF translation, 3-DOF hybrid motion or complete 6-DOF spatial motion, depending on the types of the actuation (rotary or linear) chosen for the actuators. The schematic diagram of one such design is shown in Fig. 4. The manipulator architecture completely decouples translation and rotation of the end effector for individual control. This manipulator is still in the preliminary stage of development but it has great promise for the implementation of 3-D printers. The details of the same can be found in the work of Jin et al. (2004).

Fig. 3 Prototype of the 6-DOF modular PKM (Yang et al. 2004)

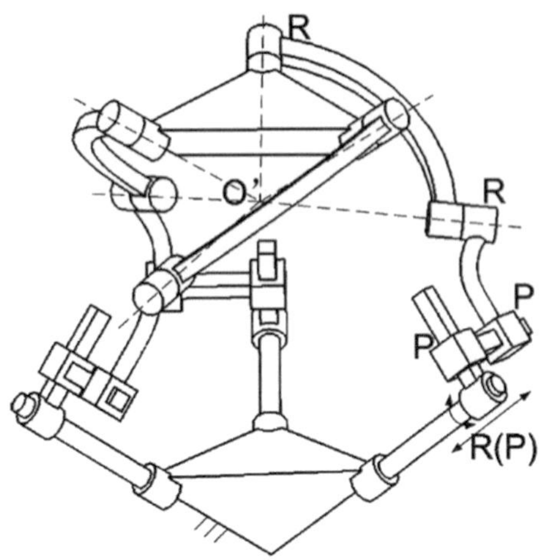

Fig. 4 Schematic diagram of SA-PM (Yang et al. 2004)

There has been considerable interest in using parallel kinematic machines such as SGP for manufacturing applications in recent years. The SGP has been successfully implemented for machining processes requiring multi-axis motions in limited workspace (Son et al. 2003). The present work proposes SGP as an alternate configuration for AM application. Song et al. (2015) proposed working prototype of SGP for workspace with size of 100 mm × 100 mm × 100 mm with pitch and roll motions in the range of −30° to +30°. This paper proposes an optimal design of the similar SGP with higher dexterity and larger workspace.

3 Kinematic Modelling of Stewart–Gough Platform

This section presents kinematic modelling of the SGP with an objective of gaining an understanding of the factors required for deriving an optimal design. The kinematic model attempts to establish the relationship between the task space and the joint space, where the task space is the Cartesian coordinates of the platform centre (and by extension end of the tool point if attached to it) and the joint space comprises the lengths of the linear actuators. The forward kinematics (Merlet 2003) describes functional relationship between length of the linear actuators and position and orientation of the platform. On the other hand, inverse kinematic analysis maps the position and orientation of the platform to the length of the linear actuators. In contrast to serial manipulators, the inverse kinematics problem for parallel manipulators is trivial due to the existence of a closed vector loop structure which provides an analytical solution. The forward kinematics problem is complex due to high degree of nonlinearity of equations leading SGP having 40 different assembly modes (Dietmaier 1998). The forward kinematics problem is typically attempted using numerical solvers. It is important to discuss the differential kinematics of SGP through definition of the Jacobian matrix. The study of the Jacobian is important as it is an index through which dexterity and isotropic behaviour of the SGP can be quantified further. It leads to the definition of kinetostatic performance indices which form the basis for optimal design of the SGP.

3.1 Inverse Kinematics

A simplified diagram of the SGP is illustrated in Fig. 2 with fixed base and movable platform connected by six-length changeable struts. The two coordinate frames are attached to the base $\{B\}$ with its origin O_B at the centre of the base circle and the platform $\{P\}$ with its origin O_P at the canter of the platform circle, respectively. The platform frame is movable and often referred to as the tool point; since orientation and position control of this point implies control of tool, it can be treated as a rigid extension of this point. The passive joints, connecting the linear actuators with the base and platform, are at the points B_i which lie on the base plane and P_i lying on the platform plane, where $i = 1, 2, 3, 4, 5, 6$. The positions of the passive joints are described using vectors $\mathbf{b}_i$ and $\mathbf{p}_i$. The lengths of the length changeable struts are denoted by l_i and their direction with the unit vector $\mathbf{s}_i$. The position and orientation of the tool point are described using position vector ${}^{B}\mathbf{t}$ and the rotation matrix ${}^{B}\mathbf{R}_P$, where the leading superscript implies the frame in which the vector is expressed and the trailing superscript implies the frame it is being transformed to. The rotation matrix ${}^{B}\mathbf{R}_P$ is used to express the orientation of frame $\{P\}$ with respect to frame $\{B\}$ and can be used to transform vectors between frames. The rotation matrix can be represented in multiple ways of which the roll–pitch–yaw (α, β, γ) representation is

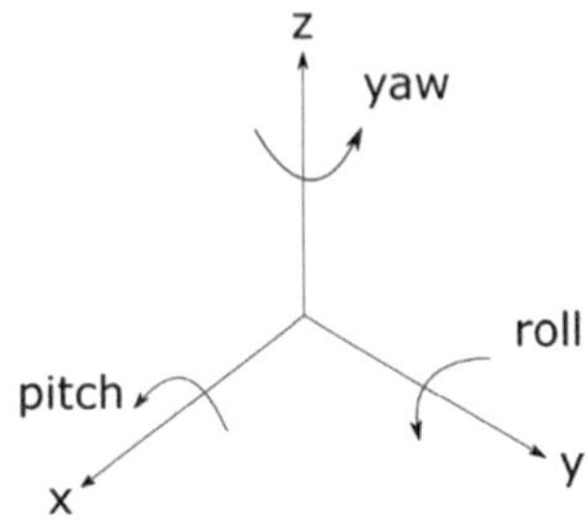

Fig. 5 Roll (β), pitch (α) and yaw (γ) rotations of a frame

found to be useful. The roll–pitch–yaw angles are illustrated in Fig. 5. The rotations along x-, y- and z-axes can be composed to form the rotation matrix in Eq. (4).

$$\begin{aligned}\mathbf{R}_{PRY}(\alpha,\beta,\gamma) &= \mathbf{R}_z(\gamma)\mathbf{R}_y(\beta)\mathbf{R}_x(\alpha)\\ &= \begin{bmatrix} c\beta c\gamma & s\alpha s\beta \varsigma\gamma - c\alpha s\gamma & c\alpha s\beta c\gamma + s\alpha s\gamma \\ c\beta c\gamma & s\alpha s\beta s\gamma + c\alpha c\gamma & c\alpha s\beta s\gamma - s\alpha c\gamma \\ -s\beta & s\alpha c\beta & c\alpha c\beta \end{bmatrix}\end{aligned} \tag{4}$$

The inverse kinematics involves the determination of strut lengths $l = [l_1\, l_2\, l_3\, l_4\, l_5\, l_6]$ given the position ${}^{B}\mathbf{t}$ and orientation ${}^{B}\mathbf{R}_P$ of the moving platform. Equation (5) can be used to describe the closed vector loop governing each limb of the SGP.

$$l_i^B\mathbf{s}_i = {}^{B}\mathbf{t} + {}^{B}\mathbf{R}_P{}^{P}\mathbf{p}_i - {}^{B}\mathbf{b}_i \tag{5}$$

The length of the struts can be found by calculating the Euclidean norm using Eq. (5) for each limb. The length of each leg is therefore obtained as Eq. (6).

$$l_i = \left\{ \left({}^{B}\mathbf{t} + {}^{B}\mathbf{R}_P{}^{P}\mathbf{p}_i - {}^{B}\mathbf{b}_i\right)^{\mathrm{T}} \left({}^{B}\mathbf{t} + {}^{B}\mathbf{R}_P{}^{P}\mathbf{p}_i - {}^{B}\mathbf{b}_i\right) \right\}^{1/2} \tag{6}$$

The negative values of the Euclidean norm are not considered as it implies a negative strut length which is not possible physically. This ensures a unique inverse kinematic solution for the system.

3.2 *Differential Kinematics—The Jacobian*

The differential kinematics deals with understanding of relationship between joint velocities and end effector velocities/tool velocities. This is done through the study of the Jacobian which is the transformation of the linear velocities of the actuators into spatial velocity of the platform. Furthermore, the Jacobian can be used to measure the motion accuracy of the platform, predicting *singular configurations* of the manipulator and evaluation of dexterity of the platform. The differential kinematic equations relating the two velocity spaces can be formulated by differentiating Eq. (5) with respect to time and can be represented as

$$\dot{l}_1{}^B\mathbf{s}_i + l_i{}^B\dot{\mathbf{s}}_i = {}^B\mathbf{v}_P + {}^B\dot{\mathbf{R}}_P{}^P\mathbf{p}_i + {}^B\mathbf{R}_P{}^P\dot{\mathbf{p}}_i - {}^B\dot{\mathbf{b}}_i \tag{7}$$

where ${}^B\mathbf{v}_P$ is the linear velocity of the tool point described with respect to the base frame. The time derivatives of vectors ${}^P\mathbf{p}_i$ and ${}^B\boldsymbol{b}_i$ are zero, since they are constant position vectors expressed in their respective frames. The relation between the rotation matrix rate and the angular velocity in a fixed frame is given by

$$ {}^B\boldsymbol{\Omega} = \begin{bmatrix} 0 & -\omega_z & \omega_y \\ \omega_z & 0 & -\omega_x \\ -\omega_y & \omega_x & 0 \end{bmatrix} = {}^B\dot{\mathbf{R}}_P{}^B\mathbf{R}_P^T \tag{8}$$

where is ${}^B\boldsymbol{\Omega}$ a skew-symmetric matrix containing information about the angular velocities. From Eq. (8) ${}^B\dot{\mathbf{R}}_P{}^P\mathbf{p}_i$ can be rewritten as

$${}^B\dot{\mathbf{R}}_P{}^P\mathbf{p}_i = {}^B\dot{\mathbf{R}}_P{}^B\mathbf{R}_P{}^{TB}\mathbf{R}_P{}^P\mathbf{p}_i = {}^B\boldsymbol{\omega} \times \left({}^B\mathbf{R}_P{}^P\mathbf{p}_i\right) = {}^B\boldsymbol{\omega} \times {}^B\mathbf{p}_i \tag{9}$$

where ${}^B\boldsymbol{\omega}$ is a vector describing the angular velocity of the tool point in the base frame. The differential kinematic equation is thereby simplified as

$$\dot{l}_i{}^B\mathbf{s}_i + l_i{}^B\dot{\mathbf{s}}_i = {}^B\boldsymbol{v}_P + {}^B\boldsymbol{\omega} \times {}^B\mathbf{p}_i \tag{10}$$

By the multiplication of both sides of Eq. (10) by unit vector ${}^B\mathbf{s}_i^T$, the final differential kinematic equation is expressed as Eq. (11) which can be reduced to Eq. (12).

$$\dot{l}_i{}^B\mathbf{s}_i^{TB}\mathbf{s}_i + l_i{}^B\mathbf{s}_i^{TB}\dot{\mathbf{s}}_i = {}^B\mathbf{s}_i^{TB}\boldsymbol{v}_P + {}^B\mathbf{s}_i^T\left({}^B\boldsymbol{\omega} \times^B \mathbf{p}_i\right) \tag{11}$$

$$\dot{l}_i = \mathbf{s}_i^T\boldsymbol{v}_P + (\mathbf{p}_i \times \boldsymbol{s}_i)^T\boldsymbol{\omega} \tag{12}$$

where $\dot{l}_i$ is the velocity of the ith actuator and $\boldsymbol{v}_P$ and $\boldsymbol{\omega}$ are the linear and angular velocities of the tool point, expressed in the base frame. Rearranging the expression into matrix form yields the Jacobian of the SGP as

$$\dot{\boldsymbol{l}} = \mathbf{J}\dot{\boldsymbol{q}} \leftrightarrow \dot{\boldsymbol{q}} = \mathbf{J}^{-1}\dot{\boldsymbol{l}} \tag{13}$$

where $\dot{\boldsymbol{l}}$ is the actuator velocity vector, $\mathbf{J}$ is the Jacobian and $\dot{\boldsymbol{q}}$ is the velocity vector of the tool point. Stacking the equations as matrix-vector multiplication form yields the following 6×6 Jacobian transformation matrix that is a function of the current pose of the platform ($\boldsymbol{q} = [x, y, z, \alpha, \beta, \gamma]$):

$$\dot{\boldsymbol{l}} = \begin{bmatrix} \dot{l}_1 \\ \dot{l}_2 \\ \dot{l}_3 \\ \dot{l}_4 \\ \dot{l}_5 \\ \dot{l}_6 \end{bmatrix} \mathbf{J} = \begin{bmatrix} \boldsymbol{s}_1^T & (\boldsymbol{Rp}_1 \times \boldsymbol{s}_1)^T \\ \boldsymbol{s}_2^T & (\boldsymbol{Rp}_2 \times \boldsymbol{s}_2)^T \\ \boldsymbol{s}_3^T & (\boldsymbol{Rp}_3 \times \boldsymbol{s}_3)^T \\ \boldsymbol{s}_4^T & (\boldsymbol{Rp}_4 \times \boldsymbol{s}_4)^T \\ \boldsymbol{s}_5^T & (\boldsymbol{Rp}_5 \times \boldsymbol{s}_5)^T \\ \boldsymbol{s}_6^T & (\boldsymbol{Rp}_6 \times \boldsymbol{s}_6)^T \end{bmatrix} \dot{\boldsymbol{q}} = \begin{bmatrix} \dot{x} \\ \dot{y} \\ \dot{z} \\ \dot{\omega}_x \\ \dot{\omega}_y \\ \dot{\omega}_z \end{bmatrix} \tag{14}$$

The first three columns of Jacobian are related to the translational motion and last three columns are related to the rotational motion of the SGP. Furthermore, multiplying both sides of Eq. (13) by dt results in the following expressions:

$$\mathrm{d}t\dot{\boldsymbol{l}} = \mathbf{J}\mathrm{d}t\dot{\boldsymbol{q}} \tag{15}$$

$$\dot{\boldsymbol{l}} = \mathbf{J}\mathrm{d}\boldsymbol{q} \leftrightarrow \mathrm{d}\boldsymbol{q} = \mathbf{J}^{-1}\mathrm{d}l \tag{16}$$

The above expressions imply that the Jacobian describes the accuracy of the robot, i.e. how a small change in the displacement of the actuators translates to a change in the position of the tool point.

4 Kinetostatic Performance Indices

The Jacobian provides an important quantitative insight into the force and motion transmission capabilities of the robotic manipulator from joint space to task space and vice versa. In this section, a general discussion on manipulability and dexterity is conducted concluding in the definition of kinetostatic performance indices namely the Global Conditioning Index (GCI) and the Global Manipulability Index (GMI) which are used to evolve an optimum design of the SGP further.

4.1 Manipulability Ellipsoid and Singular Configurations

Manipulability may be defined as the capacity of change in position and orientation of the end effector of a robot given a joint configuration/pose. In mathematical terms, it can be considered as an ellipsoid in n-dimensional Euclidean space. This ellipsoid symbolizes the motion capacity of the robot at a given pose, and hence it is known as the Manipulability Ellipsoid (ME). It is used to map the change in direction and magnitude of the output vector of actuator velocities given an input vector of tool point velocities. This space is generally represented as a Euclidean norm but analysed as the square of the Euclidean norm as it is easier to study and behaviours of both are quite similar. **J** can be considered as scaling the input joint velocities to produce the output tool or end effector velocities. The Singular Value

Decomposition (SVD) is a common tool used to characterize this linear transformation. It describes transformation as a product of a pure rotation and scaling followed by another pure rotation. In this process, the lengths of the principal axes of the ME are obtained (also called the singular values) which are given by

$$\sigma_i(\mathbf{J}) = \sqrt{\lambda_i(\mathbf{J}\mathbf{J}^T)} \quad i = 1, 2, \ldots, 6 \tag{17}$$

where λ_i is the ith eigenvalue of the matrix $\mathbf{J}\mathbf{J}^T$. The lengths of the axes are directly proportional to the velocity developed in the direction. The concept is illustrated using Fig. 6, where σ_1 and σ_2 are the lengths of the minor and major axes of the ellipse. In the context of the ME, the singular configuration implies that one of the axes has zero length and the ellipsoid collapses and become degenerate. Physically, it indicates that velocities and forces cannot be transmitted in this direction. Therefore, manipulability measure is given by

$$\mu(\mathbf{J}) = \sigma_1\sigma_2\ldots\sigma_n \tag{18}$$

Equation (18) can be interpreted as volume of the ellipsoid. By substituting Eq. (17) into Eq. (18) and rearranging the terms,

$$\mu(\mathbf{J}) = \sqrt{\lambda_1\lambda_2\ldots\lambda_n} = \sqrt{\det(\mathbf{J}\mathbf{J}^T)} \tag{19}$$

For $n \times n$ Jacobian ($\mathbf{J}$), Eq. (19) simplifies to $\mu(\mathbf{J}) = \det(\mathbf{J})$. When the determinant of the Jacobian becomes zero, the SGP is said to be in a singular configuration where it gains or loses one degree of freedom.

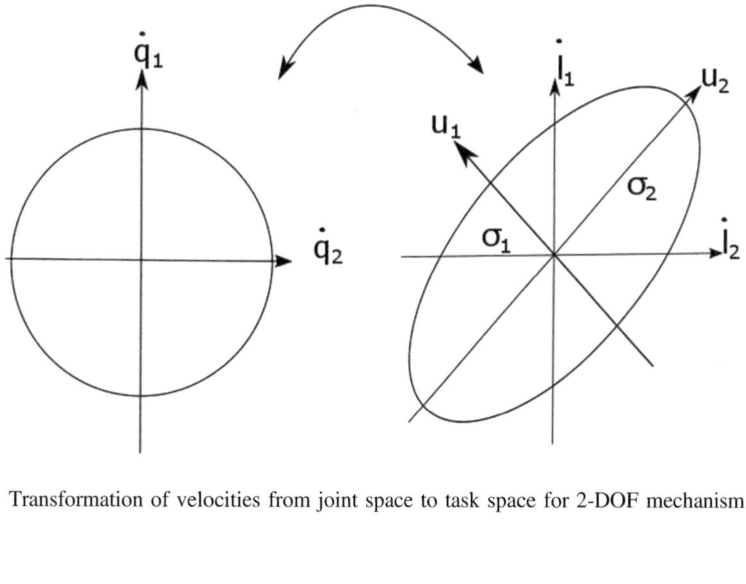

Fig. 6 Transformation of velocities from joint space to task space for 2-DOF mechanism

4.2 Kinematic Isotropy

The singular values of the Jacobian indicate the level of distortion of the unit hypersphere of tool point velocities to the unit hyper-ellipsoid of the joint rates. When all the singular values are identical, the unit-joint-rate hyper-ellipsoid collapses to a uniform hypersphere which is either deflated or inflated. The Jacobian at such a pose is said to be in an isotropic configuration and it indicates that the mobility of the platform is same in all directions, i.e. the ability of the platform to transfer velocities is uniform in all directions. Therefore, for optimal design, the Jacobian should be isotropic. But since the Jacobian is a function of the current pose, isotropy cannot be achieved over the entire workspace without changing the dimensions of the platform which is not viable. Hence, a measure of the degree of isotropy is defined as the condition number given by

$$\kappa(\mathbf{J}) = \|\mathbf{J}\|\|\mathbf{J}^{-1}\| \tag{20}$$

If the Euclidean norm is used, the condition number is the ratio of the maximum singular value to the minimum singular value of the Jacobian matrix. It can be expressed as

$$\kappa(\mathbf{J}) = \frac{\sigma_{\max}}{\sigma_{\min}} \tag{21}$$

The condition number is a quantitative measure of the distortion of the unit hypersphere where distortion is directly proportional to the value of the condition number. A condition number of 1 indicates a perfect sphere and an isotropic configuration. When the condition number approaches infinity, one of the axes of the unit hypersphere collapses indicating a singular configuration. The condition number is usually evaluated as the reciprocal $1/\kappa$ to ensure that it is bounded between 0 and 1 where 0 indicates a singular configuration and 1 indicates an isotropic configuration.

4.3 Global Conditioning Index (GCI)

The condition number as discussed previously is a function of the current pose and hence provides a measure of the local dexterity of the SGP. For a given continuous workspace, w is a measure of global dexterity and it can be defined by the global conditioning index presented in Eq. (22) (Angeles and Gosselin 1991).

$$\mathrm{GCI} = \frac{\int_w \frac{1}{\kappa}\,\mathrm{d}w}{\int_w \mathrm{d}w} \tag{22}$$

GCI can also be expressed as Eq. (23) for total number of evaluated discrete poses in the workspace.

$$\mathrm{GCI} = \frac{1}{n}\sum_{i=1}^{n}\frac{1}{\kappa} \tag{23}$$

The GCI is also bounded between 0 and 1, with 1 indicating pure isotropic behaviour over the entire workspace. Therefore, optimal design for dexterity dictates that the GCI should be as close to 1 as possible.

4.4 Global Manipulability Index

Another metric commonly used to characterize the performance of the SGP is the Global Manipulability Index (GMI) which can be defined as (Lara-Molina et al. 2011)

$$\mathrm{GMI} = \frac{\int_w \mu(\mathbf{J})\mathrm{d}w}{\int_w \mathrm{d}w} \tag{24}$$

where μ is the local manipulability of the SGP at a given pose and it is given using Eq. (25).

$$\mu(\mathbf{J}) = \sqrt{\det\left(\mathbf{J}\mathbf{J}^T\right)} \tag{25}$$

The GMI can be considered as ratio of the volumes of the ME to the unit hypersphere. It is essentially a scaling factor indicating the degree of amplification of tool point velocities to actuator velocities. Hence, the GMI should be very small as small velocities in actuators should translate to large platform velocities for rapid motion of the SGP. For a discrete workspace, the GMI can be written as

$$\mathrm{GMI} = \frac{1}{n}\sum_{i=1}^{n}\mu(\mathbf{J}) \tag{26}$$

5 Kinematic Optimization of SGP

This section deals with applying the kinetostatic performance indices defined in Sect. 4.3 for optimal design of the SGP. A high GCI value indicates dextrous behaviour in the workspace, which is a requirement for smooth trajectory generation for the printer. For higher speed of motion, the configuration can be further optimized for low GMI. The GMI and GCI are opposing objectives and optimizing for both can give a poor resultant value for GCI especially for the larger workspaces that the robot attempts to cover. Therefore, the design of the SGP is formulated as a single-objective constrained optimization problem in the present study where GCI is used as a single evaluation metric to be maximized. The objective function for the optimization problem can be written as

$$\text{Maximize: GCI} \rightarrow f_1(\mathbf{x}) = -\text{GCI}. \tag{27}$$

5.1 Geometric Design Space of the SGP

The set of free variables which can be controlled by the designer are known as the design variables of the optimization problem. These variables together constitute a vector space commonly known as design space, the dimensionality of which is dependent on the number of the variables. The higher dimension design space usually means more freedom in design but also imply complexity of the optimization problem. It is also associated with large amounts of time and computational resource requirements to solve the problem. Figure 7 shows the geometric design of the Semi-Regular Stewart–Gough platform (henceforth called the SRSGP), which has been adopted further in the design of multi-axis 3-D printer. The SRSGP can be completely articulated by the use of five design variables. These variables are listed as follows:

0. r_B, radius of the base circle.
1. r_P, radius of the platform circle.
2. ϕ_B, spacing angle between a set of base passive joints B_1–B_2, B_3–B_4, B_5–B_6.

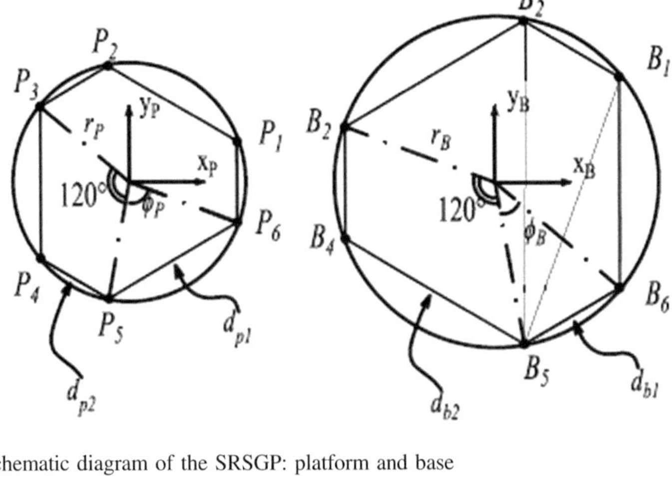

Fig. 7 Schematic diagram of the SRSGP: platform and base

3. ϕ_P, spacing angle between a set of base passive joints P_1–P_2, P_3–P_4, P_5–P_6.
4. h, distance between the origins of the base and platform frames when the platform is at neutral position. The neutral position is defined when the base and the platform are parallel with the x–y coordinates coinciding each other. In other words, the platform is at the centre of the defined workspace.

The design space can thus be represented by the vector

$$\mathbf{x} = [r_B r_P \phi_B \phi_P h] \tag{28}$$

The position vectors of the ends of the limbs can be redefined in terms of the design variables as

$$\boldsymbol{b}_i = \begin{bmatrix} r_b \cos(\varphi_{B,i}) \\ r_b \sin(\varphi_{B,i}) \\ 0 \end{bmatrix}, \boldsymbol{p}_i = \begin{bmatrix} r_p \cos(\varphi_{P,i}) \\ r_p \sin(\varphi_{P,i}) \\ 0 \end{bmatrix}$$

Here,

$$\begin{aligned} \varphi_{B,i} &= \frac{i\pi}{3} - \frac{\varphi_B}{2}, \varphi_{P,i} = \frac{i\pi}{3} - \frac{\varphi_P}{2} \qquad i = 1,3,5 \\ \varphi_{B,i} &= \varphi_{B,i} + \varphi_B, \varphi_{P,i} = \varphi_{P,i-1} + \varphi_P \qquad i = 2,4,6 \end{aligned} \tag{29}$$

The next step in the optimization process is to define bounds for these variables which allows the algorithm to search for solution in a feasible domain. The design considerations used for defining bounds on various design variables are as follows:

(1) Maximum floor space of the platform cannot exceed 1.5 m × 1.5 m and hence the maximum radius for base and platform is 0.75 m.
(2) The minimum radius of the base depends on the minimum space required for the plasticizer unit which may be enclosed inside the structure.
(3) The minimum radius of the platform depends on the radius of the barrel of the extruder which is to be mounted on the platform.
(4) The range of values for ϕ_B and ϕ_P are chosen to be 0°–120° to prevent leg crossover.
(5) The height of the platform is constrained again by components of the system enclosed by the structure but bounds on the height are only a formality required for the genetic algorithm to work effectively. The height is indirectly constrained by the length of the struts which are imposed as active constraints on the system.

5.2 Formulation of Indirect Constraints

These constraints are different from the bounds as they represent relationships between the input variables and do not appear as simple upper or lower limits on the design variables. These constraints arise from physical or performance requirements from the system. The constraints are represented as linear or nonlinear functions of the design vector. The limbs of the platform should not share the same passive joint as the physical construction of such a joint is difficult and furthermore would heavily limit the motion of the platform due to collisions. Setting the two spacing angles ϕ_B and ϕ_P greater than 0° and less than 120° does not ensure that any two joints coincide as it does not account for the finite radius of the joints. To account for the same, a parameter has been defined which is called as spacing distance henceforth. It is defined as the minimum distance between two passive joints $d_{\min}$. The symmetry of the platform ensures that there exist only two such distances for the base d_{b1} and d_{b2} as illustrated in Fig. 7. They can be defined by the cosine rule as Eqs. (30)–(32).

$$d_{b1}^2 = r_B^2 + r_B^2 - 2r_B r_B \cos\phi_B \tag{30}$$

$$d_{b1} = \sqrt{2r_B^2 - 2r_B^2 \cos\phi_B} \tag{31}$$

$$d_{b2} = \sqrt{2r_B^2 - 2r_B^2 \cos(120^\circ - \phi_B)} \tag{32}$$

The spacing distance should be greater than 0.15 m to allow for a passive joint of radius 75 mm, which is assumed to be sufficient. The constraints can be therefore expressed as

$$d_{b1} \geq 0.15\,\text{m} \tag{33}$$

$$0.15\,\text{m} - d_{b1} \leq 0 \tag{34}$$

$$0.15\,\text{m} - d_{b2} \leq 0 \tag{35}$$

$$d_{b1} \geq 0.15\,\text{m} \tag{36}$$

$$0.15\,\text{m} - d_{b1} \leq 0 \tag{37}$$

$$0.15\,\text{m} - d_{b2} \leq 0 \tag{38}$$

Similarly, the spacing distance for the moving platform can be expressed as

$$0.15\,\text{m} - d_{p1} \leq 0 \tag{39}$$

$$0.15\,\mathrm{m} - d_{p2} \leq 0 \tag{40}$$

It will be subject to the following additional constraints:

$$d_{P1}^2 = r_P^2 + r_P^2 - 2r_P r_P \cos\phi_P \tag{41}$$

$$d_{P2} = \sqrt{2r_P^2 - 2r_P^2\cos(120^\circ - \phi_B)} \tag{42}$$

The kinetostatic performance indices are bulk indices which describe the mean kinematic performance of the SRSGP and cannot be used to determine whether it encounters singularity configurations in its workspace. Hence, condition number at every point in the workspace should be above a certain value to avoid singularities. This value is set to 0.01 for the present design and the constraint is expressed as

$$0.01 - \min\left(\frac{1}{\kappa}\right) \leq 0 \tag{43}$$

5.3 Practical Evaluation of Kinetostatic Performance Index

The GCI cannot be evaluated analytically due to complex nonlinear nature of the integral involved and hence it is evaluated numerically over the workspace discretized into a finite number of points also called poses of the SRSGP. The workspace defined with respect to the FDM process is a $100\,\mathrm{mm} \times 100\,\mathrm{mm} \times 100\,\mathrm{mm}$ cube centred at $(0, 0, h)$ over which the platform can roll, pitch and yaw at $(-45, 45)^\circ$. The Cartesian workspace can then be defined in global reference frame $\{B\}$ using relationships

$$\begin{aligned} x &\in (-0.15, 0.15)\,\mathrm{m} \\ y &\in (-0.15, 0.15)\,\mathrm{m} \\ z &\in (-0.15 + h, 0.15 + h)\,\mathrm{m} \end{aligned} \tag{44}$$

The corresponding orientation workspace defined by Euler angles can be written as

$$\begin{aligned} \alpha &\in (-45, 45)^\circ \\ \beta &\in (-45, 45)^\circ \\ \gamma &\in (-45, 45)^\circ \end{aligned} \tag{45}$$

The level of discretization of the workspace can be controlled by two indices n_{CW} and n_{EW} which represent the number of subdivisions of x, y and z in the

Cartesian workspace and the number of divisions of α and β in the Euclidean workspace. The total number of poses generated can thus be calculated as

$$N_{\text{total}} = n_{\text{CW}}^3 n_{\text{EW}}^3 \tag{46}$$

An attempt to increase the number of poses generated results into increased precision of the GCI but it also yields larger computation time. Thus, the number of poses generated can be treated as a trade-off between precision and computation time. A discretization of $n_{\text{CW}} = 5$ and $n_{\text{EW}} = 5$ is chosen for the optimization problem in the present study as it ensures convergence and precision still maintaining fairly lower number of poses to be determined.

5.4 Optimization Using the Genetic Algorithm

The final optimization problem now can be formulated as follows:

$$\left.\begin{array}{ll}
\text{Maximize: GCI} \rightarrow \quad f_1(\mathbf{x}) = -\text{GCI} & \\
\text{Over:} \quad \mathbf{x} = [r_B r_P \phi_B \phi_P h] & \\
\text{Subject to:} & \\
g_i = L_{\min,i} - 1.1 \leq 0 & \text{for } i = 1,2,3,4,5,6 \\
g_{6+i} = L_{\max,i} - 2L_{\min,i} + 0.3 \leq 0 & \text{for } i = 1,2,3,4,5,6 \\
g_{13} = 0.15 - d_{b1} \leq 0 & \\
g_{14} = 0.15 - d_{b2} \leq 0 & \\
g_{15} = 0.15 - d_{p1} \leq 0 & \\
g_{16} = 0.15 - d_{p2} \leq 0 & \\
g_{17} = 0.01 - \min\left(\frac{1}{\kappa}\right) \leq 0 &
\end{array}\right\} \tag{47}$$

The above problem is single-objective, bounded problem with nonlinear constraints. It lacks an explicit analytical form. Furthermore, it is not continuous and the gradient is not well defined; therefore, standard gradient-based solvers cannot be used for such problems. Typically, evolutionary algorithms are used to solve such problems as they are not gradient information dependent. There are few research attempts related to optimal design of SRSGP based on the Jacobian which can be found from (Boudreau and Gosselin 1999; Lara-Molina et al. 2011; Kelaiaia et al. 2011). Genetic Algorithm (GA) has been found to be particularly robust for such problems although it does not guarantee the global optimum solution. The algorithm requires well-defined search space which is decided on the basis of bounds discussed in the previous section. Table 1 summarizes bounds for various design variables.

Table 1 Bounds for constraints

r_B (m)	r_P (m)	ϕ_B	ϕ_P	h (m)
0.05–0.5	0.03–0.125	0°–60°	0°–60°	0.1–0.25

Table 2 GA parameters used in the program

Population size	Maximum generations	Crossover probability	Mutation probability	Distribution index for crossover	Distribution index for mutation
50	500	0.8	0.1	20	20

Table 3 Values of design variables derived from GA

GCI	r_B (m)	r_P (m)	ϕ_B	ϕ_P	h (m)
0.375	0.4	0.117	1.972	0.1	0.48

The proposed optimization problem has been implemented in the form of computational program to determine optimal design variables using GA-based optimization algorithm available in MATLAB. The following parameters are used to initiate the optimization algorithm for the problem (Table 2).

It has been observed that algorithm converges after 30 generations and results are recorded. Table 3 summarizes output of GA after 30 generations. The convergence is achieved when the improvements in the fitness of the solution are less than a threshold value which is set as 1×10^{-6} in the present study. The GCI of the system at optimal design parameters is determined as 0.375. The corresponding geometric design values of the variables are listed in Table 3.

The values of geometric variables listed in Table 3 are used in determining dimensions for the linear actuators required for the SGP. The actuator dimensions are $l_{\min} = 0.26\,\text{m}$; $l_{\max} = 0.90\,\text{m}$ $l_{\text{stroke}} = 640\,\text{mm}$. The GCI of the proposed design is slightly lower than conventional designs referred in the literature which have a range of 0.4–0.7 (Angeles and Gosselin 1991). This is due to larger Cartesian as well as orientation workspace required to be covered by the manipulator. The result indicates that the design of the manipulator is a trade-off between workspace and dexterity.

Figure 8 shows workspace of the configuration for various orientations when a range of motion of 70° is imposed on the universal and spherical joints. It can be seen that the workspace begins to differ from the ideal cube as the orientation angle of extruder is increased. After 30° orientation, the workspace begins to deteriorate and reaches a limit at 41° beyond which the mechanism does not have a workspace. Therefore, the actual working range of the platform largely depends on the range of motion of the universal joints and ball joints which act as limiting factors.

Figures 9 and 10 show variations of the reciprocal of condition number $(1/\kappa)$ for different heights and roll and pitch angles. It is evident from Fig. 10 that the condition number is mainly affected by roll and pitch angles. The values of zero roll and pitch angles are best from the point of condition number, whereas condition

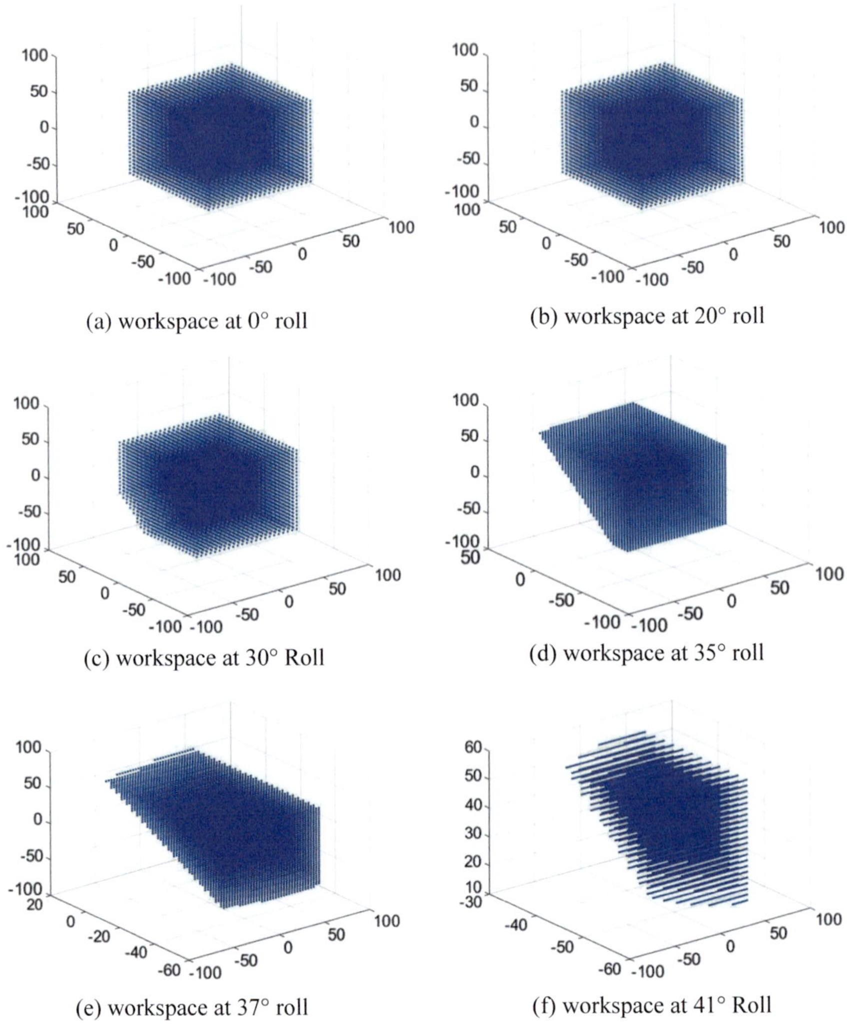

(a) workspace at 0° roll

(b) workspace at 20° roll

(c) workspace at 30° Roll

(d) workspace at 35° roll

(e) workspace at 37° roll

(f) workspace at 41° Roll

Fig. 8 Variation of workspace with different orientations

number increases ($1/\kappa$ decreases) with increase in values of roll and pitch angles. It can be seen that the condition number is greater than 0.01 for all the cases. This indicates that the robot configuration does not reach singularity condition in the entire workspace.

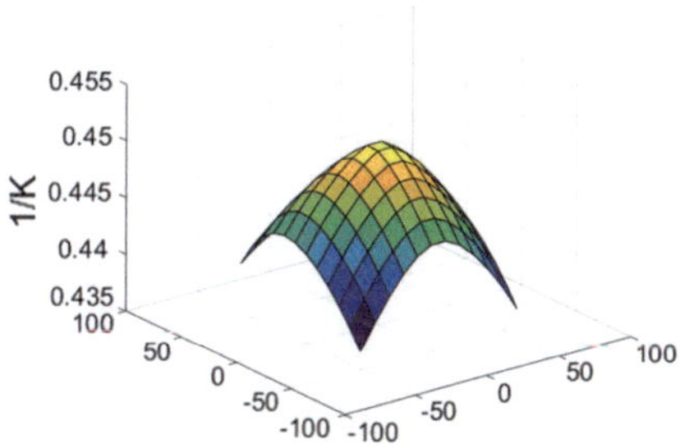

(a) Variation of condition number at height 465mm

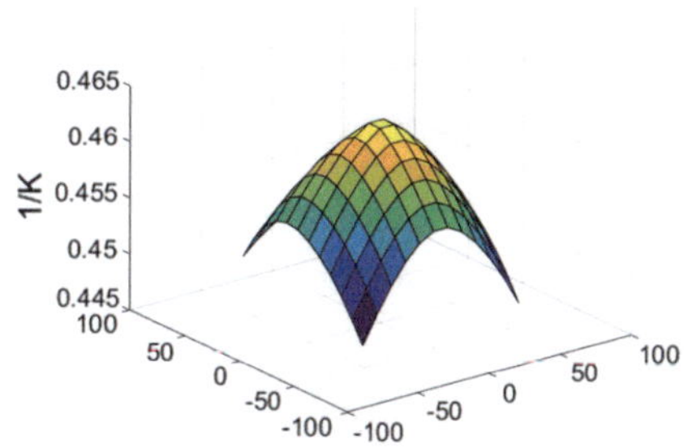

(b) Variation of condition number at height 480mm

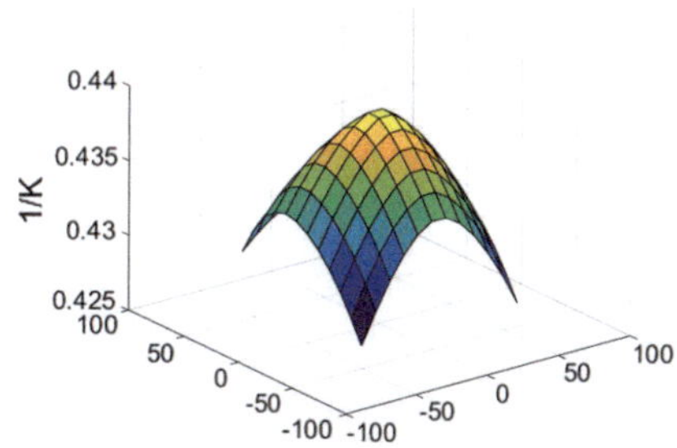

(c) Variation of condition number at height 495mm

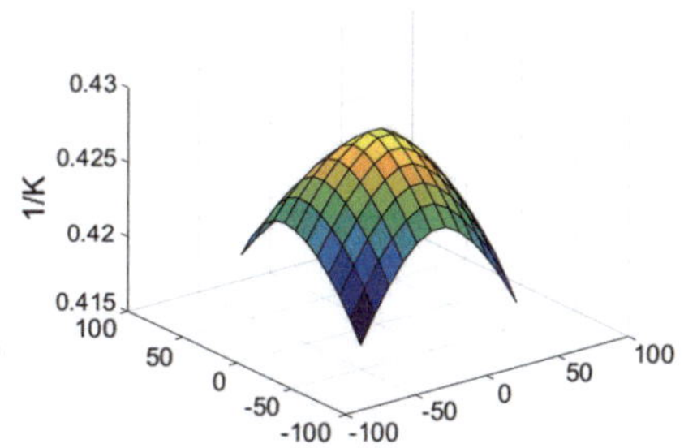

(d) Variation of condition number at height 510mm

Fig. 9 Variation of $1/\kappa$ with increasing height and constant roll and pitch

The conceptual model of the platform is shown in Fig. 11. The dimensions of linear actuator, top and bottom platform/frame, and universal joints are determined from optimal design algorithm presented in this chapter. The extruder unit for FDM is to be mounted on the bottom platform. The further work will focus on development of multi-axis 3-D printer based on these dimensions and investigate aspects related to building of the components.

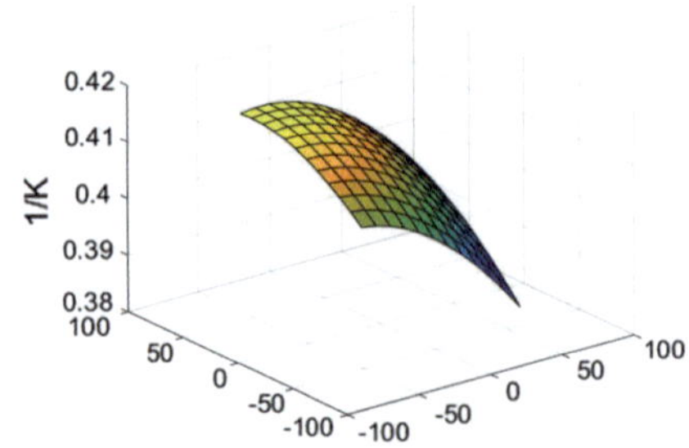

(a) Variation of condition number roll= 10° and pitch = 10°

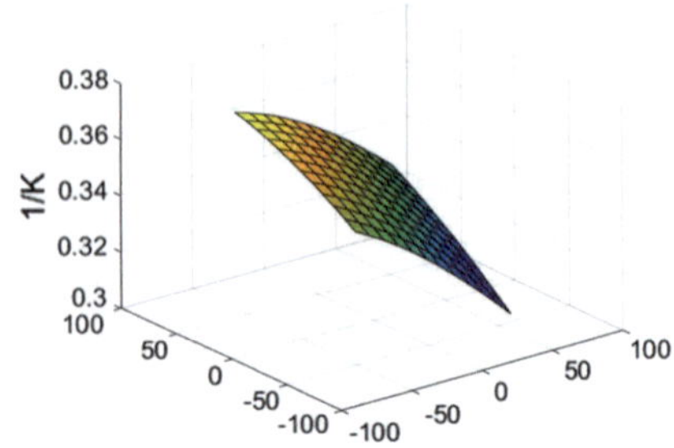

(b) Variation of condition number roll = 20° and pitch = 20°

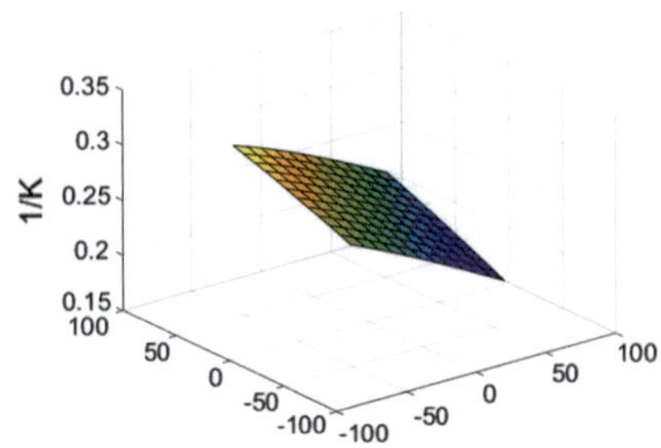

(c) Variation of condition number roll = 30° and pitch = 30°

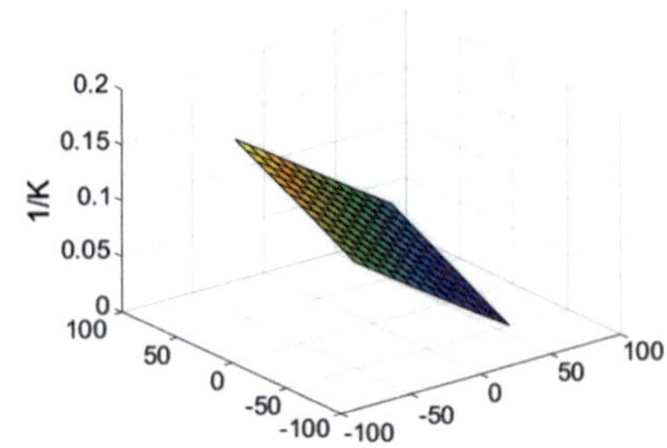

(d) Variation of condition number roll = 45° and pitch = 45°

Fig. 10 Variation of $1/\kappa$ with increasing roll and pitch and constant height = 480 mm

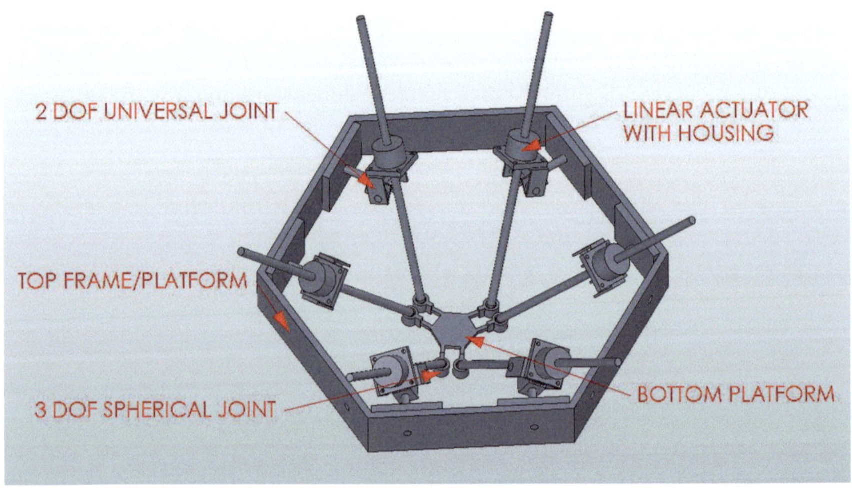

Fig. 11 Conceptual model of SGP

6 Conclusion

This chapter presented a conceptual design of a parallel kinematics machine to carry out multidirectional 3D printing. The existing 3-D printing technologies are layer based with linear motions in three Cartesian axes. This machine configuration has limitations in the form of poor part accuracy at angles oblique to the build direction, slower build speeds and additional support structure requirements. The present study addresses some of these issues by allowing six-axis motions between tool and base. To envisage this, a number of robotic architectures are discussed in the chapter and a qualitative criterion is used to compare them for the use in 3D printing application. One such popular architecture, viz. SGP, is explored in detail and its capability as a viable platform for 3D printing is illustrated. The design of SGP for multidirectional 3D printing is realized by formulating it as a single-objective optimization problem. The details of the optimization formulation and the consequent results are discussed at length. The proposed design of the multidirectional AM will not only minimize post-processing requirements but also allow for faster build-up of the part. Additionally, it is expected to offer advantages such as building around inserts, simplified slicing mechanism and control of isotropic material properties. The future work would focus on realizing the design conceived in this work for 3D printing. Further, refinement of design using global manipulability index as a second objective function will also be studied.

References

Angeles, J., and C. Gosselin. 1991. A global performance index for the kinematic optimization of robotic manipulators. *ASME Journal of Mechanical Design* 113 (3): 220–226.

Bandyopadhyay, S., and A. Ghosal. 2008. An algebraic formulation of kinematic isotropy and design of isotropic 6-6 Stewart platform manipulators. *Mechanism and Machine Theory* 43 (5): 591–616.

Bi, Z.M., and Y. Jin. 2011. Kinematic modeling of exechon parallel kinematic machine. *Robotics and Computer Integrated Manufacturing* 27 (1): 186–193.

Boer, C.R., L. Molinari-Tosatti, and K.S. Smith. 1999. *Parallel kinematic machines*. London: Springer.

Boudreau, R.R., and C.M. Gosselin. 1999. The synthesis of planar parallel manipulators with a genetic algorithm. *ASME Journal of Mechanical Design* 121 (4): 533–537.

Cajal, C., J. Santolaria, J. Velazquez, S. Aguado, and J. Albajez. 2013. Volumetric error compensation technique for 3D printers. *Procedia Engineering* 63: 642–649.

Dasgupta, B., and T.S. Mruthyunjaya. 2000. The Stewart platform manipulator: A review. *Mechanism and Machine Theory* 35 (1): 15–40.

Dietmaier, P. 1998. The Stewart-Gough platform of general geometry can have 40 real postures (Book Chapter). In *Advances in robot kinematics: Analysis and control*, 7–16.

Gao, W., Y. Zhang, D. Ramanujan, K. Ramani, Y. Chen, C.B. Williams, C. Wang, Y.C. Shin, S. Zhang, and P.D. Zavattieri. 2015. The status, challenges and future of additive manufacturing in engineering. *Computer Aided Design* 69: 65–89.

Gausemeier, J., N. Echterhoff, M. Kokoschka, and M. Wall. 2011. Thinking ahead the future of additive manufacturing—analysis of promising industries (online print). http://www.innovations-wissen.de/uploads/tx_publicationscrud/pdfs/Thinking_ahead_the_Future_of_Additive_Manufacturing_-_Future_Applications_2_pdf. Accessed on 10th Aug 2017.

Huang, T., M. Li, X. Zhao, J. Mei, D. Chetwynd, and S. Hu. 2005. Conceptual design and dimensional synthesis for a 3-DOF module of the TriVariant—a novel 5-DOF reconfigurable hybrid robot. *IEEE Transactions: Robotics* 21 (3): 449–456.

Husty, M.L. 1996. An algorithm for solving the direct kinematics of general Stewart-Gough platforms. *Mechanism and Machine Theory* 31 (4): 365–379.

Jin, Y., I.M. Chen, and G. Yang. 2009. Kinematic design of a family of 6-DOF partially decoupled parallel manipulators. *Mechanism and Machine Theory* 44: 912–922.

Jin, Y., I. Chen, and G. Yang. 2004. Kinematics analysis of a 6-DOF selectively actuated parallel manipulator. *IEEE Conference on Robotics, Automation and Mechatronics* 1: 231–236.

Kataria, A., and D.W. Rosen. 2001. Building around inserts: Methods for fabricating complex devices in stereolithography. *Rapid Prototyping Journal* 7 (5): 253–262.

Kelaiaia, R., A. Zaatri, and O. Company. 2011. Multi-objective optimization of parallel kinematic mechanisms by the genetic algorithms. *Robotica* 30: 783–797.

Lara-Molina, F., J. Rosario, and D. Dumur. 2011. Multi-objective optimization of Stewart-Gough manipulator using global indices. In *IEEE/ASME international conference on advanced intelligent mechatronics (AIM)*, 79–85.

Low, A. 2002. Design and development of 2-DOF module and calibration fixture for modular robots. Senior Project Report, Nanyang Technology University, Singapore.

Mason, A. 2006. Multi-axis hybrid rapid prototyping using fusion deposition modeling. Master's thesis, Ryerson University, Toronto, Ontario, Canada.

Merlet, J.P. 2003. *Parallel robots*. London: Springer.

Narahara, H., and K. Saito. 1995. Study on the improvement of surface roughness of complex model created by three dimensional photo-fabrication proposal of lift up irradiation method. *Journal of Japan Society of Precision Engineering* 61 (2): 233–237.

Oropallo, W., and L. Piegl. 2016. Ten challenges in 3D printing. *Engineering with Computers* 32: 135–148.

Pan, Y., X. Zhao, C. Zhou, and Y. Chen. 2012. Smooth surface fabrication in the mask projection based stereolithography. *SME: Journal of Manufacturing Processes* 14 (4): 460–470.

Pandey, P.M., N.V. Reddy, and S.G. Dhande. 2003. Improvement of surface finish by staircase machining in fused deposition modelling. *Journal of Materials Processing Technology* 132 (1): 323–331.

Rehsteiner, F., R. Neugebauer, S. Spiewak, and F. Wieland. 1999. Putting parallel kinematics machines (PKM) to productive work. *Annals of CIRP* 48 (1): 345–350.

Ruan, J., K. Eiamsaard, and F.W. Liou. 2005. Automatic process planning and toolpath generation of a multi-axis hybrid manufacturing system. *Journal of Manufacturing Processes* 7 (1): 57–68.

Sager B. 2006. SLA characterization for surface finish improvement: Inverse design methods for process planning. Ph.D. thesis, Georgia Institute of Technology, Atlanta, GA, USA.

Sager, B., and D.W. Rosen. 2008. Use of parameter estimation for stereolithography surface finish improvement. *Rapid Prototyping Journal* 14 (4): 213–220.

Son, S., T. Kim, S.E. Sarma, and A. Slocum. 2003. A hybrid 5-axis CNC milling machine. *Precision Engineering* 33 (4): 430–446.

Song, X., Y. Pan, and Y. Chen. 2015. Development of a low-cost parallel kinematic machine for multidirectional additive manufacturing. *ASME: Journal of Manufacturing Science and Engineering* 137 (2): 1–13.

Warnecke, H.J., R. Neugebauer, and F. Wieland. 1998. Development of hexapod based machine tool. *Annals of CIRP* 47 (1): 337–340.

Yang, G., I. Chen, W. Chen, and W. Lin. 2004. Kinematic design of a six-DOF parallel kinematic machine with decoupled motion architecture. *IEEE Transactions on Robotics* 20 (5): 876–887.

Zhang, D. 2010. *Parallel robotic machine tools*. London: Springer.
Zhang, G.Q., W. Mondesir, C. Martinez, X. Li, T.A. Fuhlbrigge, and H. Bheda. 2015. Robotic additive manufacturing along curved surface—a step towards freeform fabrication. In 2015 *IEEE International conference on robotics and biomimetics (ROBIO)*, Zhuhai, 721–726. (https://doi.org/10.1109/robio.2015.7418854).

Development of a Flexible Assembly System Using Industrial Robot with Machine Vision Guidance and Dexterous Multi-finger Gripper

Atul Mishra, I. A. Sainul, Sudipta Bhuyan, Sankha Deb, Debashis Sen and A. K. Deb

Abstract In today's era of mass customization, assembly automation systems should be designed with necessary production flexibility to cope with the growing product varieties to adapt to diverse customer requirements, yet the production costs should not be significantly different from those of comparable products made by mass production. In order to cope with this product variety-cost trade-off, robotics offers a flexible automation technology for turning assembly systems into efficient and flexible systems. Despite their great potential for high flexibility, there is a range of issues which must be addressed for its successful implementation. This chapter examines some of these key issues and challenges, reviews the results of previous research and describes our ongoing research on development of a flexible assembly system for mechanical products, using an industrial robot with machine vision guidance and dexterous multi-finger gripper. As part of the research work reported in this chapter, a Sexual Genetic Algorithm (SGA)-based approach for generation of optimal assembly sequence, a knowledge-based system for generating the robot task-level plan, a multi-finger robot gripper for flexible assembly based on a tendon-driven mechanism and an impedance control algorithm, and finally a strategy for implementation of robotic assembly under machine vision guidance have been presented.

A. Mishra · S. Deb (✉)
FMS and CIM Laboratory, Department of Mechanical Engineering, IIT Kharagpur, Kharagpur, India
e-mail: sankha.deb@mech.iitkgp.ernet.in

I. A. Sainul · S. Bhuyan
Advanced Technology Development Centre, IIT Kharagpur, Kharagpur, India

D. Sen
Department of Electronics & Electrical Communication Engineering, IIT Kharagpur, Kharagpur, India

A. K. Deb
Department of Electrical Engineering, IIT Kharagpur, Kharagpur, India

S. S. Pande and U. S. Dixit (eds.), *Precision Product-Process Design and Optimization*, Lecture Notes on Multidisciplinary Industrial Engineering,
https://doi.org/10.1007/978-981-10-8767-7_2

Keywords Flexible assembly system · Assembly sequence optimization
Robot task-level planning · Multi-finger gripper for flexible assembly
Vision-guided robotic assembly

1 Introduction

In the past, automation of assembly operations in industries had been successfully accomplished by employing fixed automation equipment for mass production, resulting in lower prices of products due to economies of scale. The reasons for its success can be attributed to primarily stable business environments, in which customers did not demand product differentiation. However, in today's era of mass customization, mass-produced products do not satisfy customers, which are designed largely to fulfil the average needs of customers. Therefore, flexibility is needed in assembly production systems with the necessary agility and adaptability to diverse customer requirements and capability of mass customization. Yet, the production costs of mass customized products should be low enough so that their prices are not significantly different from those of comparable products made by mass production. In order to cope with this product variety-cost trade-off, robotics offers a flexible automation technology for turning assembly systems into efficient and flexible systems (Deb and Deb 2010). Reconfiguring the robotic assembly system for new products involves a change in the software program (which is accomplished by simply reprogramming the same robot) as opposed to major hardware changes that may be required in case of hard automation. This would permit quick changeover of the assembly system in response to customer demands. Thus, several products can be made using a single robotic workcell, resulting in saving costs associated with multiple installations. Despite their great potential for high flexibility, there is a range of challenging issues which must be addressed for successful implementation of flexible assembly systems using robots. This chapter examines a number of these issues, reviews the results of previous research and describes our ongoing research on development of a flexible assembly system for mechanical products, using an industrial robot with machine vision guidance and dexterous multi-finger gripper.

2 Review of Previous Research

2.1 *Assembly Planning and Optimization*

Before programming the robot motions for carrying out assembly, it is first necessary to plan the order or the sequence in which the components need to be assembled. Further, it is important to identify the optimal assembly sequence to minimize the assembly time and cost. Various approaches have been developed as reported in the literature for automatic generation of optimal sequence. A few of them include application of mathematical algorithms, graph theoretic approaches, and more recently various evolutionary soft computing based optimization techniques which include GA, memetic algorithm, ant colony optimization, artificial immune systems, cuckoo search algorithm, flower pollination algorithm, harmony search algorithm, particle swarm optimization, and hybrid of the above. A brief review of these approaches is given below.

Chen et al. (2004) used various graph-based methods such as above graph, Assembly Precedent Diagram (APD), and relational model graph to model the relative positions, precedence constraints and relationships between different parts of an assembly. For the generation and evaluation of a feasible assembly sequence, two methods namely a mathematical model based on penalty index and Revised Minimum Spanning Table (RMST) methods were used. To solve the assembly sequence optimization problem, Bonneville et al. (1995) developed a Genetic Algorithm (GA)-based approach, starting from feasible assembly sequences provided by experienced assembly planners. To generate the offspring from the parent population, crossover and mutation operators were used followed by evaluation and selection of resulting offspring. The feasibility of operations was checked by the Liaisons and geometric constraints. Chen and Liu (2001) developed an adaptive GA (where rules were used to vary the genetic-operator probabilities accordingly) for obtaining the global best or near best assembly sequences efficiently. Marian et al. (2006) also used GA for assembly sequence planning optimization where guided search was used to generate the solutions taking both extrinsic and intrinsic precedence relations into consideration. Stochastic sampling selection, guided search with supplementary conditions type crossover and modified guided search mutation operators were used. Choi et al. (2009) solved the assembly sequence optimization problem by an approach based on GA in which precedence preservative type crossover and swap mutation had been used. Mishra and Deb (2016a) proposed a GA-based assembly sequence optimization approach in which a partially matched crossover and a swap mutation were used. Wang et al. (2005) proposed an Ant Colony Optimization (ACO) approach for optimization of assembly sequences based on minimizing the reorientations during the assembly processes. The concept of assembly by disassembly was adopted by the authors, and disassembly matrix was used to ensure the feasibility of sequences. However, the ACO algorithm has a lot of parameters which needs to be varied in order to obtain the best convergence rate of the algorithm. A binary-coded GA was proposed by Mishra and Deb (2016b) to optimize the parameters of the ACO algorithm. Cao and Xiao (2007) developed the Immune Optimization Algorithm (IOA) for generation of optimal assembly plan. It worked on the bionic principles of artificial immune systems. Attributes such as degree of feasibility, base component location in the sequence, and number of tool changes and reorientations required were used to evaluate the assembly sequences. Lv and Lu (2010) showed an application of discrete Particle Swarm Optimization (PSO) in assembly sequence optimization. Number of tool changes, number of reorientation required, number of changes in operation type and number of interferences during the product assembly were considered as optimization criteria. The particle positions and velocities in PSO were updated using subtraction, addition and multiplication operators. Gao et al. (2010) proposed a memetic algorithm based assembly sequence optimization approach on where assembly sequences are treated as chromosomes that consist of genes containing the component number and its assembly direction. Number of reorientation required during assembly and assembly sequence feasibility as the objective function were considered. Generation of new offspring was done using PMX-type crossover and swap mutation along with the local search operator. Li et al. (2016) proposed an Improved Harmony Search (IHS) for assembly sequence optimization. The paper demonstrated novel features like an initial Harmony Memory (HM) using the Opposition-Based Learning (OBL) method, a way to improvise a new harmony and a strategy for local search. Number of reorientations required during assembly, number of tool changes, and stability criteria in the objective function were considered. Mishra and Deb (2016c) developed a discrete Flower Pollination Algorithm (FPA)-based assembly sequence optimization approach for generating not only the global best assembly sequence but as many unique optimum solutions as possible. Mishra and Deb (2017) proposed a discrete cuckoo search algorithm to generate global best assembly sequences considering minimization of orientation changes, tool/gripper changes, assembly stability and base component location. It was integrated with an upstream assembly product database to extract information on assembly directions, contact details, assembly precedence constraints and tool/grippers. Zhou et al. (2011) used a hybrid bacterial chemotaxis and GA-based approach for assembly sequence optimization. The chromosomes were the assembly sequences, wherein gene of the chromosome was the bacterium. Length of longest sub-sequence, number of reorientations required and number of tool changes were considered in fitness function. Xing and Wang (2012) developed a combination of PSO and GA approach based on graph theory for compliant assembly sequence optimization. Liaison graph and adjacency matrix were employed to evaluate the geometry of the compliant assemblies. The string of components represented the assembly sequences, whose length is equal to number of assembly components. Evaluation of assembly sequences was based on assembly variation due to dimensional tolerance. Karthik and Deb (2017) proposed and implemented a methodology for assembly sequence optimization using a hybrid Cuckoo Search Genetic Algorithm (CSGA).

2.2 Robot Task-level Planning

After assembly sequence planning, a robot task-level plan containing a sequence of intermediate steps to achieve the robot task-level goals has to be drawn up from the assembly sequence. Different approaches have been discussed in the literature for automating this task. A few of them include use of agent-based techniques, object-oriented approaches, knowledge-based systems, etc. A brief review of these approaches is given below.

Ramos (1992) presented an intelligent robotic assembly system, incorporating a link with computer vision and an efficient task-level planning methodology based on a cooperative agent-based approach and the implementation of the intended task using a robot manipulator. The various agents are MODELS (representing the models of objects), VISION, world descriptor that converts numerical data given by VISION in symbolic relationships and constraints, task-level planner, task executor which is responsible for the execution and monitoring of the tasks. Osuna et al. (2003) developed an intelligent task planning system that was built upon a cooperative web-based environment to integrate the product and assembly system design processes. For intelligent task planning, object-oriented and machine learning techniques were integrated. Cambon et al. (2004) developed a planner, aSyMov. They considered symbolic and geometric constraints at each step of the planning process. The representations used by a symbolic task planner and the representations used by a realistic motion and manipulation planning library had been shown to be effectively linked. Cho et al. (2010) proposed the implementation methods of robotic task planning and execution based on the description logic knowledge base. Their implementation used a description logic knowledge base which had been expanded to encompass the description of task's goal and behaviours. A representation scheme for knowledge had been investigated and a simple algorithm was realized. A complete process chain was presented in Thomas and Wahl (2010) that starts with initial specification of assembly tasks using assembly sequence planning and finishes with task planning and execution. Their system demonstrated how to generate robot programs automatically from CAD data. Backhaus (2013) introduced and discussed an approach to simplify the use of task-oriented programming for assembly systems. The system in the output provided the device-specific code based on the input as description of assembly systems and devices (namely robot, gripper, conveyor belt, etc.). Recently, Alatartsev et al. (2015) surveyed several papers and presented their findings detailing robotic task sequencing problem considering collision-free path planning, production scheduling, multi-robot task planning, task-level planning, combination of task-level planning and path planning, online control-based planning, and manipulation planning.

2.3 *Design of Robotic Grippers and Their Controller Design*

In a flexible assembly system, dexterous robot grippers in the form of multi-finger anthropomorphic hands have a crucial role to play to support the grasping and manipulation of objects of diverse geometric shapes. A number of multi-finger robot gripper designs can be found in the literature. These grippers can be classified according to the nature of actuation mechanism employed, i.e. either fully actuated grippers (i.e. same number of actuators as the number of DOFs) or under-actuated grippers (i.e. lesser number of actuators than the number of DOFs). A growing trend in gripper designs is the use of under-actuated mechanisms based on linkages and gear trains (LiCheng et al. 2009; Tlegenov et al. 2014). However, the limitations of these mechanisms, especially the increase in overall size and weight, and also their lack of compliance, have prompted researchers to think of other alternative options including use of Tendon-Driven Mechanisms (TDMs), which have the potential to overcome the above drawbacks. Tendons, or more generally speaking, cables, had been widely used earlier in many mechanical devices since the nineteenth century. The use of tendons for robotic applications to develop grippers is a more recent development and has been studied since the early 1980s. Several tendon-actuated robot grippers have been developed all over the world, worth mentioning among them are the Stanford/JPL hand by Salisbury and Roth (1983) and Loucks et al. (1987), the Utah/MIT hand by Jacobsen et al. (1984, 1986), the multi-jointed finger hand by Okada (1986), the Barrett gripper by Townsend (2000), the Belgrade/USC hand by Bekey et al. (1990), DLR hand developments by Butterfass et al. (1998, 2001), Gao et al. (2003) and Liu et al. (2008), the LMS hand by Gazeau et al. (2001), the NASA Robonaut hand by Lovchik and Diftler (1999) and Ambrose et al. (2000) and JU hand by Pal et al. (2008). Despite the aforementioned developments, it has been generally found that the industrial applications of tendon-driven robot grippers are more limited as compared to other types of mechanical robotic grippers. This is so because there are some challenges in its implementation particularly when it comes to control of the system as it cannot be directly controlled by conventional control techniques. Ozawa et al. (2014) developed a tendon-driven robotic hand and implemented different types of controller. Compliant behaviour was incorporated by adding springs to the passive tendons in the fingers. Abdallah et al. (2012) developed two-tier architecture of position and force control in two feedback loops which improves the overall performances of using only position or force controller. Impedance control approach is preferred while system interacting with the environment, as it gives great flexibility by allowing regulation of both motion and contact force. Diftler et al. (2011) designed dual-priority based impedance controller for Robonaut hand.

2.4 Vision-Guided Robotic Assembly

Once the task-level plan for the required assembly is generated, a robot manipulator guided by a machine vision system can be deployed to automatically assemble the constituent parts. A few investigations in this regard are discussed below. Pena-Cabera (2005) proposed a method of object recognition and their localization for assembly components. Using a Fuzzy ARTMAP neural network model, assembly components were accurately recognized. Golnabi and Asdapour (2007) described the role and importance of the machine vision systems in industrial applications. System design methodology and a generic machine vision model were discussed. The justification for utilizing machine vision had been considered on the basis of economic and logistic considerations. Zhang et al. (2011) introduced a vision-guided alignment method that utilized a camera space manipulation control process and relied on the components' CAD model. To achieve the high accuracy alignment for the final assembly, a local calibration method was used. Kobari et al. (2013) presented an error resilient vision-based motion control method for parts assembly. They considered those assembly parts which get deformed during assembly. Visual information obtained from a camera reading the deformation of the parts is used to determine the force applied. Chen et al. (2015) reviewed papers on vision system working in combination with force control integrated assembly and demonstrated a system using a high accuracy assembly process. In a semi-structured environment, industrial robot can perform assembly more accurately by taking both vision and force control data. The vision system was not calibrated carefully as it provided only rough position data of parts. In addition to vision, to deal with errors, a local searching method was used which was based on force–torque control.

3 Proposed Methodology

The following sections discuss the methodologies that have been proposed in this work for (1) assembly sequence planning and optimization, (2) robot task-level plan generation, (3) design of a multi-finger robot gripper for flexible assembly, and finally (4) implementation of the robotic assembly system under the guidance of machine vision.

3.1 Assembly Sequence Planning and Optimization

In the following sections, the key issues and challenges in assembly sequence planning and optimization, a brief problem description and the methodology proposed in this work and the results and discussions will be presented.

3.1.1 Key Issues and Challenges

Before the robot is programmed for performing assembly of a product, it is necessary to decide the order or sequence in which the components need to be inserted, such that no component interferes with others, thereby implying that they must comply with certain precedence constraints. A significant amount of human expertise and experiential knowledge is needed to accomplish the above. Moreover, an assembly product may be assembled by following number of distinct sequences, out of which the optimal sequence must be identified. Further, as the part count (i.e. the number of components in the assembly) increases, the number of feasible assembly sequences also growths exponentially, which makes assembly sequence optimization arduous and time consuming, if it is done manually. Keeping the above in mind, in this work, a Sexual Genetic Algorithm (SGA)-based approach is proposed for automating the generation of optimum assembly sequence. First, a brief problem description is given below.

3.1.2 Brief Problem Description

The problem of assembly sequence optimization involves determining the best feasible assembly sequence based on one or more criteria such as minimization of number of direction changes and tool changes, maximizing the stability of the components/sub-assemblies while performing the assembly, etc. It is obvious that changing the direction of assembly of the consecutive components increases the handling time resulting in increase in overall cost of the assembly. Equation (1) shows how to compute the total number of direction changes (n_{d}).

$$n_{\mathrm{d}} = \sum_{i=1}^{n-1} (\mathrm{dir\_change}_{i,i+1}) \tag{1}$$

where

$$\mathrm{dir\_change}_{i,i+1} = \begin{cases} 0, & \text{if assembly\_dir}_i = \text{assembly\_dir}_{i+1} \\ 1, & \text{otherwise} \end{cases} \tag{2}$$

$\mathrm{dir\_change}_{i,i+1}$ represents the change in direction of assembly for two successive assembly operations and n represents the number of components in the assembly, $\mathrm{assembly\_dir}_i$ is the direction of assembly for component number i (the direction of assembly for components may be one of the following namely, $\pm x$ or $\pm y$ or $\pm z$).

At times, several tools and grippers (in case of robotic assembly) are required for performing the assembly or manipulation. The change in tool/gripper is also a non-productive task as it consumes handling time leading to increase in overall assembly cost. Hence, minimizing the tool changes is also one of the important criteria to reduce the overall assembly time and cost. Equation (3) shows how to compute the number of tool changes (n_{t}).

$$n_{\mathrm{t}} = \sum_{i=1}^{n-1} (\mathrm{tool\_change}_{i,i+1}) \tag{3}$$

where

$$\mathrm{tool\_change}_{i,i+1} = \begin{cases} 0, & \text{if tool\_number}_i = \text{tool\_number}_{i+1} \\ 1, & \text{otherwise} \end{cases} \tag{4}$$

$\text{tool\_change}_{i,i+1}$ represents the change in assembly tool/gripper for two successive assembly operations and n represents the number of components in the assembly; tool_number_i represents the tool number required for handling/insertion of component number i (a unique tool number is given to each tool in the tool database).

The base component in an assembly sequence should be the first component in assembly order. This base component information is provided by the user. The location of the base component in the sequence is represented by B, whose value is calculated as follows.

$$\text{Location of base component}, B = \begin{cases} 1, & \text{if the base component is in first position in the sequence} \\ 0, & \text{otherwise} \end{cases} \tag{5}$$

The connections between the components can be categorized into stable connection (SC), conditional stable connection (CSC) and unstable connection (USC). As the name suggests, SC signifies inseparable contact between components. Sometimes between components, there is CSC, which signifies that the connections are not autonomously stable, but components may separate from each other spontaneously. For unstable connections, fixtures are necessary to hold the components in place and maintain contact. This classification scheme and its details can be found in the paper by Li et al. (2016). To calculate the stability index of an assembly sequence, a stability (connection) matrix is used, which is a square matrix, having rows and columns equal to number of assembly components. The value of each element of the stability matrix can be 0, 1 or 2 depending on whether the connection type is unstable, conditionally stable or stable, respectively. The stability index (SI) can be found out as follows.

$$\text{SI} = \sum_{i=2}^{n} S_i | 0 \le \text{SI} \le 2n - 2 \tag{6}$$

where S_i is the stability of the component "i" and n is the number of components in the assembly.

The feasibility violations (FV) are the number of precedence violations because of certain component number violating the precedence. This can be understood from the example explained below, say, an assembly has four components and one of the infeasible sequences for this assembly is [4, 1, 2, 3]. Suppose that the Precedence Matrix (PM) of the assembly is as follows.

PM =

Component Number	1	2	3	4
1	0	0	0	0
2	1	0	0	0
3	1	1	0	0
4	1	1	1	0

(7)

This PM conveys the information as to which component has precedence with which component. The dimension of this matrix is equal to the number of assembly components. Here, 0 implies that no precedence constraint exists between the components and 1 implies that a precedence constraint exists. Now, for calculating the number of feasibility violations, from an infeasible solution, the first component is selected and checked to determine with which components it has precedence, and whether it is following the precedence or not. In this example, component '4' has precedence with components '1', '2' and '3', since the value in the PM is 1 here for each of PM(4,1), PM(4,2) and PM(4,3). So because of location of component '4' in the assembly sequence, it has a total of three feasibility violations. Next, the precedence is checked for the second component of the assembly sequence, i.e. component '1'. It reveals that it does not require any other component to be assembled before it, and hence it has no feasibility violation. In a similar manner, each component of the assembly sequence is checked to see if it satisfies the precedence criterion or not. For the solution [4, 1, 2, 3], components 2 and 3 also do not have any feasibility violations. Hence, it can be concluded that the number of feasibility violations for the infeasible solution [4, 1, 2, 3] is 3. It is obvious that for a feasible assembly sequence, the number of feasibility violations would be zero.

3.1.3 Proposed Approach Based on Sexual Genetic Algorithm

The Genetic Algorithm (GA) tries to mimic the various processes of natural selection involved in the intricate process of biological evolution. One such process is mate choice, or sexual selection that counteracts as well as enhances natural selection. Nowadays, much of the biological diversity and complexity is ascribed to this force. The Sexual Genetic algorithm (SGA) can successfully counter to a large extent the general tendency of a GA of slow convergence and getting stuck in a suboptimal solution. Hence, a SGA-based strategy has been adopted in this work. The entire population in SGA is divided into male and female chromosomes. Reproduction is done sexually to emulate male vigour and female choice; hence, in crossover, any female chromosome will mate only with a male chromosome having better average fitness (Goh et al. 2003). The rest of the process is similar to that of a GA.

The steps involved in the SGA are stated in Fig. 1. The genetic operators modified to suit the particularities of the assembly sequence planning problem are described here as well. A flowchart of the SGA is given in Fig. 1.

Selection

A selection operator called roulette wheel selection has been used that selects the assembly solutions to form the mating pool. All the solutions are evaluated using the fitness function, and the ranges for the solutions are created according to their fitnesses. Random numbers are generated multiple times, which equals the size of the population. Depending on the ranges within which these random numbers fall in, corresponding solutions are selected. The selected solutions are then put into the

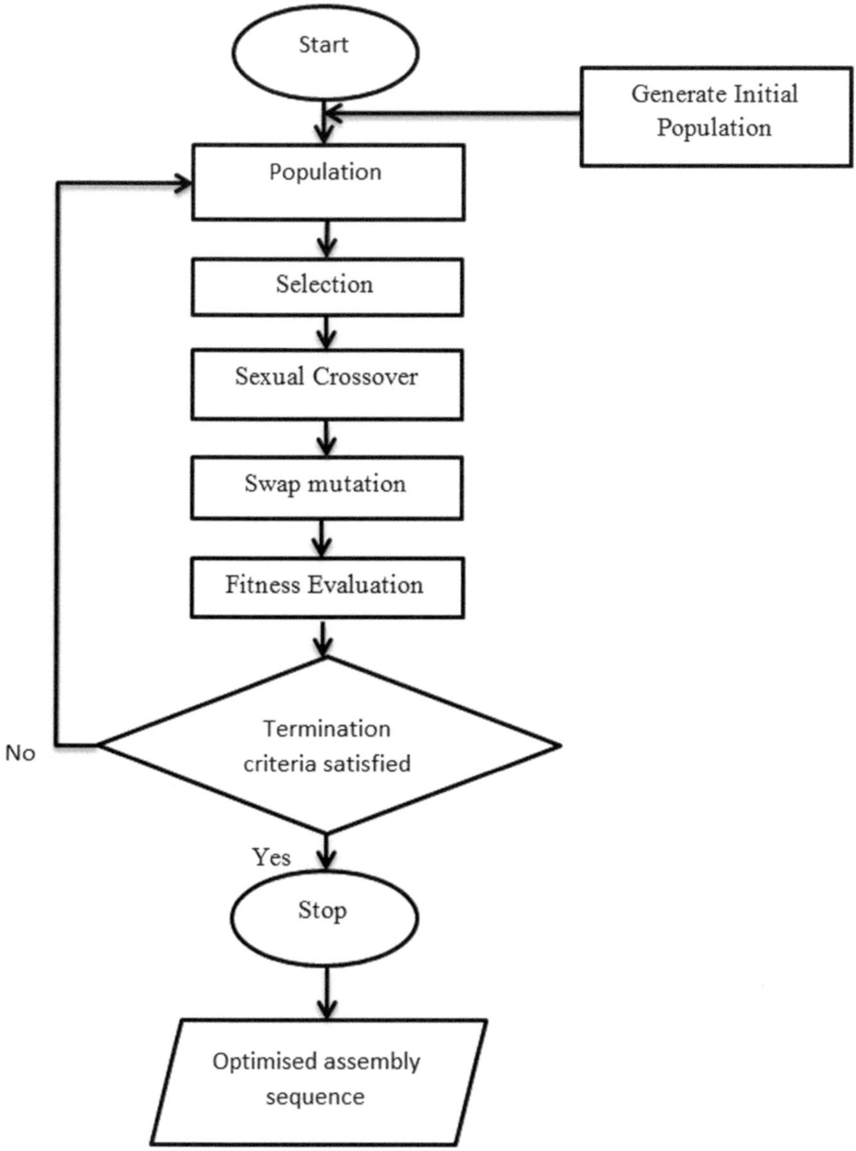

Fig. 1 The SGA flowchart for finding the optimal assembly sequence

mating pool on which special crossover operation is carried out. The size of the mating pool is equal to population size of the GA.

Sexual Crossover

Sexual crossover is the main genetic operator that differentiates SGA from a generic GA. In the process, a female chromosome is selected to mate with a male partner of a better general fitness (best among 4–5 male chromosomes). It is ensured that the off-springs are of different genders so that the sex ratio balance is maintained. The constrained nature of assembly sequence planning and necessity to maintain feasibility at every step calls for a modified crossover technique. The steps involved are described as below:

(a) The population of chromosomes is first randomly divided into two: male population and female population, each containing equal number of chromosomes.
(b) Select each female chromosome in a consecutive order (without replacement), i.e. each female will only get to be selected to mate once.
(c) Select a male chromosome of a better general fitness (best among 4–5 male chromosomes).
(d) Apply the Partially Matched Crossover (PMX) operator as follows. Generate an offspring by selecting a part of a solution (i.e. number of successive genes) from one parent and conserving the order and the positions of as many genes as possible from the other parent. A two-point crossover is applied where the part of a solution serves as boundaries for the swapping operations. This crossover operator takes care of the problem of repetitions of part numbers in the assembly sequence.

Swap Mutation

The mutation operator called swap mutation is used in this work. The genes are swapped arbitrarily inside a solution in case of swap mutation.

Fitness Evaluation of the Population

For the given problem of assembly sequence optimization, a random population of individuals (i.e. chromosomes representing assembly sequences) is generated which can be feasible or infeasible. This population is evaluated using a fitness function (FF) given in Eq. 8, consisting of number of tool changes (n_t) direction changes (n_d), base component location in assembly sequence (B) and stability index (SI). To ensure feasibility of final individuals, a feasibility criterion is incorporated that is measured in terms of number of feasibility violations (FV) due to components' locations in the sequence.

$$\mathrm{FF} = w_1 \frac{(n-1-n_\mathrm{d})}{(n-1)} + w_2 \frac{(n-1-n_\mathrm{t})}{(n-1)} + w_3 * B + w_4 * \frac{\mathrm{SI}}{(2n-2)} + \frac{1}{\left(\frac{FV}{n}+1\right)} \tag{8}$$

where n represents number of assembly components and w_1, w_2, w_3 and w_4 are weight coefficients for direction changes, tool changes, base component location and stability index, respectively, with their corresponding values as 0.25, 0.25, 0.1 and 0.4.

3.1.4 Results and Discussions

To demonstrate the working of the proposed system, a 14-component product assembly shown in Fig. 2 has been used. Figure 3a, b shows an extract of the Precedence Matrix (PM) and Stability Matrix (SM) of the above assembly, respectively. Table 1 gives the information on the assembly directions and tools/grippers required for the assembly.

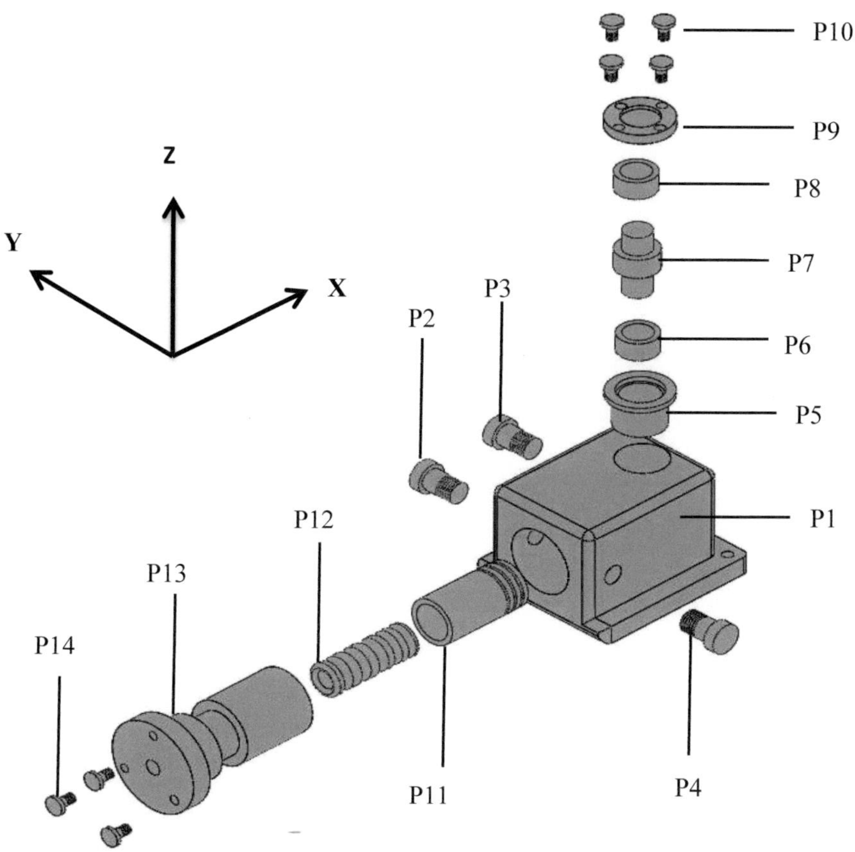

Fig. 2 14-component assembly

(a)

```
PM = [0 0 0 0 0 0 0 0 0 0 0 0 0 0;
      1 0 0 0 0 0 0 0 0 0 0 0 0 0;
      1 0 0 0 0 0 0 0 0 0 0 0 0 0;
      1 0 0 0 0 0 0 0 0 0 0 0 0 0;
      1 0 0 0 0 0 0 0 0 0 0 0 0 0;
      1 0 0 0 1 0 0 0 0 0 0 0 0 0;
      1 0 0 0 1 1 0 0 0 0 0 0 0 0;
      1 0 0 0 1 1 1 0 0 0 0 0 0 0;
      1 0 0 0 1 1 1 1 0 0 0 0 0 0;
      1 0 0 0 1 1 1 1 1 0 0 0 0 0;
      1 0 0 0 0 0 0 0 0 0 0 0 0 0;
      1 0 0 0 0 0 0 0 0 0 1 0 0 0;
      1 0 0 0 0 0 0 0 0 0 1 1 0 0;
      1 0 0 0 0 0 0 0 0 0 1 1 1 0];
```

(b)

```
SM = [0 2 2 2 0 0 0 0 0 0 0 0 0 0;
      1 0 0 0 0 0 0 0 0 0 0 0 0 0;
      0 0 0 0 0 0 0 0 0 0 0 0 0 0;
      1 0 0 0 0 0 0 0 0 0 0 0 0 0;
      0 0 0 0 0 1 0 0 0 0 0 0 0 0;
      0 0 0 0 0 0 0 0 0 0 0 0 0 0;
      0 0 0 0 0 0 0 1 0 0 0 0 0 0;
      0 0 0 0 0 0 0 0 1 0 0 0 0 0;
      0 0 0 0 0 0 0 0 0 2 0 0 0 0;
      0 0 0 0 0 0 0 0 0 0 0 0 0 0;
      1 2 0 0 0 0 0 0 0 0 0 0 0 0;
      0 0 0 0 0 0 0 0 0 0 0 0 0 0;
      0 0 0 0 0 0 0 0 0 0 0 0 0 2;
      0 0 0 0 0 0 0 0 0 0 0 0 0 0];
```

Fig. 3 **a** Precedence matrix and **b** stability matrix for the assembly shown in Fig. 2

Table 1 Information on assembly directions and tools/grippers required for the assembly shown in Fig. 2

Component number	Assembly direction	Tool/gripper number
1	*Z*−	1
2	*Y*−	2
3	*Y*−	2
4	*Y*+	2
5	*Z*−	3
6	*Z*−	3
7	*Z*−	4
8	*Z*−	5
9	*Z*−	5
10	*Z*−	6
11	*X*+	5
12	*X*+	5
13	*X*+	5
14	*X*+	6

The proposed SGA has been used to generate the best assembly sequence for the above 14-component product assembly. The main parameters affecting the performance of SGA algorithm are population size, crossover probability (p_c) and mutation probability (p_m) used in generating new eggs and number of iterations. To study their effects on performance of the SGA, a sensitivity analysis has been carried out and the optimum set of parameters found to be able to provide the optimal/near optimal assembly sequence in reasonable time is population size of 20, crossover probability of 0.90 and mutation probability of 0.05. After the SGA is run using the above-mentioned parameters, the optimal assembly sequence obtained is [1, 2, 3, 4, 5, 6, 7, 8, 9, 10, 11, 12, 13, 14] with fitness value of 1.5269, requiring seven tool changes, four reorientation changes and stability index of nine, having the base component at the first location of the assembly sequence, and most importantly the sequence was found to have no feasibility (precedence) violations. The efficiency of the proposed SGA is evaluated by comparing it with previously developed GA (Mishra and Deb 2016). The best sequence generated by the GA algorithm is also found to be [1, 2, 3, 4, 5, 6, 7, 8, 9, 10, 11, 12, 13, 14] with fitness value of 1.5269, requiring seven tool changes, four reorientation changes and stability index of nine, having base component at the first location of the assembly sequence and no feasibility violations. Figure 4 shows comparison between the

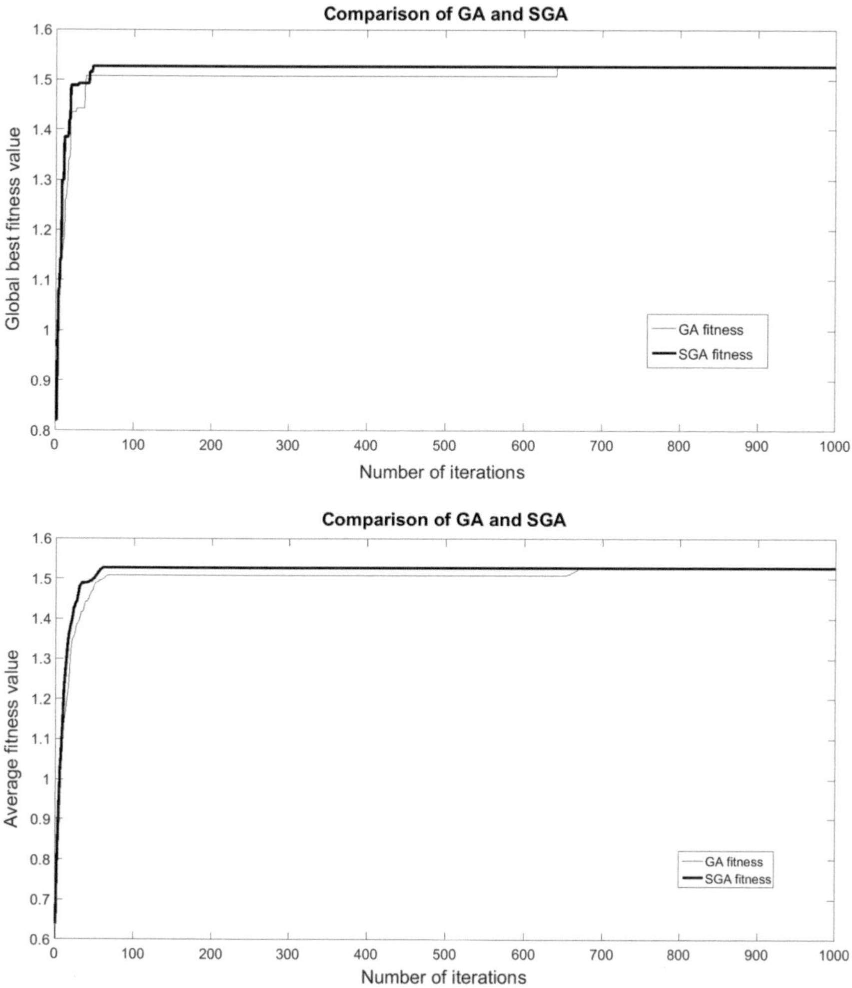

Fig. 4 Comparison between the convergence graphs of the proposed GA algorithm and SGA algorithm

Table 2 Comparison between the results of the GA and SGA

	GA	SGA
Best fitness value	1.5269	1.5269
Mean of the best fitness values for 10 runs	1.5015	1.5138
Number of runs out of 10 independent runs in which convergence to global best fitness was achieved	3	6
% of obtaining the optimal assembly sequence	30	60

convergence graphs of the GA and SGA algorithms from which it can be concluded that the proposed SGA clearly outperforms GA in terms of convergence speed as it is able to reach the global best fitness value in lesser number of iterations. The SGA reached the global best solution in 19 iterations, while the GA took 669 iterations to reach the global best solution. Table 2 gives a summary of the results of comparison between the GA and the SGA algorithms. It is found that in 6 times out of 10 runs, the SGA could find the optimal solution of 1.5269, which gives a 60% probability of finding the best solution or assembly sequence. In contrast, the probability of finding the best solution by GA is only 30%.

From the above, it is evident that the proposed SGA-based approach is able to generate the optimum solution. Furthermore, the algorithm is also more consistent than the previously developed GA-based approach. It should be noted that the optimal assembly sequence provided by the SGA requires the least number of orientation changes and tool/gripper changes, it is the most stable and most importantly it does not have any feasibility violations.

3.2 Knowledge-Based System for the Generation of Task-level Assembly Plan

The following sections present the key issues and challenges in generation of task-level assembly plan, architecture of the proposed knowledge-based system and the results and discussions.

3.2.1 Key Issues and Challenges

After assembly sequence planning, a robot task-level plan has to be drawn up from the assembly sequence. The task-level plan contains a sequence of intermediate steps to achieve the robot task-level goals such as, for example, for performing assembly of two parts *A* and *B*, obtain first the initial location coordinates of part *A* using a suitable sensor like an overhead vision camera, move the robot over part *A*, perform grasping of part *A* by closing the gripper, obtain the location coordinates of its destination on part *B* in the assembly jig by vision camera, move the part

A over part *B*, open gripper to insert part *A* inside part *B*, etc. For a product requiring large number of components to be assembled, the traditional method of manually generating the aforementioned task-level plan is very tedious and time consuming. Although this task is usually performed by experts, any inadvertent mistake can be very costly and result in loss of productivity. Hence, it will be beneficial if this task is automated, for which a knowledge-based expert system approach has been proposed in this work.

A knowledge-based system has been developed to generate the task-level plan for robotic assembly from a given feasible and optimum assembly sequence that is generated by the assembly sequence planner described in the previous section. The above knowledge-based system has been implemented using the expert system shell, CLIPS.

3.2.2 Architecture of the Proposed Knowledge-Based System

The 'C Language Integrated Production System', abbreviated as 'CLIPS', is an expert system shell that was first developed at NASA (Giarratano and Riley 2002). It basically consists of a set of knowledge-based rules, supported by facts in order to fire or activate a rule that are stored in a database. An executable CLIPS rule basically consists of two parts with an 'IF' and a 'THEN' statement. In order for the rule to fire successfully to give a desired output, the syntax of its 'IF' part must match with the syntax of the facts provided in the database. If this condition is satisfied, the 'THEN' part of the rule comes into action. Both the facts and the IF-THEN rules must be defined first, before being used. The CLIPS system consists of mainly three components namely, a database, a knowledge base and an inference engine. In addition, the CLIPS system also has a user interactive interface. The architecture of the developed knowledge-based system using CLIPS has been shown in Fig. 5.

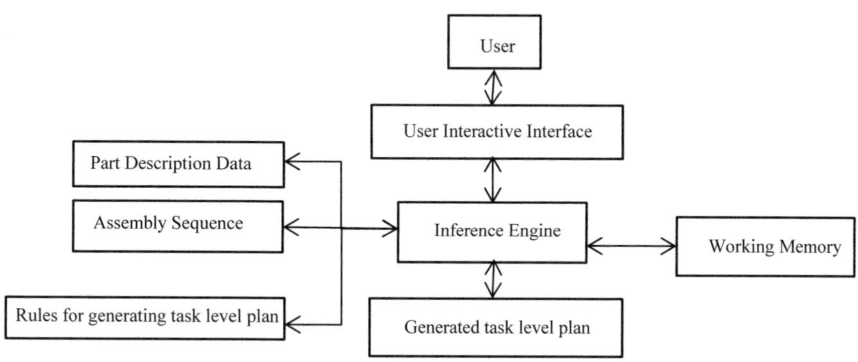

Fig. 5 Architecture of the developed knowledge-based system

Database

The database consists of a list of facts about the problem being solved, which forms the input to the expert system and must be entered by the user. The facts are represented in the CLIPS expert system shell using the concept of template. A template is an organized structure of input data that is used to define and store values for a fact description in a field called, the 'slot'. For the given problem which is to generate a task plan for the robot, the input data required are detailed information about the assembly components, such as component number, its name, type (i.e. functional part or fastener), part count, components with which it has contact, assembly direction, any special assembly requirement, etc., which are to be provided by the user. All these data are stored in the database. An operational template in CLIPS format is presented in Fig. 6a, along with a sample of input database facts given in Fig. 6b.

Knowledge Base

The knowledge base contains domain knowledge encoded in the form of rules, to be used by the expert system. Sample rules for the knowledge base are presented

(a)

```
; DEFTEMPLATE 1
        (deftemplate MAIN::component
        (slot number (type INTEGER)(default ?NONE))
        (slot name (type SYMBOL))
        (slot component_type(type SYMBOL)(allowed-symbols functional_part fastener))
        (slot part_count(type NUMBER)(default 1))
        (multislot contact_with(type NUMBER)(default 0))
        (slot thread_present(type SYMBOL)(allowed-symbols yes no)(default no))
        (slot tool_required(type SYMBOL)(allowed-symbols screwing_machine .......))
        (slot assembly_direction (type SYMBOL)(allowed-symbols x+ y+ z+ x- y- z-))
        (slot assembly_requirement (type SYMBOL)(allowed-symbols pre_requirement ......)))

; DEFTEMPLATE 2
        (deftemplate MAIN::assembly_method
        (slot assembly_method (type SYMBOL)(allowed-symbols manual robotic hybrid)))

;DEFTEMPLATE 3
        (deftemplate MAIN::sequence
        (multislot in_order_of (type INTEGER)))
```

(b)

```
(deffacts MAIN::components_data
        (component(number 1)(name base)(component_type functional_part)(part_count 1)(contact_with 2 3 4 5 7 9
        10 11 13 14)(thread_present no)(assembly_direction z-))
        ...........................))

(deffacts MAIN:assembly_method
                (assembly_method (assembly_method robotic)))

(deffacts MAIN::seqeunce_data
                (sequence (in_order_of 1 2 3 4 5 6 7 8 9 10 11 12 13 14)))
```

Fig. 6 **a** Templates for entering the assembly information in databases, **b** assembly information facts stored in the database, **c** a rule for generating task-level plan

(c)

```
; DEFRULE 1a
        (defrule MAIN::rule_1a
        (component(number ?nu1)(name ?na1)(component_type functional_part)(part_count 1))
        (sequence (in_order_of ?nu1 $?))
        (assembly_method (assembly_method robotic))
        =>
        (open "assembly_plan.txt" assembly_plan "w")
        (format assembly_plan "obtain coordinates of %s(%d) by vision%n" ?na1 ?nu1)
        (format assembly_plan "MOVE robot in joint mode above %s(%d)%n" ?na1 ?nu1)
        (printout assembly_plan "open gripper" crlf)
        (format assembly_plan "MOVE robot in linear mode to %s(%d)%n" ?na1 ?nu1)
        (printout assembly_plan "close gripper" crlf)
        (printout assembly_plan "obtain coordinates  of assembly fixture" crlf)
        ............................
        ............................
        (close))

; DEFRULE 3a
        (defrule MAIN::rule_3a
        (sequence (in_order_of $? ?nu1 ?nu2 $?))
        (component (number ?nu1)(name ?na1)(component_type functional_part)(assembly_direction ?ad1))
        (component (number ?nu2)(name ?na2)(component_type functional_part)(part_count ?pc2)
        ............................))
        (test (eq ?ad1 ?ad2))
        =>
        (open "assembly_plan.txt" assembly_plan "a")
        (printout assembly_plan "-----------------------------" crlf)
        (format assembly_plan "obtain coordinates of %s(%d) by vision%n" ?na2 ?nu2)
        (format assembly_plan "MOVE robot in joint mode above %s(%d)%n" ?na2 ?nu2)
        (format assembly_plan "MOVE robot in linear mode to component %s(%d)%n" ?na2 ?nu2)
        (printout assembly_plan "close gripper" crlf)
        (printout assembly_plan "obtain coordinates  of assembly fixture" crlf)
        (printout assembly_plan "MOVE robot in joint mode above fixture" crlf)
        (printout assembly_plan "MOVE robot in linear mode to fixture" crlf)
        (format assembly_plan "open gripper and insert the component %s(%d)%n" ?na2 ?nu2)
        (printout assembly_plan "MOVE robot in linear mode above fixture" crlf)
        (printout assembly_plan "MOVE robot in joint mode to safe point" crlf)
        (format assembly_plan "REPEAT steps of block %d times%n" ?pc2)
        (close))
```

Fig. 6 (continued)

later in this section for illustration. For the given problem of generating the task-level plan, a set of production rules have been developed using domain knowledge for task-level planning to efficiently program a robotic assembly system. These rules are stored in the knowledge base of the expert system. An extract from knowledge base rules and database facts in CLIPS format are shown in Fig. 6b, c, respectively.

Inference Engine

The inference engine in CLIPS serves as the brain of the knowledge-based system. It is a computer program, used for making inferences from the given data facts about a problem, through a process of reasoning. In CLIPS, the knowledge being stored in the form of rules, the inference engine makes inference by deciding, as to which of the rules from the knowledge base are satisfied by the database facts, and then it prioritizes the order of firing the satisfied rules.

User Interface

The user interactive interface in CLIPS consists of a display mechanism, to monitor each and every processing step during the execution of a program. It can be customized, by selecting or deselecting the options for viewing the status of database, rule activations, etc. during execution of the program in real time, which is sometimes useful for debugging purpose.

3.2.3 Results and Discussions

Example of an eight-component assembly as shown in Fig. 7 is given to demonstrate the results of the proposed knowledge-based system for generation of the task-level plans. Figure 8 shows the corresponding task-level plan generated by the knowledge-based system.

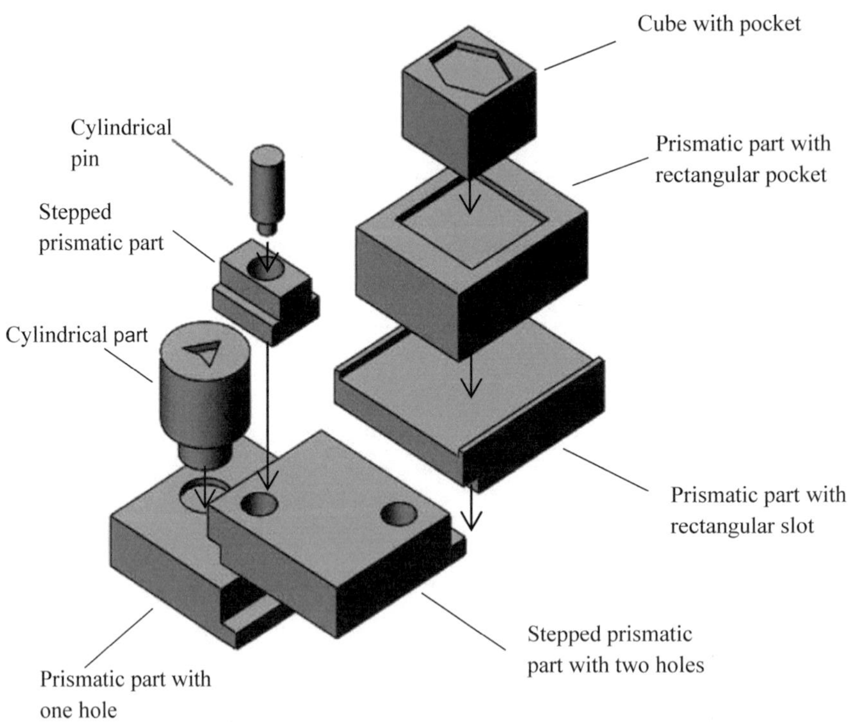

Fig. 7 An eight-component assembly

```
obtain coordinates of prismatic_part_with_one_hole(1) by vision
MOVE robot in joint mode above prismatic_part_with_one_hole (1)
open gripper
MOVE robot in linear mode to prismatic_part_with_one_hole (1)
close gripper
check if assembly jig is empty
obtain coordinates  on assembly jig where prismatic_part_with_one_hole is to be placed
MOVE robot in joint mode above jig
MOVE robot in linear mode to jig
open gripper
MOVE robot in linear mode above jig

obtain coordinates of stepped_prismatic_part_with_two_holes(2) by vision
MOVE robot in joint mode above stepped_prismatic_part_with_two_holes (2)
MOVE robot in linear mode to component stepped_prismatic_part_with_two_holes (2)
close gripper
obtain the coordinates on assembly jig where stepped_prismatic_part_with_two_holes is to be placed
MOVE robot in joint mode above jig
MOVE robot in linear mode to jig
open gripper and insert the component stepped_prismatic_part_with_two_holes (2)
MOVE robot in linear mode above jig
MOVE robot in joint mode to safe point
REPEAT steps of block 1 times

obtain coordinates of prismatic_part_with_rectangular_slot(3) by vision
MOVE robot in joint mode above prismatic_part_with_rectangular_slot (3)
MOVE robot in linear mode to component prismatic_part_with_rectangular_slot(3)
close gripper
obtain the coordinates on assembly jig where prismatic_part_with_rectangular_slot is to be placed
MOVE robot in joint mode above jig
MOVE robot in linear mode to jig
open gripper and insert the component prismatic_part_with_rectangular_slot (3)
MOVE robot in linear mode above jig
MOVE robot in joint mode to safe point
REPEAT steps of block 1 times
………………………….
………………………….
```

Fig. 8 Extract of the task-level plan for the eight-component assembly

3.3 *Proposed Multi-finger Robot Gripper for Flexible Assembly*

In the following sections, the key issues and challenges in design of multi-finger robot gripper for flexible assembly, the proposed gripper design and the results and discussions will be presented.

3.3.1 Key Issues and Challenges

As manufacturing is increasingly shifting away from high-volume, low-mix production to high-mix, low-volume production, flexible assembly cells are steadily gaining importance. Such assembly cells commonly require parts of diverse geometric shapes and sizes to be handled to perform various tasks like picking, fixing, placing, mating, etc. The mechanical grippers routinely used in industrial robots to perform above assembly tasks are simple two-finger grippers or parallel jaw-type grippers, employing actuation mechanisms based on various types of linkages. They must be, however, custom engineered according to specific application requirements of grasping. For example, to hold objects whose basic shape is cuboidal having flat faces, a gripper design with flat fingertips is preferred. On the other hand, in case of objects whose basic shape is cylindrical or spherical, a different gripper design with fingertips having V-grooves must be employed to provide a larger area of contact with the curved surfaces of the object. Thus, these two-finger or parallel jaw-type grippers are very effective for handling operations in fixed assembly cells, where part shapes are either only cuboidal or only cylindrical. But they lack the grasping flexibility like human hands to adapt to objects of diverse geometric shapes and therefore if such type of two-finger grippers were to be employed for flexible assembly, multiple grippers will be needed. Additionally, frequent gripper changes in the robot will be necessary, resulting in loss of productivity. An alternative solution could be to design a single multi-finger robot gripper, inspired by a human being's hand that will be capable of performing grasping and manipulation of variety of different object shapes without the need for changing the gripper. However, such type of gripper design may be too complex to engineer in practice and not economically viable for industrial applications. Thus, it is necessary to develop simple multi-finger robotic gripper designs that will be able to overcome the limitations of existing two-finger gripper designs and will have necessary flexibility like that of a human hand to adapt to different geometric shapes of objects.

Further, the grippers actuated using linkages lack the compliance required for parts mating and insertion operations during assembly. This drawback can be handled by introducing a compliant behaviour into the gripper. There are two ways of ensuring compliant behaviour: either by a passive compliance or by an active compliance. Passive compliant behaviour can be achieved using a tendon-driven mechanism. The active compliance of the gripper can be ensured by a purposely designed control system, e.g. impedance control, which is an indirect method of force control via motion control without explicit force feedback.

Keeping the above in mind, a tendon-driven, multi-finger gripper is being proposed in this work that will have necessary flexibility like that of a human hand to adapt to different geometric shapes of objects and a control algorithm with necessary compliance to perform parts mating and insertion.

3.3.2 Proposed Multi-finger Gripper

Figure 9 shows the proposed model of a three-finger gripper. Ulrich et al. (1988) presented a more detailed description of the gripper mechanism. Sainul et al. (2016) discussed finger actuation mechanism for a three-fingered hand. The gripper consists of three identical fingers with each finger having three links namely knuckle, middle and distal links. Base of the thumb finger is fixed on the palm of the gripper. The other two opposing fingers have revolute joints at the base which helps fingers to spread sideways. Further, knuckle-middle and middle-distal links are connected using two more revolute joints. These two revolute joints help finger to accomplish closing and opening motion during grasping.

Proposed Actuation Mechanism for the Gripper

The articulated fingers of the gripper are actuated by a tendon-driven mechanism where remotely placed actuators at the palm generate grasping force to the gripper. Pulleys are used to join the knuckle-middle links and the middle-distal links. Non-stretchable tendons run over the pulleys and routing points which are used to guide the tendons as shown in Fig. 10. Spring-loaded tendons or spiral/torsional springs at the joints are used for extension motion of the fingers. The spreading of the opposing fingers is achieved by a DC motor and worm gear system placed inside the palm. One DC motor for each finger is used to drive the tendon–pulley system. A total of four DC brushless motors are used to actuate all the eight joints of the gripper.

An end of a tendon is attached to the DC motor and the other end is fixed at the distal link. The tendon force generates torque at the middle and distal joints. Pulling tendons can only generate force in one direction (tendons only can be pulled, not push), so two tendons are required to control a single joint, one tendon for each

Fig. 9 Model of the robot gripper

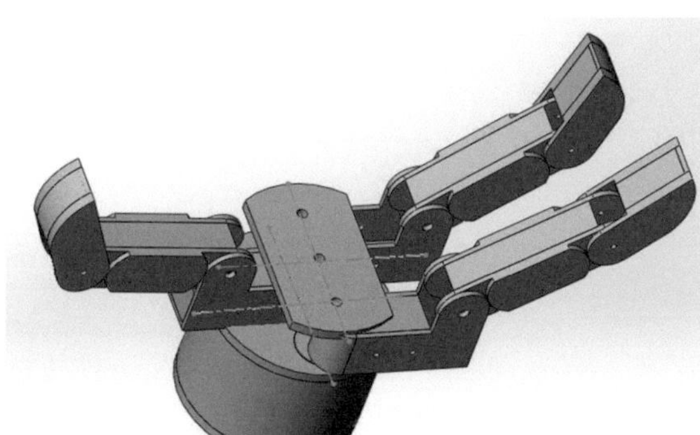

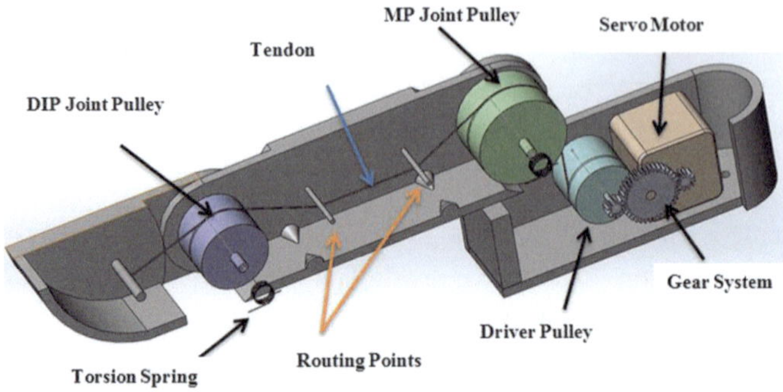

Fig. 10 Model of the tendon-driven finger showing the tendon and pulley arrangement

direction, i.e. a finger having n joints requires a total of $2n$ number of tendons for controlling all the joints independently. Although a minimum of $n+1$ tendons can independently control the joints with limited control capabilities which is implying that a finger having two joints needs at least three tendons, here, one tendon is used for finger flexion and two spring-loaded tendons are used for finger extension and they are called active and passive tendons, respectively.

Let n is the total number of joints, m is the total number of active and passive tendons, where $m = n + 1$, $l \in R^m$ be the tendon displacements, and $q \in R^n$ be the joint displacements. Then, the relation between the two displacements is as follows:

$$\boldsymbol{l} = \boldsymbol{J}_j \boldsymbol{q} + \boldsymbol{l}_0 \tag{9}$$

where $J_j \in R^{m \times n}$ is the Jacobian matrix and $l_0 \in R^m$ be the initial tendon displacements vector.

Let $\tau \in R^n$ be the joint torques and $f_t \in R^m$ be the tendon forces. Then, the relation between them is as follows.

$$\tau = J_j^T f_t \tag{10}$$

Let r_1 and r_2 be the pulley radii. Then, the Jacobian matrix for tendon arrangement as shown in Fig. 11 can be written as follows:

$$\boldsymbol{J}_j = \begin{bmatrix} r_1 & r_2 \\ r_1 & -r_2 \\ -r_1 & 0 \end{bmatrix} \tag{11}$$

DC motors control the tendon displacements or the tendon tension forces of the active tendons. The other two passive tendons are not actively controlled. Use of only one active tendon to control two joints of a finger makes the system under-actuated. Here, the use of passive tendons ensures that the joint displacements are uniquely determined for given tendon displacements and makes the

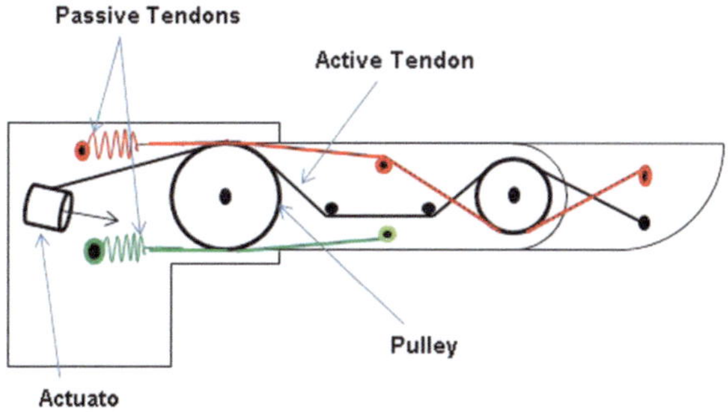

Fig. 11 Internal view of tendons run over the pulleys and routing points

system controllable in tendon space, i.e. tendon displacements and tendon forces. As the force in the passive tendons cannot be actively controlled, a limited set of joint displacements or torques can be achieved.

Control System of the Gripper

In recent times, the control of under-actuated system has drawn a lot of attention in the robotics field due to various advantages of under-actuated system. Under-actuated system requires less number of actuators compared to the full-actuated system which reduces the overall weight. Although the control system for under-actuated system is lot more challenging than the conventional full-actuated system, object grasping task is divided into two stages, pre-shaping of the fingers before grasping and gripper closure. In pre-shaping stage, the gripper adjusts its finger position by spreading the opposing fingers according to the various types of grasp which depend on the shape of object. During the pre-shaping, fingers make no interaction with the object. Gripper closure involves under-actuated control of the last two joints. The dynamic equation of a finger with three joints is as follows:

$$\boldsymbol{M}\ddot{\boldsymbol{q}} + \boldsymbol{H}(\dot{\boldsymbol{q}}, \boldsymbol{q}) + \boldsymbol{G}(\boldsymbol{q}) = \boldsymbol{\tau} \tag{12}$$

where $M \in R^{3\times3}$ be the inertia matrix, $H \in R^3$ is the joint velocity and position dependent Coriolis term, and $G \in R^3$ is the nonlinear gravity term.

A subsystem of the dynamic Eq. (12) with last two joints is considered in the subsequent section, as tendons are used only for the last two joints of a finger. Coefficients with bar line are used to differentiate with derived system from the original system. The joint torque and tendon force relation is as follows:

$$\bar{\boldsymbol{\tau}} = \boldsymbol{J}_j^T \boldsymbol{f}_{\mathrm{t}} \tag{13}$$

However, force in the passive tendons cannot be controlled as they are spring loaded. The force vector f_{t} is partitioned to separate the active component f_{a} from the passive component f_{p}. Then, the control input is as follows:

$$\bar{\tau} = J_j^T f_a + J_p^T f_p \tag{14}$$

by, putting Eq. (14) in Eq. (12) and rearranging the passive component, the dynamic equation of motion is obtained as follows:

$$\bar{M}\ddot{\bar{q}} + \bar{H}(\dot{\bar{q}}, \bar{q}) + \bar{G}(\bar{q}) - J_p^T f_p = J_j^T f_a \tag{15}$$

where $\overline{M}$ be the inertia matrix, $\overline{H}$ be the Coriolis force, $\overline{G}$, be the gravity force, $\bar{q}$ and $\bar{\tau}$ are joint displacement and joint torque of the last two joints, respectively. The relation between the tendon forces and tendon displacements for the spring-loaded tendons is as follows:

$$f_p = Kl = KJ_p\bar{q} \tag{16}$$

where K be the spring stiffness matrix of the spring-loaded tendons.

Impedance Control of the Gripper
Diameter of the pulleys is chosen in such a way that generated torque at the middle joint is greater than the distal joint, i.e. $\tau_2 > \tau_3$. Such design makes sure that the applied tendon force moves middle joint first, once it is stopped by touching object or joint limit, then only distal joint starts to move. Figure 12 shows finger closure operation, where initially only first link moves, and then second link starts to move after first link touches the object.

During the grasping tasks, fingers make physical interaction with the environment. The generated contact forces due to interaction with the object cause a deviation from the desired finger trajectory. Impedance control is preferred for tasks involving interaction with the environment which regulates motion and contact forces without explicit force feedback. It is not possible to control all the joints independently due to the under-actuation mechanism (i.e. less number of control inputs than the total number of joints in a finger), so inspired by the work of Arai

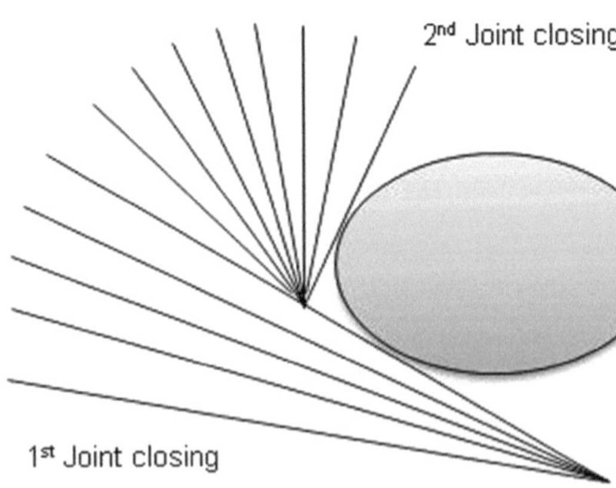

Fig. 12 Finger closure operation

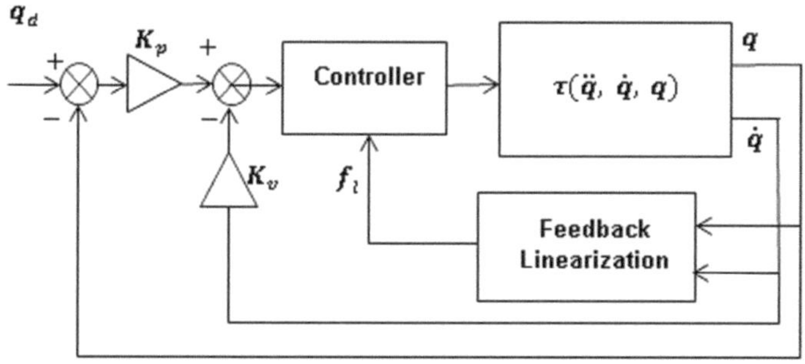

Fig. 13 Impedance-based grasp controller in joint space

and Tachi (1991); a two-phase control approach is proposed for controlling the joint displacements as well as the contact forces exerted by the links in joint space. In the first phase, tendon force controls only the middle joint. Then, in next phase, tendon force controls the distal joint.

Figure 13 shows the impedance-based grasp controller in joint space. In the presence of contact force, the dynamic equation of motion is as follows:

$$\overline{M}\ddot{\overline{q}} + \overline{H}(\dot{\overline{q}}, \overline{q}) + \overline{G}(\overline{q}) - J_{\mathrm{p}}^{T} K J_{\mathrm{p}} \overline{q} + \tau_{\mathrm{c}} = J_{j}^{T} f_{\mathrm{a}} \tag{17}$$

where τ_{c} is the generated torque at the joints due to the contact force.

From Eq. (17), the dynamic equation can be rewritten as

$$m_{22}\ddot{q}_2 + m_{23}\ddot{q}_3 + v_2 - t_2 + \tau_{\mathrm{c}}^{1} = r_1 f_{\mathrm{a}} \tag{18}$$

$$m_{32}\ddot{q}_2 + m_{33}\ddot{q}_3 + v_3 - t_3 + \tau_{\mathrm{c}}^{2} = r_2 f_{\mathrm{a}} \tag{19}$$

where $t = J_{\mathrm{p}}^{T} K J_{\mathrm{p}} \overline{q}$ and $v = \overline{H}\left(\dot{\overline{q}}, \overline{q}\right) + \overline{G}(\overline{q})$.

The following are the impedance control laws with feedback linearization.

First phase: Control law is

$$f_{\mathrm{a}} = K_{\mathrm{p}}\left(q_2 - q_2^{d}\right) - K_v \dot{q}_2 + \frac{1}{r_1}\left((v_2 - t_2) + \frac{m_{23}}{m_{33}}(t_3 - v_3)\right) \tag{20}$$

At equilibrium, generated torque at the middle joint due to contact force is as follows:

$$\tau_{\mathrm{c}}^{1} = r_1 K_{\mathrm{p}}\left(q_2 - q_2^{d}\right) \tag{21}$$

Second phase: Control law is

$$f_{\mathrm{a}} = K_{\mathrm{p}}\left(q_3 - q_3^{d}\right) - K_{\mathrm{v}} \dot{q}_3 + v_3 - t_3 \tag{22}$$

At equilibrium, generated torque at the distal joint due to contact force is as follows:

$$\tau_c^2 = r_2 K_p \left(q_3 - q_3^d \right) \tag{23}$$

3.3.3 Results and Discussions

The impedance control has been implemented on the proposed three-finger gripper. Table 3 shows all the parameters of each finger of the gripper, where (l_1, l_2, l_3) and (m_1, m_2, m_3) are the link lengths and masses of the knuckle, middle and distal links, respectively.

Figure 14 shows the joint displacement responses for implementation of impedance control laws as discussed in previous section. A time step of 10^{-3} s, proportional gain of $K_p = 10$ and derivative gain of $K_v = 0.1$ are used for the

Table 3 Finger parameters of the gripper

l_1 (mm)	l_2 (mm)	l_3 (mm)	m_1 (g)	m_2 (g)	m_3 (g)	r_1 (mm)	r_2 (mm)
60	60	40	50	50	40	10	8

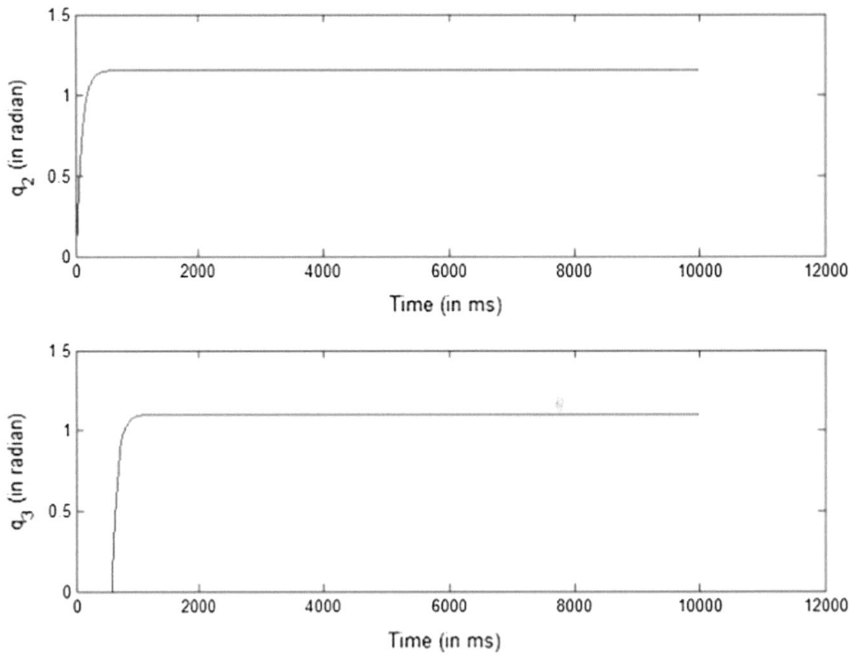

Fig. 14 Results of impedance control

implementation. From the figures, it is clearly seen that the middle link starts to move first and once the desired joint values are reached, then only the distal link starts to move.

Figures 16 and 17 show the grasping poses for cylindrical and cuboidal objects. The snapshots show the intermediate stages of pre-shaping and fingers closure. Figure 15a shows the pre-shaping stage where the two opposing fingers spread sideways. Figure 16b, c shows the implementation of the proposed two-phase control approach for gripper closure, Fig. 15b shows the closure of middle link and Fig. 15c shows the distal link closure, once the middle link is stopped by touching the object. Pre-shaping operation is not required for grasping of object shown in Fig. 16.

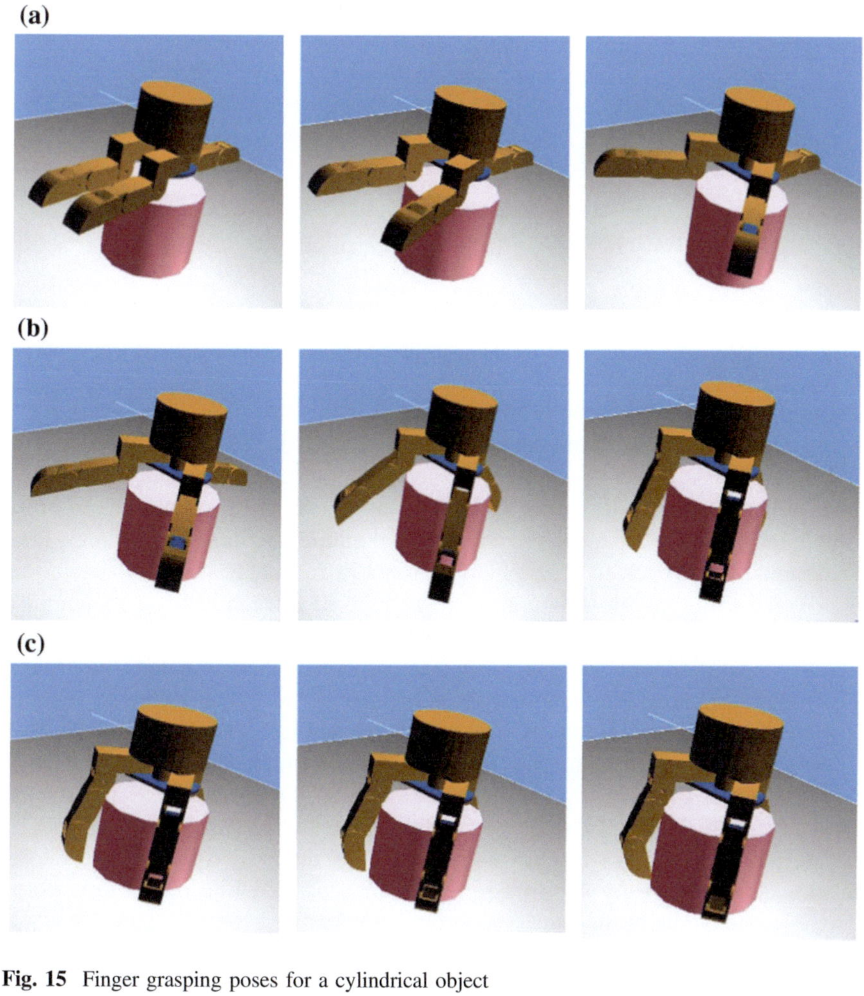

Fig. 15 Finger grasping poses for a cylindrical object

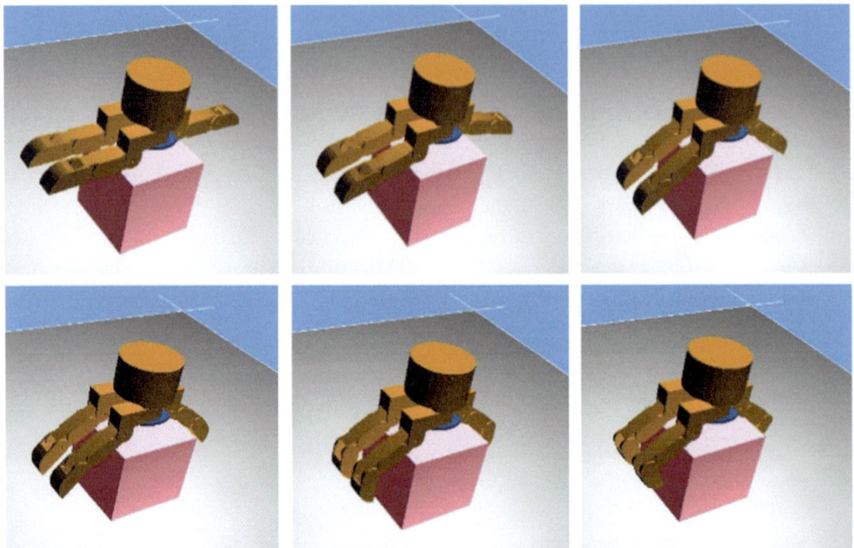

Fig. 16 Finger grasping poses for a cuboidal object

3.4 *Vision-Assisted Robotic Assembly*

In the following sections, the key issues and challenges in implementation of the assembly task-level plan using a vision-guided robotic system will be presented.

3.4.1 Key Issues and Challenges

In accordance with the task-level plan, the robot needs to be programmed to perform the assembly operations. The conventional online programming requires, first, the desired motion cycle to be taught to the robot by manual lead-through technique or powered lead-through technique with the help of a teach pendant to carefully define robot end-effector positions and orientations during and at the end of the motion. Then, it requires the programmer to prepare textual commands, describing the desired motion of the manipulator, and finally the program is entered into the controller memory for playback. Obviously, the online programming method is arduous and time consuming. The quality of motions and the end-effector positions taught relies on the skill level of the programmer. This exercise may have to be repeated again for robotic assembly of a new product even with slight variations in design, and thus it lacks the flexibility and reusability. To overcome the above drawbacks, a strategy for automatic programming of industrial robots to perform assembly operations by machine vision guidance has been proposed in this work. Furthermore, as inspection is important in any assembly process to rule out the possibility of erroneous robotic assembly, a strategy for automated assembly inspection based on machine vision has been also proposed. The aforementioned proposed system has been demonstrated by performing a mechanical assembly using a vision-assisted robotic assembly system, comprising of an industrial robot manipulator, an overhead CCD camera and the National Instrument (NI)'s LabVIEW graphical programming environment.

3.4.2 Implementation of Assembly Task Plan by Vision-Guided Robot

After the task-level plan is generated using knowledge-based system described in Sect. 3.2, it is implemented by a vision-guided robotic system that has been developed. Figure 17 shows the system setup. It consists of a robotic system comprising of a six-axis industrial robot manipulator (model Yaskawa Motoman MH5) equipped with a two-finger pneumatic gripper and a robot controller (model FS100) with open software architecture and a robot control interface (Digimetrix), a vision system comprising of an overhead-mounted CCD camera and NI vision assistant utility for image processing, and NI LabVIEW graphical programming environment for implementing the overall system. The storage bin containing the parts that are to be assembled and the assembly jig on which the assembly is to be built up are both placed within the field of view of the overhead camera and within the robot work envelope.

Recognition of Parts, and Estimation of Position and Orientation by Machine Vision

To carry out assembly using the robot, it is necessary to first of all recognize all the parts involved from the storage bin and determine their initial locations and orientations, which are accomplished by the machine vision system. The vision system

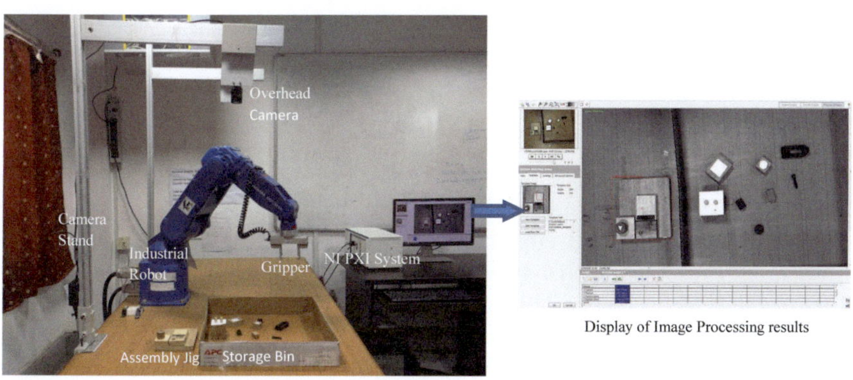

Fig. 17 Setup for vision-guided robotic system for assembly

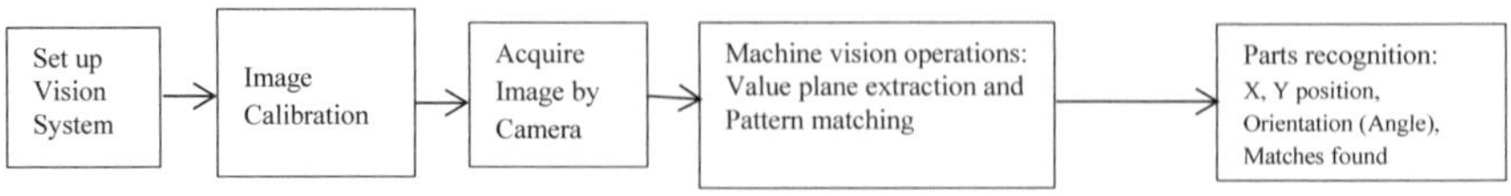

Fig. 18 Flowchart of operations performed by machine vision system for part recognition and determining position and orientation

is used to further determine the target hole (cavity) locations and their orientations on the base part of the assembly or sub-assembly into which each part has to be inserted. The vision assistant utility of LabVIEW has been used to accomplish these tasks. The vision system hardware consists of an overhead camera used to acquire the image of the parts placed inside the robot work envelope and a PC with data acquisition board. The camera used is Basler acA1300-22gc GigE camera with Sony ICX445 CCD sensor of 1.3 MP resolution, 22 fps and with Edmund Optics lens of 6 mm focal length.

Figure 18 shows the operations performed by the machine vision system.

Following are the steps for recognition of parts and determining their locations and orientations. First, the camera is to be calibrated to map the pixels of the image to real world coordinates of the robot. The calibration is done by the point coordinates calibration routine of LabVIEW by feeding the real world coordinates of any four different pixels from an image taken by the camera in the given workspace. After the calibration, the setup is kept fixed and images of the workspace, containing the parts to be assembled and the assembly jig, are captured using the camera for use in robotic manipulation. The captured images are colour images containing the (*R*)ed, (*G*)reen and (*B*)lue channels at every pixel, and hence, every pixel can be represented in the RGB space. As changes in the brightness of an image are more distinguishable than that only in colour (Wang et al. 2001), a perceived brightness measure from the image simply called the value has been considered. This is obtained by converting a pixel in an *RGB* space to that in *HSV* (Hue–Saturation–Value) space, where *H* and *S* are the chromatic components and *V* is the value. *V* is simply computed as (Solomon and Breckon 2011)

$$V = \max\left(\frac{R}{255}, \frac{G}{255}, \frac{B}{255}\right), \quad R, G, B \in [0, 1, 2, \ldots, 254, 255] \tag{24}$$

In the next step, a pattern matching approach is applied on the value plane of the image, to detect the presence of a particular part. This system matches available templates of parts to the various regions of the image to identify a part in it. The well-known normalized 2D cross-correlation value (Haralick and Shapiro 1992) between templates and similar-sized local regions in the captured image is maximized (best fit above a threshold) to find a part. To make the matching rotation invariant, finite number of rotations (between 0° and 360°) of a given template is considered. Scale invariance of the pattern matching is also important and hence multi-scale processing is performed during the pattern matching. Multi-level

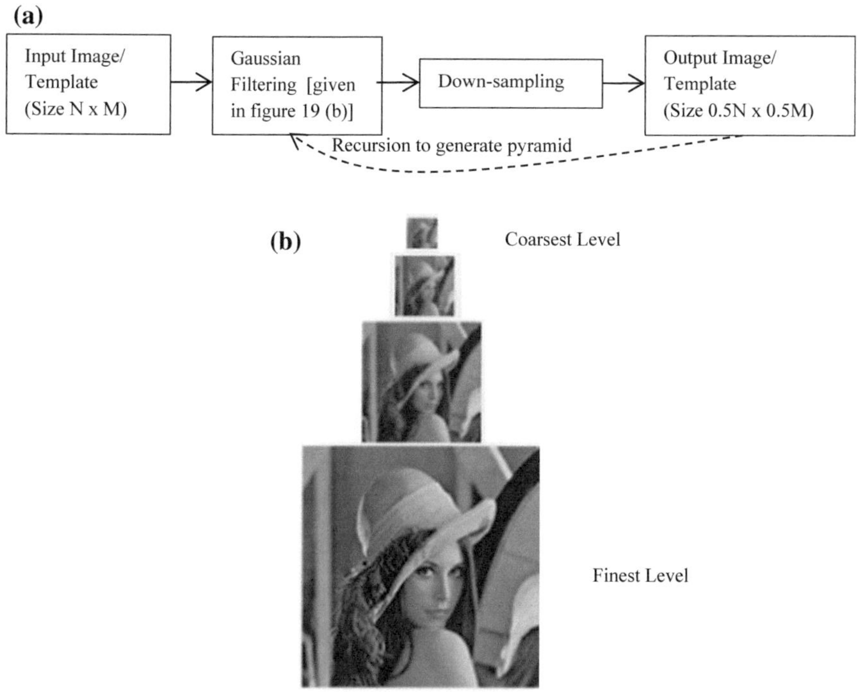

Fig. 19 **a** Multi-level Gaussian pyramid decomposition, **b** the pyramid structure. *Source* The USC-SIPI image database (http://sipi.usc.edu/database)

Gaussian pyramid decomposition (Gonzalez and Woods 2008), shown in Fig. 19, of both the template and the image at hand, is considered.

The matching is done across all the scales provided by the pyramid decomposition of both the template and the image. The number of levels in the pyramid has been considered as 4. The matching operation simply considers the real values from the value planes of the images (and templates) and works fine even in the presence of intricate textures with dense edges. It is found that the NI vision assistant utility's pattern matching described above is able to accurately locate objects that vary in size and orientation (0°–360°). To reduce complexity of the processing, a coarse-to-fine matching is considered. This helps immensely as all possible rotations and locations are considered only at the coarsest level, and then for the finer level, only a subset of possible rotations and locations are considered based on the best match orientation in the coarsest level. This avoids exhaustive search at every level of the pyramid, and actually makes the processing faster than that without pyramid decomposition, and this is another advantage of using the decomposition other than the intended scale invariance through multi-scale processing.

Once the pattern matching operation is completed, we have the estimated locations and orientations of all the parts identified. The *X*, *Y* position is then easily mapped to the world coordinates as per the initial calibration.

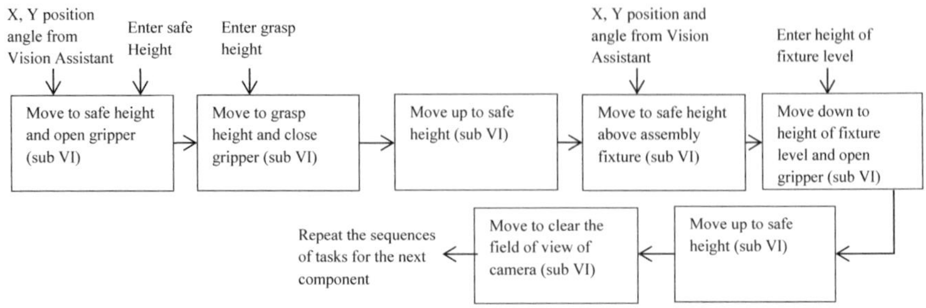

Fig. 20 Flowchart of the strategy for robotic assembly

Strategy for Robotic Assembly Under Machine Vision Guidance

After the assembly parts are recognized and their positions and orientations determined by machine vision system, the decisions regarding motion of the robot manipulator to perform the assembly operations are implemented in accordance with the assembly task plan generated earlier by the robot task-level planner. The robot task-level plan lists sequentially the handling tasks that are to be performed by the robot to pick up the parts one by one from their initial locations and orientations in the storage bin, and also the insertion tasks that are to be performed by the robot to assemble them into the respective positions and orientations of the holes (cavities) in the base part on the assembly jig. The destination coordinates on the assembly jig are also obtained from the vision system. The task-level plan also contains details of the robot motion sequences necessary for handling each part in a proper order as shown in Fig. 20, starting from moving the robot to a safe height above the recognized part before picking it up and ending with actuating the opening of the gripper fingers to release the part on the base part before moving up to a safe height. After these motion sequences, the manipulator needs to move away to clear the field of view of the camera and await further instructions for handling and insertion of the next part. For this, machine vision system is employed repeatedly as explained earlier. Each of the above-mentioned sequences has been implemented by developing a sub-VI in LabVIEW. If any part to be assembled is missing from the bin, the robot will automatically stop the assembly process after getting a “No match found” signal from the vision system.

Strategy for Assembly Inspection by Machine Vision

In any assembly process, inspection plays a vital role as there is always a possibility of erroneous robotic assembly due to missing parts. This defect may result from accidental slipping of a part from the gripper during handling before the robot reaches the assembly jig. In that case, the missing part from the assembly can be detected by first taking an image of the intermediate stage of assembly with the sub-assembled parts, each time after the robot completes an assembly operation,

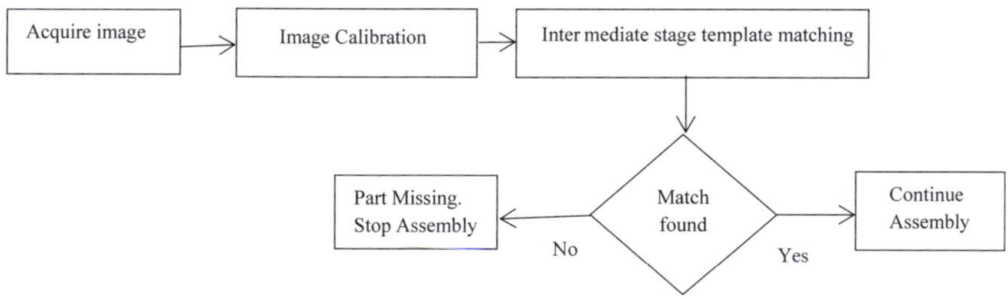

Fig. 21 Flowchart of operations performed by machine vision for assembly inspection

and then performing template matching of that sub-assembly image with the previously stored correct intermediate stage image. If the match is found to be perfect, a signal 'Continue assembly' is sent to the robot controller to proceed for picking the next part. Otherwise, robot stops after receiving a 'Part missing, stop assembly' signal. This same process is continued at each stage of the assembly to check for missing parts. Flowchart of the inspection process is shown in Fig. 21.

3.4.3 Results and Discussions

The eight-component assembly shown in Fig. 7 is used to demonstrate the results of the developed vision-assisted robotic assembly system. Figure 23a–c, shows the initial, intermediate and the final states of the task environment, respectively, indicating the positions of the jig as well as other parts and the robot's movement for performing the assembly. All the above parts were initially scattered randomly in the task environment of the robot but placed in a bin within the field of view of the overhead camera as shown by the initial state of the task environment in Fig. 22a. Table 4 summarizes the information about the estimated initial positions

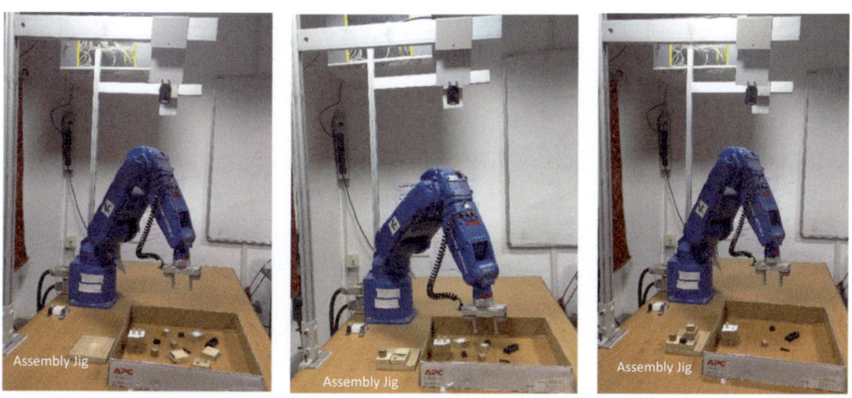

(a) Initial state (b) Intermediate state (c) Final state

Fig. 22 Initial, intermediate, final states of the task environment showing the robot's movement for performing assembly

Table 4 Summary of information of estimated initial positions

S. No.	Assembly part	Real world coordinates (mm)		Angle in deg.	Score (out of 1000)	No. of matches	Destination coordinate on assembly fixture (*X*, *Y*, in mm)
		X	*Y*				
1	Prismatic part with one hole	455.876	−453.262	131.384	970.814	1	(82.967, −487.149)
2	Stepped prismatic part with two holes	366.85	−488.54	87.52	900.36	1	(136.215, −493.232)
3	Prismatic part with rectangular slot	396.238	−438.93	210.591	973.521	1	(136.914, −443.458)
4	Cylindrical part	311.134	−431.381	88.0231	787.212	1	(71.8167, −488.361)
5	Stepped prismatic part with one hole	367.907	−393.178	48.0247	916.051	1	(110.927, −496.193)
6	Cylindrical pin	229.033	−493.784	11.1204	947.648	1	(368.281, −388.704)
7	Prismatic part with rectangular pocket	398.583	−319.615	230.692	953.062	1	(125.933, −444.452)
8	Cube with pocket	303.292	−363.209	312.238	956.559	1	(121.456, −442.341)

(a) Stored template (b) Screenshot display of Image Processing result (c) Calibrated matches

Fig. 23 Output from the vision assistant utility of LabVIEW

of all assembly parts including the holes (cavities) on the jig that were extracted using the vision assistant utility of LabVIEW. Figure 23 presents an extract of a typical output from the vision assistant for one of the assembly parts, showing the stored template, recognized part, estimated initial *X*, *Y* positions, angle, score, etc. Finally, Fig. 23c shows the final state of the task environment after successful completion of the assembly by the robot.

4 Summary and Conclusions

In an attempt to support the ongoing efforts for developing flexible robotic assembly system with necessary agility and adaptability to satisfy diverse customer requirements and capability of mass customization, this chapter has examined some of the crucial issues and challenges that need to be addressed for its successful implementation. Some of the significant contributions of the reported research are as follows.

- A Sexual Genetic Algorithm (SGA)-based approach for assembly sequence planning and optimization has been developed and implemented. The information on assembly directions of the parts, precedence relationships, various tools/grippers needed and location of the base component in the assembly is inputted. It is capable of automatically generating the feasible and optimal sequences based on minimizing the number of reorientations and tool changes, and maximizing the stability of the components/sub-assemblies.
- Further, a robot task-level planning system has been proposed and its implementation shown using a vision-guided industrial robot manipulator for performing mechanical assemblies. To accomplish task-level planning, a knowledge-based system is developed and executed using the expert system shell CLIPS, which is capable of automatically generating all the assembly tasks that are to be performed by the robot from a given optimal assembly sequence generated by the above sequence planner.

- Moreover, a multi-finger, tendon-driven robotic gripper for flexible assembly has been proposed that has necessary flexibility like that of a human hand to adapt to different geometric shapes of objects and an impedance control algorithm with necessary compliance needed to perform parts mating and insertion during assembly.
- Finally, a strategy for robotic assembly under machine vision guidance has been also developed, in which the vision system comprising an overhead-mounted CCD camera and NI vision assistant utility for image processing has been used to guide a Motoman industrial robot in performing mechanical assembly operations in a task environment, where the parts are initially scattered randomly. Further, a strategy for automated assembly inspection based on machine vision has also been proposed to rule out any possibility of erroneous robotic assembly. The above system is found to be successfully working with minimal human intervention. Research work is presently ongoing for use of multiple sensors in conjunction with machine vision to guide the robot in performing more complex assembly tasks.

Acknowledgements The authors gratefully acknowledge the funding support received from Indian Institute of Technology Kharagpur under the Centre for Excellence in Robotics grants.

References

Abdallah, M.E., R. Platt, B. Hargrave, and F. Permenter. 2012. Position control of tendon-driven fingers with position controlled actuators. In *IEEE international conference on robotics and automation*, RiverCentre, Saint Paul, Minnesota, USA, 14–18 May 2012.

Alatartsev, S., S. Stellmacher, and F. Ortmeier. 2015. Robotic task sequencing problem: A survey. *Journal of Intelligent and Robotic Systems: Theory and Applications* 80 (2): 279–298.

Ambrose, R.O., H. Aldridge, R.S. Askew, R.R. Burridge, W. Bluethmann, M. Diftler, C. Lovchik, D. Magruder, and F. Rehnmark. 2000. Robonaut: NASA's space humanoid. *IEEE Intelligent System*, 57–63, July/Aug 2000.

Arai, H., and S. Tachi. 1991. Position control of a manipulator with passive joints using dynamic coupling. *IEEE Transactions on Robotics and Automation* 7 (4), Aug 1991.

Backhaus, J., and G. Reinhart. 2013. Efficient application of Task oriented programming for assembly systems. In *IEEE/ASME international conference on advanced intelligent mechatronics (AIM)*, Australia, 750–755.

Bekey, G.A., R. Tomovic, and I. Zeljkovic. 1990. *Control architecture for the Belgrade/USC hand*. New York: Springer.

Bonneville, F., C. Perrard, and J.M. Henrioud. 1995. A genetic algorithm to generate and evaluate assembly plans. In *Proceedings of INRIA/IEEE symposium on emerging technologies and factory automation*, 231–239, Paris, France.

Butterfass, J., G. Hirzinger, S. Knoch, and H. Liu. 1998. DLR's multisensory articulated part I: Hard- and software architecture. *IEEE International Conference on Robotics and Automation*.

Butterfass, J., M. Grebenstein, H. Liu, and G. Hirzinger. 2001. DLR-hand II: Next generation of a dextrous robot hand. In *IEEE international conference on robotics and automation*, vol. 1, 109–114.

Cambon, S., F. Gravot, and R. Alami. 2004. A robot task planner that merges symbolic and geometric reasoning. In *European conference on artificial intelligence (ECAI)*, 1–5, Apr 2004.

Cao, P.B., and R.B. Xiao. 2007. Assembly planning using a novel immune approach. *International Journal of Advanced Manufacturing Technology* 31 (7): 770–782.

Chen, S., and Y.J. Liu. 2001. An adaptive genetic assembly sequence planner. *International Journal of Computer Integrated Manufacturing* 14 (5): 489–500.

Chen, R.S., K.Y. Lu, and P.H. Tai. 2004. Optimizing assembly planning through a three-stage integrated approach. *International Journal of Production Economics* 88 (3): 243–256.

Chen, H., B. Zhang, and G. Zhang. 2015. Robotic assembly. In *Handbook of manufacturing engineering and technology*, 2347–2400.

Cho, J., H. Kim, and J. Sohn. 2010. Implementing automated robot task planning and execution based on description logic KB. Lecture Notes in Computer Science (including subseries Lecture Notes in Artificial Intelligence and Lecture Notes in Bioinformatics), 6472, 217–228.

Choi, Y.K., D.M. Lee, and Y.B. Cho. 2009. An approach to multi-criteria assembly sequence planning using genetic algorithms. *International Journal of Advanced Manufacturing Technology* 42: 180–188.

Deb, S.R., and S. Deb. 2010. *Robotics technology and flexible automation*. New Delhi: Tata McGraw Hill.

Diftler, M.A., J.S. Mehling, M.E. Abdallah, N.A. Radford, L.B. Bridgwater, A.M. Sanders, R.S. Askew, D.M. Linn, J.D. Yamokoski, F.A. Permenter, B.K. Hargrave, and R. Platt. 2011. Robonaut 2—the first humanoid robot in space. In *2011 IEEE international conference on robotics and automation*, Shanghai, China, 9–13 May 2011.

Gao, X.H., M.H. Jin, L. Jiang, Z.W. Xie, P. He, L. Yang, Y.W. Liu, R. Wei, H.G. Cai, H. Liua, J. Butterfass, M. Grebenstein, N. Seitza, and G. Hirzinger. 2003. The HIT/DLR dexterous hand: Work in progress. In *IEEE international conference on robotics & automation* Taipei, Taiwan, 14–19 Sept 2003.

Gao, L., W.R. Qian, X.Y. Li, and J.F. Wang. 2010. Application of memetic algorithm in assembly sequence planning. *International Journal of Advanced Manufacturing Technology* 49 (9–12): 1175–1184.

Gazeau, J.P., S. Zeghloul, M. Arsicualt, J.P. Lallemand. 2001. The LMS hand: Force and position controls in the aim of fine manipulation of objects. In *IEEE International conference on robotics and automation*, 2642–2648.

Giarratano, J., and G. Riley. 2002. *Expert systems principles and programming*, 3rd ed. Thomson Learning and China Machine Press. ISBN 7-111-10844-2.

Goh, K.S., A. Lim, and B. Rodrigues. 2003. Sexual selection for genetic algorithms. *Artificial Intelligence Review* 19: 123–152.

Golnabi, H., and A. Asdapour. 2007. Design and application of industrial machine vision systems. *Robotics and Computer-Integrated Manufacturing* 23: 630–637.

Gonzalez, R.C., and R.E. Woods. 2008. *Digital image processing*, 3rd ed. Pearson.

Haralick, R.M., and L.G. Shapiro. 1992. *Computer and robot vision*, vol. II, Addison-Wesley.

Jacobsen, S.C., J.E. Wood, D.F. Knutti, and K.B. Biggers. 1984. The UTAH/M.I.T. dextrous hand: Work in progress. *The International Journal of Robotics Research.*

Jacobsen, S., E. Iversen, D. Knutti, R. Johnson, and K. Biggers. 1986. Design of the Utah/M.I.T. dextrous hand. In *1986 IEEE international conference on robotics and automation*, vol. 3, 1520–1532, Apr 1986.

Karthik, G.V.S.K., and S. Deb. 2017. A methodology for assembly sequence optimization by hybrid Cuckoo-Search Genetic Algorithm. *Journal of Advanced Manufacturing Systems* (in press).

Kobari, Y., T. Nammoto, J. Kinugawa, and K. Kosuge. 2013. Vision based compliant motion control for part assembly. In *IEEE/RSJ international conference on intelligent robots and systems (IROS)*, Japan, 293–298, Nov 2013.

LiCheng, W., G. Carbone, and M. Ceccarelli. 2009. Designing an underactuated mechanism for a 1 active DOF finger operation. *Mechanism and Machine Theory* 44: 336–348.

Li, X., K. Qin, B. Zeng, L. Gao, and J. Su. 2016. Assembly sequence planning based on an improved harmony search algorithm. *International Journal of Advanced Manufacturing Technology* 84: 2367–2380.

Liu, Hong, P. Meusel, G. Hirzinger, Minghe Jin, Yiwei Liu, and Zongwu Xie. 2008. The modular multisensory DLR-HIT-hand: Hardware and software architecture. *IEEE/ASME Transactions on Mechatronics* 13 (4).

Loucks, C., V. Johnson, P. Boissiere, G. Starr, and J. Steele. 1987. Modeling and control of the Stanford/JPL hand. In *International conference on robotics and automation*, vol. 4, 573–578, Mar 1987.

Lovchik, C.S., and M.A. Diftler. 1999. The robonaut hand: A dexterous robot hand for space. In *IEEE international conference on robotics and automation*, 907–912.

Lv, H.G., and C. Lu. 2010. An assembly sequence planning approach with a discrete particle swarm optimization algorithm. *International Journal of Advanced Manufacturing Technology* 50 (5–8): 761–770.

Marian, R.M., L.H.S. Luong, and K. Abhary. 2006. A genetic algorithm for the optimisation of assembly sequences. *Computers & Industrial Engineering* 50 (4): 503–527.

Mishra, A., and S. Deb. 2016a. An intelligent methodology for assembly tools selection and assembly sequence optimisation. In *Proceedings of CAD/CAM, robotics and factories of the Future*, India, 323–333, Jan 2016.

Mishra, A., and S. Deb. 2016b. A GA based parameter meta-optimization of ACO algorithm for solving assembly sequence optimization. In *Proceedings of international conference on computers & industrial engineering CIE 46*, Tianjin, China, 29–31 Oct 2016.

Mishra, A., and S. Deb. 2016c. Assembly sequence optimization using a Flower Pollination Algorithm-based approach. *Journal of Intelligent Manufacturing*. https://doi.org/10.1007/s10845-016-1261-7.

Mishra, A., and S. Deb. 2017. Robotic assembly sequence planning and optimization by cuckoo search algorithm. In *Proceedings of international conference on production research 24th ICPR*, Poznan, Poland, July 30–Aug 3 2017.

Okada, T. 1986. Computer control of multi-jointed finger system for precise object handling. *International Trends in Manufacturing Technology Robot Grippers*, 391–417.

Osuna, R.V., T. Tallinen, J.L.M. Lastra, and R. Tuokko. 2003. Assembly and task planning in a collaborative web-based environment based on assembly process modeling methodology. In *Proceedings of IEEE international symposium on assembly and task planning*, 79–84, Jan 2003.

Ozawa, R., K. Hashirii, Y. Yoshimura, M. Moriya, and H. Kobayashi. 2014. Design and control of a three-fingered tendon-driven robotic hand with active and passive tendons. *Autonomous Robots* 36: 67–78.

Pal, S., S. Chattopadhyay, and S.R. Deb. 2008. Design and development of a multi-degree of freedom dexterous instrumented Robot Gripper. *Sensors and Transducers Journal* 87 (1): 63–73.

Pena-Cabrera, M. 2005. Machine vision approach for robotic assembly. *Assembly Automation* 25 (3): 204–217.

Ramos, C., and E. Oliveira. 1992. Planning, execution and sensor-based reaction for assembly robotic tasks, Technical paper, Faculdade de Engenharia da Universidade do Porto, Portugal.

Sainul, I.A., S. Deb, and A.K. Deb. 2016. A three finger tendon driven robotic hand design and its kinematics model. In *International conference on CADCAM, robotics and factories of the future*, Kolaghat, India, Jan 2016.

Salisbury, K.S., and B. Roth. 1983. Kinematics and force analysis of articulated mechanical hands. *Journal of Mechanisms, Transmissions and Actuation in Design* 105: 35–41.

Solomon, C., and T. Breckon. 2011. *Fundamentals of digital image processing*. Wiley & Blackwell.

Tlegenov, Y., K. Telegenov, and A. Shintemirov. 2014. An open-source 3D printed underactuated robotic gripper. In *IEEE/ASME international conference on mechatronic and embedded systems and applications*, Senigallia, 10–12 Sept 2014.

Townsend, W.T. 2000. MCB—industrial robot feature article-barrett hand grasper. *Industrial Robot: An International Journal* 27 (3): 181–188.

Thomas, U., and F.M. Wahl. 2010. Assembly planning and task planning—two prerequisites for automated robot programming. *Robotic Systems for Handling and Assembly, STAR* 67: 333–354.

Ulrich, N., R. Paul, and R. Bajcsy. 1988. A medium-complexity complaint end effector. In: *IEEE international conference on robotics and automation*.

Wang, Y., J. Ostermann, and Y.Q. Zhang. 2001. *Video processing and communications*. Prentice Hall.

Wang, J.F., J.H. Liu, and Y.F. Zhong. 2005. A novel ant colony algorithm for assembly sequence planning. *International Journal of Advanced Manufacturing Technology* 25: 1137–1143.

Xing, Y.F., and Y.S. Wang. 2012. Assembly sequence planning based on a hybrid particle swarm optimisation and genetic algorithm. *International Journal of Production Research* 50 (24): 7303–7312.

Zhang, B., J. Wang, G. Rossano, C. Martinez, and S. Kock. 2011. Vision-guided robot alignment for scalable, flexible assembly automation. In *IEEE international conference on robotics and biomimetics*, Thailand, 944–951, Dec 2011.

Zhou, W., J.R. Zheng, J.J. Yan, and J.F. Wang. 2011. A novel hybrid algorithm for assembly sequence planning combining bacterial chemotaxis with genetic algorithm. *International Journal of Advanced Manufacturing Technology* 52 (5–8): 715–724.

A Piezoelectric Actuator Based Compact Micro-manipulation System for Robotic Assembly

R. K. Jain, Surajit Saha and Bhaskar Ghosh

Abstract In this chapter, a novel design of miniature wireless mobile micro-manipulation system (WMMS) for robotic assembly is proposed where a three-piezoelectric actuator fingers' based compact gripper is developed for handling the small objects. The piezoelectric actuator has the potential of generating displacement in micron range and produces high force after applying voltage for miniature objects. In order to perform pick-and-place operation of the object from one to another position in desired workspace, the kinematics for WMMS is carried out and throughput analysis is performed using ADMAS software. The simulations are obtained and verified by developing a physical prototype. It is also demonstrated that the compact WMMS shows handiness during handling and manipulation of lightweight objects without destructing it in a robust manner during robotic assembly.

Keywords Piezoelectric actuator · Micro-gripper · Wireless mobile micromanipulation system (WMMS) · Handling

1 Introduction

Robotics technology has been improved significantly during the last few decades. There are so many fields where robots are used every time. Starting from defense system to medical science everywhere, we are involved to use robots such as micromanipulation, nanotechnology, scanning tunneling, atomic force microscopy, micro-electromechanical system (MEMS), etc. They need reliable, fast, and precise actuator- and sensor-based systems which give more impact on day-to-day life (Mok et al. 2000; Huang and Tsai 2005; Valdastri et al. 2006). In this aspect, the conventional industrial manipulators are fixed at some places and can only execute tasks within a limited workspace. In order to execute tasks during locomotion in a

R. K. Jain (✉) · S. Saha · B. Ghosh
Surface and Field Robotics Group, CSIR-CMERI, Durgapur, India
e-mail: rkjain@cmeri.res.in

S. S. Pande and U. S. Dixit (eds.), *Precision Product-Process Design and Optimization*, Lecture Notes on Multidisciplinary Industrial Engineering,
https://doi.org/10.1007/978-981-10-8767-7_3

wide area, a vehicle-mounted mobile manipulator is more effective as compared to fixed-type manipulator (Oriolo and Mongillo 2005). During development of a vehicle-mounted mobile micro-manipulator, it needs proper movement of the manipulator to grasp an object without hitting anything from one position to another. In order to improve micro-manipulator performance, the reliable, fast, and accurate sensor- and actuator-based system is required that can provide precise positioning and accuracy during manipulation of the object. By fulfilling this kind of requirement, the design and development of the micro-assembly system becomes more potent in terms of precise handling of miniature and small-scale parts in a robust and economical way toward batch assembly. Regarding this, a few micro-manipulation systems for robotic assembly have been developed (Zaeh et al. 2003; Clévy and Chaillet 2006; Probst et al. 2009; Bolopion and Régnier 2013). The advantage of compliance behavior for ionic polymer metal composite (IPMC), shape memory alloy (SMA), and piezoelectric-based smart materials are used for robotic assembly (Deole et al. 2008; Jain et al. 2009, 2011). IPMC has proven its potential of handling and manipulating various flexible as well as rigid parts during different types of assembly and handling tasks due to its compliant behavior. In this aspect, Jain et al. (2013a, b) have developed IPMC-based compliant micro-grippers for different applications of micro-manipulation such as handling of parts in micro-factory test bed concept and SCARA robot. The IPMC micro-gripper is actuated with small voltages (0–3 V) and produces large displacement (10–16 mm), whereas SMA actuated micro-grippers for manipulation have been developed by Kohl et al. (2002) and Lin et al. (2009). Unfortunately, IPMC and SMA suffer from long-term actuation problem (Hall 2001; Bhandari et al. 2012). Therefore, another approach is that the piezoelectric actuator can be used for developing the micro-gripper. Compared to IPMC and SMA, the piezoelectric actuator has some benefits such as high micro/nanoscale displacement, large force generation within microseconds for developing the micro-gripper. The only limitation of piezoelectric actuator is that it suffers from the nonlinear behavior and shows hysteresis. To reduce this nonlinearity of piezoelectric actuator with operating voltage, several researchers have studied various control methods (Sun et al. 2004; Song et al. 2005). These methods also have some drawbacks like measuring the force/actuation data for a long time through the operating voltage in order to develop the micro-gripper. Tracking control of a piezoelectric actuator for compensating hysteresis has been introduced by Rakotondrabe and Ivan (2011), and Zhang et al. (2012) have incorporated the vision-based system with self-sensing method for attaining electromechanical behavior of piezoactuator with controlled voltage toward development of micro-gripper in micro-assembly. These methods have also shown some limitations like thermal insulation characteristic and packaging aspect problems of piezoactuators for real-time application when the operating voltage varies from 0 to 60 V. This is encountered by Jain et al. (2013a, b) in the design of piezoelectric actuator based micro-gripper along with micro-manipulator with three degree of freedom which can perform the manipulation task within a small workspace. For further improvement, a novel design of wireless mobile micro-manipulation system with complaint piezoelectric actuator based micro-gripper is proposed which is capable to handle the object without destructing it and performs robotic micro-assembly in a robust manner.

The objectives of this research work focus on the following points:

(a) A new design of a piezoelectric actuator based compliant micro-gripper along with mobile manipulation system for assembly;
(b) Analysis of mobile micro-manipulation system for finding the throughput speed for manipulation and handling the object;
(c) Development of a miniature wireless mobile micro-manipulation system and evaluation of its performance for robotic micro-assembly.

Relevant research work reported in these areas is highlighted here.

In the past, several researchers have carried out research work on various designs, modelings, and control methods and fabrication of piezo-actuators for micro-gripper and micro-manipulation. Gao and Swei (1999) have developed a six-degree-of-freedom (DOF) micro-manipulator based on piezoelectric translators for a dexterous high-precision manipulation where high-performance piezoelectric actuation system is used for translator along with a monolithic leaf spring and a preload mechanism. Sitti et al. (2001) have designed, fabricated, and characterized unimorph actuators (PZT-5H and PZN-PT) toward designing micro-aerial flapping mechanism. The characterization of the unimorphs is performed using optical measurement system in quasi-static and dynamic mode, and the performance of the unimorphs is compared with desired design specifications. Hristov et al. (2002) have focused on using stack actuators for simple liver system and proposed a mathematical model of the mechanism with Lagrangian method. Popa et al. (2003) have introduced dynamic model of bimorph MEMS actuators similar to piezoactuator. The operating principle is based on differential thermal expansion induced by Joule heating process. Dadfarnia et al. (2004) have proposed a Lyapunov-based control strategy for the regulation of a Cartesian robot manipulator. The arm base motion is controlled utilizing a linear actuator, while a piezoelectric (PZT) patch actuator is bonded to the surface of the flexible beam for suppressing residual beam vibrations. Agnus et al. (2005) have attempted on a design of micromanipulation station based on a microprehensile microrobot on chip (MMOC) micro-gripper. The microprehensile is a kind of seizing or grasping mechanism by wrapping around which microrobot on chip (MMOC) micro-gripper is actuated via independently controlled currents or magnetic field. Pérez et al. (2006) have presented a fabrication technique for the integration of a force sensor with the silicon technology in a micro-gripper actuated by a piezoactuator system. The force sensor consists of a silicon box with an output beam. Fuchiwaki et al. (2008) have focused on the development of the multi-axial micromanipulation organized by versatile micro-robots using micro-piezoelectric tweezers. This device has 21-DOF with less than 100 nm resolution and demonstrated the flexible handling of minuscule glass spheres with a diameter of 20 μm. Liu and Liu (2010) have described a micro-manipulation mechanism with a converse piezoelectric effect where the micro-step mechanism based on converse piezoelectric effects is designed.

A micro-gripper actuated by piezoelectric cantilever is also developed for micro-parts' assembly. Tamadazte et al. (2011) have attempted on a full automatic micromanipulation and micro-assembly where a visual feedback method has been proposed. This method is based on a mono-view and multiple-scale image-based visual servo using dynamic zoom. Chen et al. (2012) have presented a micromanipulation method for handling the micro-objects, which is based on dynamic adhesion control with compound vibration. Surface and bulk micromachining technologies are employed to fabricate the micro-gripper and a piezoelectric ceramics module used in the system. Yang et al. (2012) have presented the design and analysis of a handheld manipulator for vitreoretinal microsurgery and other biomedical applications. The design involves a parallel micromanipulator utilizing six piezoelectric linear actuators, combining compactness with a large range of motion and relatively high stiffness. Li et al. (2013) have established the kinematics and dynamic modeling of 2-DOF micromanipulator where the two scales of level amplifier principle are adopted for designing the flexure hinge to replace the traditional hinge. Boudaoud and Regnier (2014) have presented an overview of gripping force measurement using two-fingered micromanipulation systems. The challenges about measuring forces below the micro-Newton for the manipulation of highly deformable materials and embedding force sensors within micro-grippers toward measurement of high resolution are described. Xiao and Li (2014) have designed a flexure-based 3-DOF micro-positioning stage driven by piezoelectric-based actuators which are controlled by a sliding mode-based controller and applicable in micro/nanomanipulation. Amari et al. (2014) have presented a robust master/slave control strategy of the dual-stage micro- and nano-manipulator which is designed using a laser beam tracking system to predict the beam location for performing the efficient tracking and following control approach. Further, Jain et al. (2015a, b) have also attempted on micro-manipulation system where two piezoelectric fingers based micro-gripper is developed for robotic micro-assembly.

In this chapter, a novel design of three piezoelectric fingers based micro-gripper along with mobile manipulation system is proposed for robotic micro-assembly for handling the lightweight and miniature object. This is a novel part of this paper.

This chapter is organized as follows: a new design of mobile manipulation system along with a three piezoelectric actuator fingers based micro-gripper for robotic micro-assembly is discussed in Sect. 2. Further, kinematic analysis of micro-manipulation system is carried out in Sect. 2.1. The simulation results are presented. A prototype is developed and robotic micro-assembly is carried out as discussed in Sect. 3. The results toward robotic assembly are discussed in Sect. 4. Conclusions are drawn in Sect. 5.

2 Design of Piezoelectric Actuators Based Compact Miniature Wireless Manipulation System (WMMS)

Under this section, a novel design of piezoelectric actuators based miniature wireless manipulation system (WMMS) is discussed. Further, throughput analysis is carried out using ADAMS software, and angular speeds are found and different cases are analyzed.

2.1 Design and Working Principle of the Manipulation System

In order to execute the robotic micro-assembly in a large workspace, a novel multi-DOF based mobile micro-manipulation system along with piezoelectric actuators based micro-gripper is designed as shown in Fig. 1. This system consists of a mobile wheels based chassis, articulated robotic arm, link arm along with a wrist, and three piezoelectric actuator fingers based micro-gripper. The mobile wheels based chassis in manipulator has a strong aluminum (Al) base which has four-wheel driving system. This driving system provides motion to the manipulator in desired workspace. At the top of mobile wheels based chassis, one end of an articulated robotic arm is fixed where one rotational motion is provided for up-down motion during manipulation of object. The other end of articulated arm is connected to one end of arm link which has one rotational motion also. This provides the articulation and other end of arm link is connected to wrist. This end of link arm has rotational motion for adjusting the vertical alignment of the object. The other end of wrist has also rotational motion for rotating the object at desired

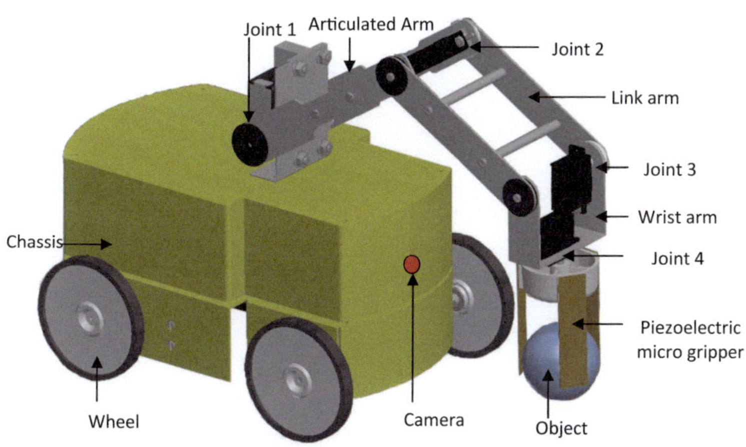

Fig. 1 CAD model of three fingers based micro-manipulator using piezoelectric actuators

orientation. At this end, a three piezoelectric actuator fingers based micro-gripper is integrated for holding and grasping the object. These three fingers are designed using bimorph piezoelectric actuators which are held in a plastic-based cylindrical clamp. These piezoelectric-based fingers (each size 45 mm × 11 mm × 0.6 mm) are activated through 0–60 V. Through bi-directional behavior of bimorph piezoelectric fingers, this grasps the object and by movement of other arms the object can be placed at desired position. A camera is also mounted on the chassis in order to visualize the object within desired work. This manipulation system can be held by the different sizes and shapes of the objects like circular, triangular, etc.

2.2 Kinematic Analysis of the Micro-manipulation System

In order to attempt the manipulation task, identification of workspace is one of the important parameters for robotic assembly. This defines the possible position that can be achieved by its end effector relative to workspace. Here, a kinematics of mobile micro-manipulation system is discussed. For developing the kinematics of mobile micro-manipulation system, one mobile vehicle arrangement along with articulated arm, link arm, wrist, and end effector as a three piezoelectric fingers based micro-gripper are analyzed together. In order to find the end effector position, the schematic diagram of micro-manipulation system is shown in Fig. 2. The two wheels of mobile mechanism are denoted as W_1 and W_2. The base is mounted on mobile mechanism at Joint J_1 where one end of articulated arm is fixed at distance of L_1. This arm rotates about this joint with angle of θ_1. The other end of articulated arm with length L_2 is attached with one end of link arm at joint J_2 and this arm also rotates about J_2 with the angle of θ_2. The other end of link arm with length of L_3 is

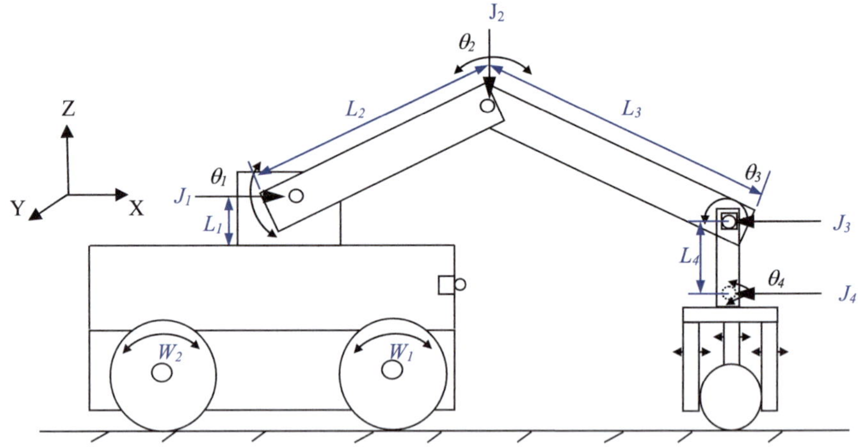

Fig. 2 Schematic representation for kinematic analysis of micro-manipulation system

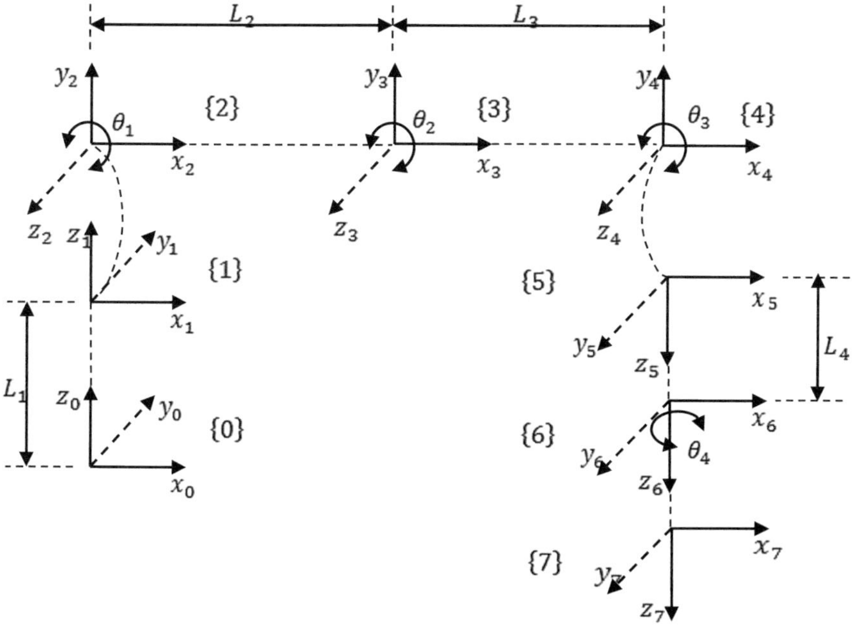

Fig. 3 Schematic diagram for forward kinematics of manipulation system with local axes

connected with the one end of wrist at joint J_3 and this also rotates about same joint (J_3) with angle of θ_3. The end effector is integrated at the end of wrist at joint J_4 with length of L_4. The wrist can be twisted with angle of θ_4 in order to attempt the twisting motion.

In order to develop the kinematics, the location of each link relative to its neighbors is shown in Fig. 3. The positions of frames {0} and {1} are available at joint 1. When one articulated arm rotates about *Z*-axis with angle of θ_1, the frame {2} is developed. Similarly, the frames {3} and {4} are established for providing the moment of other end of articulated arm and link arm, respectively. The movements of these arms are provided through θ_2 and θ_3. Afterward, the frames {5} and {6} are the positions of wrist where wrist is rotated through θ_4. The frame {7} is the position of end effector where three piezoelectric fingers based micro-gripper will be integrated. By applying the principle of forward kinematics according to Denavit–Hartenberg (D-H) method, the position of end effector or micro-gripper can be found for picking and placing of the object because the manipulation system rotations take place about local axis. The D-H parameter is also mentioned in Table 1.

After applying the D-H method, straightforward kinematics is rotated about all rotational movement about respective axes of reference frame (${}^{0}T_1$, ${}^{1}T_2$, ${}^{2}T_3$, ${}^{3}T_4$, ${}^{4}T_5$, ${}^{5}T_6$, and ${}^{6}T_7$), respectively. The transformation matrix of each link with respect to joint can be written as

Table 1 D-H parameters for finding the position of micro-gripper

a_i	d_i	α_i	θ_i
0	L_1	0	0
0	0	90°	0
L_2	0	0	θ_1
L_3	0	0	θ_2
0	0	90°	θ_3
0	L_4	0	0
0	0	0	θ_4

$$T_1^0 = \begin{bmatrix} 1 & 0 & 0 & 0 \\ 0 & 1 & 0 & 0 \\ 0 & 0 & 1 & L_1 \\ 0 & 0 & 0 & 1 \end{bmatrix} \tag{1}$$

$${}^1T_2 = \begin{bmatrix} 1 & 0 & 0 & 0 \\ 0 & 0 & -1 & 0 \\ 0 & 1 & 0 & 0 \\ 0 & 0 & 0 & 1 \end{bmatrix} \tag{2}$$

$${}^2T_3 = \begin{bmatrix} \cos\theta_1 & -\sin\theta_1 & 0 & L_2\cos\theta_1 \\ \sin\theta_1 & \cos\theta_1 & 0 & L_2\sin\theta_1 \\ 0 & 0 & 1 & 0 \\ 0 & 0 & 0 & 1 \end{bmatrix} \tag{3}$$

$${}^3T_4 = \begin{bmatrix} \cos\theta_2 & -\sin\theta_2 & 0 & L_3\cos\theta_2 \\ \sin\theta_2 & \cos\theta_2 & 0 & L_3\sin\theta_2 \\ 0 & 0 & 1 & 0 \\ 0 & 0 & 0 & 1 \end{bmatrix} \tag{4}$$

$${}^4T_5 = \begin{bmatrix} \cos\theta_3 & -\sin\theta_3 & \sin\theta_3 & 0 \\ \sin\theta_3 & \cos\theta_3 & -\cos\theta_3 & 0 \\ 0 & 1 & 0 & 0 \\ 0 & 0 & 0 & 1 \end{bmatrix} \tag{5}$$

$${}^5T_6 = \begin{bmatrix} 1 & 0 & 0 & 0 \\ 0 & 1 & 0 & 0 \\ 0 & 0 & 1 & L_4 \\ 0 & 0 & 0 & 1 \end{bmatrix} \tag{6}$$

$${}^6T_7 = \begin{bmatrix} \cos\theta_4 & -\sin\theta_4 & 0 & 0 \\ \sin\theta_4 & \cos\theta_4 & 0 & 0 \\ 0 & 0 & 1 & 0 \\ 0 & 0 & 0 & 1 \end{bmatrix} \tag{7}$$

The total transformation matrix ${}^{0}T_{7}$ can be written as

$$ {}^{0}T_{7} = {}^{0}T_{1} \times {}^{1}T_{2} \times {}^{2}T_{3} \times {}^{3}T_{4} \times {}^{4}T_{5} \times {}^{5}T_{6} \times {}^{6}T_{7} \quad (8) $$

or in terms of joint variable is

$$ {}^{0}T_{7} = \begin{bmatrix} \cos(\theta_1+\theta_2+\theta_3+\theta_4) & \sin(\theta_1+\theta_2+\theta_3+\theta_4) & \sin(\theta_1+\theta_2+\theta_3) & L_2\cos\theta_1 + L_3 cos(\theta_1+\theta_2) + L_4\sin(\theta_1+\theta_2+\theta_3) \\ -\sin\theta_4 & -\cos\theta_4 & 0 & 0 \\ \sin(\theta_1+\theta_2+\theta_3+\theta_4) & \cos(\theta_1+\theta_2+\theta_3+\theta_4) & -\cos(\theta_1+\theta_2+\theta_3) & L_1 + L_1\sin\theta_1 + L_2\sin(\theta_1+\theta_2) - L_4\cos(\theta_1+\theta_2+\theta_3) \\ 0 & 0 & 0 & 1 \end{bmatrix} \quad (9) $$

For finding the throughput/speed, the basic model of micro-manipulator is developed in ADAMS software as shown in Fig. 4. During development of model, the chassis is constructed of aluminum sheet material where four-wheel drive is made for controlling the chassis. This chassis is controlled by three servomotors. These motors are placed at revolute joints of wheels. By controlling the motors, the chassis moves in specified area. After that, an articulated link is placed at the top of chassis with a base. The base is fixed on the top of chassis. One end of articulated link is attached with base with help of a revolute joint and motor. This motor provides an articulation of arm. In the other end, another motor is mounted with revolute joint for articulating the arm and connection of one end of link arm. The other end of link arm is attached with wrist where a motor is also mounted with revolute joint. A motor is also attached to the end effector for providing the rotational motion during assembly. The end effector is designed using three fingers

Fig. 4 ADAMS model of mobile micro-manipulation system

Table 2 Different dimensional parameters of micro-manipulation system

Name of parts	Material properties of each mechanism				
	Material	Young's modulus (N/m^2)	Density (Kg/m^3)	Poisson's ratio	Weight (g)
Chassis	Al	6.9×10^{10}	2240	0.29	272
Base	Al	6.9×10^{10}	2240	0.29	31
Articulated arm	Al	6.9×10^{10}	2240	0.29	16
Link arm	Al	6.9×10^{10}	2240	0.29	11
Wrist	Al	6.9×10^{10}	2240	0.29	14
Piezoelectric micro gripper	Piezoelectric actuator	4.5×10^{10}	7500	0.30	10

based micro-gripper. These three fingers are constructed using bimorph piezo-electric actuators. These are operated at 0–60 V. An object is also placed at ground so that this micro-manipulator will pick the object and placed in identified work.

The different material properties of each link mechanism are given in Table 2. In order to attempt, the pick-and-place operation of robotic assembly, the following steps are carried out:

I. At first, manipulator is initialized at home position. After initialization of the manipulator, it will ask the position of object and scan the work area for identifying the object.
II. In the second step, the manipulator will search for the shortest path for reaching the desired position of the object.
III. In the third step, four-wheeled system will run through the motor for left-right movement.
IV. Subsequently, the manipulator reaches near to the object and then the articulated and link arms will move downward at top position of the object accordingly. If minor orientation will be required during holding the object, this will be adjusted by movement of wrist.
V. After that, piezoelectric micro-gripper will be activated through 0–60 V for holding the object.
VI. After holding the object, the articulated and link arms will move upward accordingly for placing the object at different positions. Further, manipulator follows the same steps from II to V for placing the object at different positions.

In order to find the relation of different link positions, simulation has been carried out and the three cases are considered. In the Case-I, the object is closest to the chassis of the mobile manipulator and when it picks the object, the angles between different links change from initial position. This is indicated as projected angles between different links. The changes of values of these angles between different links (i.e., articulated arm, link arm, and wrist) are 20°, 10°, and 5°, respectively. In the Case-II, the object is little far from the chassis of the mobile

Table 3 Projected angles of different links of the wireless mobile manipulation system

S. No.	Projected angles of links (°)		
	Articulated arm	Link arm	Wrist
Case-I	20°	10°	5°
Case-II	16°	8°	4°
Case-III	14°	7°	3.5°

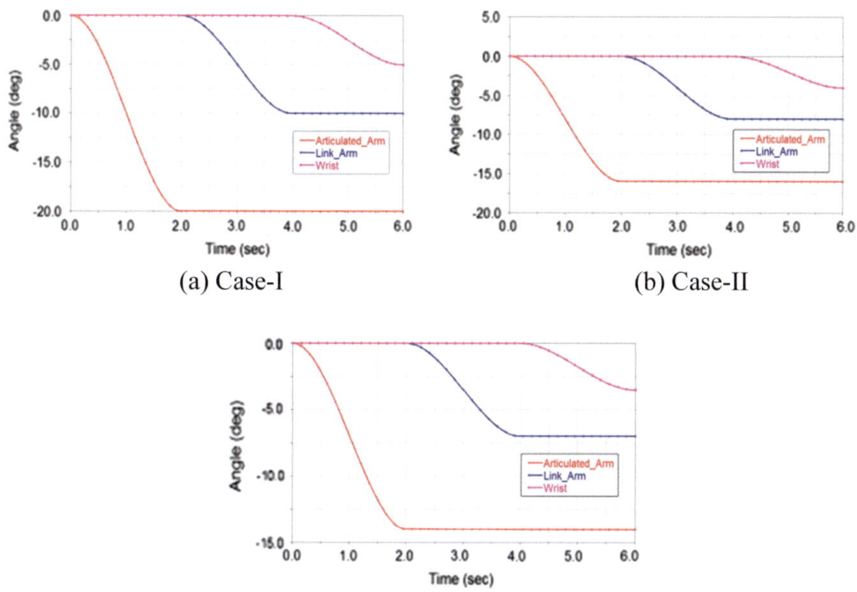

(a) Case-I (b) Case-II

(c) Case-III

Fig. 5 Simulation results of different case studies for pick-and-place operation

manipulator and the projected angles for articulated arm, link arm, and wrist are 16°, 8°, and 4°, respectively. In the Case-III, the object is far from the chassis of the mobile manipulator and projected angles are 14°, 7°, and 3.5°, respectively, as given in Table 3.

From the above case study, it is clearly shown that if the object attempts the pick-and-place operation from one to another position, the projected angles for articulated arm, link arm, and wrist are in the ratio of 4:2:1. The simulation results of these cases are shown in Fig. 5. It is envisaged that the behaviors of each arm are similar in each case. The angular velocity of each case is analyzed as shown in Fig. 6. It is observed that the angular velocities of each case are also in same ratio (4:2:1) and the values of articulated arm, link arm, and wrist for all cases are given in Table 4.

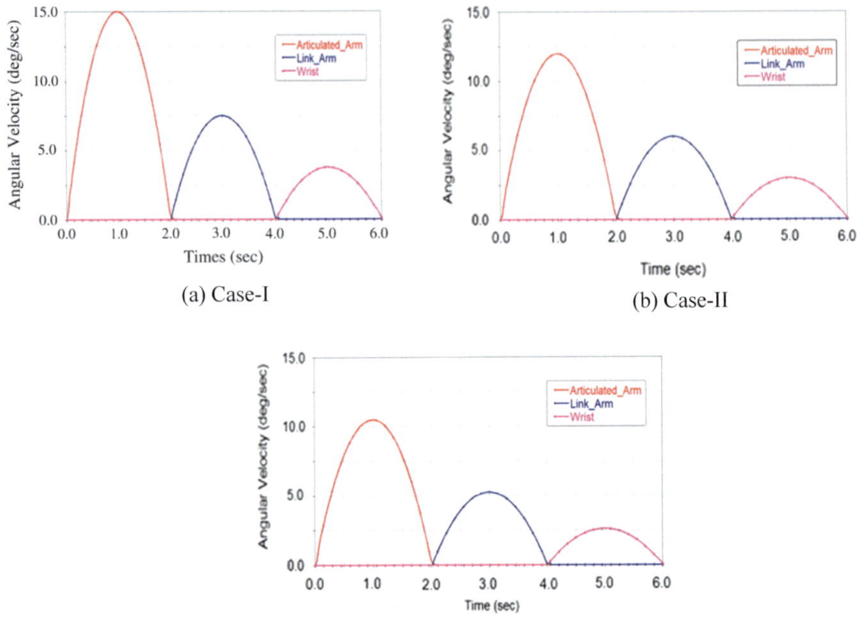

(a) Case-I

(b) Case-II

(c) Case-III

Fig. 6 Simulation results of angular velocity in different case studies

Table 4 Angular velocity of different links of the wireless mobile manipulation system

S. No.	Angular velocity of links (deg/s)		
	Articulated arm	Link arm	Wrist
Case-I	15.0	7.5	3.75
Case-II	12.0	6.0	3.0
Case-III	10.5	5.25	2.625

3 Development of Control System for Wireless Micro-manipulation System

In order to develop the novel piezoelectric actuator based miniature wireless mobile manipulator, a mobile wheels based chassis is developed as shown in Fig. 7. During development, the manipulation system is developed with two mechanisms (mobile and manipulation units) while performing the assembly operation. The mobile mechanism is used for movement in large workspace for handling operation and manipulation mechanism is developed by using 3-DOF based manipulator with a piezo-based gripper. The gripper is fitted at the extreme end which is capable to orient itself according to the alignment of the object with the help of the nano-servo motor attached to it which is mounted on the mobile arrangement. The mobile arrangement has four-wheel drive system which gives the manipulator motion to

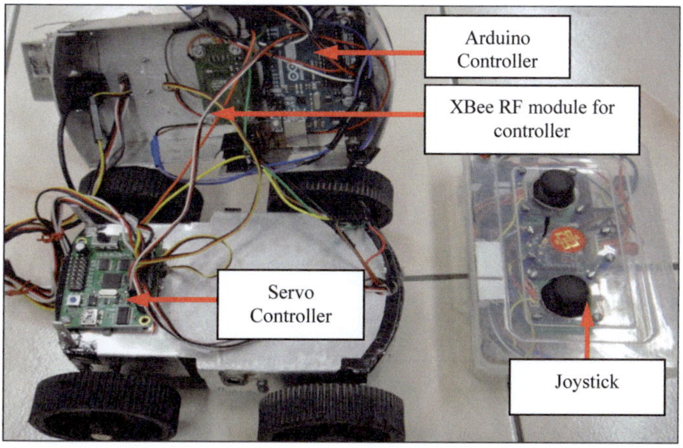

Fig. 7 Integration of wireless controller in manipulation system

move around anywhere. Each wheel has dimension of 48 mm diameter and 5 mm width. The rear two wheels are responsible for the driven speed, and the front two wheels provide the left-right motion with help of a steering mechanism. In the manipulation mechanism, three linkages arms (articulated, link, and wrist arm) and an end effector are developed. The end effector (gripper) is developed using three piezoelectric bimorph actuators. The gripper consists of a holder and three piezoelectric bimorph actuators where it is used as an active finger and operated through 0–60 VDC. Therefore, these fingers can compensate the misalignment itself during holding and pick-and-place operations of robotic assembly. During development of the wireless manipulation system, the control system has been developed using two modules, i.e., manipulation module and control panel module as shown in Fig. 8.

In manipulation module, one miniature servo controller for steering, arm rotation, gripper orientation, mobile mechanism, and piezoelectric controller for gripping are assembled and integrated with Arduino Uno microcontroller board (Make: Arduino, Italy) which makes decision as requirement. This is operated by two-axis joystick control system. In this module, an XBee RF module for wireless communication is also assembled, and another module of XBee is integrated on control panel module for proper wireless communication. This XBee is also powered by Arduino and Arduino is operated with 6 V Ni-MH battery. Two 2-axis joysticks (Make: iTEAD studio) are supplied with 5 VDC from the Arduino board and the *X*- and *Y*-axis pins are connected to analog input pins of Arduino, whereas the button pins are connected to the digital input pins. Deflection of the piezoelectric actuator varies from −500 to 500 μm while varying the voltage from −30 to 30 VDC. In the range of 60 V DC, total 1000 μm deflection is possible for the piezoelectric actuator. The control input to the 60 V circuit from the Arduino can be discretized into 255 discrete levels with inbuilt 8-bit DAC of Arduino Uno controller board which yields 255 discrete voltage levels in the range of 60 V. Hence, the deflection resolution for the piezoelectric actuator used in the gripper is

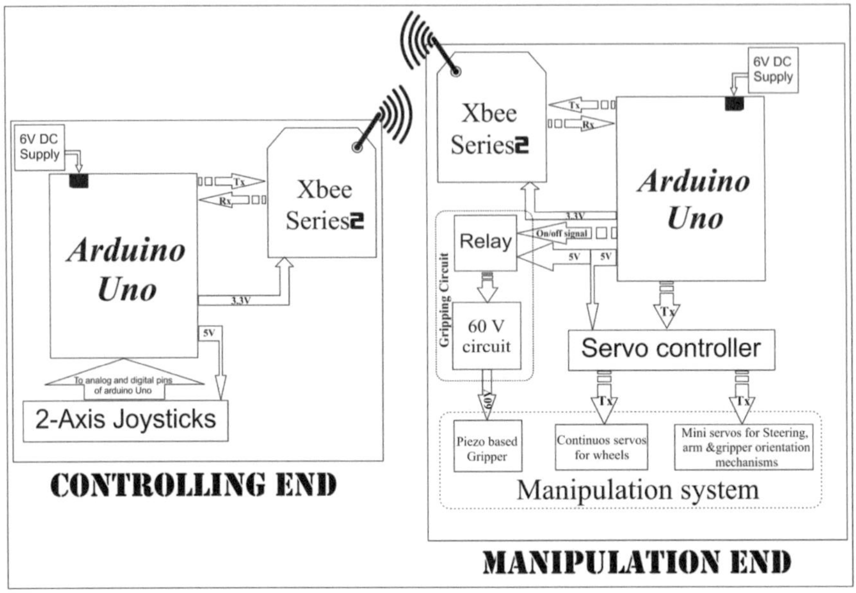

Fig. 8 Connection diagram of controller for wireless mobile manipulation system

1000/255 = 3.9215 μm. The gripper is being turned on/off using a relay breakout board (Make: iTEAD studio) which turns on/off the supply of 60 V to the gripper fingers. All the servos are controlled with a servo controller (Model: Propeller USB Servo controller; Make: Parallax, USA). According to the joystick values received remotely through the XBee at the manipulator module, the servo controller is instructed to function as per it by the Arduino Uno controller board on the manipulator side. The range of remote operation is 100 m. In order to operate the WMMS, the joystick module and manipulator module are initialized at first. In the joystick module, the default joystick values are set with ($X = 0$, $Y = 0$). This is one-time initialization after powering on this module. Afterward, current joystick values (new *X*, new *Y*) are taken through analog input pins and button values of joysticks (*B*) through digital input pins of Arduino Uno board. Subsequently, these are transmitted wirelessly through XBee RF module. Then, after a small time pause, it is checked whether there is an acknowledgement signal is received or not. The scanning of joystick values and sending them wirelessly will be continued until data are received from the other end (i.e., manipulation end). After receiving the data, a time pause is taken by this module and then it is checked whether the operation of joystick module is required or not. It will go through all the above steps from the scanning of joystick values again and again until the termination of this module by user as shown in Fig. 9. In the manipulation module, initially all the servos for wheels, steering, and arm are set to its default positions. This is also one-time initialization after powering on this module. After the initialization, a small time pause is taken before receiving the joystick values wirelessly. The data are received

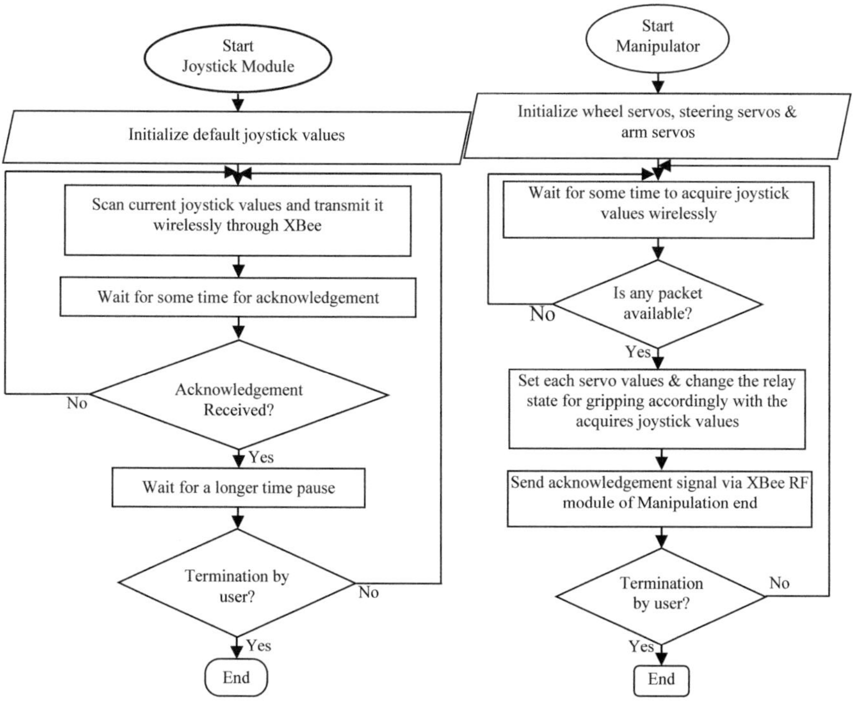

Fig. 9 Flowchart for controlling of mobile manipulation system

by the manipulation module, and acknowledgement is sent after performing the operation. After getting the packet holding the joystick values, the Arduino Uno on the manipulator side instructs all the servos to work as desired fashion by setting proper position values of all individual servos. It is not only the servos whose states may be changed but the state of the gripper is also changed according to the joystick value received. After doing all the required tasks, an acknowledgement signal is transmitted wirelessly via the XBee at the manipulator end. Then, it is checked whether the service of manipulator module is required or not. It will go through all the above steps from acquiring the joystick values again and again until the termination of this module by user.

4 Results and Discussions

In this section, the deflection behavior of piezoelectric actuator is discussed for gripping the object. Further, a WMMS is developed. The prototype is tested and its experimental performance for gripping is evaluated toward pick-and-place operation of assembly. The performance of handling capability of different objects is demonstrated.

4.1 Deflection Behavior of Piezoelectric Actuator

In order to examine the deflection characteristic with the voltage signal of bimorph piezoelectric actuator, the piezoelectric actuator is fixed with a holder which is connected to a high-voltage (HV) amplifier for providing the voltage. According to the input in microcontroller, the control signal is sent to HV amplifier. The amplified voltage activates the bimorph piezoelectric actuator properly. This controlled voltage signal provides the bending to the piezoelectric actuator as shown in Fig. 10. When the polarity of the voltage is reversed, then bending occurs in opposite direction. To study the bending behavior of the actuator, voltage is applied and it attains the maximum deflection of piezoelectric actuator up to 0.5 mm (500 μm) as shown in Fig. 11. When the voltage is varied in the reverse way, then the actuator does not follow the same path to return to its initial position. It shows the hysteresis error in deflection. This is minimized through microcontroller where relays are set accordingly. The deflection error is minimized up to 84%. Then, this voltage is sent to micro-gripper during assembly.

During gripping the objects, the piezoelectric actuator is actuated at 30 VDC which yields quiet high average force of 104.9558 mN. The tip force values of piezo-finger at different trials for different voltages are presented in Table 5. From the trials at the gripping voltage of 30 VDC, the distribution is also plotted as shown in Fig. 12. The repeatability is calculated from the distribution as 0.3298 mN.

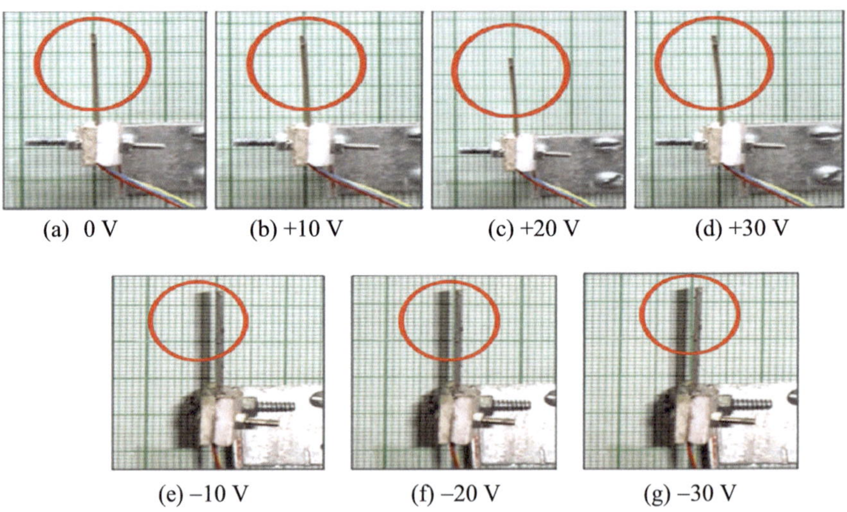

Fig. 10 Successive steps for testing of bimorph piezoelectric actuation

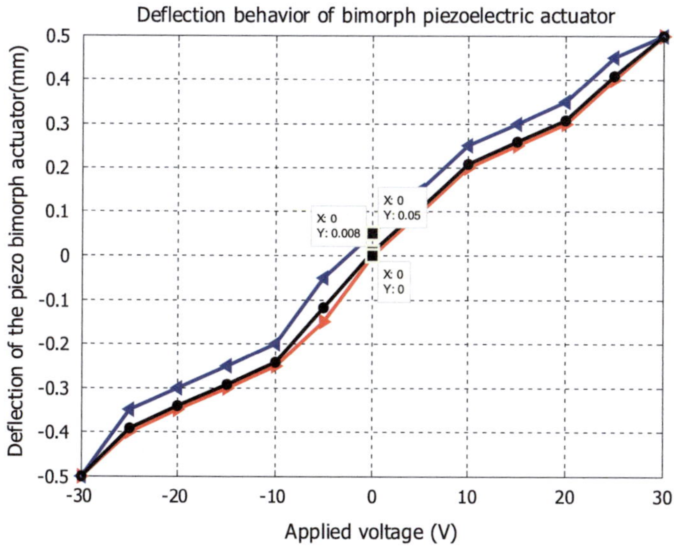

Fig. 11 Deflection behavior of bimorph piezoelectric actuator

Table 5 Experimental data for force analysis of piezoelectric actuator

Volt	Force value in mN					Average force (mN)
	F1	F2	F3	F4	F5	
0	0	0	0	0	0	0
10	65.39	65.282	65.527	65.458	65.085	65.348
20	91.955	91.906	92.671	92.563	92.946	92.408
30	104.959	105.042	104.924	104.904	104.953	104.9558
Standard deviation at 30 V						0.3298 mN

4.2 *Behavior of Wireless Mobile Micro-manipulation System During Assembly*

In order to execute the robotic assembly, the prototype of mobile micro-manipulation system is developed as shown in Fig. 13. The mobile manipulation system can move in any direction with the help of wireless joystick arrangement because mobile mechanism is operated through continuous servo motors with help of a servo controller and Arduino Uno controller. The PWM pulses are sent to individual servos as commanded by the central commanding element in the controller. The steps of operation for manipulation in stationary condition during the arm movement are given below:

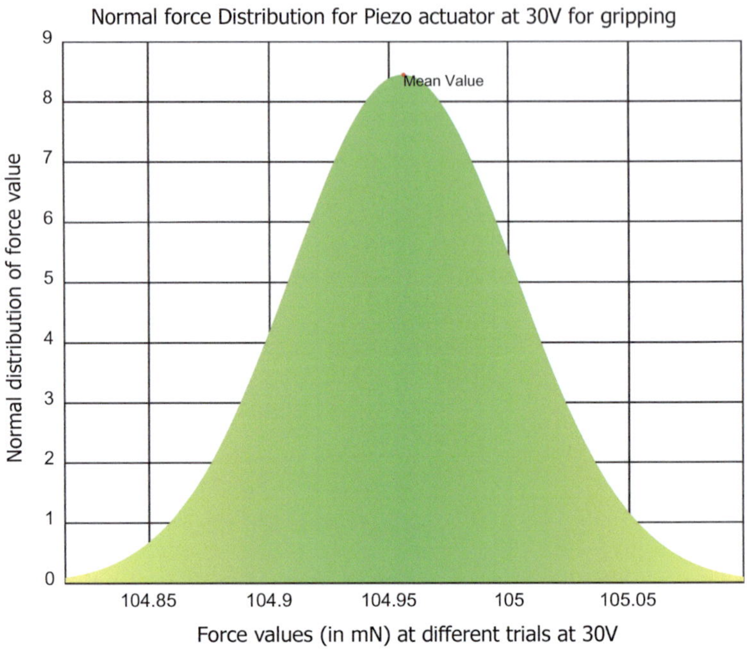

Fig. 12 The normal distribution curve for force of piezoelectric actuator

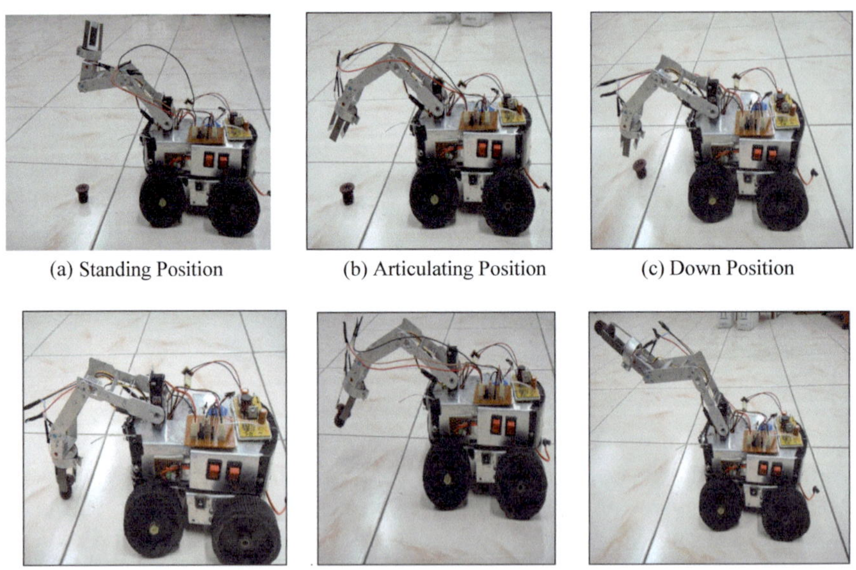

(a) Standing Position (b) Articulating Position (c) Down Position

(d) Grasping Position (e) Up Position (f) Lifting Position

Fig. 13 Development of prototype of wireless mobile manipulation system

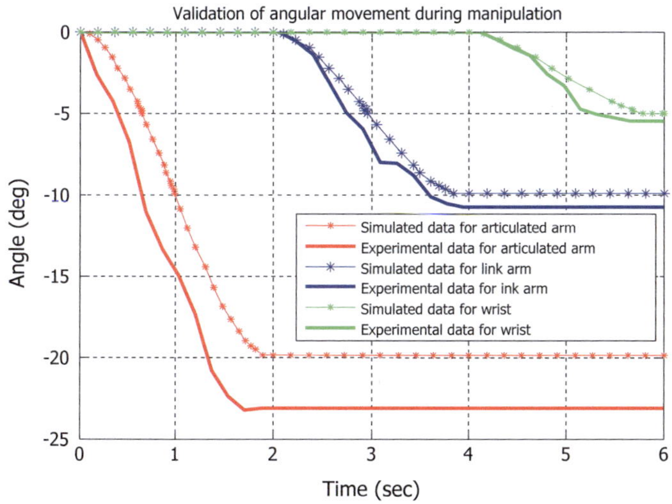

Fig. 14 Validation of different arm behaviors during manipulation

(i) At first, the manipulator is initialized where all the three arms are kept at a predefined set position.
(ii) Afterward, the wrist arm is operated through command with Arduino via the servo controller and it is rotated downward.
(iii) Subsequently, all three arms (articulated, link, and wrist) are operated synchronously through command to attain the position of the object.
(iv) After the three piezo-fingers based gripper reaches the object, it is activated to grip the object through 0–60 V DC which is controlled by the Arduino and a relay.
(v) At that time, the gripper holds the object. Then, all arm mechanisms are again operated synchronously in upward direction for lifting the object.
(vi) Finally, the end arm is moved further to attain the final position.

In order to evaluate the performance of different link positions, experiments are carried out where link position angles are set at angle of 20°, 10°, and 5° (i.e., articulated arm, link arm, and wrist), respectively. The performance is shown in Fig. 14. It shows that the experimental behavior is similar to that of the simulation data.

In order to evaluate the performance of micro-manipulation system, three markers (red, blue, and green) are placed at different locations like link joint, articulated joint, and arm joints. A camera (Make: Sony, Model: DSC TX-9) is mounted on the front of micro-manipulation system for acquiring the videos of movements in real time. A program in MATLAB software is written for converting

Table 6 Experimental data for grasping time

Time (s)	Distance from origin (mm)										
	Trial1	Trial2	Trial3	Trial4	Trial5	Trial6	Trial7	Trial8	Trial9	Trial10	Average
0	362.3	365.3	361.8	360.6	362.5	364.2	361.5	360.6	362.7	363.5	362.5
2	313.9	319.7	325.3	326.3	324.6	321.3	320.1	318.5	322.4	320.9	321.3
4	271.7	279.3	278.9	286.5	279.6	274.6	275.7	281.7	279.9	283.1	279.1
6	253.4	255.4	255.1	262.4	254.5	253.2	257.5	258.5	258.5	256.5	256.5
8	234.4	237.8	236.5	237.4	238.3	235.6	234.6	237	238.2	235.2	236.5
10	231.7	234.5	225.4	228.4	229.9	230.6	231.3	226.2	228.6	232.4	229.9

Table 7 Experimental data for lifting time

Time (s)	Distance from origin (mm)										
	Trial1	Trial2	Trial3	Trial4	Trial5	Trial6	Trial7	Trial8	Trial9	Trial10	Average
10	232.8	234.15	224.4	228.3	229.6	231.5	232.5	227.45	228.1	229.7	229.85
12	255.6	268	261	256	267.2	262.5	254.6	258.15	257.2	262.25	260.25
14	291.9	304.5	297.5	291.3	294.4	297.5	298.3	292.6	296.8	297.4	296.22
16	324.6	327.6	321.7	320.3	321.2	326.6	322.3	324.8	323.3	325.6	323.8
18	346	345.3	344.4	338.2	337.6	339.8	344.5	336.8	346.6	345.5	343.47
20	361	362	362.9	361.6	363.3	360.5	364.6	359.5	362.2	361.9	361.95

the videos into different images/templates using template matching method. From these different images/templates, *X*-*Y* coordinate data of different color markers are generated by conducting different trials as given in Tables 6 and 7.

From these experimental data, the performance is evaluated by plotting the average data as shown in Fig. 15a. It shows that the manipulation system takes 8 s downtime for grasping and within 2 s the manipulation system grasps the objects. Further, the manipulator lifts the object within 10 s and ready to carry the object at different locations. In order to find the repeatability during grasping, the experiments have been carried out and data are collected as given in Table 8. After conducting the different trials, it is found that the normal distribution is 0.1420 and the repeatability of manipulation mechanism is 99.85%. The normal distribution curve is also plotted as shown in Fig. 15b.

4.3 Different Handling Capabilities of Mobile Micro-manipulation System During Assembly

The handling capabilities of different types of objects are tested as shown in Fig. 16. The load capabilities are summarized in Table 9. By conducting different trials, it is observed that using bimorph piezoelectric actuator based micro-gripper

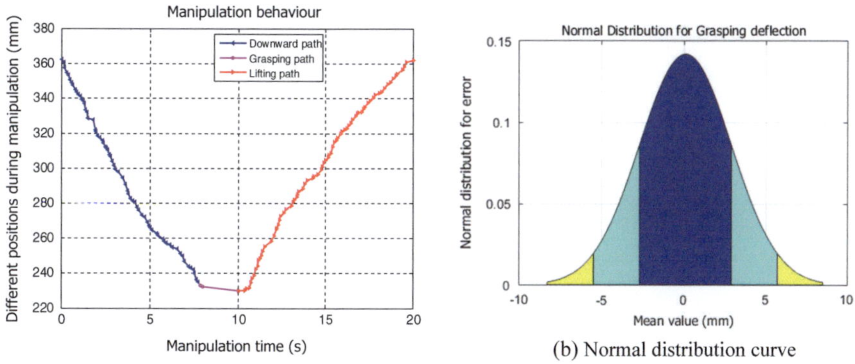

(a) Grasping time

(b) Normal distribution curve

Fig. 15 Behavior of mobile manipulation system during handling

Table 8 Experimental data for repeatability analysis

Trial number	Trial1	Trial2	Trial3	Trial4	Trial5	Trial6	Trial7	Trial8	Trial9	Trial10
Actual distance (mm)	231.7	234.5	225.4	228.4	229.9	230.6	231.3	226.2	228.6	232.4
Desired distance (mm)	230									
Error	−1.7	−4.5	4.6	1.6	0.1	−0.6	−1.3	3.8	1.4	−2.4
Mean	0.1000									
Standard deviation	2.8083									
Normal deviation	0.1420									

provides the dexterous handlings of object where the bimorph piezoelectric fingers are actuated individually and this also provides adjust misalignment during operation automatically. These kinds of handling exhibited acceptable handiness and handling capability of various millimeter-scale components. These devices can also be used in different kinds of assemblies where placement of object is required for the precise positioning and misalignment ability during robotic assembly.

(a) Handling of Nut

(b) Handling of DC motor

(c) Handling of thermocol

Fig. 16 Handling capability of different objects

Table 9 Testing of different types of object

S. No.	Type of object	Object size	Weight (g)
1	Thermocol ball	Diameter 22.2 mm	0.130
2	Plastic nut	Head diameter 22.15 mm, length 22.20 mm	3.250
3	DC motor	Diameter 25.0 mm	22.0

5 Conclusion

In this paper, a novel design of wireless piezoelectric actuator based mobile micro-manipulation system is proposed where error minimization and repeatability analysis of piezoelectric actuator based compliant micro-gripper is discussed. In robotic microassembly, compliant piezoelectric actuator based micro-gripper helps for accommodating the misalignment of peg when assembly operations like pick-and-place and peg-in-hole are performed by 3-DOF based mobile micro-manipulation system. During development of piezoelectric actuator based micro-gripper, bimorph piezoelectric actuator is used as an active actuator. The bending behavior and high mass force generation techniques are utilized for handling purpose. By constructing a piezoelectric actuator based micro-gripper, it is established that use of bimorph piezoelectric strips helps in manipulation task where holding and lifting rigid/flexible miniature components without utilizing any conventional motor can be performed. By developing the prototype of mobile micro-manipulation system, it is demonstrated that 3-DOF mobile micro-manipulator with bimorph piezoelectric actuator based compliant micro-gripper is capable of handling the peg during peg-in-hole assembly. This capability shows the utilization of piezoelectric actuator during manipulation in robotic micro-assembly. The future scope of research will be explored in the area of inspection and surveillance like small pipeline inspections and mobile surveillance for domestic and other applications.

Acknowledgements The authors are grateful to the Director, CSIR-CMERI, Durgapur, West Bengal, India for granting the permission to publish this paper. This work is the part of entitled project “Development of piezo actuator based micro manipulation system” under SINP on “Intelligent Devices and Smart Actuator” (Project No. ESC-203/10) for 12th 5-year plan which is financially supported by Council of Scientific Industrial Research (CSIR), Govt. of India, New Delhi, India.

References

Agnus, J., P. Nectoux, and N. Chaillet. 2005. Overview of microgrippers and design of a micromanipulation station based on a MMOC microgripper. In *Proceedings of International Symposium on Computational Intelligence in Robotics and Automation*, 117–123.

Amari, N., D. Folio, and A. Ferreira. 2014. Motion of a micro/nanomanipulator using a laser beam tracking system. *International Journal of Optomechatronics* 8 (1): 30–46.

Bhandari, B., G.Y. Lee, and S.H. Ahn. 2012. A review on IPMC material as actuators and sensors: fabrications, characteristics and applications. *International Journal of Precision Engineering and Manufacturing* 13 (1): 141–163.

Bolopion, A., and S. Régnier. 2013. A review of haptic feedback teleoperation systems for micromanipulation and microassembly. *IEEE Transactions on Automation Science and Engineering* 10 (3): 496–502.

Boudaoud, M., and S. Regnier. 2014. An overview on gripping force measurement at the micro and nano-scales using two-fingered microrobotic systems. *International Journal of Advanced Robotic Systems* 11: 45.

Chen, T., M. Pan, Y. Wang, J. Liu, L. Chen, and L. Sun. 2012. Manipulation of micro objects based on dynamic adhesion control. *International Journal of Advanced Robotic Systems* 9: 89.

Clévy, C., and N. Chaillet. 2006. Micromanipulation and micro-assembly systems. In *Proceedings of International Advanced Robotics Programm, IARP'06*, Paris, France, sur-CD.

Dadfarnia, M., N. Jalili, B. Xian, and D.M. Dawson. 2004. A Lyapunov-based piezoelectric controller for flexible Cartesian robot manipulators. *Journal of Dynamic Systems, Measurement, and Control* 126 (2): 347–358.

Deole, U., R. Lumia, M. Shahinpoor, and M. Bermudez. 2008. Design and test of IPMC artificial muscle microgripper. *Journal of Micro-Nano Mechatronics* 4 (3): 95–102.

Fuchiwaki, O., A. Ito, D. Misaki, and H. Aoyama. 2008. Multi-axial micromanipulation organized by versatile micro robots and micro tweezers. In *Proceedings of IEEE International Conference on Robotics and Automation*, 893–898.

Gao, P., and S.M. Swei. 1999. A six-degree-of-freedom micro-manipulator based on piezoelectric translators. *Nanotechnology* 10 (4): 447.

Hall, D.A. 2001. Review nonlinearity in piezoelectric ceramics. *Journal of Materials Science* 36 (19): 4575–4601.

Hristov, K., F. Ionescu, and K. Kostadinov. 2002. Modeling, procedure and development of piezo actuated mechatronic systems. *Journal of Problems of Engineering Cybernetics and Robotics* 53: 76–79.

Huang, S.J., and J.P. Tsai. 2005. Robotic automatic assembly system for random operating condition. *International Journal of Advanced Manufacturing Technology* 27 (3–4): 334–344.

Jain, R.K., U.S. Patkar, and S. Majumder. 2009. Micro gripper for micromanipulation using IPMCs (ionic polymer metal composites). *Journal of Scientific & Industrial Research* 68 (1): 23–29.

Jain, R.K., S. Datta, S. Majumder, and A. Dutta. 2011. Two IPMC fingers based micro gripper for handling. *International Journal of Advanced Robotics Systems* 8 (1): 1–9.

Jain, R.K., S. Majumder, and A. Dutta. 2013a. SCARA based peg-in-hole assembly using compliant IPMC based micro gripper. *Robotics Autonomous Systems* 61 (3): 297–311.

Jain, R.K., S. Saha, and S. Majumder. 2013b. Development of piezoelectric actuator based compliant micro gripper for robotic peg-in-hole assembly. In *IEEE International Conference on Robotics and Biomimetics (ROBIO-2013)*, 1562–1567.

Jain, R.K., S. Majumder, B. Ghosh, and S. Saha. 2015a. Development of mobile micro manipulation system with a compliant piezoelectric actuator based micro gripper. *Journal of Manufacturing Systems* 35: 76–91.

Jain, R.K., S. Majumder, and B. Ghosh. 2015b. Design and analysis of piezoelectric actuator for micro gripper. *International Journal of Mechanics and Materials in Design* 11: 253–276.

Kohl, M., B. Krevet, and E. Just. 2002. SMA microgripper system. *Sensors and Actuators A: Physical* 97: 646–652.

Li, Y., Z. Wu, and X. Zhao. 2013. Optimal design and control strategy of a novel 2-DOF micromanipulator. *International Journal of Advanced Robotic Systems* 10: 162.

Lin, C.M., C.H. Fan, and C.C. Lan. 2009. A shape memory alloy actuated microgripper with wide handling ranges. In *Proceedings of IEEE/ASME International Conference on Advanced Intelligent Mechatronics*, 12–17.

Liu, T.J., and Q.Y. Liu. 2010. The design of micromanipulator based on piezo actuator. *Key Engineering Materials* 426: 529–531.

Mok, S.M., C.H. Wu, and D.T. Lee. 2000. A system for analyzing automatic assembly and disassembly operations. In *Proceedings of IEEE International Conference on Robotics and Automation, ICRA'00*, vol. 4, 3695–3700.

Oriolo, G., and C. Mongillo. 2005. Motion planning for mobile manipulators along given end-effector paths. In *IEEE International Conference on Robotics and Automation*, 2154–2160.

Pérez, R., N. Chaillet, K. Domanski, P. Janus, and P. Grabiec. 2006. Fabrication, modeling and integration of a silicon technology force sensor in a piezoelectric micro-manipulator. *Sensors and Actuators A: Physical* 128 (2): 367–375.

Popa, D.O., B.H. Kang, J.T. Wen, H.E. Stephanou, G. Skidmore, and A. Geisberger. 2003. Dynamic modeling and input shaping of thermal bimorph MEMS actuators. In *Proceedings of IEEE International Conference on Robotics and Automation-2003 (ICRA'03)*, vol. 1, 1470–1475.

Probst, M., C. Hürzeler, R. Borer, and B.J. Nelson. 2009. A micro assembly system for the flexible assembly of hybrid robotic MEMS devices. *International Journal of Optomechatronics* 3 (2): 69–90.

Rakotondrabe, M., and I.A. Ivan. 2011. Development and force/position control of a new hybrid thermo-piezoelectric micro gripper dedicated to micromanipulation tasks. *IEEE Transactions on Automation Science and Engineering* 8 (4): 824–834.

Sitti, M., D. Campolo, J. Yan, and R.S. Fearing. 2001. Development of PZT and PZN-PT based unimorph actuators for micromechanical flapping mechanisms. In *Proceedings of IEEE International Conference on Robotics and Automation (ICRA)*, vol. 4, 3839–3846.

Sun, L., C. Ru, and W. Rong. 2004. Hysteresis compensation for piezoelectric actuator based on adaptive inverse control. In *Fifth World Congress on Intelligent Control and Automation (WCICA 2004)*, vol. 6, 5036–5039.

Song, G., J. Zhao, X. Zhou, and J.A. De Abreu-García. 2005. Tracking control of a piezoceramic actuator with hysteresis compensation using inverse Preisach model. *IEEE/ASME Transactions on Mechatronics* 10 (2): 198–209.

Tamadazte, B., N.L.F. Piat, and S. Dembélé. 2011. Robotic micromanipulation and microassembly using monoview and multiscale visual servoing. *IEEE/ASME Transactions on Mechatronics* 16 (2): 277–287.

Valdastri, P., P. Corradi, A. Menciassi, T. Schmickl, K. Crailsheim, J. Seyfried, and P. Dario. 2006. Micromanipulation, communication and swarm intelligence issues in a swarm microrobotic platform. *Robotics and Autonomous Systems* 54 (10): 789–804.

Xiao, S., and Y. Li. 2014. Dynamic compensation and H∞ control for piezoelectric actuators based on the inverse Bouc-Wen model. *Robotics and Computer-Integrated Manufacturing* 30 (1): 47–54.

Yang, S., R.A. MacLachlan, and C.N. Riviere. 2012. Design and analysis of 6 DOF handheld micromanipulator. In *IEEE International Conference on Robotics and Automation (ICRA)*, 1946–1951.

Zaeh, M.F., D. Jacob, M. Ehrenstrasser, and J. Schilp. 2003. Hybrid micro-assembly system for teleoperated and automated micromanipulation. *Ziegert, JC: Machines and Processes for Micro-scale and Meso-scale Fabrication, Metrology and Assembly, Florida. North Carolina: Proceedings of ASPE* 28: 119–124.

Zhang, Y.L., M.L. Han, M.Y. Yu, C.Y. Shee, and W.T. Ang. 2012. Automatic hysteresis modeling of piezoelectric micromanipulator in vision-guided micromanipulation systems. *IEEE/ASME Transactions on Mechatronics* 17 (3): 547–553.

Numerical Modeling and Experimental Validation of Machining of Low-Rigidity Thin-Wall Parts

Gururaj Bolar and Shrikrishna N. Joshi

Abstract In the present work, a realistic three-dimensional thermomechanical finite element method (FEM) based model is developed to simulate the complex physical interaction of helical cutting tool and workpiece during thin-wall milling of an aerospace grade aluminum alloy. Lagrangian formulation with explicit solution scheme is employed to simulate the interaction between helical milling cutter and the workpiece. The behavior of the material at high strain, strain rate, and the temperature is defined by Johnson–Cook material constitutive model. Johnson–Cook damage law and friction law are used to account for chip separation and contact interaction. Experiments are carried out to validate the results predicted by the developed 3-D numerical model. Four case studies are conducted to test the capability of developed 3-D numerical model. It is noted that the milling force and wall deformation predicted by the developed model match well with the experimental results. Overall, this work provides a useful tool for prior study of the precision machining of low-rigidity thin-wall parts.

Keywords Thin-wall machining · Thermomechanical analysis
Finite element method · Milling force · Wall deflection · Aluminum 2024

Abbreviations

2-D	Two-dimensional
3-D	Three-dimensional
ALE	Arbitrary Lagrangian–Eulerian
CAE	Computer-aided engineering
CNC	Computer numerical control

G. Bolar · S. N. Joshi (✉)
Department of Mechanical Engineering, Indian Institute of Technology Guwahati, Guwahati 781039, Assam, India
e-mail: snj@iitg.ac.in

G. Bolar
e-mail: gururaj@iitg.ac.in

S. S. Pande and U. S. Dixit (eds.), *Precision Product-Process Design and Optimization*, Lecture Notes on Multidisciplinary Industrial Engineering,
https://doi.org/10.1007/978-981-10-8767-7_4

FEA	Finite element analysis
FEM	Finite element method
J-C	Johnson–Cook
MOGA	Multi-objective genetic algorithm
VMC	Vertical machining center

Nomenclature

A, B, c, m, n	Johnson–Cook (J-C) material model coefficients
$[C]$	Viscous damping matrix
$[\bar{C}]$	Capacitance matrix
c	Damping coefficient
C_p	Specific heat
D	Scalar damage parameter
D_1-D_5	Johnson–Cook (J-C) damage constants
E	Elastic modulus
F	External force vector
F_C	Resultant milling force
f_f	Fraction of heat energy conducted into the chip
f_t	Feed per tooth
F_x	Milling force in X direction
F_y	Milling force in Y direction
F_z	Milling force in Z direction
G_f	Hillerborg's fracture energy
h_c	Convection coefficient
k	Stiffness coefficient
$[K]$	Stiffness matrix
$[\bar{K}]$	Time-dependent conductivity matrix
K_c	Fracture toughness
k_chip	Shear flow stress
k_t	Thermal conductivity
$[M]$	Mass matrix
n_s	Spindle speed
p	Shorter length of two edges
$\dot{q}_\mathrm{conv}$	Convection coefficient
$\dot{Q}$	Total heat generation rate
$\dot{Q}_f$	Volumetric heat flux due to friction
$\dot{Q}_\mathrm{pl}$	Volumetric heat flux generated due to inelastic plastic deformation
r_d	Radial depth of cut
t	Plate thickness
T	Nodal temperature vector
$\dot{T}$	Derivative of the temperature with respect to time
T_c	Cutting temperature
T_melt	Melt temperature

T_{room}	Room temperature
$\ddot{u}$, $\dot{u}$, u	Nodal acceleration, velocity, and displacement vectors
$\bar{u}$	Equivalent plastic displacement
$\bar{u}_f$	Equivalent plastic displacement at failure
α	Thermal expansion
β	Mean friction angle
γ	Shearing strain
$\dot{\gamma}$	Shear stain rate
$\dot{\gamma}_s$	Slip rate
$\bar{\varepsilon}$	Equivalent plastic strain
$\dot{\varepsilon}$	Equivalent plastic strain rate
$\bar{\varepsilon}_{\text{f}}$	Equivalent plastic strain at failure
ε_{fi}	Equivalent strain at fracture initiation
$\dot{\varepsilon}_o$	Reference plastic strain rate
$\bar{\varepsilon}_{0i}$	Plastic strain at damage initiation
μ	Friction coefficient
v	Poisson ratio
ρ	Material mass
ρ_m	Material density
$\bar{\sigma}_{jc}$	Equivalent flow stress
σ_n	Normal stress
σ_y	Yield stress of the material
τ_{criti}	Critical frictional stress
τ_f	Frictional stress
τ_s	Shear stress

1 Introduction

Due to homogeneity and excellent strength-to-weight ratio, monolithic thin-wall components are widely used in aerospace, marine, electronics, and automobile industry. Airframe structures of modern commercial and military aircraft contain hundreds of unitized monolithic metal structural components, which comprise of thinner ribs and webs. In aero-engine construction, about 90% parts are thin-walled. An impeller blade can be an example of thin-wall asymmetric-open component while engine-casing can be considered as thin-wall axisymmetric-closed components (Geng et al. 2011). Panels or structural parts and heat sinks of processors of electronics products such as laptop computers are also made up of thin-wall parts. Thin-wall machining is essential in the manufacture of dies and molds required to produce thin-wall plastic parts. It is also used in power engineering applications such as turbine blades, housings, and enclosures.

The conventional thin-wall structural components are manufactured in parts and then assembled together using riveting or welding operations. The process involves high part cost, and it is time-consuming and laborious (Campbell 2011). Machining thin-wall parts eliminate the need for different setups and processes; however, it consumes a lot of power because of machining of about 90–95% bulk material. Today's manufacturing and tool room industry are striving to reduce the component cost and to improve the product quality in terms of surface finish and dimensional accuracy. To fulfill these requirements, it is imperative to focus research attention on improving the product quality and overall productivity during machining of thin-wall components.

In general, machining of parts having thin sections is called as thin-wall machining. Peripheral milling is a commonly used in machining of thin-wall parts. During the machining process, low-rigidity thin sections deflect under the influence of milling force and the heat generated by severe plastic deformation. Various milling parameters, viz., feed per tooth, axial depth of cut, radial depth of cut, tool diameter, spindle speed, and tool angles influence the magnitude of the milling force, which in turn affects the accuracy of the machined component.

According to Kennedy and Earls (2007), a wall of thickness in the range of 1–2.5 mm can be considered as a thin-wall. Yang (1980) defined the thin plates and thick plates for plate bending theory as:

$$\left.\begin{array}{l}\text{Thick plate} = t/p > (1/5) \\ \text{Thin plate} = (1/100) \leq t/p \leq (1/5)\end{array}\right\}, \quad (1)$$

where p is the shorter length of two edges of the plate and t is the plate thickness.

The main difference between thin-wall machining and normal machining is that in thin-wall machining, there is very less amount of material left to support against the milling force. This leads to lower stiffness of the workpiece and results in vibrations and deformation. Consequently, the effect of the milling parameters on thin-wall machining gets amplified compared to that of a normal machining operation.

2 Challenges in Thin-Wall Machining

Most of the thin-wall components used in the industry are machined from aluminum and titanium blocks. Aluminum is widely used because of its low yield stress and good machinability rating. Moreover, good fatigue resistance makes it favorable in aerospace and automobile applications. In the field of thin-wall machining, high-speed thin-wall machining is gaining popularity due to its merits such as low milling force, low cutting temperature, reduced machining time, and generation of the better quality surface (Davim et al. 2008). High-speed machining

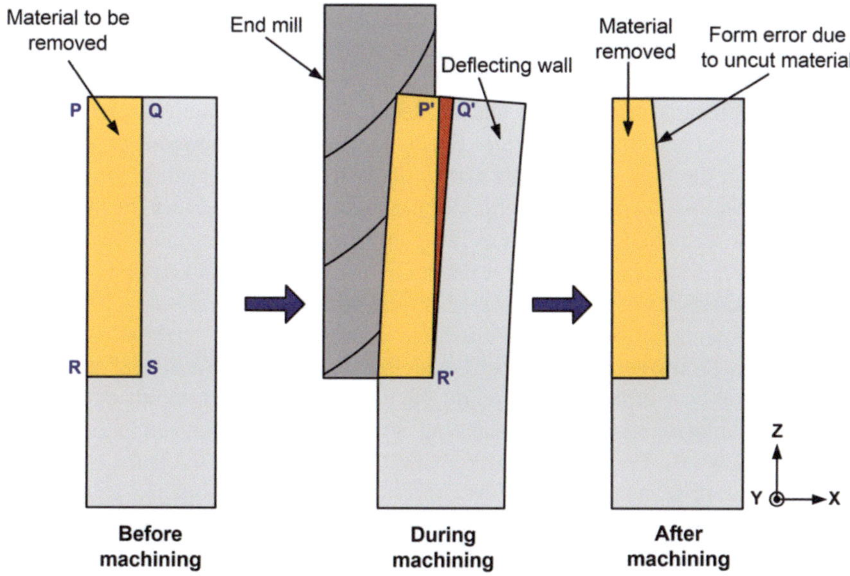

Fig. 1 Wall deflection and form error produced in machining thin-wall feature

utilizes machine tools having very high spindle power rating and thus it is very expensive. It requires high capital investment, which may not affordable by the small-scale industries. Therefore, it is essential to explore the employment of conventional low to medium duty computer numerical control (CNC) machining centers for manufacturing of quality thin-wall structures. Figure 1 shows a schematic representation of thin-wall machining process.

For the chosen milling parameters, the material to be cut is PQRS, however, under the action of milling force, the wall deflects causing the material $P'Q'R'$ to remain uncut. As the cutter moves away in the feed direction (Y direction), the wall recovers elastically and the material remains uncut which results in thicker top and thinner base. The thin-wall parts are always machined on computer numerically controlled (CNC) machines. Despite this, the process of thin-wall machining is not devoid of problems. This is because the process control by CNC is based on idealized geometry and does not take into account the deformation of the parts. As a result, there is a significant deviation between the desired part profile and the manufactured one. To reduce the form error generated due to the deflection and to improve the surface finish of work part, it is essential to set the milling parameters at their optimum levels. In view of this, if the deflection is predicted beforehand, effective countermeasures can be undertaken to obtain the desired process performance, that is, manufacturing of a dimensionally accurate finished component with good surface quality.

Another important aspect related to machining of aluminum thin-wall components is the surface roughness. Aluminum alloys possess a comparatively low modulus of elasticity, which causes the workpiece to spring back. This spring back action often results in deflection and chatter. Chatter affects the material removal rate (MRR) and leads to poor surface finish, part rejection and loss of productivity.

As seen in the Fig. 1, thin walls deflect under the action of milling force, which in turn affects the quality and accuracy of the work part. A need was thus identified to develop a realistic numerical model to predict the process responses, which will help for better control of the process parameters during the actual cutting operation.

Literature reports esteemed research articles and reports published by international universities, industry, and research organizations on various aspects of thin-wall machining such as analytical modeling, numerical modeling, experimental investigations into influence of milling parameters, viz., feed, depth of cut, tool geometry parameters, tool path planning strategies and optimization of milling process parameters. From the reported literature, it was concluded that a significant amount of work is reported on 2-D as well as 3-D simulations of the bulk milling operation (Bacaria et al. 2001; Soo et al. 2004; Davim et al. 2008; Özel and Ulutan 2012; Özel et al. 2010, 2011; Maranhão and Davim 2010; Maurel-Pantel et al. 2012; Jiao et al. 2015). Extensive research has been reported on simulation of orthogonal metal cutting process using 2-D FEM methods. Thin-wall milling using a helical end-milling cutter is a complex operation and is difficult to comprehend and analyze in a 2-D domain. Research attempts have been reported on 3-D FEM modeling and simulation of machining of low-rigidity thin-wall parts to study the milling forces, deflection, stresses, cutting temperature, and chip morphology obtained during the process (Ning et al. 2003; Wan et al. 2005; Gang 2009; Rai and Xirouchakis 2008, 2009; Li et al. 2015; Cheng et al. 2015). Most of them carried out 3-D simulation of metal cutting for only one rotation of milling cutter hence lack applicability because in reality, it is essential to simulate the complete pass of the milling cutter over the specified length of the work part to obtain the instantaneous wall deflection.

Keeping this in view, in the present work, a Lagrangian-based 3-D FEM-based numerical model has been developed to simulate the deflection of workpiece wall during thin-wall machining of aluminum 2024-T351 alloy. The developed model was aimed to accurately compute the milling forces, temperature, stress distribution and to study the chip formation in work part for chosen process parameters by considering realistic material behavior, friction consideration, damage model, heat generation and by employing realistic geometry of the milling tool. Subsequently, experimental trials were carried out to validate the developed model. Details of the model development are presented in the sections to follow.

3 Overview of the Process Model Development

To understand and improve the thin-wall machining process, it is important to develop a mathematical model to establish a realistic relationship between input and output process parameters. The primary objective of this work was to analyze the metal cutting phenomenon in thin-wall milling process by carrying out a thermo-mechanical analysis. The work mainly includes the following steps. These steps are schematically depicted in Fig. 2.

- Design of geometric models of helical cutting tool, workpiece and selection of suitable mesh element.
- Choosing the material models and material failure criteria.
- Specify governing equation and boundary conditions.
- Selection and application of required process conditions.
- Solution of the problem using finite element analysis (FEA) tool in nonlinear mode.
- Determination of stress–strain, temperature profiles in the workpiece-cutting tool interaction zone, and computation of process performance parameters such as milling force, stress distribution, cutting temperature, part deflection, and chip morphology.
- Validation of computed responses with the experimental results.
- Evaluation of computational efficiency of developed 3-D numerical simulation approach.

These steps are discussed at length in the following sections.

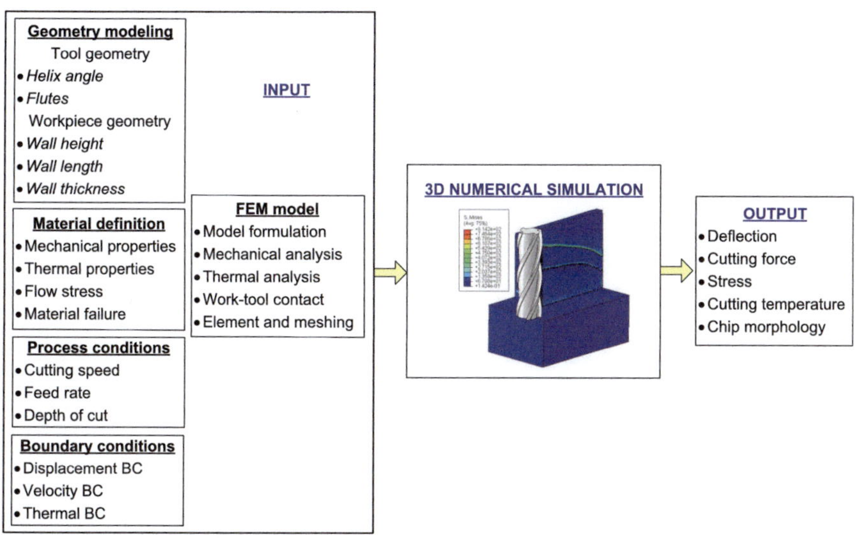

Fig. 2 Overall approach and development of 3-D numerical model

4 Thermomechanical Modeling of End-Milling Operation

In the end-milling operation, the material is removed from the workpiece using a rotating tool. The thin-wall sections are deformed due to the influence of forces developed during tool-workpiece interaction and the heat generated during cutting. The milling force developed depends upon various process parameters; therefore, the accuracy of the component depends upon the proper selection of the process parameters during machining. In the machining process, heat is generated due to plastic deformation and frictional contact between the workpiece and the tool. The heat generated directly influences the material properties. Therefore, it is essential to couple the mechanical and thermal responses to obtain the solution for such a problem. In this work, a fully coupled thermomechanical analysis of thin-wall milling operation has been attempted.

4.1 Assumptions

Numerical modeling and simulation of thin-wall machining involve complex interaction between the tool and workpiece, contact modeling, material properties. The following assumptions were made in the present thermomechanical modeling:

- The helical milling cutter has sharp cutting edges.
- The milling cutter is stiffer than the workpiece hence it is modeled as a rigid body. It has translation, rotation, and thermal degrees of freedom.
- Workpiece material is deformable. It is isotropic and homogeneous in nature.
- The workpiece material is free of residual stresses.
- Initial temperatures for both workpiece and the tool are set at 25 °C.
- Workpiece and tool surfaces lose the heat generated during the machining process to the environment by convection. The convection coefficient, h_c is assumed as 20 W/m^2 °C (Kiliçaslan 2009).
- The workpiece has uniform thickness along the height and width. There are no deformations present in the workpiece at the commencement of the machining simulation.
- It is considered as 90% of the total energy generated due to plastic deformation process converts into the heat energy (Li et al. 2002).
- It is also considered that the entire energy that generates during metal cutting operation transforms into heat energy and half of the heat dissipates into the chip (Li et al. 2002).

4.2 Workpiece, Tool Geometry, and Finite Element Meshing

Initially, the geometric models of cutting tool and workpiece have been developed. The thin-wall workpiece was modeled as an inverted cantilever structure with bottom portion being constrained, while the other three ends were free. The tool geometry influences the surface properties of the machined workpiece, so importance has been given to accurately model the cutting tool. A 3-D model of the end mill with actual tool geometry parameters was designed using a CAE tool and then imported in Abaqus/Explicit as a solid 3-D homogeneous part. Figure 3 shows the workpiece, cutting tool, and relevant geometric parameters.

The influence of vibrations of thin-wall structure that occurs during the machining was not considered in this work. The parameters related to cutting tool geometry are listed in Table 1.

Thin-wall machining process involves nonlinear and complex interaction between the tool and workpiece. It also involves large deformations. It is thus essential to employ proper element type for the modeling process. The workpiece is meshed with 3-D solid elements C3D8RT to carry out the thermomechanical analysis. It is a 3-D displacement-temperature coupled 8-noded solid element with reduced integration and hourglass control. As shown in Fig. 4 the entire geometry is discretized into very small finite elements. For this particular case, the tool and the workpiece were discretized into 12,650 and 293,745 elements, respectively. Also, it can be seen that element size on the work material varies at different regions

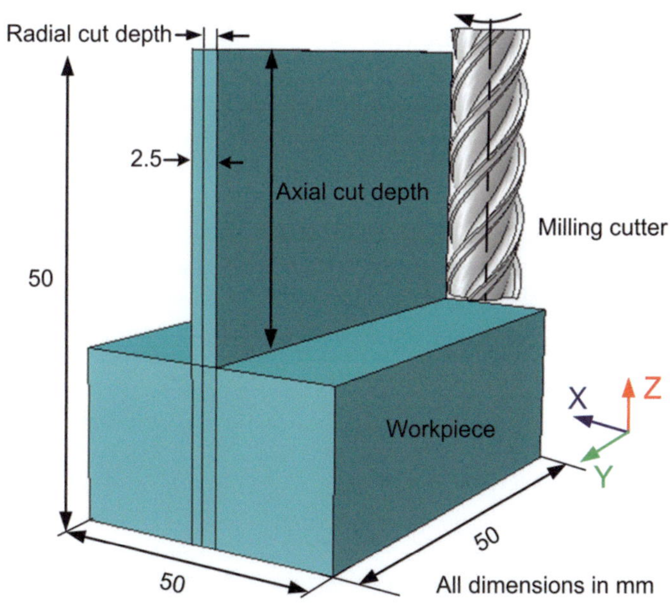

Fig. 3 Workpice geometry and dimensions

Table 1 Cutting tool geometry parameters

Tool diameter (mm)	8
Tool rake angle (°)	8
Tool helix angle (°)	45
Tool clearance angle (°)	15
Number of flutes	4

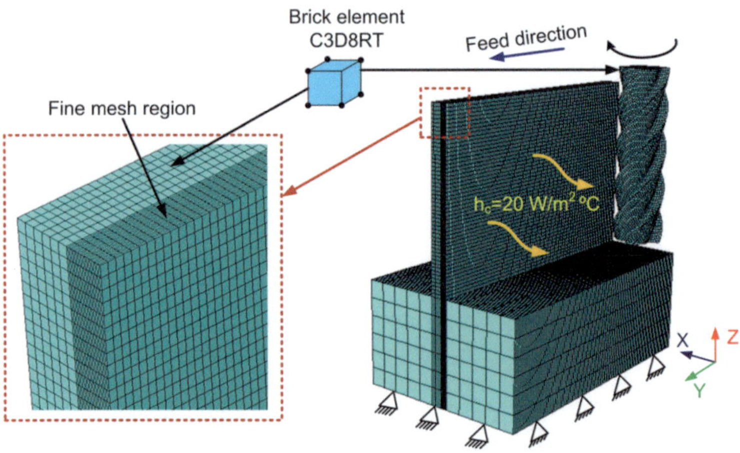

Fig. 4 Workpiece-tool meshing and boundary conditions

of the workpiece. The mesh density was kept higher at the workpiece-cutter interaction region, which helps in better simulation of chip formation at the cutting region. Remaining regions of the workpiece were coarsely meshed to reduce the number of elements in the non-active areas of the machining process.

Figure 4 also depicts the boundary conditions that applied on workpiece and tool geometries. The workpiece was constrained at the bottom to imitate the clamping action during machining. The end-milling cutter was given two motions, namely linear motion along feed direction and rotational motion about its own axis. The initial temperature of the tool and workpiece was set at room temperature. Heat loss to the environment from the tool and workpiece interface was considered primarily by the convection.

4.3 Material Properties

In this work, numerical simulations have been performed for milling of aluminum 2024-T351 aerospace grade commercial alloy. It possesses high strength and good fatigue resistance and is widely used in aircraft wing and fuselage structures. The

Table 2 Workpiece and cutting tool material properties (Mabrouki et al. 2008)

Properties	Workpiece	Tool
Density, ρ_m (kg/m^3)	2700	11,900
Elastic modulus, E (GPa)	73	534
Poisson ratio, v	0.33	0.22
Fracture toughness, K_c (MPa$\sqrt{m}$)	37	–
Specific heat, C_p (J/kg °C)	$= 0.557\ T + 877.6$	400
Thermal expansion, α (10e−6/°C)	$= 8.9 * 10^{-3} T + 22.2$	30
Thermal conductivity, k_t (W/m °C)	$25 \leq T \leq 300 = 0.247\ T + 114.4$ $300 \leq T \leq T_{\text{melt}} = 0.125\ T + 226$	50
Melt temperature, T_{melt} (°C)	520	–
Room temperature, T_{room} (°C)	25	–

workpiece material is considered to be an elastic-plastic type. The end mill is considered to be made of uncoated tungsten carbide (WC).

Tungsten carbide otherwise known as cemented carbide composes of tungsten carbide powder (85–95%) cemented with a binder material namely cobalt or nickel. It has many desirable qualities such as high resistant to abrasion, erosion, wear, compression, and heat (Juneja 2003). The effect of machining process parameters on the cutting tool wear is not considered of the present study; therefore, the tool is to be considered as a rigid body throughout the simulation. Table 2 shows the mechanical and thermal properties of the workpiece as well as the tool material (Mabrouki et al. 2008).

4.4 *Material Constitutive Model*

During the metal cutting process, the material deforms plastically and is subjected to high strains, strain rates, and temperature conditions. To describe the thermo-mechanical behavior of a material undergoing deformation at such high strains and strain rate conditions, Johnson and Cook developed a constitutive material model in 1983. As per their model, the equivalent flow stress $\bar{\sigma}_{jc}$ can be computed as

$$\bar{\sigma}_{jc} = (A + B\bar{\varepsilon}^n)\left[1 - c\ln\left(\frac{\dot{\varepsilon}}{\dot{\varepsilon}_o}\right)\right]\left(1 - \left(\frac{T_c - T_{\text{room}}}{T_{\text{melt}} - T_{\text{room}}}\right)^m\right) \quad (2)$$

Table 3 lists the J-C material model constants for aluminum 2024-T351 alloy (Mabrouki et al. 2008).

Table 3 J-C material coefficients for A2024-T351 (Mabrouki et al. 2008)

A (MPa)	B (MPa)	n	c	m
352	440	0.42	0.0083	1

4.5 Material Damage Model

In the metal cutting process, chips are formed as a result of excessive (large) material deformation at the cutting tool-work material interface under the action of applied force. The material is said to have failed when it loses its load carrying capacity. In machining processes, the prediction and control of the material failure is a critical issue. In order to investigate the surface finish and integrity of the produced parts, it is essential to simulate the damage and fracture of the material under the action of applied loads.

In this work, the ductile failure of work material was simulated under the high strain, strain rate conditions. Effects of temperature on the deformation were also considered. For this purpose, Johnson and Cook (1985) shear failure criterion was used to model the chip detachment. Deformation occurring in the final stage of chip evolution was computed using the evolution law which is based on fracture energy principle (Liu et al. 2014). Damage in the element is initiated by scalar damage coefficient D. It is a sum of ratios of increments in equivalent plastic strain $\Delta\bar{\varepsilon}$ to the equivalent strain at fracture initiation ε_{fi}. It is given in Eq. (3). The damage initiates when scalar damage parameter D exceeds unity.

$$D = \Sigma \frac{\Delta\bar{\varepsilon}}{\varepsilon_{fi}} \tag{3}$$

The equivalent strain at failure initiation ε_{fi} is

$$\varepsilon_{fi} = \left[D_1 + D_2 \exp\left(D_3 \frac{P}{\bar{\sigma}}\right)\right]\left[1 + D_4 \ln\left(\frac{\dot{\varepsilon}}{\dot{\varepsilon}_o}\right)\right]\left(1 + D_5\left(\frac{T_c - T_{\text{room}}}{T_{\text{melt}} - T_{\text{room}}}\right)\right) \tag{4}$$

The J-C damage constants (D_1-D_5) are shown in Table 4. It can be observed that coefficient D_5 is zero. This indicates that temperature has no effect on the damage initiation during machining of the aluminum alloy 2024-T351. Stress triaxiality and strain rate effects are considered to be main factors for damage initiation (Mabrouki et al. (2008)).

When the damage of a ductile material initiates, the stress–strain relationship no longer represents the material behavior during deformation. After the damage

Table 4 Johnson–Cook failure parameters for A2024-T351 (Mabrouki et al. 2008)

D_1	D_2	D_3	D_4	D_5
0.13	0.13	−1.5	0.011	0

initiation, the use of stress–strain relation causes strong mesh dependency which is based on strain localization. This reduces the energy dissipation as the mesh becomes smaller. Influence of mesh dependency on energy dissipation can be reduced using the Hillerborg's fracture energy principle. According to this principle, the fracture energy can be computed as

$$G_f = \int_{\bar{\varepsilon}_{0i}}^{\bar{\varepsilon}_f} L\bar{\sigma}_Y \mathrm{d}\bar{\varepsilon} = \int_0^{\bar{u}_f} \bar{\sigma}_Y \mathrm{d}\bar{u} \tag{5}$$

The damage evolution parameter (ω_s) is given as

$$\omega_s = \frac{L\bar{\varepsilon}}{\bar{u}_f} = \frac{\bar{u}}{\bar{u}_f} \tag{6}$$

The equivalent plastic displacement at failure can be expressed by

$$\bar{u}_f = \frac{2G_f}{\sigma_y} \tag{7}$$

where G_f is the fracture energy dissipation. It can be determined by:

$$G_f = \left(\frac{1-v^2}{E}\right)K_c^2 \tag{8}$$

where K_c is the fracture toughness of the material.

4.6 Workpiece-Milling Tool Contact Model

Study of the effect of friction between the end-milling tool and workpiece is important as it affects the cutting forces, temperature at the cutting zone and chip, product quality and tool surface wear. In this work, workpiece and milling tool contact has been considered by employing the modified Coulomb friction model. According to this model, the contact between chip and tool rake surface region is divided into two regions, viz., sliding and sticking region (see Fig. 6).

During metal cutting, the friction due to sticking occurs very near to the contact of cutting edge with the workpiece, while the friction due to sliding occurs far away from the contact area. The sliding region follows the Coulomb friction law. At the

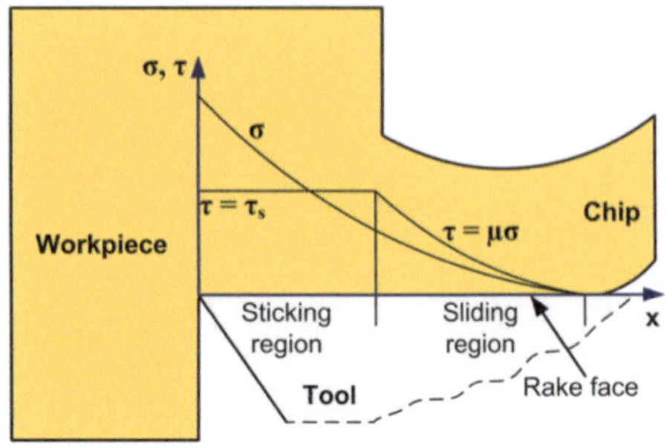

Fig. 6 Distribution of normal and shear stress at chip-tool interface

sticking region, the frictional stress τ_f is equal to the critical frictional stress τ_{criti}. Accordingly the frictional stress can be expressed as

$$\tau_f = k_{\text{chip}} \text{ when } \mu\sigma_n > k_{\text{chip}} \quad \text{(Sticking region)} \tag{9}$$

$$\tau_f = \mu\sigma \text{ when } \mu_n\sigma < k_{\text{chip}} \quad \text{(Sliding region)} \tag{10}$$

In the present work, the coefficient of friction μ is considered as 0.17 (Liu et al. 2014). In what follows, the solution methodology is presented.

4.7 *Solution Methodology*

After the development of geometric models of workpiece and tool, material properties, material damage law, and friction law were applied and thermomechanical analysis of thin-wall end milling was carried out. The mathematical equations employed in the present work are given below.

4.7.1 Mechanical Analysis

The differential equation of motion that governs the mechanical displacement is given by

$$\rho\ddot{u} + c\dot{u} + ku = F \tag{11}$$

Equation (11) is rewritten in matrix form as

$$[M]\ddot{u} + [C]\dot{u} + [K]u = F \tag{12}$$

Nodal acceleration at the beginning of time increment can be obtained by rewriting Eq. (12) as

$$\ddot{u}_i = M^{-1}(F - C\dot{u}_i - Ku_i) \tag{13}$$

In the present work, explicit formulation was employed which uses central difference scheme to discretize the equations. The acceleration equation can be written as

$$\ddot{u}_i = \frac{\left(\dot{u}_{i+1/2+} - \dot{u}_{i-1/2}\right)}{(\Delta t_{i+1} + \Delta t_i)/2} \tag{14}$$

Velocity change is calculated by integrating acceleration term using central difference method. The velocity at the middle of the current step can be computed as

$$\dot{u}_{i+1/2} = \left(\frac{\Delta t_{i+1} + \Delta t_i}{2}\right)\ddot{u}_i + \dot{u}_{i-1/2} \tag{15}$$

Displacement is calculated by integrating velocity through time, which is then used to obtain the displacement at the end of a time step. It is given by

$$u_{i+1} = u_i + \Delta t_{i+1}\dot{u}_{i+1/2} \tag{16}$$

4.7.2 Thermal Analysis

The governing equation for the transient heat conduction during the machining process is written as

$$\rho C_p \frac{\partial T}{\partial t} = \frac{\partial}{\partial x}\left[k_x \frac{\partial T}{\partial x}\right] + \frac{\partial}{\partial y}\left[k_y \frac{\partial T}{\partial y}\right] + \frac{\partial}{\partial z}\left[k_z \frac{\partial T}{\partial z}\right] + \dot{Q} \tag{17}$$

Equation (17) can be rewritten in matrix form as

$$[\bar{C}]\dot{T} + [\bar{K}]T = \dot{Q} \tag{18}$$

Solving the nodal temperature rate from the above equation yields

$$\dot{T}_i = C^{-1}\left(\dot{Q} - KT_i\right) \tag{19}$$

After applying the forward difference integration scheme on Eq. (19), the nodal temperature rate is given by

$$\dot{T}_i = \frac{T_{i+1} - T_i}{\Delta t_{i+1}} \tag{20}$$

Equation (20) can be rewritten as

$$T_{i+1} = (\Delta t_{i+1})\dot{T}_i + T_i \tag{21}$$

The explicit expression for nodal temperature rate can finally be written as

$$T_{i+1} = (\Delta t_{i+1})C^{-1}\left(\dot{Q} - KT_i\right) + T_i \tag{22}$$

Total volumetric heat generation rate is due to the heat generated during plastic deformation and friction at the work-tool interface. The generation of heat due to plastic deformation is expressed as

$$\dot{Q}_{\text{pl}} = \eta_p \sigma \dot{\varepsilon}_p \tag{23}$$

By considering that most of the energy of plastic deformation converts into heat energy, in this work, η_p is taken as 0.9 (Li et al. 2002). During the metal cutting operation, friction at the workpiece-tool contact region generates significant heat. The volumetric heat flux due to friction is given by

$$\dot{Q}_f = f_f \cdot \eta_f \cdot \tau_f \cdot \dot{\gamma}_s \tag{24}$$

During the metal cutting operation, the tool and workpiece surfaces dissipate the heat to the environment. This phenomenon is described by using Newton's law of convection as

$$\dot{q}_{\text{conv}} = h_c(T_c - T_{\text{room}}) \tag{25}$$

The milling process involves large deformations and continuously changing contact between cutting and workpiece. With the above-stated Eqs. (11)–(25), Abaqus/Explicit, the FEA solver computed the responses such as milling force, wall deflection, and cutting temperature. In the present analysis, explicit time integration scheme has been used to solve the transient problem. It was originally developed to solve high-speed dynamic problems those were difficult to simulate using the implicit method. This scheme is simple and can handle problems which involve high nonlinearity, large deformation, complex friction contact conditions, thermo-mechanical coupling and fragmentation. The formulation using explicit solution can be of Eulerian, Lagrangian, or Arbitrary Lagrangian–Eulerian (ALE) type. Figure 7 shows a comparison between Lagrangian, Eulerian, and ALE formulations. In the Lagrangian formulation, the finite element mesh is attached to the material and

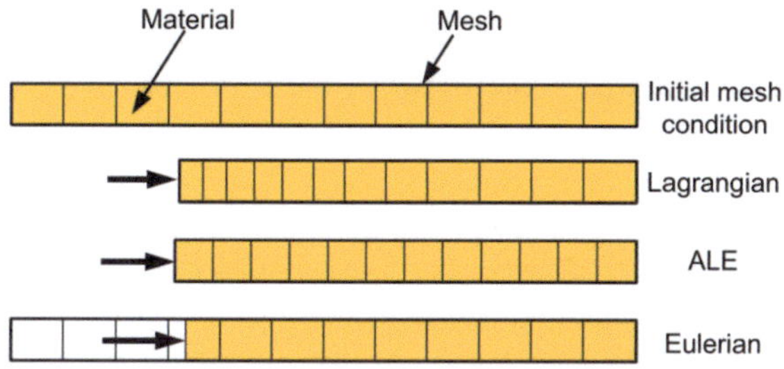

Fig. 7 Comparison of motion of mesh and material with Lagrangian, Eulerian, and ALE formulation

follows the mesh deformation. The Lagrangian formulation is used to analyze transient problems which undergo large deformations. It is widely used due to its ability to form chips and to determine the chip geometry as a function of cutting parameters, plastic deformation process, and material properties. This enables the evolution of the cut material from its nascent stage to a stable state without any predetermined material geometrical boundary conditions. The development of chip is entirely a function of physical deformation process, machining parameters, and the input material properties.

In the Eulerian formulation, the finite element mesh is fixed in the space and the material flows through the control volume that eliminates the possibility of element distortion during the process. It reduces the computation time as fewer elements required for the analysis. These models do not need separation criteria for simulating the material failure. The major drawback of the Eulerian formulation is that it needs prior knowledge of the chip geometry, chip-tool contact length, chip thickness, and contact conditions to simulate the chip formation. Thus in this formulation, it is needed to keep the conditions of chip thickness, tool-chip contact length and contact conditions constant. Therefore, the Eulerian formulation is not suitable for simulation of workpiece deformation in metal cutting. To overcome these drawbacks, researchers have developed a new iterative procedure which combines the best features of Lagrangian and Eulerian formulations. It is called Arbitrary Lagrangian–Eulerian (ALE) approach. According to this approach, the mesh follows the flow of material. The displacements are computed in Lagrangian steps. For velocity computation, the mesh is repositioned and the problem is solved in Eulerian step. The combined formulation avoids the severe element deformation which is a typical problem often associated with the Lagrangian approach. However, in view of simplicity and high computational efficiency, most of the 3-D FEM-based simulations of metal cutting operations have used the Lagrangian formulation. Therefore in the present work, the same approach has been used.

In the present work, Abaqus/Explicit, a commercial finite element solver has been used to carry out 3-D numerical simulations of machining of thin walls. A computer system of 3.9 GHz 4 GB RAM processor has been employed. Extensive trials have been carried out to fine tune the FEM solver parameters. In general, a typical simulation of material removal during a complete pass of the milling tool along the wall length of 50 mm took about 340–350 h.

5 Experimental Validation of FEM-Based Simulation of Thin-Wall Machining

After the development of numerical model, experimental validation of the responses predicted by the numerical model was carried out. For this purpose, an experimental setup was developed on a three-axis computer numerically controlled vertical machining center (CNC-VMC). Figure 8 depicts the details of the experimental setup developed. The workpiece was clamped in a workpiece holder, which was fixed firmly on a piezoelectric sensor based three-component dynamometer (make: Kistler 9272B). The dynamometer is capable of measuring three components of milling force (F_x, F_y and F_z). The dynamometer was mounted firmly onto the base plate of the machine tool. For the data acquisition, the dynamometer was connected to a computer through a force measurement multichannel charge amplifier (type 5070A). The data sampling rate was set to 500 Hz per channel. The component F_x is normal to the machined wall surface. The component F_y is oriented along the direction of feed movement and the F_z component is along the tool's axis. The workpiece deflection was measured by Solartron linear variable differential transformer (LVDT) sensor. Experiments were repeated thrice for each set of process conditions.

By using the developed numerical model, the responses such as milling force and wall deflection were measured during thin-wall machining process. In addition, the efficiency of the developed model in terms of the computational time was also analyzed. Milling conditions used for simulation studies are listed in Table 5. The results obtained during the numerical simulation of machining thin-wall aluminum 2024-T351 alloy part are discussed in the following sections.

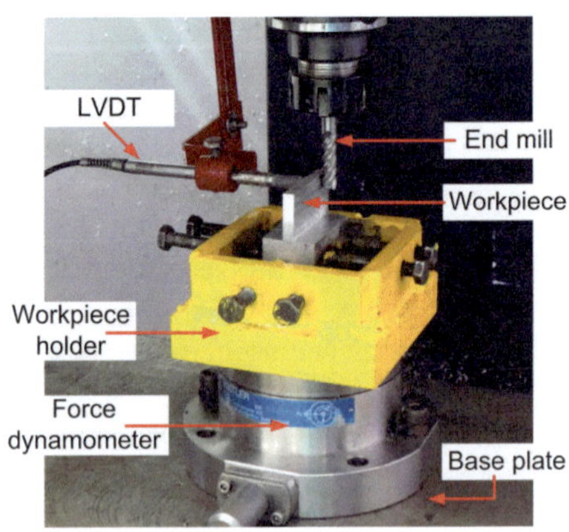

Fig. 8 Experimental setup for thin-wall milling

Table 5 Milling conditions employed for thin-wall machining simulation

Test case	n_s (rev/min)	f_z (mm/z)	r_d (mm)	Workpiece length (mm)
1	4500	0.1	1	40
2	4500	0.1	1	50
3	3500	0.1	0.625	40
4	3500	0.1	0.625	50

5.1 *Milling Force*

Using the developed experimental setup, force components, namely F_x, F_y and F_z, were recorded for a typical process condition (Test case 1) as mentioned in Table 5. For the same process condition, a numerical simulation using the developed mathematical model was carried out. Figure 9 shows the comparison between the simulation predictions and the experimental results.

The dotted lines represent the moving average values of components of milling force recorded in the stable region during machining experiments, while the solid lines represent the moving average values of simulation results. It can be noted that the forces computed by the numerical model are in good agreement with the experimental results. Average prediction errors were noted to be 13.93, 20.82, and 32.68% in F_x, F_y and F_z directions, respectively. Numerical predictions are found to be lower than that of experimental results. This is attributed to the uncontrollable factors such as material inhomogeneity, errors due to tool and workpiece setting, tool vibration and tool run-out. It can be observed that the prediction error of F_z is quite high in comparison with that of F_x and F_y. It may be due to fact that the magnitude of F_z is smaller and it is more sensitive than the other force components. It is also to be noted that the factors such as re-cutting of chips and ploughing of cutting tool edges on the work surface result in the generation of fluctuating force values.

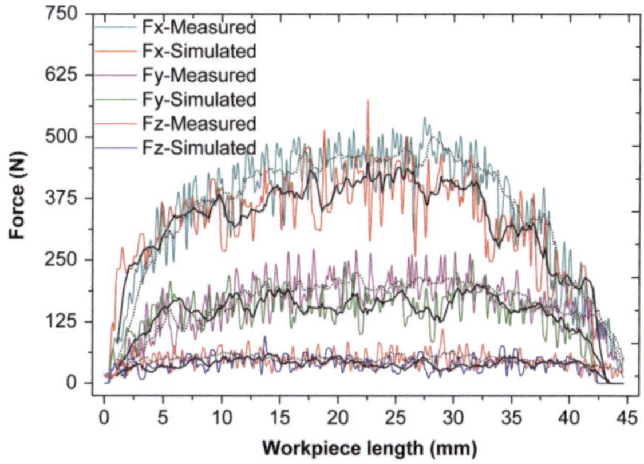

Fig. 9 Comparison of simulated and measured milling force components F_x, F_y, and F_z

From the plots, it can be seen that the trends of variations of experimental and numerical forces are similar. However the forces are fluctuating, which may be due to the fact that during the cutting process, the material softens due to the rise in temperature which reduces the force values. As the magnitude of force decreases, the heat production also reduces which in turn affects the metal softening effect that further leads to rise in the milling forces.

5.2 *Wall Deflection and Form Error*

In thin-wall milling operation, the low-rigidity thin-wall part deflects due to the application of milling forces. During the experiments, it was observed that maximum deflection occurs at top portion of the wall in comparison to that its base. It is due to the low rigidity of top portion of the workpiece and the base has sufficient rigidity as it is firmly supported by the bulk material. Figure 10 depicts the variation of deflection along the workpiece length measured perpendicular to the feed direction. It is noted that the deflections at free ends of the workpiece are higher in comparison to that of the middle portion of the wall. The two ends are less stiff and deflect readily under the action of milling force, whereas the center portion has sufficient rigidity as it is supported by the material all around. It can also be observed that experimentally obtained deflection values are slightly on the higher side than those obtained in the simulations. The mean absolute error between the experimental and numerically obtained results was noted to be 11%. The trends of variations of predicted results were matching well with respective experimental measurements. Figure 11 shows the comparison of wall sections obtained in numerical simulation and experimental study.

It can be seen that during the thin-wall machining, machined wall base is thinner than that of the top potion. This is because, due to low rigidity, instantaneous deflection of top end is more, which leads to lower cutting of the top portion in

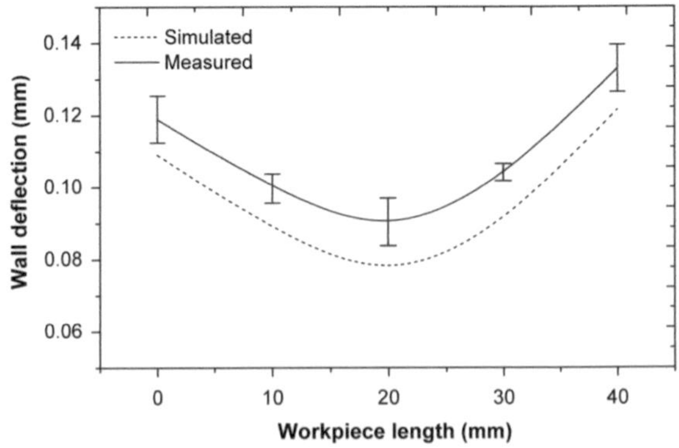

Fig. 10 Comparison of the deflection at top portion of the wall (2 mm below the top edge): numerical and experimental results

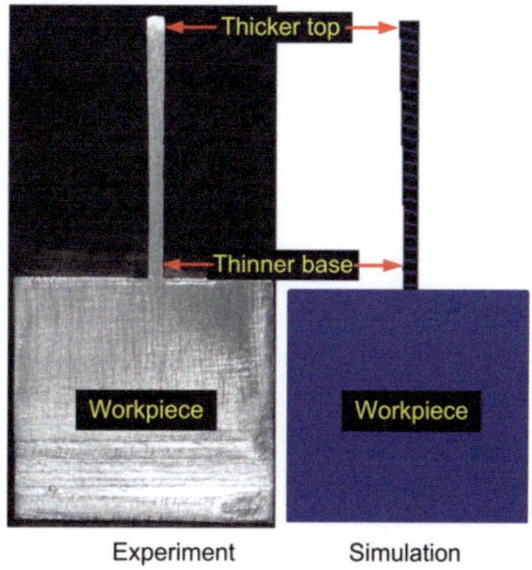

Fig. 11 Wall cross sections obtained during experiment and numerical simulation

Table 6 Comparison of the simulated and measured responses for test cases listed in Table 5

Test case		Milling force (*N*)			Wall deflection (mm)
		F_x	F_y	F_z	
1	Predicted	294.3	132.44	37.04	0.096
	Measured	341.8	167.28	55.10	0.107
	Absolute error (%)	13.93	20.82	32.68	11.00
2	Predicted	289.6	127.11	38.31	0.133
	Measured	323.1	152.9	51.64	0.124
	Absolute error (%)	10.53	16.86	25.81	6.780
3	Predicted	252.74	175.02	38.49	0.091
	Measured	287.50	204.32	52.10	0.104
	Absolute error (%)	12.09	14.34	26.13	12.50
4	Predicted	246.98	168.16	36.67	0.0956
	Measured	274.40	194.37	45.72	0.106
	Absolute error (%)	9.9	13.5	19.79	9.81
	Mean error (%)	11.61	16.38	26.1	10.02

comparison with that of the desired depth of cut. The developed numerical model could successfully simulate this phenomenon.

Similar to test case 1, experiments were conducted for three other test cases. Table 6 summarizes the values of measured and simulated milling force components, wall deflection, and workpiece temperature for all the four test cases.

It can be seen that the milling force component values predicted by the developed model match well with the values that are obtained by experiments for all the test cases. Mean prediction errors for F_x, F_y, F_z, and wall deflection were noted to be 11.61, 16.38, 26.1, and 10.02%. Overall a very good agreement between the simulated and experimentally measured responses has been noted which demonstrates the capability of the developed model to predict the process responses accurately. Thus, the present 3-D FEM-based numerical model has predicted important process performance parameters such as milling force and wall deflection quite accurately and easily. A prior knowledge about these parameters will certainly help the process engineers to tune up the process parameters to achieve the desired process performance. Predicted milling forces can be used to estimate the energy requirement. This model is capable of predicting the instantaneous wall deflection, which will be useful in correcting the tool path to minimize the form error that occurs during the thin-wall milling process. Thus, it can be said that numerical modeling and simulation provides a useful tool to the engineers and scientists to carry out a prior detail study of the machining process. Further, this model can be used to carry out parametric analysis of thin-wall machining, i.e., a study on influence of process parameters, viz., axial depth of cut, radial depth of cut, feed, and cutting speed, tool geometry parameters, viz., tool diameter, and helix angle on the performance parameters such as part accuracy, milling forces, and power consumption. This model can also be employed to study the thin-wall machining operation of important and difficult-to-cut materials, viz., titanium alloys, Inconel, etc.

6 Conclusion

This chapter presented, in details, the development of a numerical (FEM) model for the thin-wall machining process. A realistic three-dimensional thermomechanical finite element based FEM model has been developed to simulate the complex physical interaction of helical cutting tool and workpiece during thin-wall milling of an aerospace grade aluminum alloy. Lagrangian formulation with explicit solution scheme was employed to simulate the interaction between helical milling cutter and the workpiece. The behavior of the material at high strain, strain rate and the temperature was defined by Johnson–Cook material constitutive model. Johnson–Cook damage law and friction law were used to account for chip separation and contact interaction. Experimental work was carried out to validate the results predicted by the mathematical model. The developed model predicted the forces in radial, feed, and axial directions with errors of 11.61, 16.38, and 26.1%, respectively. The prediction error for deflection at the top portion of thin-wall was 10.02%. It was noted that maximum deflections occur at the free ends of the wall as compared to that at the center. It was also observed that due to the deflection of the wall, some material at the top of the wall remained uncut that further leads to geometric form error in the workpiece. The results can be used to set the process parameters to obtain the desired process performance in terms of product accuracy. This model can further be improved by considering the material anisotropy, inhomogeneity, and prestresses in the work part.

References

Bacaria, J.L., O. Dalverny, and S. Caperaa. 2001. A three-dimensional transient numerical model of milling. *Proceedings of the Institution of Mechanical Engineers, Part B: Journal of Engineering Manufacture* 215 (8): 1147–1150.

Campbell Jr., F.C. 2011. *Manufacturing technology for aerospace structural materials*. Amsterdam: Elsevier.

Cheng, Y., D. Zuo, M. Wu, X. Feng, and Y. Zhang. 2015. Study on simulation of machining deformation and experiments for thin-walled parts of titanium alloy. *International Journal of Control and Automation* 8 (1): 401–410.

Davim, J.P., C. Maranhão, M.J. Jackson, G. Cabral, and J. Gracio. 2008. FEM analysis in high speed machining of aluminium alloy (Al7075-0) using polycrystalline diamond (PCD) and cemented carbide (K10) cutting tools. *The International Journal of Advanced Manufacturing Technology* 39 (11): 1093–1100.

Gang, L. 2009. Study on deformation of titanium thin-walled part in milling process. *Journal of Materials Processing Technology* 209 (6): 2788–2793.

Geng, Z., K. Ridgway, S. Turner, and G. Morgan. 2011. Application of thin-walled dynamics for advanced manufacturing solutions. *IOP Conference Series: Materials Science and Engineering* 26 (1): 012011.

Jiao, L., X. Wang, Y. Qian, Z. Liang, and Z. Liu. 2015. Modelling and analysis for the temperature field of the machined surface in the face milling of aluminium alloy. *The International Journal of Advanced Manufacturing Technology* 81 (9–12): 1797–1808.

Johnson, G.R., and W.H. Cook. 1983. A constitutive model and data for metals subjected to large strains, high strain rates and high temperatures. In *Proceedings of the 7th International Symposium on Ballistics*, vol. 21, no. 1, 541–547.

Johnson, G.R., and W.H. Cook. 1985. Fracture characteristics of three metals subjected to various strains, strain rates, temperatures and pressures. *Engineering Fracture Mechanics* 21 (1): 31–48.

Juneja, B.L., and G.S. Sekhon. 2003. *Fundamentals of metal cutting and machine tools*. 2nd ed. New age International (P) Ltd., ISBN 0852265190.

Kennedy, B., and A.R. Earls. 2007. Wall smart: Thin-wall milling isn't for the faint of heart, but techniques exist for performing it efficiently. Cutting Tool Engineering 59 (2).

Kiliçaslan, C. 2009. Modelling and simulation of metal cutting by finite element method. M.Sc. thesis, Graduate School of Engineering and Sciences of Izmir Institute of Technology, December 2009, İZMİR, Turkey.

Li, K., X.L. Gao, and J.W. Sutherland. 2002. Finite element simulation of the orthogonal metal cutting process for qualitative understanding of the effects of crater wear on the chip formation process. *Journal of Materials Processing Technology* 127 (3): 309–324.

Li, J., Z.L. Wang, P. Xi, and Y. Jiao. 2015. Analysis of 45 steel rectangular thin-walled parts milling deformation. *Key Engineering Materials* 667: 22–28.

Liu, J., Y. Bai, and C. Xu. 2014. Evaluation of ductile fracture models in finite element simulation of metal cutting processes. *Journal of Manufacturing Science and Engineering* 136 (1): 011010-1–011010-11.

Mabrouki, T., F. Girardin, M. Asad, and J.F. Rigal. 2008. Numerical and experimental study of dry cutting for an aeronautic aluminium alloy (A2024-T351). *International Journal of Machine Tools and Manufacture* 48 (11): 1187–1197.

Maranhão, C., and J.P. Davim. 2010. Finite element modelling of machining of AISI 316 steel: numerical simulation and experimental validation. *Simulation Modelling Practice and Theory* 18 (2): 139–156.

Maurel-Pantel, A., M. Fontaine, S. Thibaud, and J.C. Gelin. 2012. 3D FEM simulations of shoulder milling operations on a 304L stainless steel. *Simulation Modelling Practice and Theory* 22: 13–27.

Ning, H., W. Zhigang, J. Chengyu, and Z. Bing. 2003. Finite element method analysis and control stratagem for machining deformation of thin-walled components. *Journal of Materials Processing Technology* 139 (1): 332–336.

Özel, T., and D. Ulutan. 2012. Prediction of machining induced residual stresses in turning of titanium and nickel based alloys with experiments and finite element simulations. *CIRP Annals-Manufacturing Technology* 61 (1): 547–550.

Özel, T., M. Sima, A.K. Srivastava, and B. Kaftanoglu. 2010. Investigations on the effects of multi-layered coated inserts in machining Ti–6Al–4 V alloy with experiments and finite element simulations. *CIRP Annals-Manufacturing Technology* 59 (1): 77–82.

Özel, T., I. Llanos, J. Soriano, and P.J. Arrazola. 2011. 3D finite element modelling of chip formation process for machining Inconel 718: Comparison of FE software predictions. *Machining Science and Technology* 15 (1): 21–46.

Rai, J.K., and P. Xirouchakis. 2008. Finite element method based machining simulation environment for analyzing part errors induced during milling of thin-walled components. *International Journal of Machine Tools and Manufacture* 48 (6): 629–643.

Rai, J.K., and P. Xirouchakis. 2009. FEM-based prediction of workpiece transient temperature distribution and deformations during milling. *The International Journal of Advanced Manufacturing Technology* 42 (5): 429–449.

Soo, S.L., D.K. Aspinwall, and R.C. Dewes. 2004. 3D FE modelling of the cutting of Inconel 718. *Journal of Materials Processing Technology* 150 (1): 116–123.

Wan, M., W. Zhang, K. Qiu, T. Gao, and Y. Yang. 2005. Numerical prediction of static form errors in peripheral milling of thin-walled workpieces with irregular meshes. *Transactions-American Society of Mechanical Engineers Journal of Manufacturing Science and Engineering* 127 (1): 13–22.

Yang, G. 1980. *Elastic and plastic mechanics*. People Education Published Inc., PRC.

Modelling of SLM Additive Manufacturing for Magnesium Alloy

Ashish Kumar Mishra and Arvind Kumar

Abstract Light-weight materials, such as magnesium and its alloys, have excellent mechanical properties, such as high specific strength, high specific modulus, good damping ability, thermal conductivity, etc. These properties make them a potential candidate for aerospace and automotive applications. The inability of the conventional manufacturing processes to manufacture precise and high-quality magnesium parts drives the exploring of new manufacturing techniques. In this study, the feasibility of additive manufacturing of AZ91D magnesium alloy using Selective Laser Melting (SLM) process has been investigated. Numerical modelling and simulations are performed to study the melt pool shape and size evolution, temperature and velocity fields, temperature gradients and cooling rate during the SLM of AZ91D powder, and the influence of the input laser power is described. The model considers conduction heat transfers and radiation heat losses from the powder bed surface, melting and solidification and convection in the melt pool. The simulation results provide preliminary insights of the complex physical phenomena occurring during SLM of AZ91D by describing the interaction between the laser source and the magnesium powder. The temperature and the velocity field were found playing a significant role in deciding the melt pool dimensions and geometry. An attempt has been made to describe the melt pool stability by calculating the ratios of melt pool length to width and melt pool width to depth. The spatiotemporal variation of temperature shows that very large temperature gradients and cooling rates (of the order of 10^7 K/m and 10^6 K/s, respectively) can be achieved. The high cooling rates are likely to result in the development of finer microstructure and so improved mechanical properties.

A. K. Mishra · A. Kumar (✉)
Department of Mechanical Engineering, Indian Institute of Technology Kanpur, Kanpur 208016, India
e-mail: arvindkr@iitk.ac.in

S. S. Pande and U. S. Dixit (eds.), *Precision Product-Process Design and Optimization*, Lecture Notes on Multidisciplinary Industrial Engineering,
https://doi.org/10.1007/978-981-10-8767-7_5

1 Introduction

There is a growing interest for the use of magnesium alloys in automotive and aerospace industries, a reason behind which is their very low density compared to aluminium alloys (2/3rd of density at room temperature) and steel (2/9th of density at room temperature). The low density imparts them high specific strength and modulus, which along with their excellent thermal properties and damping characteristics make them a potential candidate for low-temperature structural applications in aerospace and automotive industries (Easton et al. 2008; Mishra 2013). However, creep susceptibility and poor surface properties limit their use as structural materials. To overcome these deficiencies, alloying elements like aluminium, zinc, manganese, etc., are usually added (Zhang et al. 2009; Zhang et al. 1998).

Casting and forming are the two conventional manufacturing techniques generally used for manufacturing magnesium alloy parts. Forming provides better mechanical properties than casting but suffers from oxidation and induced anisotropy. Castings too suffer from defects such as high porosity and gas entrapment, which deteriorate mechanical properties of the manufactured part. Moreover, difficulties in fabricating high precision parts, oxidation and huge wastage have encouraged the search for more efficient manufacturing processes. In this regard, Selective Laser Melting (SLM) based additive manufacturing can be a potential manufacturing process for magnesium alloy parts (Wei et al. 2014).

SLM is a widely used powder bed fusion based metal additive manufacturing (MAM) process. It uses a focused high energy laser beam to fully melt the powder layer particles. Full melting is commonly used for processing of pure metals and alloys, where almost the entire region of powder layer material under laser irradiation is melted along with some melting occurring in the previously solidified layer or the base layer (known as remelting). The remelting ensures proper bonding between the layers to create well-bonded high-density structures. A schematic diagram of the SLM process is shown in Fig. 1. The inherent advantages of SLM process include direct production from CAD model, excellent process capabilities and high cooling rates inside the melt pool resulting in the refined microstructure of produced parts (Wei et al. 2014), and much more. SLM fabricated metal parts have mechanical properties comparable to those of bulk materials.

SLM involves complex coupled thermal and physical phenomena, like laser–matter interaction, thermal transport, rapid melting and solidification; all significantly controlled by temperature and flow fields within the melt pool. Therefore, the knowledge of thermal behaviour during SLM is extremely important (Li and Gu 2014). Numerical modelling and simulations are extensively used in this regard, with numerous studies investigating thermal behaviour in SLM reported for common structural materials like steel (Childs and Hauser 2005; Gusarov and Smurov 2010), aluminium (Loh et al. 2015) and titanium alloys (Aggarwal et al. 2016; Li and Gu 2014).

The SLM process of magnesium alloys is challenging, due to the low melting and boiling temperatures as well as high reactivity towards oxygen and nitrogen in the powder and the molten state. Due to this, only a few experimental studies are

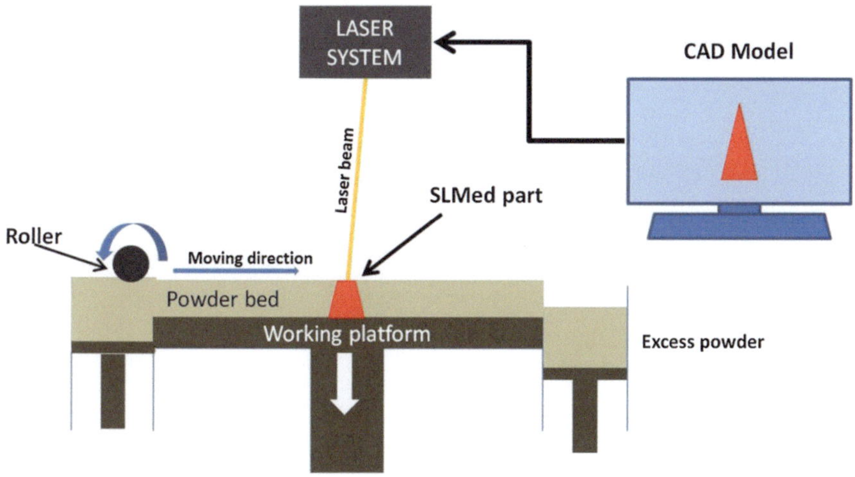

Fig. 1 Schematic diagram of SLM process

reported for SLM of magnesium alloys (Ng et al. 2011; Wei et al. 2014). Thus, numerical modelling and simulations can be used as an effective tool for detailed analysis of SLM of magnesium alloys.

In this work, a numerical model of SLM of AZ91D magnesium alloy is developed based on appropriate mathematical descriptions for the heat source, governing transport equations and boundary conditions. AZ91D is a magnesium alloy containing Al and Zn as alloying elements. Using the numerical model, simulations are performed for different laser power and results are presented for the temperature and velocity field evolution within melt pool. The melt pool characteristics (such as width, length, depth and volume), temperature gradients and cooling rate are discussed.

2 Mathematical Modelling

The model has following assumptions:

(a) The powder bed is considered as a homogeneous domain. The effect of porosity is accounted using suitable mathematical relations.
(b) Incompressible laminar flow within the melt pool.
(c) The density variation in liquid phase is approximated using the Boussinesq approximation.
(d) Material evaporation is not considered.

Simulation is performed only for half of the domain (as shown in Fig. 2) for reducing the CPU time by taking the advantage of the symmetry on the middle plane (*x*-*z* plane). The laser beam, with surface Gaussian beam profile, is applied to

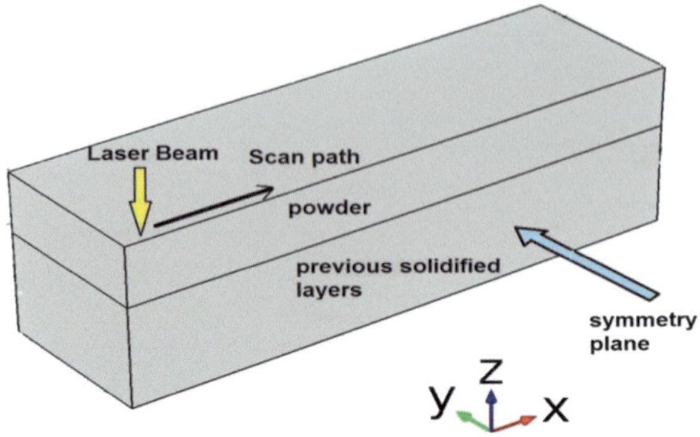

Fig. 2 Schematic of computational domain

Table 1 Process parameters for simulations

Computational domain dimensions (μm)	500 × 150 × 160
Layer thickness (μm)	60
Laser spot size (m)	100
Laser power (W)	100, 125, 150, 175, 200
Scan velocity (m/s)	1.0
Porosity (φ)	0.475

Table 2 Properties of AZ91D alloy (Friedrich and Mordike 2006)

Solidus temperature (T_S)	743 K
Liquidus temperature (T_L)	868 K
Specific heat capacity c_{PS} c_{PL}	1014 J/kg K (solid, 293 K) 1230 J/kg K (liquid)
Thermal conductivity k_S (Solid) k_L (Liquid)	72 W/m K 82.9 W/m K
Density (solid, 20°C) ρ_S (Liquid) ρ_L	1810 kg/m^3 1656 kg/m^3
Latent heat of fusion L	373 kJ/kg
Dynamic viscosity μ	3×10^{-3} Pa s
Thermal expansion coefficient	2.6×10^{-5} K^{-1}
Volumetric thermal expansion coefficient (β_T)	9.541×10^{-5} K^{-1}
Temperature coefficient for surface tension ($\partial\gamma/\partial T$)	-2.13×10^{-4} N/m K
Emissivity (ε)	0.18

the top surface of the domain. The domain is divided into two layers. The top layer is considered as powder bed, while the bottom as the previously solidified layer. Both layers are of the AZ91D alloy. The process parameters used for the simulations are listed in Table 1. The thermophysical properties of the AZ91D alloy are listed in Table 2.

(A) Powder bed properties
The effect of porosity in the powder bed is accounted in thermal conductivity and density of powder bed. These quantities are calculated using following equations:

$$\rho_{\text{powder}} = \rho_{\text{s}}(1 - \varphi) \tag{1}$$

$$k_{\text{powder}} = k_{\text{s}}(1 - \varphi) \tag{2}$$

(B) Governing transport equations.

(1) Continuity equation:

$$\vec{\nabla} \cdot \rho\vec{u} = 0 \tag{3}$$

(2) Momentum conservation equation:

$$\frac{\rho\partial\vec{u}}{\partial t} + (\rho\vec{u}\cdot\nabla)\vec{u} = -\nabla p + \nabla\cdot\left(\mu\left(\nabla\vec{u} + (\nabla\vec{u})^{T}\right)\right) + \vec{F}, \tag{4}$$

where $\vec{F}$ is a source term used to take account of the buoyancy forces and the mushy region flow resistance. The symbols ρ, $\vec{u}$, μ and p represent density, velocity field, dynamic viscosity and dynamic pressure, respectively.

$$\vec{F} = \overrightarrow{F_{\text{n}}} + \overrightarrow{F_{\text{s}}} \tag{5}$$

$$\overrightarrow{F_{\text{s}}} = \frac{C(1 - f_{\text{L}})^{2}}{b + f_{\text{L}}^{3}}\vec{u}, \tag{6}$$

where f_{L} represents the liquid fraction, C and b are constants used for flow within the two-phase zone.
The source term $\overrightarrow{F_{\text{n}}}$ is given as

$$\overrightarrow{F_{\text{n}}} = \rho\vec{g}\beta_{\text{T}}(T - T_{\text{ref}}) \tag{7}$$

(3) Energy conservation equation:

$$\rho c_{\text{P}}\frac{\partial T}{\partial t} + \rho c_{\text{P}}\vec{u}\cdot\nabla T = \nabla\cdot(k\nabla T), \tag{8}$$

where ρ, c_P and k are given by following relations,

$$\rho = (1 - f_L)\rho_S + f_L\rho_L \tag{9}$$

$$c_P = \frac{1}{\rho}\{(1 - f_L)\rho_S c_{PS} + f_L\rho_L c_{PL}\} + \frac{L\partial\alpha_m}{\partial T} \tag{10}$$

$$k = (1 - f_L)k_S + f_L k_L \tag{11}$$

$$\alpha_m = \frac{1}{2}\frac{(f_L\rho_L - (1 - f_L)\rho_S)}{(f_L\rho_L + (1 - f_L)\rho_S)} \tag{12}$$

In Eqs. (9), (10), (11) and (12), the properties k_S and ρ_S are taken as given in Eqs. (1, 2) for the powder domain and for the solid domain as given in Table 2, while c_{PS} is taken as constant for both the powder and the solid domain, as given in Table 2.

(C) Boundary conditions
At top surface

$$k\frac{\partial T}{\partial z} = q_{in} - h_c(T - T_\infty) - \varepsilon\sigma\left(T^4 - T_\infty^4\right) \tag{13}$$

$$q_{in} = \frac{2\varepsilon P}{\pi R^2}\exp\left(-\frac{2\left((x - Vt)^2 + y^2\right)}{R^2}\right), \tag{14}$$

where q_{in} is input heat flux and R is the laser spot radius. The symbols ε, σ, T_∞ and h_c represent emissivity, Stefan–Boltzmann constant, ambient temperature and convection coefficient, respectively. The velocity boundary condition at the top surface is given as

$$-\frac{\mu\partial u}{\partial z} = \frac{\partial\gamma}{\partial T}\left(\frac{\partial T}{\partial x}\right) \tag{15}$$

$$-\frac{\mu\partial v}{\partial z} = \frac{\partial\gamma}{\partial T}\left(\frac{\partial T}{\partial y}\right), \tag{16}$$

where γ is surface tension.

The domain boundaries, except plane of symmetry and top boundary, are considered adiabatic.

3 Results and Discussion

Figure 3 shows the temperature and melt pool evolution with time for laser power $P = 100$ W. The colour map shows the temperature field, while the curved boundary represents the melt pool. The temperature in the domain starts increasing with time as the laser heating starts. The melting starts once the temperature exceeds the solidus temperature, forming a melt pool which travels along the scan direction following the moving laser beam. The melt pool grows rapidly during initial time, as shown in Fig. 3 at 200 and 300 μs. However, it achieves a quasi-steady state after some time.

This is shown in Fig. 3 at 700 μs where no change is observed in the melt pool size and the temperature till 900 μs. The maximum temperature in the computational domain is observed in the central region of the melt pool, where the beam intensity is the highest. This maximum domain temperature with time is plotted in Fig. 4, which gives a glimpse of temperature field evolution within computational domain. The maximum domain temperature increases rapidly (in less than 20 μs) to 1600 K and then the further increase in the temperature slows down. After $t = 500$ μs, the variations are small enough (largest variation in maximum domain temperature is 2.6% after $t = 500$ μs) to consider the process attaining a quasi-steady state. This trend of rapidly reaching the maximum temperature and then remaining stable around the same value can be attributed to the very large thermal conductivity of the AZ91D alloy which facilitates rapid thermal transport and dissipation.

A very high velocity of the order of 3.65 m/s is induced (due to Marangoni convection) in the melt pool at 100 W. Such high velocity greatly enhances lateral heat transfer and mixing in melt pool (through convection) and results in wider melt pools. Combined with the very large thermal conductivity of the alloy, it resulted in deeper melt pool. However, despite the deeper penetration of melt pool and longer growth duration, no remelting is seen at 100 W. Remelting is defined as the melting of previous deposited/base layers and is critical to ensure manufacturing of well-bonded high-density parts.

Melt pool dimensions (length, width and depth) increase with time during the initial stage, but as the melt pool becomes steady the dimensions of the melt pool become stable. The same is illustrated by the plot of melt pool volume with time, plotted in Fig. 5. Confirming to the previous discussions, the melt pool grows initially as the volume increases linearly from 100 to 200 μs. After 200 μs, the rate of growth of the melt pool volume slows down and continues till around 680 μs after which the growth stops and melt pool becomes quasi-steady. The steady-state melt pool volume is found as 3.75×10^{-13} m^3 at 100 W laser power. From Figs. 4 and 5, it is evident that the melt pool becomes quasi-steady after $t = 700$ μs, and accordingly, the temperature, the flow velocity and the melt pool characteristics do not experience any significant variation subsequently.

Figure 6 illustrates the effect of increasing laser power on the melt pool. The results are shown at 800 μs when the quasi-steady state is already achieved.

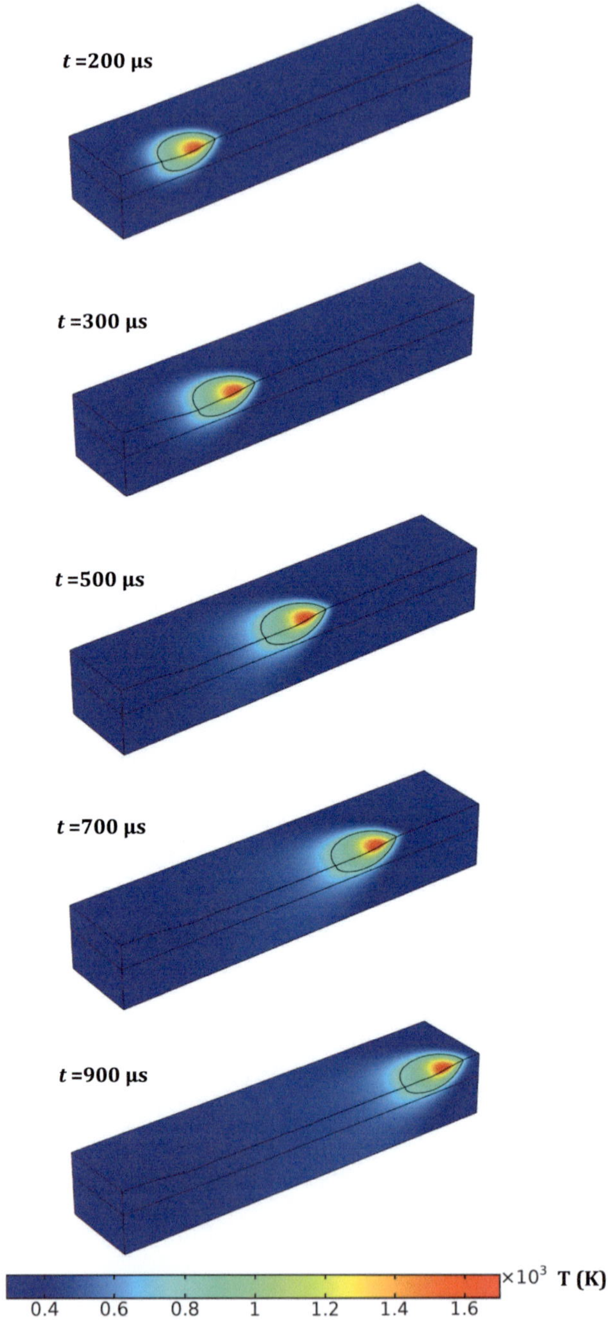

Fig. 3 Melt pool evolution with time at P = 100 W

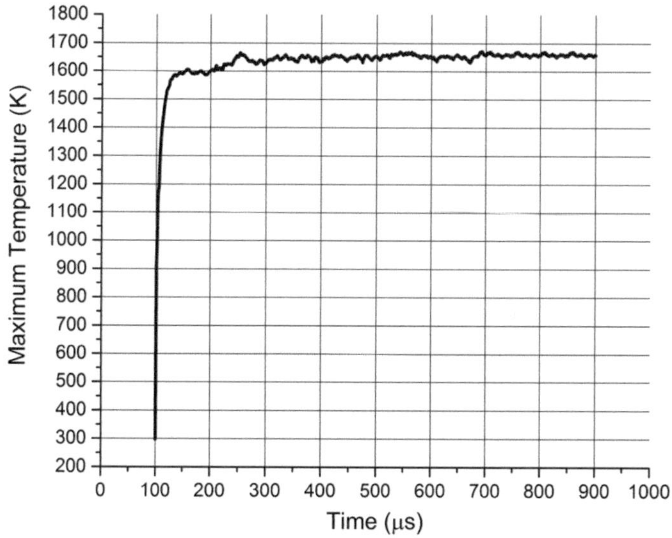

Fig. 4 Maximum domain temperature with time at P = 100 W

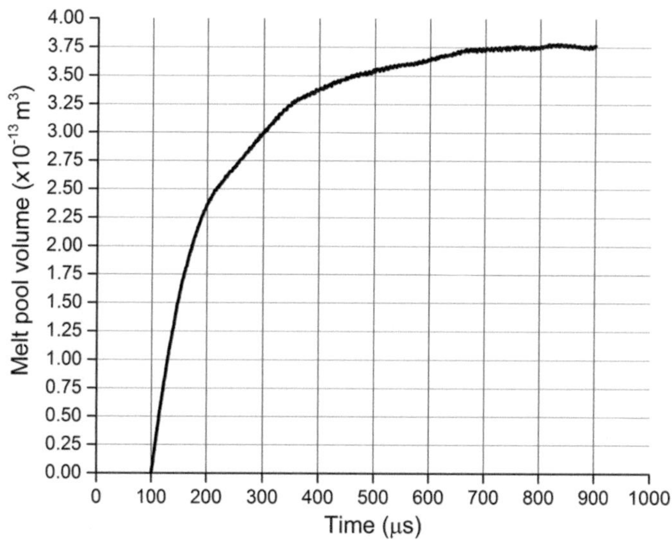

Fig. 5 Melt pool volume with time at P = 100 W

The comparison of melt pool length, width, depth, maximum temperature and velocity corresponding to these results is discussed later. The colour plot represents the temperature distribution while arrows show the velocity field. The velocity field is outwards, typical to the Marangoni flow for the molten metals. The maximum

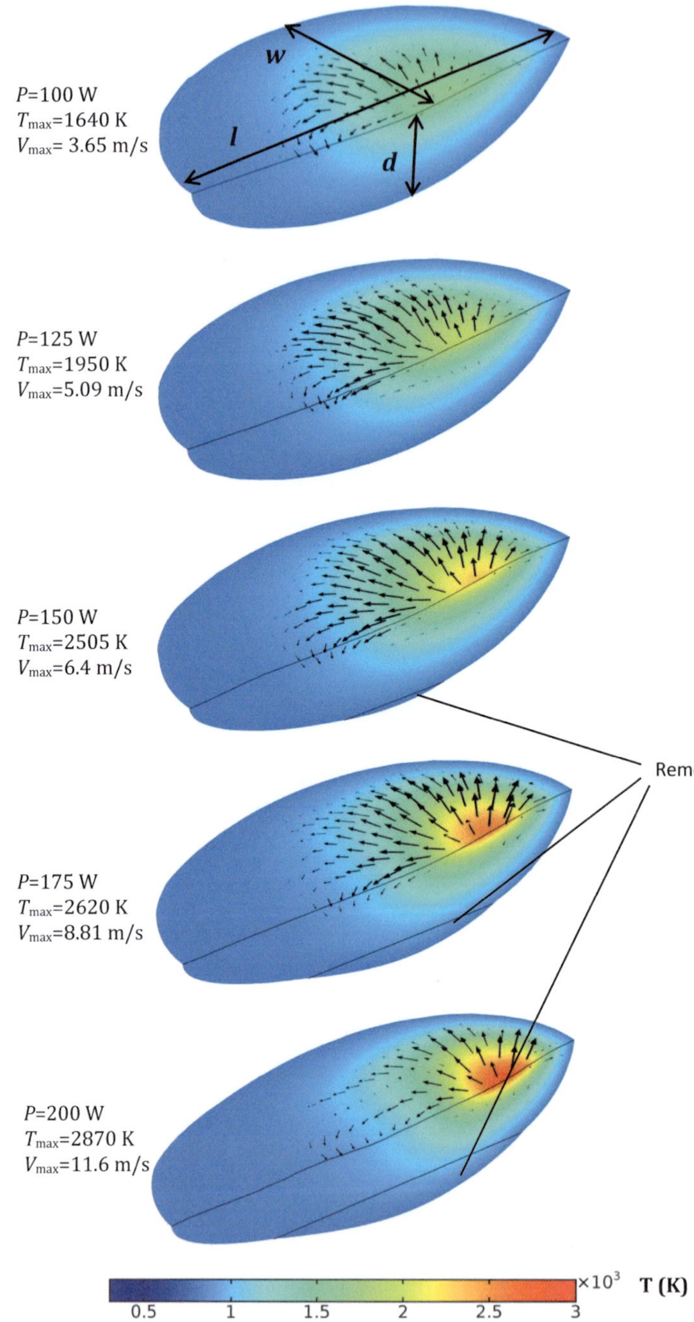

Fig. 6 Melt pool comparison for different laser power (t = 800 μs)

temperature and flow velocity values increase as the laser power increases from 100 to 200 W. The flow velocity within melt pool increases because the temperature distribution becomes steeper resulting in stronger Marangoni convection as the laser power increases. The maximum temperature and the maximum velocity in melt pool increase drastically respectively from 1640 K, 3.65 m/s at 100 W to 2870 K, 11.6 m/s at 200 W. However, such high temperatures and high velocities will not be present in reality as there will be evaporation and the region which experiences such high velocities will eventually get evaporated. Still, a maximum flow velocity of around 5 m/s is expected to be induced in reality at 200 W.

The melt pool dimensions (length, half width and depth) are found increasing as the laser power is increased (Fig. 7), owing to the higher temperatures and enhanced heat transfer at increased laser power as discussed before. The increase in the melt pool half width (w) is more linear in trend than the melt pool length (l) or the depth (d). The melt pool length increases much rapidly when the laser power is increased, showing the effect of the very high thermal conductivity of the solidified material past the melt pool resulting in more diffusion of thermal energy (both the transported energy and the evolved latent heat of solidification) over longer distances. The melt pool depth increases due to increased convection within melt pool at increased laser power, which causes the molten metal to penetrate deeper in the powder layer and eventually remelting starts to occur in the previously solidified/base layer also at 150 W. Considerable remelting is observed at laser power higher than 150 W, suggesting the use of higher laser power to obtain sufficient remelting. However, more evaporation occurs at higher laser powers and so there is a trade-off in the selection of laser power so as to prevent excess evaporation and to ensure sufficient remelting.

The same is visible in the plot of melt pool volume versus laser power, shown in Fig. 8. The melt pool volumes are plotted when they had achieved the quasi-steady state. The melt pool volume is found to increase nonlinearly with the laser power,

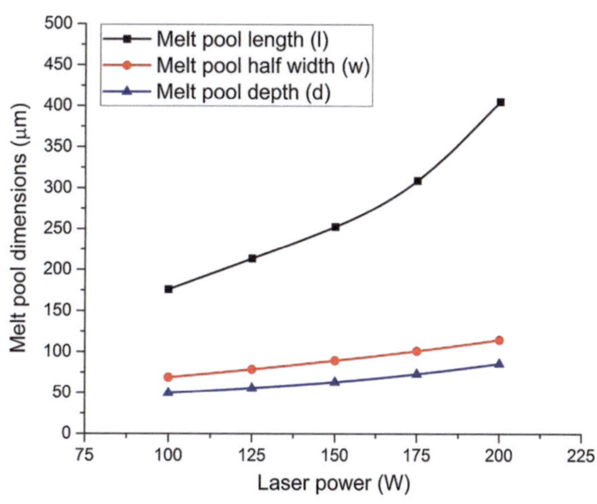

Fig. 7 Variation of melt pool dimensions for different laser power

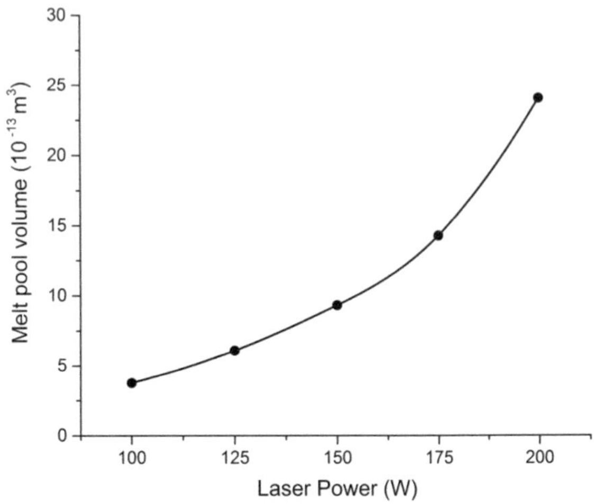

Fig. 8 Melt pool volume for different laser power at t = 800 μs

from 7.5 × 10^{-13} m^3 at 100 W to 4.8 × 10^{-12} m^3 at 200 W. An empirical relationship between the quasi-steady melt pool volume and the laser power is proposed, as given in the following

$$\text{Melt pool volume} = (0.9522 + 0.8777\exp(0.01988P)) \times 10^{-13}\text{m}^3 \quad (17)$$

where P is the value of input laser power in Watt.

The melt pool aspect ratio (length/width) and the ratio of melt pool width to depth are the key attributes for the stability of the melt pool during SLM. These are plotted in Figs. 9, 10 for different laser power. The values were calculated from the quasi-steady melt pool results. In the present case, the aspect ratio ($l/(2w)$) is found increasing from 1.28 at 100 W to 1.77 at 200 W. The melt pool width to depth ratio ($2w/d$), on the other hand, was found quite high (nearly twice the aspect ratio). It increased from 2.78 at 100 W to 2.83 at 150 W and then decreased to 2.68 at 200 W. Such high values indicate shallower melt pools typically resulted from the outwards flow of molten metal in melt pool (Marangoni flow). The decrease in the width to depth ratio after 150 W is due to the penetration of melt pool in the bulk solid material layer below the powder layer, where the very high thermal conductivity resulted in increased heat transfer and hence deeper melt pools. Higher values of these ratios should be undesirable in order to keep the melt pool stable and hence a careful selection of input process parameters is required.

An attempt is made to propose a mathematical correlation between these ratios and the input laser power through curve fitting. The melt pool aspect ratio ($l/(2w)$) versus laser power plot was found to follow a cubic polynomial of the form

$$\frac{l}{2w} = a_0 + a_1P + a_2P^2 + a_3P^3 \quad (18)$$

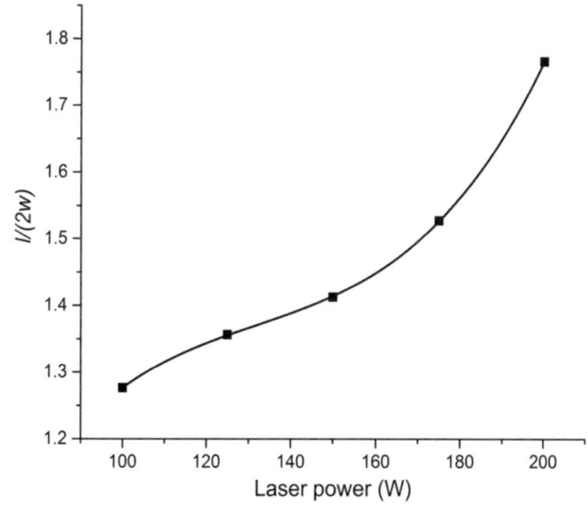

Fig. 9 Melt pool aspect ratio (*l/2w*) for different input laser power

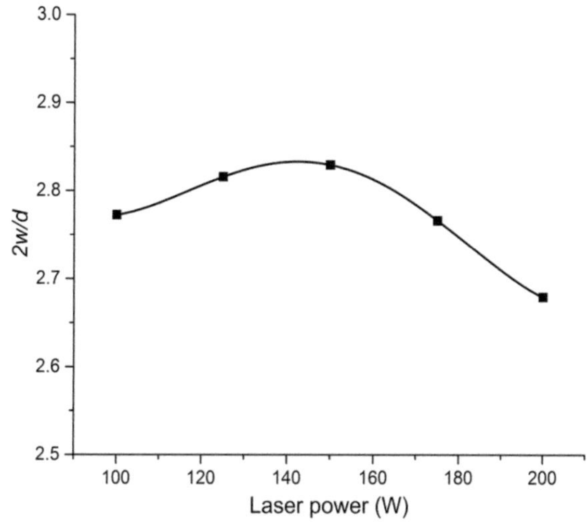

Fig. 10 Melt pool width to depth ratio (*2w/d*) for different input laser power

The melt pool width to depth ratio is found to follow a quartic polynomial of the form

$$\frac{2w}{d} = a_0 + a_1 P + a_2 P^2 + a_3 P^3 + a_4 P^4, \tag{19}$$

where P is the value of input laser power (in Watt). The value of coefficients of above polynomials is listed in Table 3.

The temperature gradient and the temperature–time history play a significant role in the development of melt pool, residual stresses and distortion in the

Table 3 Coefficients of the polynomial expressions

Coefficients	$l/(2w)$	$2w/d$
a_0	−0.7075	6.7072
a_1	0.043	−0.1214
a_2	-3.1×10^{-4}	0.00135
a_3	7.8437×10^{-7}	-6.31457×10^{-6}
a_4	–	1.0578×10^{-8}

manufactured part and hence its study has significant importance. A detailed discussion of the same is out of the scope of this work, and only a brief discussion is presented here to highlight the effect of laser power on the temperature gradient and the temperature–time history. The locations of line and point data sets used to show temperature gradient and temperature–time history are depicted in Fig. 11. Figure 12 shows temperature variation along the length, width and depth (EAB, AC and AD in Fig. 11), respectively for different laser power at time t = 500 μs. Very large temperature gradients (of the order of 10^7 K/m) are induced in the SLM process during both heating and cooling. From the plots, it is evident that the temperature gradients along all three directions (T'_x, T'_y and T'_z) increase with increasing the laser power. The temperature gradient is larger at the melting front than at the solidification front (Fig. 12a), the reason of which is the very high thermal conductivity of the bulk solid after the solidification front which results in quick heat diffusion over longer distances. Also, a notable point in Fig. 12a is the delayed temperature fall witnessed near the solidification front, shown by the inflexion in the plot towards the solidification front (region between 250 to 400 μm). The delayed fall in temperature is resulted by the evolution of the latent heat of solidification at the solidification front, which does not allow the temperature to fall continuously. The temperature distribution along the width (AC in Fig. 11) closely resembles the Gaussian distribution of incident beam energy. The temperature gradient T'_y, at a particular laser power, first increases and then decreases while moving away from the beam centre; and eventually vanishing at the outer wall of the domain. The increase in the temperature gradient components T'_x and T'_y with increasing laser power explains the increase in the melt pool flow velocity with laser power. The temperature distribution along the depth, i.e. AD in Fig. 11, shows the exponential decrease while moving away from the top surface (which is irradiated by the laser beam). The temperature gradient component T'_z too is very high near the top surface and continues to decrease exponentially along the depth.

The cooling rate is another important attribute in determining the end product microstructure and mechanical properties. Here, to determine the cooling rate, the temperature at a particular point (point ‘A’, Fig. 11) is plotted with respect to time (see Fig. 13). It is evident that during SLM, as the laser beam passes through a point

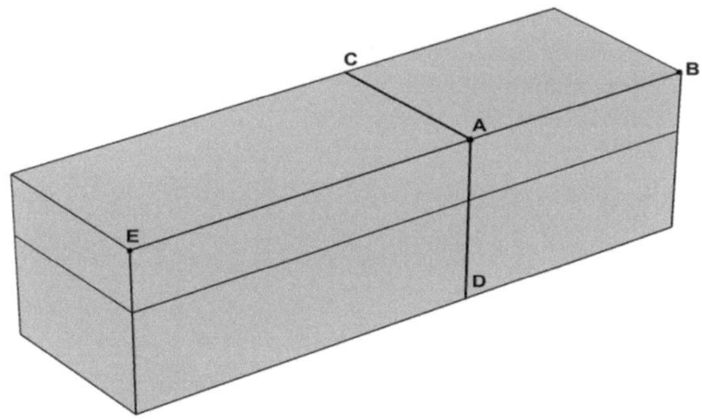

Fig. 11 Locations of line and point data sets used to show temperature gradient and temperature–time history. The location of point A is (500, 0, 160 μm)

(point ‘A’ in this case) the temperature rises and then falls rapidly resulting in very high heating and cooling rates of the order of 10^6 K/s in the melt pool. This can be attributed to the very small time of interaction of laser beam with the material (defined as the laser beam spot diameter divided by laser beam scan speed). Such high cooling rates in the melt pool give rise to the nucleation of very fine grains (rapid solidification), which along with the very small time available (interaction time) for mixing and grain growth result in finer microstructure and consequently superior mechanical properties when compared to the conventional manufacturing.

A noticeable aspect in Fig. 13 is the occurrence of prolonged arrest in the temperature drop during cooling when the melt pool is about to pass the point A, visible from the point of inflexion in the *T*-*t* curves after 600 μs. This is the effect of the evolution of the latent heat of solidification at the solidification front which prevents the temperature from falling for a longer duration. The cooling rate in the solidified material is, however, much smaller implying that further grain growth will not occur once the material is solidified. A detailed analysis of solidification microstructure and grain growth needs experimental analysis and hence is not attempted here.

Overall, it was found from the current results that the temperature and the resulting flow velocity field increase with time till the quasi-steady state is achieved, after which no significant variation was observed. These quantities were also found increasing when the value of the input laser power was increased. The melt pool dimensions and geometry were also found dependent on temperature and molten metal velocity and increased with increasing the laser power. This was also reflected in the plot of melt pool volume versus laser power which showed an exponential increase. The temperature gradients and cooling rates were found very high (of the order of 10^7 K/m and 10^6 K/s, respectively) with the latent heat evolution causing some delay in cooling in the solidification zone.

Fig. 12 Temperature distribution along **a** length (EAB), **b** width (AC) **c** depth (AD) for different laser power at t = 500 μs

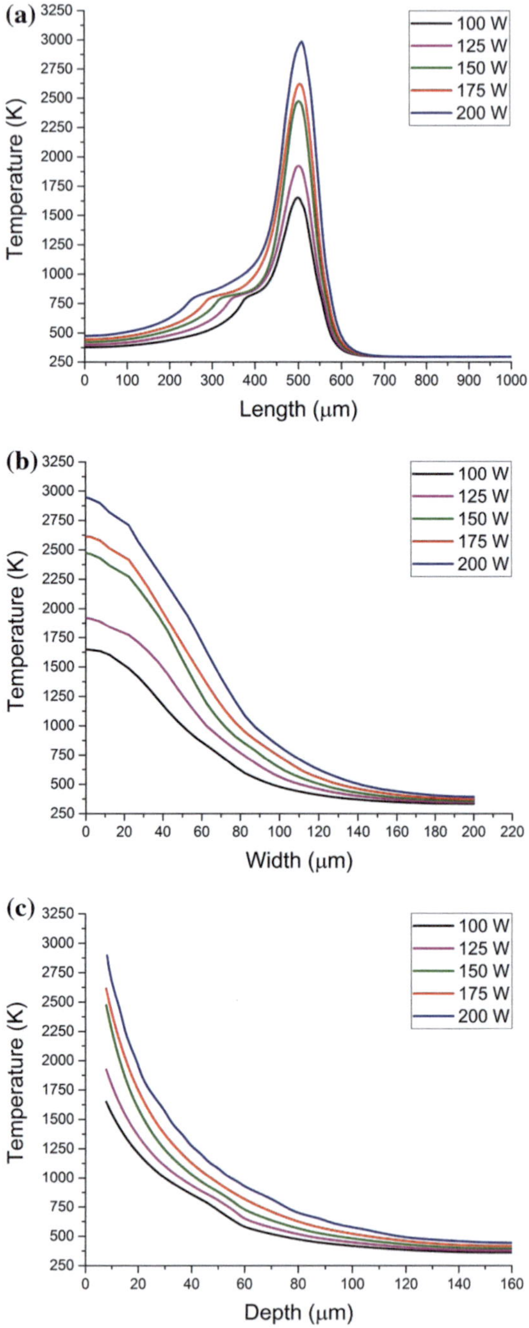

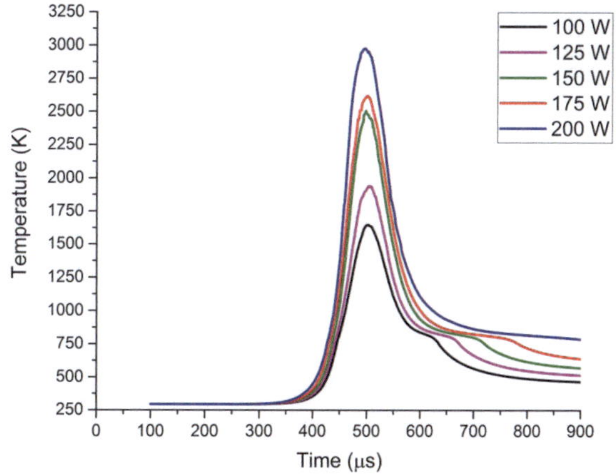

Fig. 13 Temperature–time variations at point A for different laser power

4 Conclusion

This work reports a numerical model of SLM for AZ91D magnesium alloy and simulation results for various input laser power values. Experimental studies for additive manufacturing of magnesium alloys are greatly needed and irreplaceable, then numerical simulations alone due to very little studies in this field. Nevertheless, it was found that such numerical studies can provide preliminary insights of the complex physical phenomena occurring during SLM AZ91D and their dependence on the key process parameters, such as laser power. The temperature and the resulted velocity field of molten metal were found playing a significant role in heat transfer and were governing factors in deciding the melt pool dimensions and geometry. Both the temperature and flow velocity were found increasing when the value of input laser power was increased, showing their greater dependence on the input energy. The high velocities in conjunction with the very high thermal conductivity of the material result in enhanced heat transfer in the domain which results into larger melt pool. The high velocities, though, may also cause penetration of molten metal in the powder bed and compromise the melt pool stability. The melt pool dimensions were found increasing when the laser power increased, which was reflected by the melt pool volume which increases exponentially with increase in laser power. An attempt was made to briefly study the melt pool stability by calculating the ratios of melt pool length to width and melt pool width to depth. The melt pool width to depth ratio is much higher (2.6–2.8) showing the formation of shallow melt pools. This also meant that the melt pool is prone to destabilise. The spatiotemporal variation of temperature shows that very large temperature gradient and cooling rates (of the order of 10^7 K/m and 10^6 K/s, respectively) can be achieved. Such high values can result in significant residual stresses and distortion in the manufactured part. The high cooling rates are likely to result in the development of finer microstructure and so improved mechanical properties. However, the need of a detailed experimental analysis of temperature gradients and cooling rates and their effect is felt greatly in order to support the observations made through numerical simulations.

References

Aggarwal, A., H. Khoont, and A. Kumar. 2016. Investigation of transport phenomena in selective laser melting of Ti-6Al-4 V. In *3rd Annual International Conference on Materials Science, Metal & Manufacturing*, 2016, Singapore.

Childs, T., and Hauser, C. 2005. Raster scan selective laser melting of the surface layer of a tool steel powder bed. *Proceedings of the Institution of Mechanical Engineers, Part B: Journal of Engineering Manufacture* 219: 379–384.

Easton, M., A. Beer, M. Barnett, C. Davies, G. Dunlop, Y. Durandet, S. Blacket, T. Hilditch, and P. Beggs. 2008. Magnesium alloy applications in automotive structures. *The Journal of The Minerals, Metals & Materials Society* 60 (11): 57–62.

Friedrich, H.E., and B.L. Mordike. 2006. *Magnesium technology: Metallurgy, design data, applications*. Berlin, Heidelberg: Springer.

Gusarov, A., and I. Smurov. 2010. Modelling the interaction of laser radiation with powder bed at selective laser melting. *Physics Procedia* 5: 381–394.

Li, Y., and D. Gu. 2014. Thermal behaviour during selective laser melting of commercially pure titanium powder: Numerical simulation and experimental study. *Additive Manufacturing* 1–4: 99–109.

Loh, L., C. Chua, W. Yeong, M. Mapar, S. Sing, Z. Liu, and D. Zhang. 2015. Numerical investigation and an effective modelling on the Selective Laser Melting (SLM) process with aluminium alloy 6061. *International Journal of Heat and Mass Transfer* 80: 288–300.

Mishra, S. 2013. Magnesium alloys in aerospace applications. *NCAIR newsletter* 3 (2): 7–9.

Ng, C.C., M.M. Savalani, M.L. Lau, and H.C. Man. 2011. Microstructure and mechanical properties of selective laser melted magnesium. *Applied Surface Science* 257: 7447–7454.

Wei, K., M. Gao, Z. Wang, and X. Zeng. 2014. Effect of energy input on formability, microstructure and mechanical properties of selective laser melted AZ91D magnesium alloy. *Material Science and Engineering A* 611: 212–222.

Zhang, Z., A. Couture, and A. Luo. 1998. An investigation of the properties of Mg-Zn-Al alloys. *Scripta Materialia* 39 (1): 45–53.

Zhang, E., D. Yin, L. Xu, L. Yang, and K. Yang. 2009. Microstructural, mechanical and corrosion properties and biocompatibility of Mg-Zn-Mn alloys for biomedical application. *Material Science and Engineering* C29: 987–993.

Parametric Optimization of 3D Printing Process Using MCDM Method

S. Vinodh and Priyanka Shinde

Abstract Additive Manufacturing is identified as a key emerging technology and has received much attention during recent years. Three-Dimensional Printing (3D printing) is an Additive Manufacturing (AM) method and has tremendous applications in industries. Selection of appropriate AM process for an application requires consideration of various conflicting criteria. The right AM option ensures competitive performance of manufacturing which in turn affects the quality of the parts. For achieving the best results of any manufacturing process, parametric optimization is essential which has been attempted in the case of 3D printing process using Multi-Criteria Decision-Making (MCDM) techniques. This paper represents the application of an MCDM technique, viz. Multi-Objective Optimization using Ratio Analysis (MOORA) method, to optimize the parameters of 3D printing process, which takes into account any number of criteria, both quantitative as well as qualitative, and offers a simple computational procedure. Three process parameters of FDM based 3D printer, viz. layer thickness, build pattern and fill pattern are considered in this study. Surface roughness and building time of part are taken as response parameters. Effect of each process parameter on surface roughness and build time has been studied.

Keywords Additive manufacturing · 3D printing · Fused deposition modelling
Multi-criteria decision-making · MOORA

Nomenclature

Y_i Normalized assessment value
R_a Surface roughness
I^+ Positive ideal solution

S. Vinodh (✉) · P. Shinde
Department of Production Engineering, National Institute of Technology Tiruchirappalli, Tiruchirappalli, India
e-mail: vinodh_sekar82@yahoo.com

P. Shinde
e-mail: piyashinde2792@gmail.com

S. S. Pande and U. S. Dixit (eds.), *Precision Product-Process Design and Optimization*, Lecture Notes on Multidisciplinary Industrial Engineering,
https://doi.org/10.1007/978-981-10-8767-7_6

I^- Negative ideal solution
S_i^+ Distance between PIS and normalized values
S_i^- Distance between NIS and normalized values
CC_i Closeness coefficient

1 Introduction

Additive Manufacturing (AM) is one of the revolutionary technologies that fabricates 3D objects from CAD data. Additive manufacturing has been visualized as an innovative path for future manufacturing due to its ability to create one-off custom products and capability to manufacture complex designs as a single unit. 3D printing is one of the potential AM technology in which 3D model of object is sliced in successive layers under computer control and these layers are laid down one over the other to create object. Most commonly used 3D printing technology is material extrusion process. This technology also named as Fused Deposition Modelling (FDM) uses extrusion nozzle to extrude material to create successive layers. Most commonly used materials for FDM process are thermoplastics such as Acrylonitrile Butadiene Styrene (ABS), Poly-Carbonate (PC) and Poly Lactic Acid (PLA). FDM printer is connected with a computer interface that processes STL file (Stereo lithography file format) according to which the extrusion nozzle moves both horizontally and vertically following the designed path. Build material is fed to the printer extrusion nozzle in the form of solid where it gets heated past their glass transition temperature and extruded from nozzle in the form of thin filament. These filament strings get deposited on one another to generate 3D model. By using 3D printing process, any complex geometry can be produced.

This chapter presents optimization of process parameters of a non-laser based 3D printing process working on FDM technology using MOORA method. Three process variables are considered, viz. layer thickness, build pattern and fill pattern. Layer thickness is a measure of height of each layer in additive manufacturing process measured along vertical axis (Z-axis). It is one of the important technical characteristics of printer which has impact on part quality. Build pattern is the way in which internal structure of the part is created. Fill pattern is associated with support structure style. Support structure is required to support if any overhang or cantilever is present in geometry. This affects build time and support material volume usage. Different settings of these parameters are taken and surface roughness and build time are calculated in each case. Process variables of 3D printing influence the quality of 3D printed parts (Abdullaha et al. 2015). Major quality indicator considered in this study is surface roughness of parts. Similarly, build time is also an influencing parameter that implies the time needed to build part. These two parameters have been optimized in this study to get the best setting of process variables.

1.1 Objectives of the Present Study

FDM process is the most commonly used AM technology. Parts manufactured using this process show different properties. In order to overcome certain shortcomings of this process, proper understanding of FDM process and its input parameters is essential. This study attempts to understand and analyse the FDM process thoroughly so as to make the components manufactured by this process more reliable and also to make this process more cost-efficient and environment-friendly than any other process. Taking into consideration above factors, objectives of this study are as follows:

1. One of the important objective is optimization of the printing process parameters of FDM based 3D printing. To get optimal settings of process parameters, a multi-objective optimization problem is formulated and solved using MOORA method (Multi-objective optimization using ratio analysis). Finally, the results obtained using MOORA method have been analysed using TOPSIS method and ranks obtained for each experimental run are compared.
2. A detailed study of various process parameters of FDM is carried out and their impact on part quality is investigated. Effect of each process parameter on response parameters has been studied.

2 Literature Review

Multiple studies have been conducted for optimization of process parameters for 3D printing. The studies identify how printing parameters affect various response parameters such as dimensional accuracy of 3D printed parts, surface finish and manufacturing time.

Thrimurthulu et al. (2004) determined optimal part orientation for FDM process in order to enhance surface finish and minimize build time. To obtain optimum results, genetic algorithm was used. Two case studies were carried out in this work with two parts namely axisymmetric part and a 3D part. Two contradictory objectives were achieved as minimum build time and maximum surface finish. Also, support structure minimization was implicitly done in this work. They developed a model for assessing the build time and average part surface roughness. Their contributed methodology can be used to recognize optimum part orientation for any complex part.

Wang et al. (2007) used Taguchi method with Grey Relational Analysis (GRA) to optimize FDM process parameters. Two specimens were prepared namely trapezoid test specimen and tensile test specimen. Six process parameters were taken namely layer thickness, support style, deposition style, deposition orientation in *Z*-direction, deposition orientation in *X*-direction and build location. Taguchi's L_{18} orthogonal array was applied to determine experimental runs.

Response parameters such as tensile strength, dimensional accuracy and surface roughness were taken. GRA was used to determine the optimum parameter setting. Results of GRA were verified using TOPSIS method. Results of this study showed that deposition in *Z*-direction was influencing parameter in case of tensile strength and dimensional accuracy whereas layer thickness was found to be the most influencing parameter in case of surface roughness. They proposed a methodology of integrating Taguchi method with GRA for optimizing RP processes. TOPSIS method was used to verify the resolutions of multiple quality characteristic problems.

Sood et al. (2009) investigated the influence of vital FDM parameters on dimensional accuracy of processed ABSP400 part. Process parameters studied were layer thickness, orientation angle, raster angle, air gap and raster width along with their interactions. Standard test specimens were taken and experiments were designed according to Taguchi's experimental design. Grey Taguchi method was adopted in this study in order to obtain optimum settings of process parameters. Minimization of percentage change in length, width and thickness simultaneously was achieved. It was observed from the study, that shrinkage is the dominating factor along with length and width direction of built part. They adopted grey Taguchi method for identifying common factor setting such that all the three dimensions of a fabricated part show lesser deviation from actual value. They found the optimal process parameters settings to reduce percentage change in length, width and thickness.

Anoop et al. (2011) carried out optimization study for FDM parameters using weighted principal component analysis. Process parameters taken in this study were layer thickness, orientation, air gap, raster angle and raster width. Taguchi method was used to determine experimental runs. L_{27} orthogonal array was taken in this study with five input parameters and three levels of each parameter. Tensile, flexural and impact specimen were prepared and data was collected in terms of three responses. Results of this study showed that all process parameters have significant effect on response parameters. To simultaneously optimize three responses, optimum parameter settings have been found. They concluded that factors such as layer thickness, raster angle, raster width and orientation have a great influence on the mechanical properties of FDM produced parts. They identified optimum parameter settings for optimizing three mechanical properties concurrently.

Zhang and Peng (2012) carried out Taguchi based optimization study to determine optimum parameters for FDM process. Four process parameters were selected in this study namely wire-width compensation, extrusion speed, layer thickness and filling speed. Dimensional error and warpage deformation were taken as response parameters. L_9 orthogonal array was used with four process parameters with three levels. For optimization, Taguchi method was used in combination with fuzzy comprehensive evaluation. Results showed that most significant parameter was wire-width compensation. They used Taguchi method in integration with fuzzy comprehensive evaluation for optimizing four process parameters. They found that performance index of FDM process is greatly influenced by 'wire-width compensation' followed by extrusion speed, layer thickness and filling speed.

Alhubail et al. (2013) worked on Taguchi method based optimization of FDM input parameters to get improved part quality. Process variables namely layer thickness, raster orientation, air gap, raster width and contour width were considered and impact of these parameters on quality characteristics such as tensile strength and surface roughness was studied. They concluded that setting layer thickness and raster width at lower values could minimize the surface roughness in addition to the air gap at −0.01 mm and also higher tensile strength can be obtained. They identified that tensile strength and surface roughness of a processed part is highly influenced by air gap parameter. Validation runs were done to confirm the predicted analysis.

Raol et al. (2014) studied the effect of FDM printing parameters on surface roughness. Printing parameters such as layer thickness, part built orientation and raster angle were considered. Experiments were conducted using response surface methodology and from the results of the experiments, mathematical model was developed. Experimental result analysis and surface plots concluded that part build orientation possesses most vital effect on surface roughness followed by layer thickness. Nevertheless, raster angle has the least vital influence on surface roughness.

Kumar et al. (2014) conducted study to optimize the process parameters of ABS-M30i parts built by FDM to get minimum surface roughness. Five parameters were considered in this study and Taguchi's design of experiments and ANOVA were used to analyse the effect of each parameter. It was found that in this study, not all FDM printer parameters have impact on surface roughness but vary in influence on each proposed response variables. Smooth surface construction and lower R_a were ensured with layer thickness value of 0.254 mm and negative air gap −0.01 mm or raster width of 0.508 mm. They analysed variable parameters of FDM process like laser thickness, air gap, raster width, counter width and raster orientation and their interactions. They determined that thinner layer and voids between deposited layers may minimize surface roughness.

Farzad and Godfrey (2014) optimized FDM parameters using group method for data handling and differential evolution. Relationship between FDM process parameters and tensile strength was determined. Initially, pretest was carried out considering two process parameters namely part orientation and raster angle. Results of pre-test showed that both parameters affect tensile strength. Further, parameters viz. air gap, part orientation, raster angle and raster width were taken and 16 runs were conducted. Optimal parameter settings were found using differential evolution.

Abdullah et al. (2015) investigated the impact of printing orientation and layer thickness on mechanical and topological properties of printed ABS samples. Two printing orientations (*XY* and *YZ*) with three different layer heights (0.1, 0.2 and 0.3 mm) were chosen and specimens were printed utilizing a 3D printer. ANOVA was carried out to investigate the relationship of layer height and printing orientation on tensile strength and surface roughness of the specimens. They concluded that layer height and orientation setting could be improved for better mechanical and topological properties for patient specific implant fabrication.

Manikandan et al. (2015) studied the effect of FDM parameters on flexural strength and surface roughness of PC-ABS mix using FDM 900mc machine. Parameters considered in the study were raster angle, contour style, raster width and air gap. Taguchi method was applied to design experiments and flexural strength and surface roughness were tested. They also found that contour style has the most vital effect on surface roughness of PC-ABS part made using FDM process in comparison with other parameters. They identified the best possible parameters of FDM process in order to ensure good flexural and surface roughness properties. They have shown that raster angle has great effect on flexural strength and counter style has great effect on surface roughness.

Nidagundi et al. (2015) used FDM process parameters for optimization. Process parameters considered for optimization were layer thickness, orientation angle and fill angle. Output parameters considered were ultimate tensile strength, surface roughness, dimensional accuracy and build time. Taguchi's L_9 array was used to conduct experimental runs. S/N ratio was applied to identify optimum parameter settings. They optimized the parameters of FDM process for enhancing properties namely; ultimate tensile strength; dimensional accuracy; surface roughness and manufacturing time. They validated the performance of optimum conditions of FDM by conducting verification experiment.

Rao and Rai (2016) carried out optimization study for FDM process using Teaching Learning-Based Optimization (TLBO) algorithm. FDM parameters such as air gap, layer thickness, orientation angle, raster angle and raster width were taken. Total five case studies were conducted to attain optimum combination of process parameters. Optimization algorithms used in this study were TLBO algorithm and Non-dominated Sorting TLBO (NSTLBO) algorithm. They considered three single-objective optimization problems and two multi-objective optimization problems of FDM process and solved using TLBO and NSTLBO algorithm. They concluded that TLBO algorithm shows better performance compared to GA and QPSO algorithm in terms of objective function value.

Srivastava et al. (2017) carried out optimization study for FDM process parameters using response surface methodology (RSM). Face-centred central composite design was used to conduct experiments using 86 experimental runs. In this study, contour width, orientation, raster angle, raster width, layer height and air gap were taken as process parameters. Response parameters such as build time, model material volume and support material volume for ABS were taken. Optimal parameter setting was obtained using RSM method. Developed mathematical models were tested using design expert software for accuracy.

FDM is a widely used AM technology and has tremendous role in 3D Printing. To achieve optimization of multiple input characteristics of FDM process, MOORA method is used in this study with effective experimental design, i.e. L_8 orthogonal array. Results have also been verified using TOPSIS method. Comparative analysis has been done to derive appropriate inferences. Details about process parameters, response parameters and experimental design are given in subsequent section.

Table 1 Parameter design

Process parameters	Level 1	Level 2
Layer thickness (mm)	0.2540	0.3302
Build pattern	Solid	Sparse-high density
Fill pattern	Basic	Smart

3 Experimental Setup

To determine the effect of process variables on printed parts, tensile specimens were built according to the standards ISO 527:1993 under different conditions of input parameters. Process parameters considered are (1) layer thickness, (2) build pattern and (3) fill pattern. Test specimens were manufactured using uPrint SE plus 3D printer machine and ABS plus plastic as build material. uPrint SE 3D Printer is used in this study which uses ABS plus as model material. Table 1 shows the selected values of process parameters. Two settings of layer thickness are used to build part, they are 0.254 and 0.3302 mm respectively. Build pattern is defined by 'Solid' and 'Sparse-High Density'. Two fill patterns namely 'Basic' and 'Smart' are used. Detailed description of process parameters used in this study is given in Sect. 3.1.

Two-level full factorial design is used to design the experiment runs using three process parameters and two levels providing 2^3 experiments (Hwang and Yoon 1981). Table 2 indicates experimental design according to 2^3 design. Figure 7 shows tensile specimens made by FDM 3D printing process.

3.1 Details About Process Parameters

Process parameters involved in this study are layer thickness, build pattern and fill pattern. For this study, uPrint SE Plus FDM printer is used. This printer has two settings of layer thickness, three settings of build patterns or model interior style and three settings of fill pattern or support style. The software accompanied with

Table 2 L_8 experimental design

Experiment No.	Process parameters		
	Layer thickness (mm)	Build pattern	Fill pattern
1	0.2540	Solid	Basic
2	0.2540	Solid	Smart
3	0.2540	Sparse-high density	Basic
4	0.2540	Sparse-high density	Smart
5	0.3302	Solid	Basic
6	0.3302	Solid	Smart
7	0.3302	Sparse-high density	Basic
8	0.3302	Sparse-high density	Smart

this machine is Catalyst EX 4.4. With reference to this software, detailed description about printer parameters with appropriate figures is given below (Catalyst EX 4.4 software manual):

1. *Layer thickness*: Layer thickness is a measure of the height of each successive layer of material in AM or 3D printing process. The number of layers required to create a part determines the build time required. The thinner the layers, the longer it takes to produce a part of a given height. It is the thickness of layers of material deposited by nozzle and depends on nozzle type. For this study, FDM SE plus 3D printer is used which has two settings of layer thickness, viz. 0.254 and 0.3302 mm.
2. *Build pattern*: Build pattern sometimes called as part interior style is an important parameter. Build pattern influences some important characteristics of prototypes like strength, weight, material consumption build time, etc. 3D printer has three build patterns namely solid, sparse-high density and sparse-low density. Some important features of each pattern are given below:

    a. *Solid*: This pattern has dense fill. There is no gap between adjacent rasters and rasters run perpendicular to those on the preceding layer. For this pattern, model material consumption is higher. Also, it takes longer time to build the part than other patterns. Figure 1 shows solid build pattern.
    b. *Sparse-Low density*: This pattern gives hollow interior with internal lattice for structural rigidity. Large air gaps will be there between rasters and there will be unidirectional rasters on each layer. The interior will be 'honeycombed/hatched'. This pattern results in lowest model material consumption and shortest build time. Figure 2 illustrates sparse-low density build style.
    c. *Sparse-High density*: This interior style is default and is widely applied because of lesser build times, reduced material usage and reduction of the probability of part curl for geometries with higher mass. This pattern also gives hollow interior with internal lattice for structural rigidity. It consumes slightly more model material and takes slightly longer time for building as compared to sparse-low density. Figure 3 illustrates sparse-high density build style.

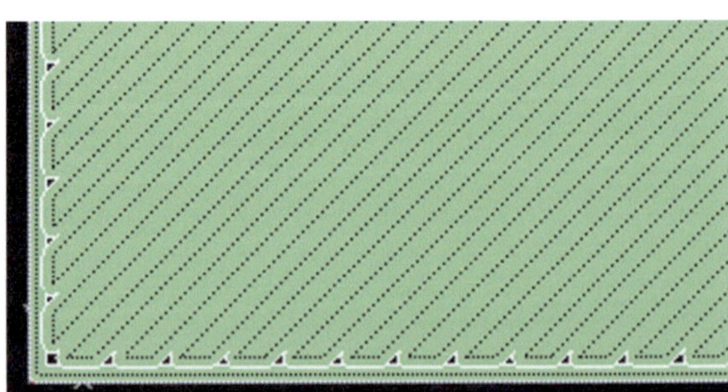

Fig. 1 Solid build pattern

Fig. 2 Sparse-low density build pattern

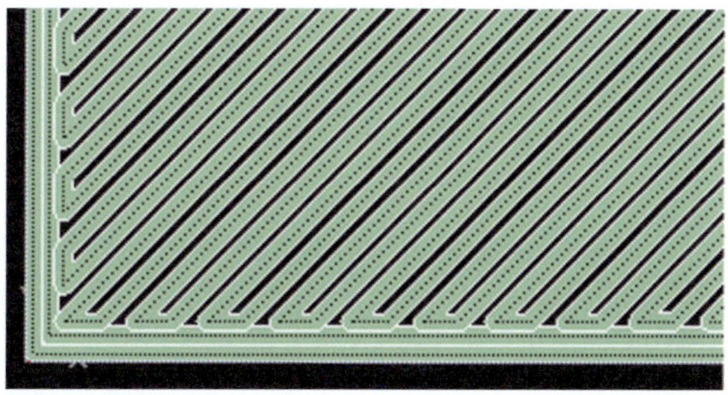

Fig. 3 Sparse-high density build pattern

3. *Fill pattern*: Fill pattern also can be taken as support style. Selection of fill pattern is important as it determines the build time required for the model. Part supports are temporary structures generated during modelling or production phase in order to enable overhang type features. Three important fill patterns are basic, smart and surround and their features are as follows:

   a. *Basic pattern*: Basic support style is the standard raster pattern support structure. It uses a consistent, narrow spacing between support raster toolpaths. This is suitable for all parts and is the default for builds using breakaway support materials. Figure 4 illustrates the basic fill pattern.
   b. *Surround pattern*: This support style encloses the entire model with support material. This style of supports is useful for tall parts with thin features that require extra support and stability during build process. Surround supports require higher build times and to be used only with soluble support material. Figure 5 shows surround support style.
   c. *Smart pattern*: This support style reduces the usage of support material, minimizes build time and enhances support removability for many parts. This style is the default style for builds using soluble supports. SMART supports can be effectively used as it reduces build time up to 14% and also reduces material consumption up to 40%. SMART supports are compatible for almost all parts, and those with large support regions. Figure 6 shows smart support style.

Fig. 4 Basic support style

Fig. 5 Surround support style

Fig. 6 Smart support style

3.2 Selection of Response Parameters

Nowadays, FDM is considered as a potential solution for manufacturing of plastic components in batch size. This process is not only considered for making prototypes or models for visualization and testing but also to make final products. Apart from commercial and customized products, implants manufactured by this process are being used in many of the biological and biomedical applications. To make this

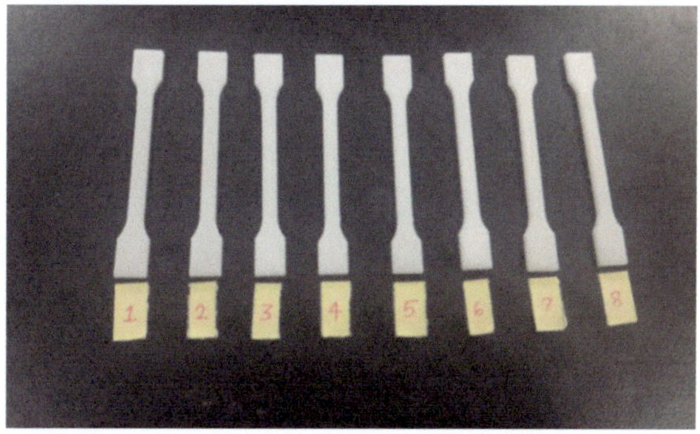

Fig. 7 Tensile specimens built using FDM 3D printer

process more environment-friendly and cost-efficient, it is essential to consider some critical parameters of FDM printers. Following response parameters are taken in this study:

a. Build time: One of the most important applications of FDM process is making prototypes for either design visualization or testing. FDM can be taken as most effective process for making prototypes rather than any other conventional process. Therefore, attention must be given to reduce the cost of prototyping. FDM process mainly influences with its build time. In order to make this process cost efficient, build time needs to be reduced.
b. Surface roughness: FDM parts in some cases are considered as final products and are being used directly. Therefore, parameters like surface finish and part interior properties are important factors to consider.

FDM printer mostly comes with different parameter settings. These printers consist of some critical parameters which directly influence the parts manufactured by this process. Input parameters like layer thickness and part orientation directly affect part surface finish and part strength. In order to take full advantage of FDM process, these input parameters need to be optimized in order to get good part quality. In this study, MOORA method is used to solve optimization problem as discussed in Sect. 4.

4 MOORA Method

MOORA method considers all attributes with their relative importance, and provides an effective assessment of the alternatives. This procedure is computationally simple, logical and robust which can concurrently include any number of quantitative and qualitative selection attributes. As it is based on simple ratio analysis, it has the least amount of mathematical calculations and is useful for practitioners also. Also, computation procedure is not influenced by the addition of any parameter.

Steps in MOORA method includes the following:

Step 1 The first step is to recognize the relevant evaluation attributes.

Step 2 A decision matrix is formulated which depicts the performance of different alternatives with reference to different criteria. The data can be represented as matrix X.

$$X = \begin{vmatrix} x_{11} & x_{12} & \cdots & \cdots & \cdots & \cdots & x_{1n} \\ x_{21} & x_{22} & \cdots & \cdots & \cdots & \cdots & x_{2n} \\ \vdots & \vdots & & & & & \vdots \\ x_{m1} & x_{m2} & \cdots & \cdots & \cdots & \cdots & x_{mn} \end{vmatrix} \quad (1)$$

where x_{ij} in above equation is measure of ith alternative on jth attribute, m and n denote the number of alternatives attributes respectively.

Step 3 The next step is to develop a ratio system. In this step, each performance on an attribute is compared to denominator which is representative for all alternatives pertaining to that attribute. Brauers et al. (2008) mentioned different ratio systems (Mandal and Sarkar 2012) and mentioned that for this denominator, the best alternative is square root of sum of squares of each alternative per attribute. Ratio is given by:

$$X_{ij}^{a} = X_{ij} / \sqrt{\sum\nolimits_{i=0}^{n} X_{ij}^{2}} \quad (j = 1, 2, \ldots n) \quad (2)$$

where x_{ij} denote dimensionless number which pertains to the interval [0, 1] indicating the normalized values of ith alternative on jth attribute.

Step 4 These normalizes performances are summed for maximization (for beneficial attributes) and subtracted for minimization (for non-beneficial attributes). Then optimization problem is:

$$Y_i = \sum_{j=1}^{g} X_{ij}^{a} - \sum_{j=g+1}^{n} X_{ij}^{a} \quad (3)$$

where g denotes number of attributes to be maximized, $(n - g)$ denote number of attributes to be minimized, and Y_i denotes normalized assessment value of ith alternative with reference to all the attributes.

Step 5 Y_i values can be positive or negative depending on totals of its maxima (beneficial attributes) and minima (non-beneficial attributes) in decision matrix (Manikandan et al. 2015). Final preferences are indicted by ordinal ranking of Y_i values. Thus, best choice possesses highest value of Y_i while worst choice possesses lowest value of Y_i. In this study, MOORA method is used for optimization. To get more accurate results, fuzzy evaluation methods can be used which deals with subjectivity and vagueness associated with data.

Application of MOORA method for optimization problem is shown in Sect. 5.

5 Optimization of Process Parameters

MOORA method is used for selecting optimized process parameters of FDM 3D printer. Parameters associated with quality of part were build time and surface roughness as depicted in Table 3. The input parameters considered are layer thickness, build pattern and fill pattern. In this study, major quality characteristics examined are surface roughness of parts. Quality of parts is good with lower surface roughness values. Also, parts which require lower build time are considered as cost-effective parts. Therefore, both surface roughness and build time correspond to lower-the-better (LB) criterion. The selected response parameters were widely applied in prior research studies (Thrimurthulu et al. 2004; Choi and Samavedam 2002). The surface roughness was measured using Mitutoyo SURF TEST SJ-301 tester. The measurement was taken in the middle of gauge length of the tensile specimen. Table 3 shows objective data obtained from experimental trials. Table 4 shows normalized assessment values of alternatives with reference to attributes, as calculated using Eq. (2).

Further, Eq. (3) has been applied to each reading and the normalized assessment scores (Y_i) of all alternatives with reference to considered attributes have been calculated. To calculate Y_i values, both build time and surface roughness are taken as non-beneficial attribute (lower values are desirable in both cases). Ranks are given according to descending assessment values and specimen having rank 1 is considered to be built with best setting of process parameters. Table 4 shows rankings of MOORA method based computations which suggest that optimum values of process parameters are layer thickness as 0.3302 mm, build pattern as Solid and fill pattern as Smart.

Table 3 Objective data

Experiment No.	Process parameters			Build time (min)	Surface roughness (R_a) (μm)
	Layer thickness (mm)	Build pattern	Fill pattern		
1	0.2540	Solid	Basic	7	10.48
2	0.2540	Solid	Smart	6	9.70
3	0.2540	Sparse-high density	Basic	6	12.09
4	0.2540	Sparse-high density	Smart	6	9.76
5	0.3302	Solid	Basic	6	8.73
6	0.3302	Solid	Smart	5	4.29
7	0.3302	Sparse-high density	Basic	5	6.43
8	0.3302	Sparse-high density	Smart	5	5.07

Table 4 Results of multi-objective analysis (normalized assessment)

Experiment No.	Squared values		Ratio		Normalized assessment value (Y_i)	Rank
	Build time	Surface roughness	Build time	Surface roughness		
1	49	109.83	0.4276	0.4254	−0.8530	7
2	36	94.09	0.3665	0.3938	−0.7603	5
3	36	146.16	0.3665	0.4908	−0.8573	8
4	36	95.25	0.3665	0.3962	−0.7627	6
5	36	76.21	0.3665	0.3544	−0.7209	4
6	25	18.40	0.3054	0.1741	−0.4795	1
7	25	41.34	0.3054	0.2610	−0.5664	3
8	25	25.70	0.3054	0.2058	−0.5112	2
	$\sqrt{268} = 16.37$	$\sqrt{606.98} = 24.63$				

6 Comparative Analysis of MOORA Results Using TOPSIS

In order to verify the ranks obtained using MOORA method, TOPSIS method is used. Ranks obtained using both the methods were compared in order to derive appropriate inferences with multiple input characteristics.

TOPSIS method

TOPSIS is a Multi-Criteria Decision-Making (MCDM) technique used to decide preference order. This method is called 'Technique for order preference by similarity to ideal solution' which was developed by Yoon and Hwang in 1981 (Hwang and Yoon 1981). Preference order is decided based on the closest alternative to the ideal solution. In this method, alternatives are graded based on the closeness to the ideal solution. The alternative which is nearer to the ideal solution is assigned the highest grade. This method follows certain steps described as below:

Step 1 *Normalization of data*

In this step, experimental data is being normalized in order to compare the parameters. Normalized values of each parameter are obtained using the following equation:

$$N_{ij} = \frac{p_{ij}}{\sqrt{\sum_{i=1}^{m} p_{ij}^2}} \tag{4}$$

where $i = 1 \ldots m$ and $j = 1 \ldots n$. p_{ij} represents the actual value of the ith value of jth experiment number and N_{ij} represents the corresponding normalized value.

Step 2 *Computation of weighted normalized matrix*

After normalizing the data, weights associated with each parameter are determined. Weighted normalized matrix is obtained by multiplying normalized value with corresponding weight. It is given by;

$$R_{ij} = W_i \times N_{ij} \tag{5}$$

where W_i represents the weights of respective parameters.

Step 3 *Computation of positive ideal solution (PIS) and negative ideal solution (NIS)*

Based on the objective, values of PIS and NIS are decided. If objective is maximization of parameters, then maximum value among each parameter obtained from weighted normalized matrix is taken as PIS (I^+) and minimum value among each parameter obtained from weighted normalized matrix is taken as NIS (I^-). If the objective is minimization, then minimum value among each parameter obtained from weighted normalized matrix is taken as PIS (I^+) and maximum value among each parameter obtained from weighted normalized matrix is taken as NIS (I^-).

PIS and NIS is calculated using the following equations:

$$I^+ = \left(P_1^+, P_2^+, P_3^+, P_4^+ \ldots\right) \text{ maximum values} \tag{6}$$

$$I^- = \left(P_1^-, P_2^-, P_3^-, P_4^- \ldots\right) \text{ minimum values} \tag{7}$$

Step 4 *Computation of distance between PIS and NIS*

$$S_i^+ = \sqrt{\sum_{i=1}^{m} \left(r_{ij} - P_i^+\right)^2} \tag{8}$$

$$S_i^- = \sqrt{\sum_{i=1}^{m} \left(r_{ij} - P_i^-\right)^2} \tag{9}$$

where S_i^+ is distance between PIS and normalized values and S_i^- is distance between NIS and normalized values.

Step 5 *Computation of closeness coefficient*

Closeness Coefficient (CC_i) is calculated using following equation:

$$CC_i = \frac{S_i^-}{S_i^+ + S_i^-} \tag{10}$$

Based on closeness coefficient value, ranking of each alternative is obtained. Highest value of closeness coefficient gives the best choice of alternative. Hence, preference order is decided based on CC_i values.

Application of TOPSIS method and results

TOPSIS method is applied in this study to compare results obtained using MOORA method. Ranks obtained by MOORA method are compared with TOPSIS ranks and results are validated. In this study, two response parameters are taken such as build time and surface roughness. Objective of this study is to minimize both response parameters and obtain optimal settings. According to TOPSIS method, these two parameters are normalized using Eq. (4). After normalization of data, weighted normalized matrix is obtained by multiplying weights of corresponding parameters with normalized value. In this study, weight for both build time and surface roughness are being taken equal to 0.5. Weighted normalized matrix is obtained using Eq. (5). Now PIS and NIS are obtained using Eqs. (8) and (9). In this study, $P^{+}_{\text{Roughness}} = 0.08706$; $P^{+}_{\text{Buildtime}} = 0.1527$; $P^{-}_{\text{Roughness}} = 0.2454$; $P^{-}_{\text{Buildtime}} = 0.2138$. Finally, closeness coefficient is calculated using Eq. (10). All results are summarized in Table 5. Table 5 shows closeness coefficient values. Based on this value, preference order is decided and the best alternative is obtained.

As shown in Table 5, first rank is obtained for experimental run 6 according to TOPSIS method. As shown in Table 4, first rank was obtained for experimental run 6 using MOORA method. Results of MOORA and TOPSIS methods are summarized in Table 6. In MOORA method, ranks are given according to descending order of ratios, i.e. first rank is given for the highest ratio. Similarly, in TOPSIS method, ranks are given based on descending order of CC_i values, i.e. highest CC_i value possesses first rank. Table 6 shows comparative ranking of MOORA and TOPSIS methods. It is clear from the table that all rankings for experimental runs using both the methods are found to be similar.

Table 5 Results obtained using TOPSIS method

Experiment No.	Normalized		Weighted normalized		S_i^+	S_i^-	CC_i	Rank
	Build time	Surface roughness	Build time	Surface roughness				
1	0.4276	0.4254	0.2138	0.2127	0.1397	0.0327	0.1897	7
2	0.3665	0.3937	0.1833	0.1969	0.114	0.0573	0.3345	5
3	0.3665	0.4907	0.1833	0.2454	0.1612	0.0305	0.1591	8
4	0.3665	0.3961	0.1833	0.1981	0.1151	0.0563	0.3285	6
5	0.3665	0.3543	0.1833	0.1772	0.0951	0.0747	0.4393	4
6	0.3054	0.1741	0.1527	0.0871	0	0.1697	1	1
7	0.3054	0.261	0.1527	0.1305	0.0434	0.1301	0.7499	3
8	0.3054	0.2058	0.1527	0.1029	0.0158	0.155	0.9075	2

Table 6 Comparison of ranks obtained using MOORA and TOPSIS methods

Experiment No.	MOORA ratio	Rank	TOPSIS CC_i	Rank
1	−0.8530	7	0.1897	7
2	−0.7603	5	0.3345	5
3	−0.8573	8	0.1591	8
4	−0.7627	6	0.3285	6
5	−0.7209	4	0.4393	4
6	−0.4795	1	1	1
7	−0.5664	3	0.7499	3
8	−0.5112	2	0.9075	2

7 Results and Discussions

In order to evaluate surface roughness and build time, each specimen is manufactured according to ISO standards. For each experiment, surface roughness and build time were measured. Build time was recorded at the time of each experimental run while surface roughness was measured with Mitutoyo SURF TEST SJ-301 tester. Table 3 shows the results of surface roughness test and build time readings under various settings of process parameters. MOORA method is applied to get the optimum setting of process parameters. In this study, MOORA method will be used for parametric optimization as this method has easy intermediate steps for calculation (Gadakh 2011). TOPSIS method is best suitable method to get preference order for alternatives (Wang et al. 2007). TOPSIS method is used to validate the results obtained using MOORA method. Table 4 shows the result of multi-objective analysis after MOORA method based computations. Also, these results are compared using TOPSIS method. As observed from Table 6, it is clear that ranks obtained by both the methods are similar for all experimental runs. The effect of printing parameters on surface roughness and build time has been studied. Results of analysis are as follows:

(a) Surface roughness: As observed from Table 3, R_a value which is the most widely used parameter to indicate the mean surface roughness is measured for each specimen. For the same layer thickness and build pattern, R_a value is lower for SMART fill pattern. However, for same values of layer thickness and fill pattern, R_a value is higher in case of SPARSE build style. Generally, as layer thickness increases surface roughness value decreases (Anitha et al. 2001).

(b) Build time: As inferred from Table 3, as layer thickness values increases, build time decreases. Also it is clear from the table that there is no significant effect of build pattern on build time. Although for SMART fill pattern, build time readings are lower. As compared to 'BASIC' fill pattern, SMART pattern will minimize the consumption of support material, reducing the build time of part and improving support removability for many parts.

8 Conclusions

The aim of this study was to carry out process parameters optimization for FDM based 3D printing process. MOORA method was applied as optimization method in this study. According to results, the following conclusions are drawn:

- Layer thickness and fill pattern are proved to be vital factors influencing part quality. It was found that build pattern and fill pattern are significant in influencing R_a value. Surface finish obtained at sparse-high density was obtained poor. There is a significant decrease in R_a value for smart fill pattern.
- Also, layer thickness was found to be an influencing parameter affecting build time. Build time was lower for higher values of layer thickness. Build time for SPARSE build style was obtained lower because it has large air gap between rasters and requires shortest build time.
- Under same layer thickness and build pattern, build time got decreased for smart fill pattern. Using MCDM method, optimum parameter setting for FDM printer was found as layer thickness 0.3302 mm, build pattern solid and fill pattern smart which gives lowest build time of 5 min and lower surface roughness value as 4.29 μm.
- Also, after comparison of MOORA method with TOPSIS method, it was observed that same ranks are obtained for all experimental runs. Hence optimal parameter settings obtained using both the methods is same i.e. experimental run 6 (Layer thickness—0.3302 mm; Build pattern—Solid; Fill pattern—Smart).
- In the present work, strength of the built parts was not measured. In future, strength could be measured. Also, tensile specimen is being used in the present work and in future, other types of specimen also could be considered. Also, in future, more comprehensive data with varying conditions could be studied.

References

Abdullaha, A., D. Mohamada, T.N. Rahim, H.M. Akil, and Z.A. Rajiona. 2015. 3d printer's parameter optimization for potential patient specific implant fabrication. *Jurnal Teknologi* 76 (7): 75–79.

Alhubail, M., D. Alenezi, and B. Aldousiri. 2013. Taguchi-based optimization of process parameters of fused deposition modelling for improved part quality. *International Journal of Engineering Research and Technology* 2 (12): 2519.

Anitha, R., S. Arunachalam, and P. Radhakrishnan. 2001. Critical parameters influencing the quality of prototypes in fused deposition modelling. *Journal of Materials Processing Technology* 118: 385–388.

Anoop kumar, S., C. Vedansh, D. Sourav, and M. Siba Sankar. 2011. Optimization of process parameters in fused deposition modeling using weighted principal component analysis. *Journal of Advanced Manufacturing Systems* 10 (2): 41–259.

Brauers, W.K.M., E.K. Zavadskas, F. Peldschus, and Z. Turskis. 2008. Multiobjective decision-making for road design. *Transport* 23: 183–193.

Catalyst Software manual.

Choi, S.H., and S. Samavedam. 2002. Modelling and optimization for rapid prototyping. *Computers in Industry* 47: 39–53.

Farzad, R., and C.O. Godfrey. 2014. Fused deposition modelling (FDM) process parameter prediction and optimization using group method for data handling (GMDH) and differential evolution (DE). *International Journal of Advanced Manufacturing Technology* 73: 509–519.

Gadakh, V.S. 2011. Application of MOORA method for parametric optimization of milling process. *International Journal of Applied Engineering Research* 1: 743.

Hwang, C.L., and K. Yoon. 1981. *Multiple attribute decision making: methods and applications*. New York, NY: Springer-Verlag.

Kumar, D., V.N. Kannan, and G. Sankaranarayanan. 2014. Parameter optimization of ABS-M30i parts produced by fused deposition modeling for minimum surface roughness. *International Journal of Current Engineering and Technology* 3: 93–97.

Mandal, U.K., and B. Sarkar. 2012. Selection of best intelligent manufacturing system (IMS) under fuzzy Moora conflicting MCDM environment. *International Journal of Emerging Technology and Advanced Engineering* 2 (9): 301–310.

Manikandan, S., A.S. Seshan Kumar, C. Sharma, V. Prabhu Raja, and A. Adhiyamaan. 2015. Investigation on the effect of fused deposition modeling process parameters on flexural and surface roughness properties of PC-ABS blend. *International Journal on Recent Technologies in Mechanical and Electrical Engineering* 2(8).

Nidagundi, V.B., R. Keshavamurthy, and C.P.S. Prakash. 2015. Studies on parametric optimization for fused deposition modelling process. *Materials Today: Proceedings* 2: 1691–1699.

Rao, R.V., and D.P. Rai. 2016. Optimization of fused deposition modeling process using teaching learning-based optimization algorithm. *Engineering Science and Technology, an International Journal* 19: 587–603.

Raol, T., K.G. Dave, D.B. Patel, and V.N. Talati. 2014. An experimental investigation of effect of process parameters on surface roughness of fused deposition modeling built parts. *International Journal of Engineering Research and Technology* 3(4).

Sood, A.K., R.K. Ohdar, and S.S. Mahapatra. 2009. Improving dimensional accuracy of fused deposition modelling processed part using grey Taguchi method. *Materials and Design* 30: 4243–4252.

Srivastava, M., M. Sachin, T.K. Kundra, and R. Sandeep. 2017. Multi-response optimization of fused deposition modelling process parameters of ABS using response surface methodology (RSM)-based desirability analysis. *Materials Today: Proceedings* 4: 1972–1977.

Thrimurthulu, K., M.P. Pulak, and N.Venkata Reddy. 2004. Optimum part deposition orientation in fused deposition modeling. *International Journal of Machine Tools and Manufacture* 44: 585–594.

Wang, C.C., Ta-Wei Lin, and Shr-Shiung Hu. 2007. Optimizing the rapid prototyping process by integrating the Taguchi method with the Gray relational analysis. *Rapid Prototyping Journal* 13 (5): 304–315.

Zhang, J., and A. Peng. 2012. Process-parameter optimization for fused deposition modeling based on Taguchi method. *Advanced Materials Research* 538: 444–447.

Intelligent PCD Tool Testing and Prediction of Performance

Ramesh Kuppuswamy and Kerry Anne Airey

Abstract The ongoing competition on cutting tools drive the tool manufactures to enhance the cutting tool quality through real-time condition monitoring and performance prediction. The digitization of tool within the manufacturing process, allows for the creation of an intelligent manufacturing system, to reduce the cycle time required for testing, and facilitate the categorization of tool performance as well as early warning signals of the change in the manufacturing processes. The system comprises two platforms such as feature extraction engine (FEE) and feature prediction engine (FPE). The FEE monitors real-time progression of operational behavior during wear with the tool–work interface signals. The accelerated tool wear testing applies a milling arrangement, incorporating a clock-testing workpiece that simulates an intermittent cutting process on hardened steel. The feature extraction engine uses the accelerated wear results to build a calibrated wear model as a reference tool for wear analysis and prediction. Flank wear lands were imaged using a Leica toolmaker's microscope and used to calibrate the wear model, correlating the digital signal feature to the tool–work interface wear behavior. The imaged wear progression, force, and acoustic emission signal features were analyzed by statistical methods including applications of Spiro-Wilks, ANOVA, and Kruskal–Wallis evaluations. This confirms the experimental accuracy and provides the baseline for wear prediction driving the development of a feature prediction engine (FPE). The results are a significant reduction in the quality control cycle time and performance prediction. The experimental results indicate that the FEE correlates accurately across sensors and progression tracking of abrasive wear in the cutting tools, clearly distinguishing between machining cuts, signal noise, and signal anomalies.

Keywords Cutting tools · PCD · Intelligent tool testing · Tool performance prediction

R. Kuppuswamy (✉) · K. A. Airey
Department of Mechanical Engineering, University of Cape Town,
Cape Town, South Africa
e-mail: Ramesh.kuppuswamy@uct.ac.za

S. S. Pande and U. S. Dixit (eds.), *Precision Product-Process Design and Optimization*, Lecture Notes on Multidisciplinary Industrial Engineering,
https://doi.org/10.1007/978-981-10-8767-7_7

Nomenclature

σ_y	Yield strength of the PCD cutting insert (GPa)
μ	Population mean of the test samples
E	Young's modulus of the PCD (GPa)
F_n	Normal force exerted on the workpiece during machining (N)
H_{PCD}	Hardness of the PCD matrix (GPa)
i	Number of data points
K_{IC}	Fracture toughness of the PCD insert (MPa $m^{1/2}$)
l_c	Cutting distance traversed (m)
l	Number of hidden units
L	Levene's statistic (L)
k	Number of sets from which the data come
m	Work hardening factor
n	Size of the sample
N	Total number of sampled cases
P	Weighted sums in a nonlinear model
s	Standard deviation of the experimental data
t	"*t*" factor
$V_{abrasive}$	Volume lost from the cutting insert (m^3)
x	Sample mean
w	Neuron weights
Z	Array of the mean and median values of the sampled case *i* from group *j*

1 Introduction

The ongoing thrust toward a high value manufacturing and services necessitates the manufacturing industries to ensure total system uptime, reliability, and efficiency, particularly for mission-critical high value assets. Conventional approaches such as scheduled preventive maintenance and reactive "fail-and-fix" methods are no longer adequate or effective to meet the increasingly higher operational availability at an affordable cost. In addition, with a paradigm shift toward "fly by the hour" business models, original equipment, lines, engines, and tools manufacturers (OEMs) are compelled to remodel the traditional ways of maintaining these resources to broaden their revenue streams by providing 100% fulfillment at all time. The problem is severe when the manufacturing processes change the product characteristics in terms of dimensional features, precision, and surface integrity. The cutting tool industries often battle to maintain a consistent quality as the manufacturing process varies continuously. Use of intelligent cutting tool testing and prediction of performance would transform manufacturing into predictive reliability so that a consistent quality can be maintained in the tool manufacturing. One such in-demand technology is the use of polycrystalline diamond (PCD) cutters, as cutting inserts brazed into oil well drill bits (National Renewable Energy Laboratory 2000). This increasing demand for PCD cutters necessitate research on PCD performance optimization, with the objective of extending tool-life cycles. The process of enforcing high quality standards and real-time modification of the manufacturing process to account for creeping manufacturing defects becomes a critical tool in the production of PCD inserts. Current quality control processes are inadequate, inefficient processes, which do not operate as real-time applications, resulting in excessive time wastage and subquality inserts that must be discarded. Research on intelligent tool testing methods, therefore, becomes a priority, emphasizing digital monitoring of PCD wear behavior and extending to appropriate predictive analysis. The intent being to establish links between the manufacturing process and the failure behavior of the PCD insert. The quality control process associated with the preemptive elimination of these undesirable features, significantly extend the manufacturing time of an insert batch and require extensive hands-on qualitative analysis by experienced personnel. Accelerating the quality control approach would enable faster rectification of nonideal failure mechanisms experienced by the PCD cutting tools. Furthermore, the application of necessary corrective actions on the manufacturing floor becomes significantly faster than current protocols allow. This, in turn, minimizes material wastage and prioritizes superior tool performance prediction strategies. Ensuring PCD tool compliance to industrial standards begins during manufacture and ends during field testing. During and immediately after manufacture, line inspection procedures assess the PCD tools, for obvious damage and dimensional compliance. In general, the tools presenting with edge chipping, micro-grooving, crack formation, and gouge marks are discarded. Minor surface defects, however, are corrected through further applications of grinding and polishing processes.

The objectives of this research are to develop an intelligent PCD tool testing system that clearly establishes the failure characteristics of the PCD cutting inserts, as well as predicting the product performance. This is implemented via (see Fig. 1):

(A) Feature extraction engine (FEE) with a dashboard interface, and
(B) A predictive tool-life model for PCD Failure.

The scope of this research extends to:

- Development of an appropriate sensor system (select, acquire, and set up the most appropriate type of sensors, placement locations, fastening methods; types, levels, and extent of sensory data to achieve balance between accuracy and analytical ability).
- Development of a data acquisition system so as to digitize the behavior of the tool–work interface.
- Testing, validation, and selection of the most appropriate time–frequency cutting signals in terms of the sensitivity and repeatability of the tool characterization at the production testing stages.

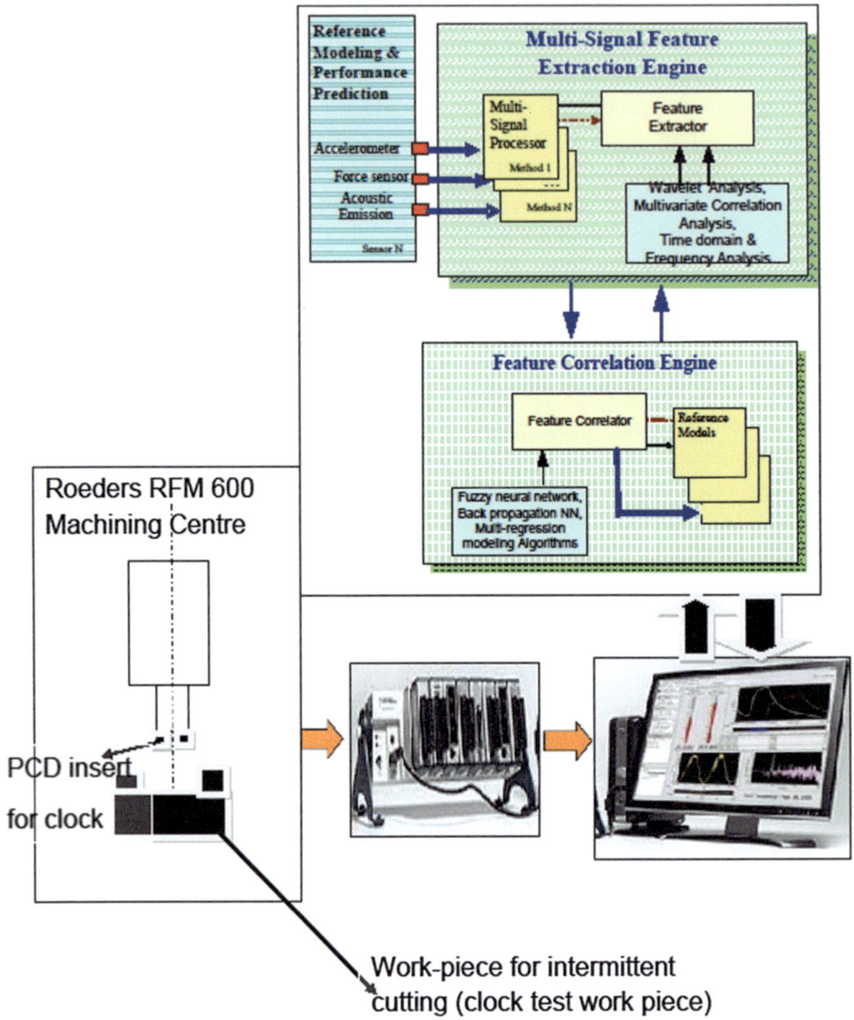

Fig. 1 Plan for development of an intelligent PCD tool testing and prediction of performance

- Testing, validation, and selection of reliable and repeatable feature extraction techniques in the selection of significant parameters for the development of tool-life predictive reference models.
- Testing, validation, and selection of effective self-learning techniques in terms of training sample size, convergence speed, and accuracy in predicting the expected life span of cutters at production testing stages.
- Testing, validation, and selection of the most effective clustering techniques in terms of the accuracy, reliability, and repeatability in mapping specific signals in

relation to known reference models for the prediction of PCD cutting insert life span at production testing stages.
- Distillation of the best sensory signals, feature extraction, learning, and clustering methods for the development of reference models.
- To test the effectiveness of the reference models for the characterization of PCD inserts in a nondestructive and dynamic manner.

2 A Review of PCD Tool Testing Methods

Ensuring PCD compliance to industrial standards begins during manufacture and ends during field testing of the inserts. The testing methods are grouped into industrial-based testing and laboratory-based testing methods as described below.

2.1 Industrial-Based Testing for PCD Tools

- **Field testing**

The field tests conducted at the industries predominantly determines the wear progression and evaluate this against performance expectations. This data is unable to isolate the performance behavior of the PCD tool against the manufacturing inputs but collectively addresses the tool, machine, and process characteristics. As a result, the tool manufacturers are unable to impart corrective actions into their production processes. Furthermore, relying on this method is time-consuming and expensive, with excessive equipment requirements and high scrappage anticipated prior to defect correction.

- **Simulated field testing**

A simulated field test is devised to eliminate third parties from the testing process. However, simulated field testing methods are time-consuming and expensive, as they tend to be associated with high levels of tools scrappage. Despite the encompassing nature of the simulated tests, several approximations are needed, and hence, the accuracy of tool testing is compromised.

- **Accelerated tool wear testing**

With the intention of reducing tool testing time accelerated tool wear (ATW) testing protocols are employed by the tool manufacturers but a time-consuming post-mortem analysis is a prerequisite to link the failure characteristics of the cutting tool to the respective manufacturing processes. A typical example of accelerated tool wear testing is shown in Fig. 2.

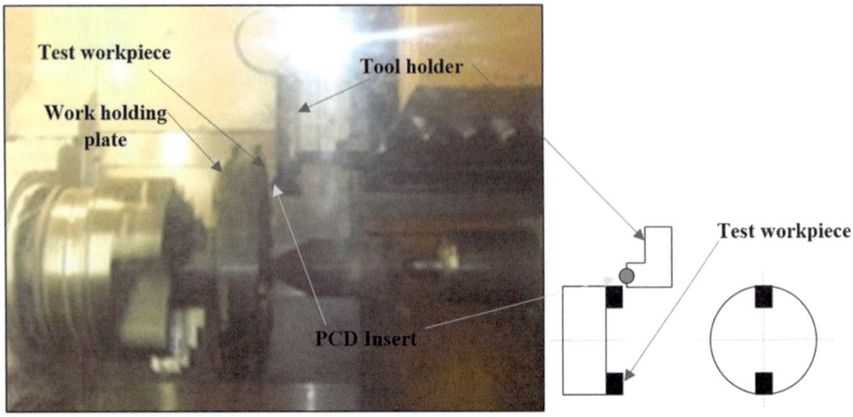

Fig. 2 Accelerated cutting tool testing employed by the cutting tool manufacturers

2.2 *Laboratory-Based Testing of PCD Tools*

The laboratory approaches to investigating cutting tool product performance are intended to develop an understanding of the failure behavior of a given cutting insert during testing.

- **Nondestructive testing (NDT)**

Nondestructive tests aim to identify and remove from circulation, defective cutting tools prior to conducting destructive tests that focus on identifying and correcting significant insert flaws.

- **Visual inspection**

Visual inspection processes in laboratory settings progress with line operators inspecting the PCD tools as they come off the production line. Scanning electron microscope was used to accurately measure the wear of the diamond-like coatings (Thorwarth et al. 2015). Material properties, including porosity and particle grain size, can be confirmed using SEM analysis, with the added advantage of inspecting the surface for the presence of micro-crack defects, likely to propagate in fracture failure. This analysis is largely unsuited for use in tracking the severity of wear between machining passes, as the time and cost commitment is high. Not all experimental or industrial environments have this equipment available for use. A more economical alternative for monitoring flank wear is a toolmaker's microscope.

- **Raman spectral analysis**

The Raman spectral analysis and applications of ultrasonic tests have had some success in identifying defects especially for the PCD matrix (Radtke 2006). Incomplete matrix mixing and nonuniform material concentration flaws can be detected, but there are limitations in their ability to locate the vertical positioning of the defect within the matrix layer. Obvious defect detection can help rectify poor quality cutting tools, but cannot assist in identifying those cutting tools without defects that will still underperform. Before and after Raman spectroscopy studies on fatigued oil and gas cutting tools mainly the PCD cutters exhibited changes in the stress field of the matrix, from general compressive stress fields to residual tensile stress fields (Vhareta et al. 2012).

On an overall note, the nondestructive evaluation confirms the cutting tool's material properties and has some success in identifying obvious external, and to a limited degree, internal defects, but is most powerful in providing baseline reference imaging of the cutting tool's surface textures.

- **Destructive testing**

Standard destructive testing protocols, like breakage tests, tend to focus on information obtained at the point of failure. Information that is more pertinent to the evaluation of wear performance is obtained in the moments prior to failure which presents the behavioral changes in the insert, indicative of imminent failure. Common apparatus for testing tool wear are lathe or milling stations. Lathes have the obvious advantage of generally mounting single cutting tools. Khidhir et al. (2015) chose to turn a single point cemented carbide cutting insert, simplifying their analysis and focusing on the creation of the prediction model for product performance. Milling machine applies a wider range of cutting tools and is capable of using multiple insert tool holders. Che et al. (2012) recommend isolating a single cutting insert, to simplify the correlation of machining signals to wear development, and eliminate noises. In this manner, a single cutting insert operating with minimal milling station features can provide a holistic signal representation of insert failure, from initiation through growth to catastrophic termination of the insert.

- **Wear growth analysis**

Most wear analysis bases itself on empirical relations tracks the progression of tool wear against the cutting distance traversed to generate that level of wear (Arsecularatne et al. 2006). Further machining with the same tool–work combination allows tool performance predictions to be calculated from this wear curve. Where new or used tool–work combinations are tested, the algebraic constant parameters developed by the Taylor relation are unlikely to exist in literature. One method of investigating tool wear is to determine these constants during testing. de Mesquita et al. (2011) have demonstrated the Taylor relation method for computing the tool life on turning ABNT 1038. The method tracks the flank wear progression against cutting time to plot a Taylor tool-life curve, and algebraic constants were computed. Using the Taylor relation, further instances of tool life were predicted to plan removal time of the cutting tool from the CNC machining operations. Evaluating these curves for wear behavior shows initially faster wear rates that stabilize to almost constant wear rates, until rapidly generated wear occurs just prior to failure, or alternatively, a catastrophic failure occurs (Kato and Adachi 2001).

Quantifying the extent and severity of wear mechanisms defines the rate of wear generation and the mechanism of insert failure.

- **Flank wear analysis**

ISO standards require that the tool maintain its cutting edge as long as the limit of flank wear not exceeding VB = 0.3 mm (Juneja 2005; Palanisamy et al. 2008), at which point the tool is declared worn and substituted for a sharp tool. Standard flank wear is measured linearly down the flank face from the cutting edge to the extremity of the wear land. A marked limitation of this approach is that the position of maximum wear can misrepresent the severity of the wear. Kuttolamadom (2012) suggested taking the average linear wear measurement as the flank wear across a series of points along the flank face, generating more realistic flank wear values. For small flank wear presentations, as with short-cycle quality control tests, this can be challenging to measure with only slight incremental wear rates experienced between passes.

- **Volumetric flank wear analysis**

A past analysis of a volumetric wear approach for the turning TI–6Al–4V with carbide cutting tools, considering the volume of material lost from the cutting tools is a more accurate indicator of the severity of wear. This allows for an accurate three-dimensional representation of the wear land through the cutting tool. Their analysis is driven by the modeling of the dominant microstructural wear mechanism, and its application to predictive modeling. This approach is highly dependent on the process, machine parameters, and tool–workpiece combination (Kuttolamadom 2012). Indeed, Adesta et al. (2010) quantified their simulated flank wear through volumetric analysis during high-speed hard turning. This approach must be customized to both the tool geometry and relative orientation of the cutting plane to describe the extent of his insert wear by integrating over both dimensions of orthogonal cutting process generating a 3D wear volume (Palmai 2014). Unique tool geometries can prove challenging to map as a 3D wear volume and small inaccuracies in the wear measurements can be magnified through calculation of the volume. Burger et al. (2009) used a Zygo NewView 7200 white light interferometer 3D imaging technique to map the worn areas of their titanium milling inserts. This allows them to characterize failure mechanisms for each of the tool edges and complements the unusual geometry of their cutting inserts that incorporate chip breaker features and clamping sites, which would otherwise make modeling the volume of the cutting inserts lost, a lengthy and problematic exercise. Unfortunately, the use of highly specialized imaging software adds time delays and costs to the data processing. 3D mapping should only be applied to those worn areas not easily determined through geometric analysis. Monitoring wear volume during machining cannot be done through imaging, as the wear face will always be in contact with the machining surface. Additional analytical approaches are needed to both categorize the wear mechanism and monitor its development during machining.

- **Tool condition monitoring**

Research by Maropoulis et al. (1996), Arsecularatne et al. (2006) and Byrne et al. (1995) illustrates this point by discussing the vast array of approaches to condition monitoring of cutting tools. Most approaches include indirectly monitoring wear via cutting forces, temperature, vibrations, surface finishes, acoustic emissions, and subsequently building relations between these signals and direct imaging of wear progression through the tool. Byrne et al. (1995), also championed the incorporation of torque, current, power, speed, touch probes, microphones, and smoke sensors. Harnessing this behavior in condition monitoring systems allows accurate, timeous quality control checks to be developed.

Condition monitoring strives for accurate real-time feedback of tool condition. Most investigations begin with dynamometer installations that offer good force/thrust tracking. This sensor offers real-time force monitoring that is independent of cutting conditions and drill geometry (El-Wardany et al. 1995). Ivester et al. (2000) used a Kistler three-axis dynamometer to record the machining forces for various rake angles in turning AISI1045 with coated and uncoated WC/CO Kennametal grade K68 turning inserts. The force comparison to wear land formation generates a calibrated database for changing rake angles. Mandal et al. (2011) used three-axis dynamometers to isolate tangential, radial, and longitudinal forces during lathe boring of Ti6Al4 V and lathe turning of zirconia toughened alumina. Wang et al. (2013) isolated the orthogonal x, y, z forces when milling stainless steel (HRC52) with ball nosed tungsten carbide cutters. Generating force data through dynamometers is an accurate, real-time indication of force increases. Correlating this data to wear generation often requires further data analysis. Wang et al. (2013) used a preliminary off-line calculation to build and calibrate their wear model, with the intent to operate the model in real time thereafter. Additional indirect wear monitoring can be achieved through acoustic emission (AE) sensors and accelerometers. Jemielniak et al. (2008) successfully compared acoustic emission signals to force signals, tracking wear development in micro-milling applications. Here, the change in amplitude from signal mean to signal peak, in both the positive and negative directions was used to “train” their system to recognize wear behavior, after each pass. The signal changes were graphed against the subsequent used tool life (where the used tool life is equal to, the current tool life (in minutes) over the total possible tool life), and approximated using a second-degree polynomial. Focusing on the change in signal was thought to eliminate baseline variations that develop due to uncontrollable variations in cutting conditions, presenting only the most relevant signals features are extracted for wear information (Jemielniak et al. 2008). Govenkar et al. (1996) discussed the applications of AE sensors to nondestructive and destructive testing protocols. AE sensor’s sensitivity to the sudden strain energy propagation of stress waves allow them to be applied to structures surrounding the machining interface, like exposing the acoustic emission sensor to the machining interface via a “jet of cooling fluid” and monitor the machining process. Govenkar et al. (1996) was able to identify the types of chips ejected from the interface, successfully differentiate between spiral chips of 160, 8, and 3 mm in length, with FFT analysis recording higher amplitudes for shorter chip lengths. Other studies have considered AE sensors mounted indirectly through the cutting fluid and directly seated on either the tool, or workpiece. Inasaki (1998) has established the effective use of AE sensor on capturing chip formation signals regardless of its placement. Furthermore, the AE sensors was effectively used on distinguishing the chip formation signals from the surrounding signals. AE monitoring of turning recorded wear signals isolated from machine vibrations and external vibrations. This is owing to the AE operating frequency range, which falls above the operating vibration frequency of the lathe. In a study comparing the ability of AE sensors, accelerometers, and spectrometric oil analysis to detect pitting wear in spur gears, the AE sensor was the only one responsive enough to distinguish pitting wear in time to remove the gear before performance suffered (Tan et al. 2007). Li (2002) attempted experimental investigations into delamination wear phenomena. The speed at which delamination events occur from the trigger event to the surface separation makes difficulty in prediction of event. The ensuing signal chaos is a secondary contributor, as the delaminated surface interacts with the machining interface, causing damage. Successful tests found that when monitoring the AE signal output in both the frequency and the time domains, it exhibits distinct amplitude drops in the signal output, immediately following delamination events. Ravindra et al. (1997) had success with using their AE sensor to track flank wear, with their rise times resembling rapid run in wear (stage 1 of flank wear), steady wear progression, and finally chipping or catastrophic tool failure. AE sensors offer real-time operation in frequencies that immediately exclude structural resonances and background noise generated by surrounding machines. AE signals are sensitive to wear fluctuations and fracture events, with non-directionality that reduces the number of sensors required. AE sensors are compact and easy to mount, but are largely dependent on the process parameters, with reference wear models requiring experimental calibration before implementation. An alternative to the AE sensor is an accelerometer. Bierman et al. (2013) successfully measured the vibrations in five-axis machining, where the frequency recorded is a multiple of the spindle rotation frequency. Comparing this accelerometer to contact-emission sensors commonly used to tune musical instruments, they found good correlation between sensors, although the positioning of the contact sensor very close to the location of the vibration was necessary to ensure accuracy. The acoustic emission sensor provides reasonably accurate results, regardless of positioning. Fang et al. (2012) found a triaxial accelerometer sensitive enough to track orthogonal vibrations in high-speed machining of Inconel 718, using wavelet analysis and comparing this output to the cutting force data to evaluate the extent of the tool wear. Touch probes are another consistent vibration monitoring option that is easy to implement and rarely requirement machine modifications. Their dependency on specific experimental materials and cutting conditions limits their application to generic wear monitoring, while their low sensitivity limit their applications to monitoring high performance tools (El-Wardany et al. 1995). Analyzing the quality of a machined surface finish is a good indicator of insert cutting performance, but is heavily influenced by tool geometry, cutting speeds, and feed rates. Colding (2004) suggested that average surface roughness values must be related to tool radius before accurate wear implications can be made. The presence of burr formations (Chen 2005) or material adhesions can indicate that the cutting interface temperature rose too high during machining, implying a worn tool was operated, or nonideal failure occurred. This artificially raises the average surface roughness values, with large peak-to-peaks expected. Ozel and Karpat (2005) used calibrated models to begin predicting the surface finish after training a neural network with experimental data and monitoring flank wear. They noted that surface finish's reliance on geometric and machining parameters make this difficult to apply to new tool–work combinations (Ozel et al. 2007). Childs et al. (2008), Khidhir et al. (2015) and Kuttolamadom et al. (2015) obtained their experimental average surface roughness parameters using various profilometers, all of which require testing to be conducted post-machining, adding time delays and increasing the personnel requirements for quality testing protocols. Palanisamy et al. (2008) used surface finish to target their optimal cutting conditions before beginning to build their tool wear prediction model. Surface finish used in this manner could help identify accelerated wear testing conditions designed to reduce the testing cycle for quality control processes. A final method of directly determining the influence that wear generation in cutting tools has on the machining process is to measure the increase in temperature at the cutting interface during each successive cutting pass and establishing relations between this temperature behavior and the quantity of wear presenting in the tool. During the turning of AISI 1045 with coated and uncoated WC/Co inserts, Ivester et al. (2000) incorporated thermocouples into the bi-conducting tool–chip interface, operating as the junction for the thermocouple and tracking the voltage readout. The results of which were lower amplitude variations than produced across force signatures and insufficient data from the coated inserts. Thermocouples are time-consuming to embed into the workpieces and cannot necessarily be mounted close enough to the machining interface. Thermal cameras providing real-time temperature evaluation must be positioned in line-of-sight of the cutting interface without forward protective barriers. Lin and Ting (1995), among others, noted that placement of sensors in the appropriate orientation and correct proximity to the cutting interface are vital to obtain precise experimental results. El-Wardany et al. (1995) commented that it would further reduce the impact of noisy signals, and false alarms generated by natural frequencies of the spindle and motor within the machine. This can present positioning challenges in close machining or drilling environments, where tools are expected to bore down into or through workpiece materials, while the monitoring system is expected to continue to operate.

One commonality through the research so far is that it mostly pertains to flank wear, although Lu and Chou (2011) considered on delamination failure events. The designer of quality control tests should be mindful that at some point, inferior tools failing by nonideal wear mechanisms will pass through the evaluation. The system must be able to handle, identify, and alert the operator to the discovery of nonideal fracture failure of the PCD tool. Table 1 summarizes the past PCD tool testing methods

Table 1 PCD tool testing methods

PCD tool testing method	Principle	Advantages	Shortfalls	Reference
Industrial approach	Field testing	Reasonably accurate	Difficult to identify the exact source of failure	Huges (2014)
Industrial approach	Simulated field tests	Reasonably accurate	Difficult to identify the exact source of failure	Huges (2014)
Visual inspection	SEM analysis	Enable to identify the grain size, porosity, and micro-cracks	Lack of information on wear resistance and strength	Thorwarth et al. (2015)
Raman spectral analysis	Optical test	Enable to identify the defects with the PCD matrix	Difficult to establish the internal defects of PCD	Radtke (2006) and Vhareta (2012)
Machining test	Continuous turning	Enable to establish the wear resistance in the form of flank wear	Consumes enormous time as the PCD tool has to undergo an image analysis to understand its failure characteristics	Khidhir et al. (2015)
Machining test	Process parameters study	Enable to establish the wear generation phenomenon	It is equivalent to a postmortem study and consumes time	Byrne et al. (1995)
Condition monitoring	Process parameters study using dynamometers	Accurate	Suitable only for the laboratory	El-Wardany et al. (1995)
Condition monitoring	Process parameter study using AE sensors	Less expensive method	Erratic results at times of delamination	Li (2002)

The PCD tool testing methods described in Table 1 are centered on in-process manufacturing tests and consume time. As a result, the scrap rate of PCD tools between the time of fault finding and corrective action on the production floor is high. Therefore, intelligent approaches were applied to estimate tool wear states. Application of neural networks (NN) and clustering techniques such as multilayer perceptron (MLP), adaptive resonance networks (ART2), support vector machine (SVM), and self-organizing maps (SOM) was attempted for establishing the tool wear maps (Si et al. 2011). Also, hidden Markov models (HMM) were successfully applied as it represents the temporary signal dynamics and reasoning in speech recognition. These methods were used for prognostic evaluation of tool wear, finding 92.5% experimental compliance with wear signals (Zhu et al. 2009). A major limitation of these methods is its lack of consideration on fracture failure mechanisms. Intelligent tool testing systems that did not handle the fracture events tend to limit the real-time quality control processes. Research on intelligent tool testing methods, therefore becomes, a priority, emphasizing digital monitoring of PCD wear behavior and extending to appropriate predictive analysis. Accelerating the quality control approach would enable faster rectification of nonideal failure mechanisms experienced by the inserts. Furthermore, the application of necessary corrective actions on the manufacturing floor becomes significantly faster than the current protocols allow. This, in turn, minimizes the material wastage and prioritizes the PCD life prediction strategies. Therefore, the objectives of this project are to develop an intelligent PCD insert clock-testing system that clearly establishes the failure characteristics of the PCD cutting inserts, as well as predicting the product performance.

3 Experimental Setup

The machining experiments were conducted on Röeders RFM 600 high-speed machining center using a PCD milling cutter head over a linear clock-testing workpiece as illustrated in Fig. 3. Throughout the experiments, the force signatures were captured and amplified, and further analysis was done using the data acquisition software Dewesoft, to understand the tool–work interface behavior. Every cutting experiment was repeated five times, and the average cutting forces of at least three tests with the clearest values were taken. The machined workpiece surface texture was examined using a scanning electron microscope (SEM) and the surface roughness was examined using a Surtronic profilometer. After every 5 m of cutting distance, the flank wear was measured using a tool maker's microscope. The AE sensor tracks machine frequencies above 10 Hz, while the dynamometer x, y-directions track forces from −10 kN up to 10 kN. The z-direction has 15 kN upper and lower limits. The temperature at the cutting interface was estimated by assessing the color of the sparks generated during testing and comparing these to the known glow point of silver steel. The accelerometer was calibrated using a Rion VE10 vibration generator.

This milling cutter head was configured to hold a PCD insert of size D16 × 12 mm at a negative rake angle and mounted on HSK 40 tool holder. The milling conditions for performing the machining test are given in Table 2.

4 Results and Discussion

The experimental results have enabled to create a quality control approach that analyzes PCD insert cutting performance, with a view to incorporating accelerated wear generation and combining this with intelligent condition monitoring in order to predict the product performance. The results include the output of extensive experimental findings and its signal correlation with the flank wear behavior.

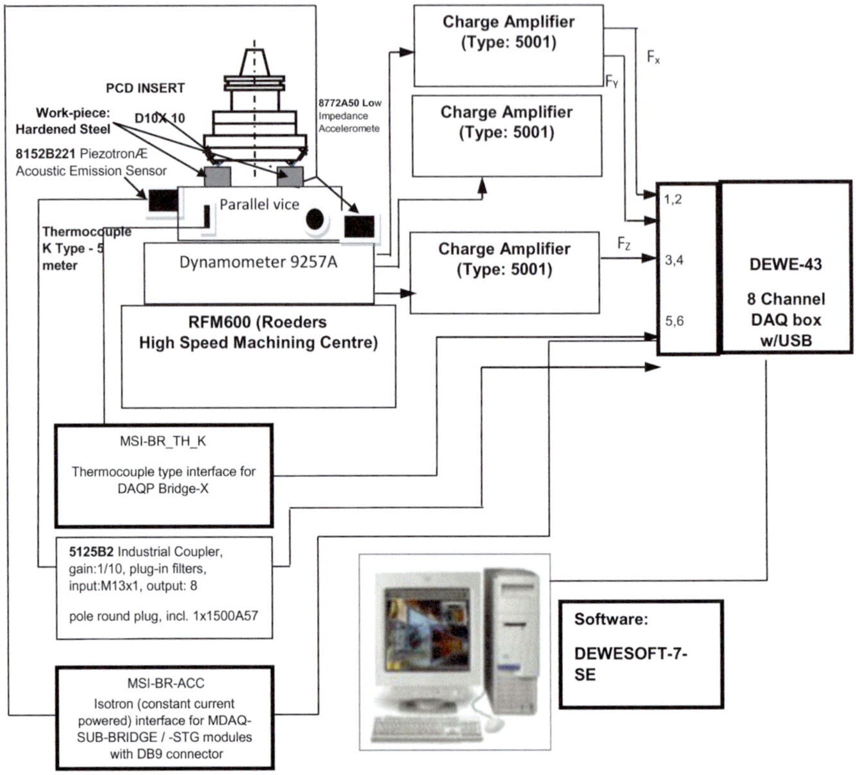

Fig. 3 Experimental setup and data acquisition arrangement for the machining experiments

Table 2 Milling conditions for the accelerated wear tests

Detail	Unit	Value
Milling type	–	Intermitted cutting (clock testing)
Cutting speed	m/min	534.2
Feed rate	mm/min	1200.00
Depth of cut	mm	0.05
Material removal rate	mm^3/min	60.00
PCD insert size	mm/mm	D16 × 12
Workpiece material and hardness	–	Medium carbon alloy steel and HRC 45

Furthermore, ANOVA analysis was performed for validating the AE sensor signatures, and neural network analysis was done to find the performance prediction of the PCD inserts.

4.1 Flank Wear

Flank wear is the gradual, consistent loss of material from a cutting tool, which degrades the cutting edge and distorts the tool geometry. Flank wear is usually characterized by abrasion, sudden catastrophic failure and brittle fracture of a PCD insert, and the growth of flank wear indirectly explains the physical properties of a PCD insert such as wear resistance and transverse rupture strength. In order to distil the effect of built-up edge formation, volumetric flank wear, and VB, accelerated wear tests were performed, and the results are shown in Figs. 4 and 5. In general, the localized cracking around diamond grains or failure of the binding matrix will also result in the smallest diamond particles becoming loose and falling out during machining. All of these effects combine in the abrasion wear of a PCD insert as shown in Fig. 6.

Using the PCD material information, the following relationship for the calculation of volumetric wear under abrasive wear mechanisms is presented by Stolarski (2000):

$$V_{\text{abrasive-theoretical}} = \frac{m^2 * \sigma_y * E * F_n^{3/2} * l_c}{K_{\text{IC}}^2 \, H_{\text{PCD}}^{3/2}} \tag{1}$$

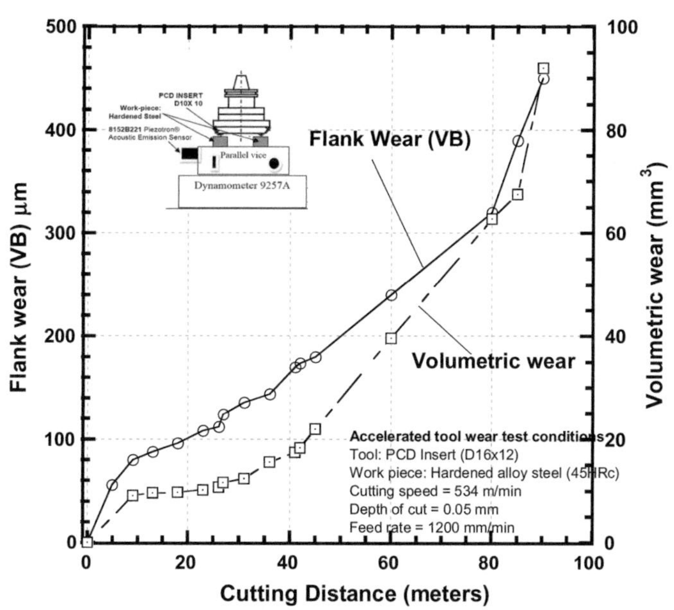

Fig. 4 Flank wear behavior of a PCD insert at accelerated tool test conditions

$V_{abrasive}$ represents the volume lost from the cutting insert (m^3), while m is the work hardening factor, equated to the workpiece hardness over cutting insert hardness. σ_y, is the yield strength (GPa) of the cutting insert, of PCD owing to its nature as a brittle material, E is Young's modulus (GPa) for PCD, F_n is the normal force (N) exerted on the workpiece during machining and l_c is the cutting distance (m) traversed. K_{IC}, denotes the fracture toughness (MPa $m^{1/2}$) experienced by the PCD material and finally H_{PCD} represents the hardness (GPa) of the PCD matrix.

Abrasion wear volume can be corroborated through direct imaging and measurement of the wear region under a toolmaker's microscope. Post-machining weigh-ins can be conducted, comparing the final worn PCD insert weight to the weight of the insert prior to machining, and calculating the volume of the lost material based on the density of the PCD material matrix. This does become slightly more complicated once the WC-Co base becomes involved in wear mechanisms, however, the use of accelerated wear testing protocol should limit wear to the PCD matrix. Geometric analysis can model the volume of PCD material lost from a theoretically ideal cutting insert, by assuming the PCD cutting inserts as a simple cylindrical shape with an inclined wear plane. In this manner, the lost volume would approximate a cylindrical wedge tracks the progression of the wear from zero to a completely used insert as the inclined machining interface progresses through a PCD insert (Fig. 7).

By considering the geometrical form of the worn portion of the PCD tool, the volumetric wear is given as

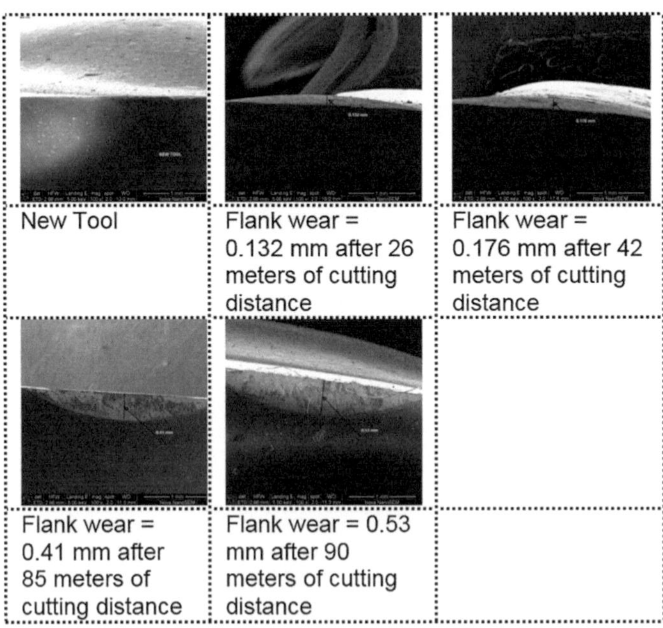

Fig. 5 Flank wear images of a PCD insert at accelerated tool test conditions

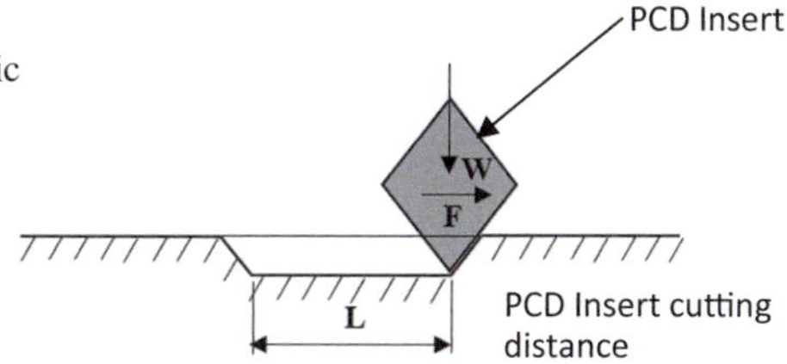

Fig. 6 Geometric arrangement of volumetric abrasion wear determination

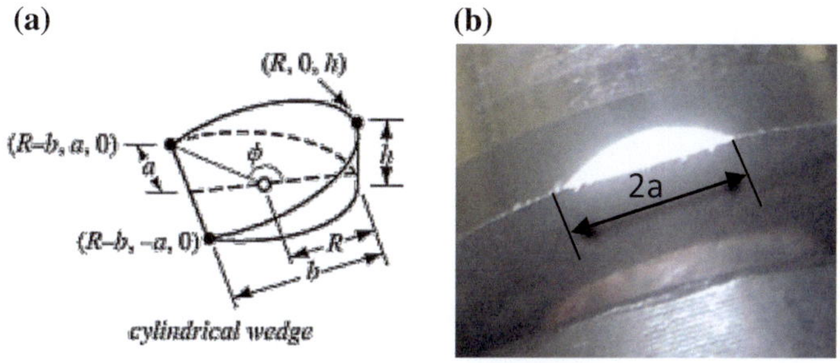

Fig. 7 Dimensional correlation for volumetric wear computations, **a** theoretical model and **b** actual wear

$$V_{\mathrm{PCD}} = \frac{hR^2}{3}\left[\frac{3\sin\theta - 3\theta\cos\theta - \sin 3\theta}{1-\cos\theta}\right] \tag{2}$$

The values for a, b, h, and θ was obtained from the PCD worn insert image as it corresponds to dimension 2a indicated on the flank wear image (Fig. 7b) (Wolfram Math World 2013) (see Appendix 1 for more details).

- **Wear modeling and signal interrelations**

The tool–work interface conditions were monitored in terms of cutting forces, acoustic emission signals, and correlated with the flank wear growth behavior of the PCD cutting inserts. Specific relationships between these signals and the cutting distance were established using the data-driven models. The developed data-driven model was applied to predict the remaining useful life of the PCD tool. Shown in Fig. 8 is the behavior of cutting forces and AE_{RMS} signals for the PCD tools at various cutting distances. The cutting forces behavior suggests that as the cutting distance increases the cutting forces increases consistently with each cutting pass until a maximum force is reached. Continuing machining resulted in an additional signal spike and beyond this signal, the failure of the PCD insert was apparently seen. Between the cutting distances from 0 to 30 m, the increase in cutting forces was found to be higher than the cutting distance range 30–80 m. AE_{RMS} has exhibited a similar trend and confirms the suitability of using AE_{RMS} signals as a measure of cutting forces and flank wear growth behavior.

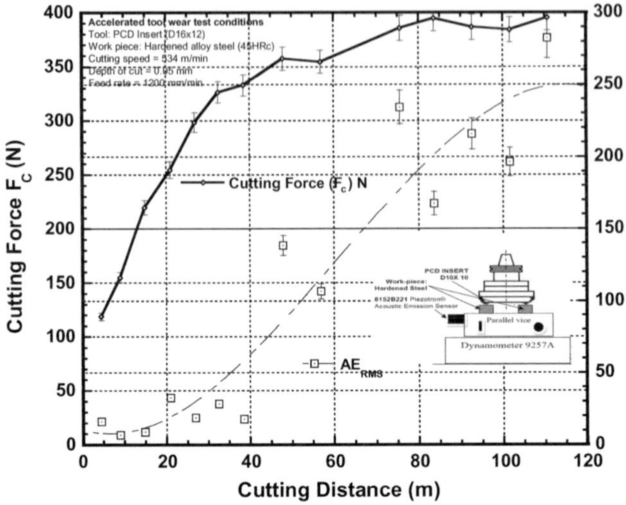

Fig. 8 Cutting forces and AE_{RMS} behavior for various cutting distances

- **ANOVA analysis**

To further demonstrate the correlation between force signals and AE sensor signals, ANOVA tests were performed. As a result, the dynamic digitized map of the tool–work interface signals of the reference samples was clearly sketched inclusive of errors and variation. Shown in Table 3 are the experimental results of cutting forces and AE_{RMS} signals for various cutting distances.

Table 3 Variables factors of the experiments and results

Cutting conditions	Cutting distance (m)	Cutting force F_c (N)	AE_{RMS} (dB)	Flank wear (μm)
Fixed cutting conditions: intermitted cutting (clock testing); cutting speed = 534.2 m/min; feed rate = 1200.00 mm/min; depth of cut = 0.05 mm; material PCD insert: D16 × 12	4.8	118.78	16.344	56
	9	154.99	6.7500	80
	13	219.96	8.9300	88
	18	254.56	32.688	96
	23	298.33	18.910	108
	27	326.22	28.370	124
	31	332.85	17.920	136
	36	357.57	138.48	144
	41	354.54	106.64	170
	45	385.61	234.38	180
	60	394.81	167.57	240
	80	386.49	215.82	320
	85	384.08	196.54	390
	90	395.45	282.44	450

The cutting force (F_c) is the resultant of the measured forces F_X, F_Y, and F_Z. AE_{RMS} signals were measured at the nearest point of machining zone as the AE sensor was mounted on the workpiece. In each machining experiment, five sample signals of F_X, F_Y, F_Z, and AE_{RMS} were taken, and F_c and AE_{RMS} values were computed. The columns F_c and AE_{RMS} shown in Table 3 indicate the mean value of F_c and AE_{RMS} of each experiment. A one-way ANOVA test was applied to compare the effect of a single factor on the different groups: cutting force (F_c) and AE_{RMS} signals. The findings of ANOVA comparison tests on (i) cutting force and flank wear, and (ii) flank wear and AE_{RMS} are shown in Tables 4 and 5, respectively.

The p-values of the parameters resulted from the ANOVA are presented in Tables 4 and 5. The smaller the p-value, the smaller the probability of making mistakes by rejecting the null hypothesis, and consequently, the larger the corresponding coefficient. By evaluating the p-values of the parameters: cutting force signatures and AE_{RMS} signatures, it is found that the cutting force with the p-values smaller than 0.05 exhibits a larger effect on the flank wear than AE_{RMS} signatures. Also, the p-values for AE_{RMS} signatures show the correlation between the cutting force and AE_{RMS} signatures and suggest that the flank wear growth can be studied by both cutting force and AE_{RMS} signatures with reasonable accuracies. Tables 4 and 5 also give the F-value which is the ratio of the groups' mean square over the error mean square. As the value on both cases was found to be more than 1, it suggests that the samples were drawn from a different population although the measurement of AE_{RMS} signals is closest to the cutting zone whereas the cutting forces are the measurement at the cutting zones. The F-values computed using the t-test for the unpaired groups (i) cutting force and flank wear (ii) AE_{RMS} and flank wear was found to be 1.775 and 1.518, respectively. Also, the homogeneity of variance was tested in SPSS by applying the Levene's statistic for nonparametric data (NIST 2014). If the significance p, of statistic L, remains above 0.05, for a 95%

Table 4 ANOVA comparison tests on cutting force and flank wear

Source of variation	Cutting force F_c (N) and flank wear				
I	DF	Seq SS	AdjMS	F	p
In between	1	113,442	113,442	9.8	0.004
Within	26	300,923	11,573		
Total	27	414,365	15,346		

Table 5 ANOVA comparison tests on AE_{RMS} and flank wear

Source of variation	AE_{RMS} and flank wear				
I	DF	Seq SS	AdjMS	F	p
In between	1	44021.0	44021.0	3.58	0.06
Within	26	319,235	12,278.2		
Total	27	363,256	13453.9		

confidence interval, indicating no significant variation in the data for all inserts, the homogeneity assumptions for comparative analysis would be satisfied. The Levene's statistic (L) equation is given as (NIST 2014)

$$L = \frac{(N-k)\sum_{i=1}^{k} N_i (Z_i - Z)^2}{(k-1)\sum_{i=1}^{k}\sum_{j=1}^{N_i} (Z_{ij} - Z_i)^2} \tag{3}$$

where k represents the number of sets from which the data come, N is the total number of sampled cases, and Z is an array of the mean and median values of the sampled case i from group j. The test values for homogeneity of variance in PCD wear data, using the Levene's statistic was found to be, 0.299, 0.070, 0.952, 0.763, and 0.564 for PCD Insert #1 to Insert #5, respectively. Analyzing the value of the p factor shows no datasets of particular significance, with all p factors consistently above 0.05 in value. As a result, the homogeneity of variance is consistent across all experimental data from PCD cutting inserts, #1 to #5, and satisfies that the data be drawn from population exhibiting equal variance. A similar Levene's test on data: cutting forces (F_c) and AE_{RMS} signals also conclude that the data were drawn from population exhibition equal variances.

- ***t*-test**

The "t" factor analysis assesses the variability of the data within each machining cycle. The distribution of the data points within each pass is normally distributed, and the "t" factor analysis can be applied directly. The "t" factor is computed using Eq. 2 as (IBM 2015)

$$t = \frac{(x - \mu)}{(s/\sqrt{n})} \tag{4}$$

where x is the sample mean, and the population mean is denoted by μ, with "s" being the standard deviation, and finally, n represents the size of the sample. The "t" factor for each distinct machining pass for the PCD tool is presented in Fig. 9

The cutting force "F_c" and "t" factor for F_c were tracked against the flank wear, which shows the general correlation between the two, and enable to quantitatively express the cutting force signatures as a wear. On a similar note, the AE_{RMS} signals and t-factor for AE_{RMS} have a correlation with the flank wear growth and doubly confirm the use of AE_{RMS} signals as a measure of flank wear. The same was also expressed quantitatively in terms $t_{AE\text{-}RMS}$ factor.

4.2 *Frequency Domain Analysis*

Between the cutting distance from 5 to 80 m the frequency signal spikes occur at $f = 48, 97$ Hz, for both the AE_{RMS} and cutting force (F_c) signals. However, beyond 80 m of cutting distance, the frequency signals spikes occur at 45, 95, and 190 Hz confirming the significant flank wear of the PCD tool.

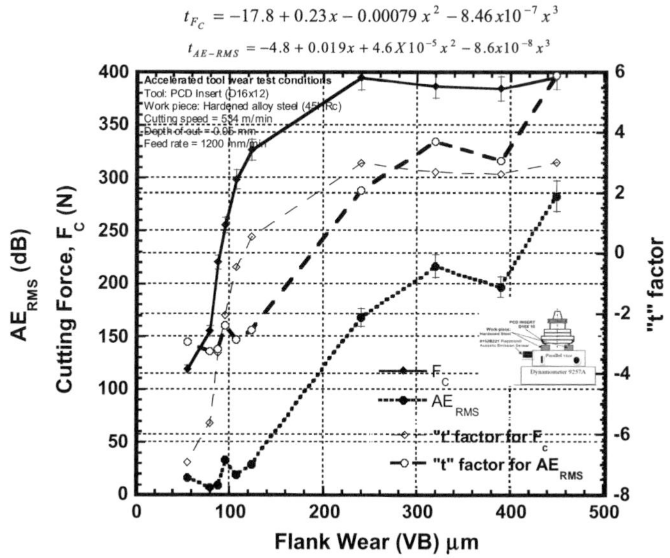

Fig. 9 Cutting force F_c and AE_{RMS} behavior in terms of "*t*" factor

4.3 Predictive Model Generation

Analyzing the signal relationships through multivariate analysis has enabled to build the neural network and a wear model using the machining variables (Liu and Jolley 2015). Shown in Fig. 10 is the PCD wear multilayer perceptron using the neural network architecture.

The hidden layer tracks the data bias between outputs and is used to evaluate the influence of the inputs through weighted sums in a nonlinear model as detailed in Eq. 3 (IBM 2015).

$$P = \sigma\left\{\sum_{i=1}^{l} \lambda_i \sigma(x.w_i)\right\} \tag{5}$$

where l equates to the number of hidden units, the sigmoidal function is given as $\sigma(x) = 1/(1 + \exp(-x))$, such that x are the input covariates and "w" refers to the neuron weights. The number of data points then becomes "i". The bias term is defined by $x.w_i \underline{\underline{\text{def}}} \sum_k x_k w_{ik} + w_{io}$. The input covariates for this model are the machining signals identified such as cutting force signatures and acoustic emission signatures.

The neural network architecture uses a single hidden layer with real-time data training to update the synaptic weight after each data record, until the stopping criteria are met. The training criteria for stopping is set as no more than a consecutive step with no error decrease before the algorithm halts itself. Table 6 shows

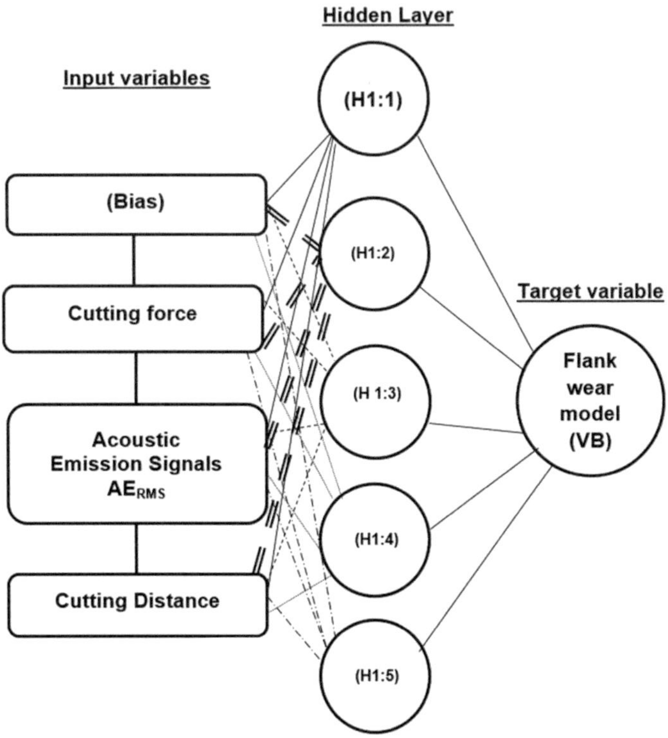

Fig. 10 PCD wear multilayer perceptron using the neural network architecture

the summary of neural network model. The assessment of the model accuracy shows a 6.472 compliance under sum-of-squares methodology creating a 0.087% relative error. The outcome of the neutral network training is an estimation of the model parameters defining the relation between the machining signals variables, the hidden layer, and the wear formation on the PCD cutting insert. These parameters are summarized in Table 7. Understanding the significance of each machining parameter on the magnitude of the wear is critical to ensure the correct emphasis of the variable. Calculating the weighted sum for each of the covariates quantifies their influence on the target variable is given by the following equation (IBM 2015):

$$w_j(x) = \sum_{i=1}^{l} \lambda_i \sigma'(x.w_i) w_{ij} \tag{6}$$

where the sigmoidal function is given as $\sigma(x) = 1/(1 + \exp(-x))$, such that x are the input covariates and w refers to the neuron weights. The number of data points then becomes i, the bias term is defined by $x.w_i \underline{\underline{\text{def}}} \sum_k x_k w_{ik} + w_{io}$, and lambda is the constant term for the data position. Furthermore, the neural network identifies that cutting distance followed by cutting force and acoustic emission signals are the

Table 6 Neural network model summary

Training	Sum-of-squares error	6.472
	Relative error	0.087
	Stopping rule used	1 consecutive step(s) with no decrease in error[a]
	Training time	0:00:00,03
Testing	Sum-of-squares error	2.165
	Relative error	0.058

a excluding the bias unit

Table 7 Neural network estimation of parameter relations: output layer parameter estimates

Predictor		H(1:1)	H(1:2)	H(1:3)	H(1:4)	Predicted hidden layer (H1:5)
Input layer	(Bias)	−0.092	0.951	0.922	−0.141	0.358
	Cutting force	−0.2	−0.522	1.054	−0.172	0.433
	Acoustic emission	0.213	−0.093	−0.037	0.397	0.439
	Cutting distance	0.161	1.086	0.265	0.07	0.409
Output layer: flank wear		−0.706	1.645	0.096	−0.236	−0.426

significant factors of weightage 100, 29.8, and 24.3%, respectively. The imaged flank wear experienced by the PCD inserts during testing are graphed against cutting distance in Fig. 3. The sensor array data signals were modeled against progressive cutting distances and the project values for the cutting force and acoustic emission signals occurring at 80 m of cutting distance and above and projected to be 410 N and 45.05 Hz, respectively.

5 Conclusions

This investigation confirms that the developed intelligent PCD clock-testing method enables to evaluate the PCD insert performance in real time. The feature extraction coupled with statistical evaluation along with the neural network training of a predictive model has become an incredibly powerful performance predictor. Once the calibrated model was developed, the DAQ was automated, with the resulting cycle time reducing to approximately 1 min and 40 s for a complete destructive test and real-time calculation of the insert performance relative to the calibrated wear models for a single polycrystalline diamond cutting insert. This represents a noticeable reduction of quality control cycle times, with the actionable feedback produced in real time to be passed to manufacturing production lines. The accelerated wear test protocol generates representative wear behaviors in the PCD insert, where abrasive flank wear and one instance of insert chipping failure mechanisms were noted without degrading the tool through graphitization mechanisms. Despite the limited fracture events, anecdotal evidence suggests that the sensor array is sensitive enough to extract meaningful signal features from these events, should they occur. Statistical analysis of the data improves the accuracy of the wear models and allows predictive modeling of the flank wear. The strength of this system lies in its ability to effect analysis in real time and its proposed cross-application portability.

Acknowledgements The authors wish to thank Mr. Jonathan Waite University of Cape Town, Cape Town, South Africa for the important contributions made to this research in the initial PCD clock-testing experiments. Further thanks go to Element 6 for the supply of PCD inserts and AE sensors. This project was supported by fund NRF Grant: Incentive Funding for Rated Researchers (IPRR)–South Africa through Reference: IFR150204113619 and Grant No: 96066.

Appendix 1: Computation of Volumetric Wear Using the Principles Cylindrical Wedge

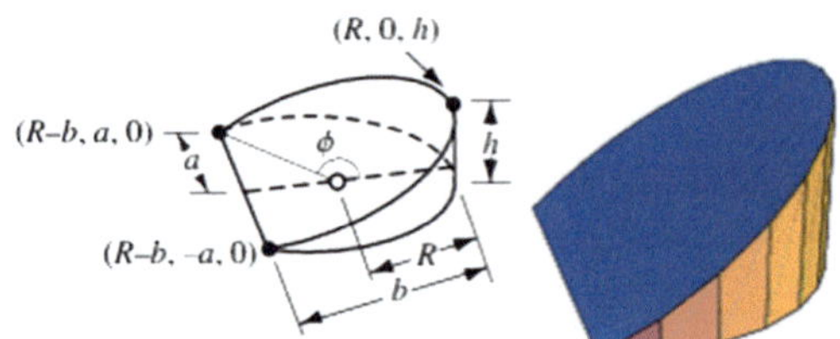

cylindrical wedge

A wedge is cut from a cylinder by slicing with a plane that intersects the base of the cylinder. The volume of a cylindrical wedge can be found by noting that the plane cutting the cylinder passes through the three points illustrated above (with $b > R$), so the three-point form of the plane gives the equation

$$\begin{vmatrix} x & y & z & 1 \\ R-b & a & 0 & 1 \\ R-b & -a & 0 & 1 \\ R & 0 & h & 1 \end{vmatrix} = h(R-b-x)+bz$$

Solving for z is given as

$$z = \frac{(h)(x-R+b)}{b}$$

The value of "a" is given as

$$a = \sqrt{(R^2)\left((b-R)^2\right)}$$
$$a = \sqrt{b(2R-b)}$$

The volume of cylindrical wedge is given as the integral of rectangular areas along the x-axis

$$z = \int Z(x)y(x)\,\mathrm{d}x$$
$$Z = \int_{R-b}^{R} \frac{h(x-R+b)}{b}\sqrt{(R^2-x^2)}\,\mathrm{d}x$$
$$z = \frac{h}{6b}\left[2\sqrt{(2R-b)b}\left(3R^2-2ab+b^2\right) - 3\pi R^2(R-b)\sin^{-1}\frac{(R-b)}{R}\right]$$
$$a = R\sin\theta; \quad b = R(1-\cos\theta); \quad b^2 = 2bR - a^2$$

Hence, $V = \frac{2}{3}hR^2$.

References

Adesta, E., M. Al Hazza, M.A. Riza, and R. Rosehan. 2010. Tool life estimation model based on simulator tool wear during high speed hard turning. *European Journal of Scientific Research* 39: 265–278.

Arsecularatne, J., L. Zhang, and C. Montross. 2006. Wear and tool life of tunsten arbide, PCBN and PCD cutting tools. *International Journal f Machine Tools and Manufacture* 46: 482–491.

Bierman, D., A. Zabel, T. Bruggemann, and A. Barthelmey. 2013. A comparison of low cost structure-borne sound measurement and acceleration measurement for detection of workpiece vibrations in 5-axis simultaneous machining. *CIRP* 12: 91–96.

Burger, U., M. Kuttolamadom, A. Bryan, and L. Mears. 2009. Volumetric flank wear characterization for titanium milling insert tools. In *Indiana: Proceedings of the 2009 ASME International manufacturing Science and Engineering Conference.*

Byrne, G., D. Dornfeld, I. I., G. Ketteler, W. König, and R. Teti. 1995. Tool condition monitoring (TCM)—the status of research and industrial application. *Annals of the CIRP* 44: 541–567.

Che, D., P. Han, P. Guo, and K. Ehmann. 2012. *Issues in polycrystalline diamond compact cutter-rock interaction from a metal machining point of view-part II: Bit performance and rock cutting mechanics*, 134. JMSE: ASME.

Chen, G.-L. 2005. Development of a new and simple quick-stop device for the study on chip formation. *International Journal of Machine Tools and Manufacture* 45: 789–794.

Childs, T., K. Sekiya, R. Tezuka, Y. Yamane, D. Dornfeld, D.-E. Lee, and P. Wright. 2008. Surface finishes from turning and facing with round nosed tools. *CIRP Annal—Manufacturing Technology* 57: 89–92.

Colding, B.N. 2004. A predictive relationship between forces, surface finish and tool-life. *CIRP Annals* 53: 85.

de Mesquita, N., J. de Oliveira, and A. Ferraz. 2011. Life prediction of cutting tool by the workpiece cutting condition. *Advanced Materials Research* 223: 554–563.

El-Wardany, T., D. Gao, and M. Elbestawi. 1995. Tool condition monitoring in drilling using vibration signature analysis. *International Journal of Machine Tools and Manufacture* 36 (6): 687–711.

Fang, N., P. Pai, and N. Edwards. 2012. Tool-edge wear and wavlet packet transform analysis in high-speed machining of Inconel 718. *Journal of Mechanical Engineering* 58: 191–202.

Govekar, E., P. Muzic, and I. Grabec. 1996. Classification of chip form based on AE analysis. *Ultrsonics* 34: 467–469.

Hughes, Baker. 2014. https://www.bakerhughes.com/products-and-services/drilling/drill-bit-systems/pdc-bits.

IBM. 2015. *IBM SPSS neural networks.*

Inasaki, I. 1998. Application of acoustic emission sensor for monitoring machining processes. *Ultrasonics* 36: 273–281.

Ivester, R., M. Kennedy, M. Davies, R. Stevenson, J. Thiele, R. Furness, and S. Athavale. 2000. Assessment of machining models: Progress report. *Machining Science and Technology* 4 (3): 511–538.

Jemielniak, K., S. Bombinski, and P. Aristimuno. 2008. Tool condition monitoring in micromilling based on hierarchical integration of signal measures. *CIRP Annuals—Manufacturing Technology* 57: 121–124.

Juneja, B. 2005. *Fundamentals of metal cutting and machine tools*, 2nd ed, 134. New Delhi: New Age International (P) Limited Publishers.

Kato, K., and K. Adachi. 2001. *7: Wear mechanisms.* Boca Raton: CRC Press LLC.

Khidhir, B., W. Al-Oqaiel, and P. Kareem. 2015. Prediction models by response surface methodology for turning operation. *American Journal of Modelling and Optimization* 3 (1): 1–6.

Kuttolamadom M. 2012. *Prediction of the wear and evolution of cutting tools in carbide/Ti–6Al–4V machining tribosystem by volumetric tool wear.* Tigerprints.

Kuttolamadom, M., M. Laine Mears, and T. Kurfess. 2015. The correlation of volumetric wear rate of turning tool inserts with carbide grain sizes. *ASME Journal of Manufacturing Science and Engineering* 137: 011015.

Li, X. 2002. A brief review: Acoustic emission method for tool wear monitoring during turning. *machine tools & manufacture* 42: 157–165.

Lin, S., and C. Ting. 1995. Tool wear monitoring in drilling using force signals. *Wear* 180: 53–60.

Liu, T.-I., and B. Jolley. 2015. Tool condition monitoring (TCM) using neural networks. *International Journal of Advanced Manufacturing Technology* 78: 1999–2007.

Lu, P., and Y. Chou. 2011. *Analysis of acoustic emission signal evolutions for monitoring diamond-coated tool delamination wear in machining*. Tuscaloosa: University of Alabama.

Mandal, N., B. Doloi, B. Mondal, and R. Das. 2011. Optimisation of flank wear using Zirconia Toughened Alumina (ZTA) cutting tool: Taguchi method and regression analysis. *Measurement* 44: 2149–2155.

Maropoulis, P., and B. Alamin. 1996. Integrated tool life prediction and management for an intelligent tool selection system. *Journal of Materials Processing Technology* 61: 225–230.

National Institute of Standards and Technology. 2014. *Levene test for equality of variances*. www.itl.nist.gov/div898/handbook/eda/section3/eda35a.htm.

National Renewable Energy Laboratory. 2000. *Diamond-cutter drill bits.* www.nrel.gov/docs/fy00osti/23692.

Ozel, T., and Y. Karpat. 2005. Predictive modeling of surface roughness and tool wear in hard turning using regression and neural networks. *International Journal of Machine Tools and Manufacture* 45: 467–479.

Ozel, T., Y. Karpat, L. Figueria, and J. Davim. 2007. Modelling of surface finish and tool flank wear in turning of AISI D2 steel *i*th ceramic wiper inserts. *Journal of Materials Processing Technology* 189: 192–198.

Palanisamy, P., I. Rajendran, and S. Shanmugasundaram. 2008. Prediction of tool wear using regression and ANN models in end-milling operation. *nternational Journal of Advance Manufacturing Technology* 37: 29–41.

Palmai, Z. 2014. A new physically defined equation to describe the wear of cutting tools. *ANNALS of Faculty Engineering HUnedoara-International Journal of Engineering* 12: 113.

Radtke, R. 2006. *New high strength and faster drilling TSP diamond cutters*. Kingwood: Technology International Inc.

Ravindra, H., Y. Srinivasa, and R. Krishnamurthy. 1997. Acoustic emission for tool condition monitoring in metal cutting. *Wear* 212: 78–84.

Si, X.S., W. Wang, C.H. Hu, and D.H. Zhou. 2011. Remaining useful life estimation—a review on the statistical data driven approaches. *European Journal of Operational Research* 213 (1): 1–14.

Si, X.S., W. Wang, C.H. Hu, M.Y. Chen, and D.H. Zhou. 2013. A wiener-process-based degradation model with a recursive filter algorithm for remaining useful life estimation. *Mechanical Systems and Signal Processing* 35 (1–2): 219–237.

Stolarski, 2000. *Tribology in machine design*, 1st ed. Oxford: Butterworth Heinemann.

Tan, C., P. Irving, and D. Mba. 2007. A comparative experimental study on diagnostic and prognostic capabilities of acoustic emission, vibration and spectrometric oil analysis for spu gears. *Mechanical Systems and Signal Processing* 21 (1): 208–233.

Thorwarth, K., G. Thorwarth, R. Figi, B. Weisse, M. Stiefel, and R. Hauert. 2015. On interlayer stability and high-cycle simulator performance of diamond-like carbon layers for articulating joint replacements. *International Journal of Molecular Sciences* 15: 10527–10540.

Vhareta, M., R. Erasmus, and J. Comins. 2012. Use of Raman spectroscopy to study fatigue type processes on polycrystalline diamond (PCD). In *Durban: 18th World Conference on Nondestructive Testing*.

Wang, J., P. Wang, and X. Gao. 2013. Tool life prediction for sustainable manufacturing. In *Berlin: 11th Global Conference on Sustainable Manufacturing*.

Wolfram Math World. 2013. *Cylindrical Wedge*. Retrieved 20 Oct 2014, from http://mathworld.wolfram.com/CylindricalWedge.html.

Zhu, K., Y. Wong, and G. Hong. 2009. Multi-category micro-milling tool wear monitoring with continuous hidden Markov models. *Mechanical Systems and Signal Processing* 23: 547–560.

Focused Ion Beam Fabrication: Process Development and Optimization Strategy for Optical Applications

Vivek Garg, Rakesh G. Mote and Jing Fu

Abstract This work presents an optimization of focused ion beam (FIB) milling process for the fabrication of optical elements. Focused ion beam is a nanofabrication tool involving ion–beam material interaction at atomic or molecular levels. The ion–solid interactions have been investigated for the fabrication of desired 3D geometries. Focused ion beam dwell time is an important parameter and determines the final geometry and accuracy of the fabrication. An algorithm has been developed to optimize the ion beam dwell time for the desired geometry fabrication maintaining high accuracy. Binary Fresnel zone plates (FZPs) have been fabricated using the optimized simulation results for demonstration. The algorithm can be used for fabrication of 3D elements and micro/nano-structures for various applications.

Keywords Focused ion beam · Nanofabrication · Optimization and simulation Micro/nano-optical elements

Nomenclature

a_{ij}	Milling coefficient
A	Matrix to measure the milling coefficients
$F(t)$	Objective function
f	Focal distance
$G(x)$	Standard normal Gaussian distribution
I_B	Ion beam current

V. Garg
IITB Monash Research Academy, Mumbai 400076, India
e-mail: vivekgarg@iitb.ac.in

R. G. Mote (✉)
Department of Mechanical Engineering, Indian Institute of Technology Bombay, Mumbai 400076, India
e-mail: rakesh.mote@iitb.ac.in

J. Fu
Department of Mechanical and Aerospace Engineering, Monash University, Clayton 3800, Australia
e-mail: jing.fu@monash.edu

S. S. Pande and U. S. Dixit (eds.), *Precision Product-Process Design and Optimization*, Lecture Notes on Multidisciplinary Industrial Engineering,
https://doi.org/10.1007/978-981-10-8767-7_8

M	Number of pixels to be machined
N	Total number of pixels
p, Q, r	Multiplication matrices
R	Radius
t	Ion beam dwell time
t^0	Initial value of dwell time
t^k	Dwell time from kth update
T	Ion beam milling time
dx, dy, dz	Milled dimensions (length, width, depth)
Y_s	Erosion rate
z^o	Desired geometry profile
λ	Wavelength
μ	Mean value
σ	Standard deviation
$\Delta r_{\mathrm{Rayleigh}}$	Rayleigh resolution limit

1 Introduction

Focused ion beam (FIB) is an important and advanced technology in the field of micro/nano-fabrication. It has wide range of capabilities from milling and deposition to imaging at micro/nano-scale. The distinct advantage of FIB lies in the fact that it enables mask-less direct fabrication on various materials, making it suitable for applications in the field of nanotechnology. Focused ion beam instrument, when combined with imaging technologies like scanning electron microscope (SEM), offers a wide range of applications for fabrication and characterization of micro/nano-structures. It allows mask-less and one-step fabrication, which makes it a very important technology in micro/nano-fabrication. This is because the FIB milling does not require any intermediate resist layer as in other photo or e-beam lithographic techniques, and thus enables direct writing of micro/nano-structures. Focused ion beam has a short wavelength with high energy density, which helps focusing the beam to a smaller spot size of 5–10 nanometre (nm). It allows control of beam shape and size for fabrication of complex 3D structures. The unique fabrication capabilities of FIB make it an excellent tool for rapid prototyping applications. A great detail of research exists in the literature for FIB principle, its applications and limitations (Melngailis 1987; Lehrer et al. 2001; Reyntjens and Puers 2001; Kim et al. 2012; Mote and Xiaomin 2017). Focused ion beam processing has also been applied for processing and characterization of biological, organic materials, etc. (Fu and Joshi 2010; Al-Abboodi et al. 2013; Adineh et al. 2015), allowing its use for multidisciplinary research.

Focused ion beam is a specialized process and involves several complexities in terms of beam control and fabrication accuracy. Accurate FIB fabrication of optical elements and 3D micro/nano-structures for specific applications require deep insight into the process. The accuracy of FIB fabricated profiles depends on a number of parameters such as scanning strategy, beam current, acceleration voltage, beam dwell time, etc. Out of these parameters, ion beam dwell time is one of the important parameters that determine the final geometry and accuracy of fabricated structures. A better understanding of the milling mechanisms and material removal at micro/nano-level is required for accurate fabrication and desirable performance of these elements. This will allow accurate fabrication of optical elements or any other 3D structure for specific applications in X-ray optics, micro/nano-photonics, etc.

In the present work, an optimization problem is formulated for FIB fabrication and has been solved in MATLAB by an iterative approach for an optimal solution. The algorithm minimizes the deviation of a simulated geometry obtained by FIB fabrication from the desired geometry. The proposed strategy can be further used to fabricate optical elements or any 3D micro/nano-structures. The ion beam dwell time has been optimized for the desired geometry fabrication, maintaining the required accuracy. This algorithm has been used to demonstrate the simulation of Fresnel zone plate (FZP) for specific design. Experiments have been carried out to validate the optimization model. Based on the optimization results, the designed FZP has been successfully fabricated.

The remaining chapter is organized in the following way. Focused ion beam micro/nanofabrication basics, process capabilities and limitations along with important FIB operating parameters and their effects have been discussed and assessed critically in Sect. 2. Section 3 presents the formulation of optimization problem and iterative solution approach for FIB nanofabrication. Simulation results have been illustrated in Sect. 4. Focused ion beam experiments and results have been included in Sect. 5. Finally, a conclusion and outlook are discussed in Sect. 6.

2 Focused Ion Beam Nanofabrication—Basics

The focused ion beam technique utilizes a finely focused beam of ions for milling, imaging, deposition, implantation, etc. A high-energy ionized beam of ions in a FIB system is produced from a liquid metal ion source (LMIS) by the application of electric field. This beam is focused onto a target surface, which results in a number of phenomena such as sputtering of constituent atoms, generation of collision cascade, surface modification, etc. Heavy ions also produce secondary electrons from the target surface, which can be used for imaging applications. Focused ion beam can also be used for deposition, implantation and lithography applications. In this section, FIB instrument, basic principle, ion–solid interactions and important parameters for FIB fabrication are discussed.

2.1 Focused Ion Beam Instrument

A typical dual beam FIB-SEM instrument has a liquid metal ion source (LMIS), an ion column, an electron column, a stage, detectors, a gas injection system (GIS), a vacuum system and a computer. An illustration of a Carl Zeiss Auriga dual beam FIB-SEM system (Compact-4558) available at IIT Bombay is shown in Fig. 1a.

The focused ion beam system is similar to a scanning electron microscope, where an electron beam is used. Both the systems employ charged particles, which can be used for imaging, etching and deposition applications. Use of ions in FIB offers several advantages over the electrons, due to the reason that ions are comparatively massive and larger. Ions have much more direct effect on the sample as compared to electrons, whereas electrons penetrate deeper due to their smaller size. Energetic ions carry higher momentum than electrons, which is used for material removal and milling applications. Another difference between the two systems lies in the beam steering and focusing optics. In SEM, magnetic fields are used to focus the electron beam due to smaller mass of electrons. On the other hand, charged ion beam requires stronger field; hence, electrostatic lenses are used. In a dual beam system, FIB and SEM columns are arranged at an angle of 52° to each other, which allows milling/deposition/etching by FIB and simultaneous observation through SEM. The sample is made normal to FIB by tilting the sample stage at 52°. A typical configuration of FIB-SEM columns and sample inside the vacuum chamber in a dual beam system is shown in Fig. 1b.

A highly focused ion beam is obtained from LMIS, which has a tungsten needle. The tungsten (*W*) needle is attached to a reservoir of source material. Gallium (Ga) is most commonly used FIB source because of the following reasons:

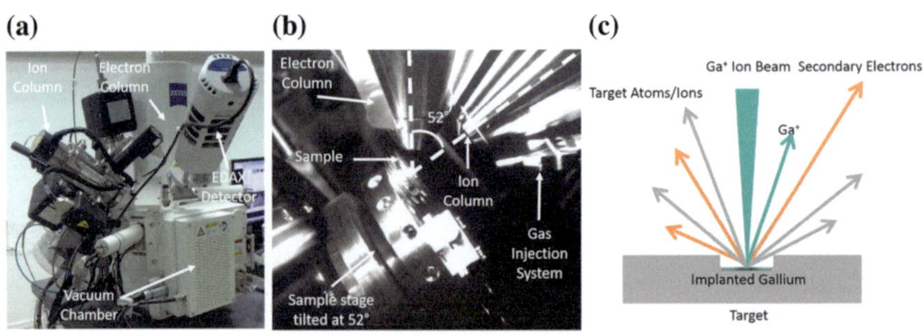

Fig. 1 Focused ion beam (FIB) system at IIT Bombay: **a** Carl Zeiss Auriga Compact-4558 FIB-SEM system, **b** the dual beam-sample configuration inside vacuum chamber and **c** schematic illustration of FIB interaction with target material: sputtering of target atoms/ions, Ga ion implantation and generation of secondary electrons from target surface

- Low melting point (29.8 °C)—It requires less heating and avoids any interaction of Ga with tungsten needle.
- Low vapour pressure—It can be used in pure form and promotes long life, avoiding the Ga evaporation.
- Gallium has suitable mechanical, electrical and vacuum properties. Gallium emission provides high angular intensity with a small spread of energy.
- Low volatility at melting point—long source life.

The heated LMIS flows and wets the tungsten needle, having a tip radius of 2–5 μm. Gallium requires less reheating and remains liquid at ambient conditions for weeks due to its supercooling properties. If an electric field (10^8 V/cm) is applied at the end of wet tungsten needle, it will cause Ga to form a point source (2–5 nm) of conical shape (Taylor cone). This shape is formed as a result of balance between surface tension of the liquid metal and the electrostatic force set up by the applied electric field. Once the cone tip is formed and is sharp enough, the extraction voltage pulls Ga ions from the tungsten tip by ionization process. The Ga ions are then passed through the ion column, where an acceleration voltage (5–50 kV) is applied. A set of apertures and lenses are used in the ion beam column to focus the ion beam of various diameters at target surface. The spot size (diameter) and the shape of the beam govern its machining and imaging capabilities.

2.2 *Ion–Solid Interactions*

The incident ion beam loses its energy to target electrons and atoms, when scanned over the sample surface. Most ions are deflected/backscattered from their path as a result of collision with target atoms and some ions penetrate over a certain distance inside the sample. As a result of collisions, sample constituent atoms are displaced from their lattice positions. The collision also results in recoil and sputtering of constituent atoms, formation of defects, electron emission and excitation, photon emission, etc. (Melngailis 1987). Mixing of constituent atoms, phase transformation, crystallization, amorphization and permanent damage might also occur as a result of thermal- and radiation-induced effects. Ion implantation on target surface due to FIB operation and sputtering results in change of surface morphology and craters are formed as a result of atoms being removed. A schematic of focused ion–beam interaction with sample surface is shown in Fig. 1c. The collision of ions with target material nucleus and electrons can be considered separately due to large difference in their masses:

- Elastic or nuclear collision—between ion and nucleus. Kinetic energy and momentum are conserved. The incident ion recoils the target atom and scattering of ion takes place.
- Inelastic or electronic collision—between ion and electrons. It results in excitation and ionization of the constituent electrons.

Following main effects take place due to ion–solid interactions (Melngailis 1987):

- Thermal heating—The sample surface gets heated due to ion–beam interactions.
- Radiation damage—The sample is damaged due to displacement of atoms from their lattice positions. The depth of ion penetration depends on the ion beam energy, material properties and operating parameters such as incidence angle. The ions are implanted in the sample.
- Sputtering—The atoms are removed from sample surface due to incident ions. Sputtering occurs at low energies.
- Electron emission—The electrons are emitted from the target surface due to ion–solid interactions. These electrons are used for imaging.
- Chemical effects—These are produced due to ion–electron interaction and ion–atom interaction. Examples are ion-assisted etching, ion-induced deposition, etc.

The ion–solid interactions for amorphous materials can be modelled using Monte Carlo simulations by SRIM/TRIM (stopping and range of ions in matter, transport of ions in matter) codes developed by Ziegler and Biersack (1985). These simulations are helpful for calculation of sputter yield, interaction volume, etc. However, these calculations are based on binary collisions, and the effects of neighbouring atoms are not taken into account.

2.3 Focused Ion Beam Governing Parameters and Important Effects

Focused ion beam nanofabrication is a complex process and requires accurate control of processing parameters for the desired fabrication. A brief introduction of important FIB governing parameters and their effects on milling are discussed in this section.

2.3.1 Acceleration Voltage

An extraction voltage is applied to pull the Ga ions from the wet tungsten needle tip. The energy and velocity of ions depend on the applied acceleration voltage in the ion column. Ions will travel faster if a higher acceleration voltage is applied, and thus, the energy imparted on the target sample will also be higher (Fu et al. 2000). An increase in the acceleration voltage does not influence the probe current. An acceleration voltage of 30 kV is generally used to get the highest possible resolution and sputtering yield (Orloff 1993). This value of acceleration voltage makes the system suitable for many applications like sputtering, implanting or imaging.

2.3.2 Beam Current

Ion beam current is an important parameter for fabrication of micro/nano-structures. It defines the resolution capability of FIB system. Beam spot size is controlled by the beam current. It is difficult to measure the spot size in practice; hence, the beam current is generally used as a process parameter. The sputtering yield is controlled by changing the beam current. At higher values of beam current (i.e. larger spot size), sputtering yield is high. The higher material removal rate is, however, at the expense of the fabrication accuracy. If a high-resolution and accurate profile is required, a low beam current must be used. During typical milling applications, a high beam current is used for bulk material removal applications, while low beam currents are used for fine polishing. The typical minimum spot size of Carl Zeiss Auriga system at MEMS department IIT Bombay is 8 nm at a beam current of 1 pA and acceleration voltage 30 kV.

2.3.3 Dwell Time

Dwell time in a FIB system is the duration of time for which the beam is stationary at a particular position (pixel). Ion beam dwell time defines the duration of time for which a particular pixel is scanned by the ion beam (Utke et al., 2008). The dwell time is generally of the order of microseconds (μs) to milliseconds (ms). Final geometry of FIB fabrication depends on the dwell time, making it an important parameter. There are two challenges associated with the optimization of the dwell time. First, excessively large dwell time results in redeposition of sputtered atoms/ions, and surface accuracy becomes poor. The fabrication depth is also more than the intended value. These effects are applicable for fabrication, where depth removed per pixel is of the order of beam diameter (Adams and Vasile 2006). Second, a very low dwell time results in undesired material removal outside the targeted area. This occurs in the fabrication of non-rectilinear profiles such as circle, ellipse, etc. A significant portion of target material is exposed to the unintended ion dose and results in an irregular profile and undesired milling (Adams and Vasile 2006). Thus, a careful choice and optimization of FIB dwell time is required for the desired accuracy of fabrication. This aspect is even more challenging for 3D features nanofabrication by FIB.

2.3.4 Ion Range

Focused ion beam with high-energy ions impacts the target surface. High-energy ions impart their energy to the surface atoms and travel a certain path inside the target material. The trajectory followed by the energetic ions is known as 'collision cascade', and the total length of the path travelled by ion is termed as 'range'. The projected values of ion range in amorphous targets can be calculated using the Monte Carlo simulation-based SRIM code developed by Ziegler and Biersack

Table 1 The ion range and sputtering yield of various materials for 30 kV Gallium ions (Glass et al. 2008)

Material	Si	Al	Cu	Ag
Ion range (nm)	27	24	10	11
Sputter yield (atoms/ion)	2.6	4.4	11	14

(1985). The ion range depends on ion–solid interactions and material properties. Ion ranges of few different materials are as presented in Table 1.

2.3.5 Sputter Yield

Sputter yield is defined as the number of target atoms ejected per incident ion at the target surface. It depends on following parameters (Yao 2007):

- Incident ions—mass, energy, dose, incidence angle, etc.
- Target material—mass, crystal orientation, crystallinity, surface binding energies, conductivity, surface curvature, etc.

There exists a threshold value of the incident ion energy for each material, below which no sputtering takes place. Table 1 shows the sputter yield for four different materials with gallium as an ion source at 30 kV.

The milling time and sputter yield can be calculated as

$$\text{Sputter Rate} = \left[\frac{\text{Volume Removed}}{\text{Ion Dosage}}\right] = \left[\frac{\text{Volume Removed}}{\text{Beam current} \times \text{Milling Time}}\right] = \left[\frac{\mathrm{d}x \times \mathrm{d}y \times \mathrm{d}z}{I_\mathrm{B} \times T}\right] \tag{1}$$

$$\text{Milling Time} = \frac{\mathrm{d}x \times \mathrm{d}y \times \mathrm{d}z}{I_\mathrm{B} \times \text{Sputter Rate}} \tag{2}$$

$$\text{Sputter Yield} = \frac{\text{Number of atoms removed from the target surface}}{\text{Number of incindent ions}} \tag{3}$$

where I_B is the ion beam current used for milling, T is the milling time and dx, dy, dz, are dimensions (length, width, depth) of the milled region.

2.3.6 Scan Mode and Pixel Spacing

The ion beam can be scanned over a target surface in two different modes—raster scan and serpentine scan (Yao 2007). In raster mode, scanning is done in the same direction. The beam always moves to initial point of next scanning line. Figure 2a shows the schematic diagram of FIB raster scanning. In this mode of scanning, the

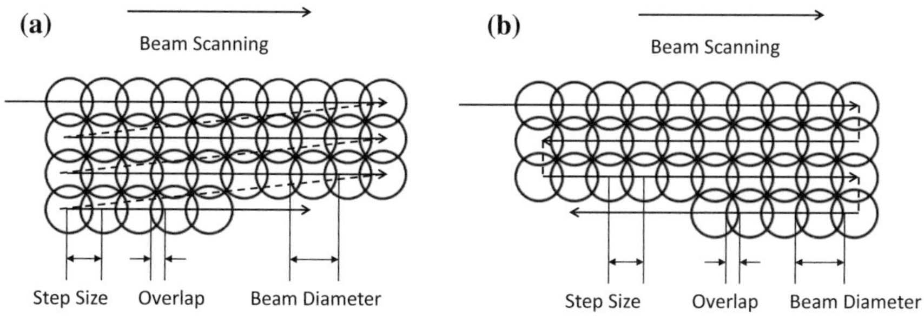

Fig. 2 Schematic illustration of FIB scanning: **a** raster scan and **b** serpentine scan

sputtered atoms continuously redeposit on the milling area. It results in loss of fabrication accuracy. Raster scanning is suitable for fabrication of V-shaped channels or inclined bottom surfaces. In serpentine scanning, ion beam moves in reverse directions for next scan line. It results in faster operation and fabrication time is less as compared to raster scan for the same area. Schematic diagram representing serpentine scanning mode is shown in Fig. 2b. The redeposition of sputtered material is less in serpentine scanning mode, as the redeposited atoms from the previously scanned line are removed in the subsequent line scan. This mode of scanning is favourable for high aspect ratio structures with vertical side-walls and flat bottom. Pixel spacing can be defined as the distance between the centres of two adjacent pixels. Milling by FIB will be uniform, if the spacing between adjacent pixels and scan lines is proper and small enough (Tseng 2005).

2.3.7 Beam Overlap

The scanning in FIB process, as illustrated in the previous section, takes place along a predefined path and ion beam is moved to subsequent pixel locations with a distance equal to step size. Focused ion beam is blanked while moving from one pixel to another. The ion beam diameter and step size, in addition to dwell time, are two important parameters that determine the profile of depth being milled. Beam overlap is defined with respect to step size (Yao 2007)

$$\text{Beam Overlap} = \frac{\text{Beam Diameter} - \text{Step Size}}{\text{Beam Diameter}} \tag{4}$$

The concept of beam overlap is represented schematically in Fig. 3. Beam overlap is 50%, if the step size is half of the beam diameter. It is a critical FIB parameter and must be optimized for different applications. Negative overlap for deposition, zero overlap for etching and positive overlap are generally used for milling applications.

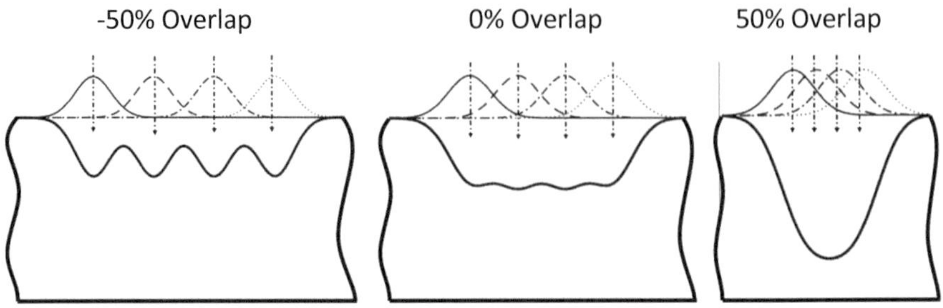

Fig. 3 Concept of beam overlap in FIB and effect on fabrication profile: negative, zero and positive overlap

2.4 Focused Ion Beam—Process Capabilities

2.4.1 Focused Ion Beam Imaging

In FIB imaging, similar to a scanning electron microscope, the beam is scanned over the sample surface. The secondary ions and electrons are generated due to the interaction between incident ions and surface atoms. These secondary ions or electrons are collected for the purpose of imaging by means of biased detectors. The detectors collect emitted ions or electrons based on the voltage applied to it. The resolution of FIB imaging depends upon the ion beam spot size (Reyntjens and Puers 2001).

Focused ion beam is proved to be a powerful tool for material analysis and characterization. The channelling contrast produced by FIB is better than the contrast provided by the electrons in an SEM. This capability is utilized for observing grain boundary orientations and crystal defects such as dislocations. However, due to ion–beam interactions with the sample surface, some damage also occurs. It limits the use of FIB for imaging of soft materials. Some of the ions from the beam penetrate into the sample and get implanted in the surface, resulting in change of surface properties. The depth of implantation depends upon the beam energy, and material damage up to an extent takes place based on the beam parameters. These effects can be reduced by using a low energy and finely focused beam of ions.

2.4.2 Focused Ion Beam Milling

Material from the sample surface can be removed precisely using a high energy and finely focused beam of ions. If the high energy beam is scanned over the surface, it results in sputtering of atoms from the sample surface. The milling process can be observed in situ and sample surface can be analysed by FIB/SEM imaging. The milling rate by FIB is generally small, as the material removal takes place atom by atom, and usually the sputtering is accompanied by redeposition of the sputtered atoms from the surface. To enhance the milling rate, an etching gas can be used in work chamber. This technique is called 'gas-assisted etching (GAE)' (Reyntjens and Puers 2001). Gas-assisted FIB etching can be regarded as a local version of plasma etching or reactive ion etching with much higher ion energy (Utke et al. 2008). Following steps are followed in gas-assisted FIB-induced etching (Yao 2007):

- A chemically neutral reactive gas through a nozzle is adsorbed on the sample surface.
- The gas reacts with the sample either in the presence or absence of ion beam.
- Volatile products are formed (desorption), which are removed from vacuum.

Ion implantation, redeposition, etc. can be reduced using gas-assisted ion beam etching. However, care must be taken to avoid any spontaneous reaction of the etching gas with the target surface.

2.4.3 Focused Ion Beam Deposition

Focused ion beam can be used for deposition of metal and insulator materials (Reyntjens and Puers 2001). The operating principle is analogous to chemical vapour deposition (CVD). Generally, deposition of platinum and tungsten is done on commercially available instruments. There are various phenomena that occur when an ion beam is scanned over target surface, where precursor gas molecules are previously desorbed:

- Desorption of the molecules,
- Dissociation of the molecules and
- Reaction of molecules with target material.

A deposition layer is sprayed over sample surface in the form of a gas. The ion beam is subsequently scanned over the sample surface and the sprayed gas is decomposed by the action of an ion beam. The desired reaction products remain on the surface as a thin film, while the volatile products are removed with the help of vacuum inside the chamber. The deposited material is not completely pure, as some organic contaminants, as well as gallium ions, are also implanted in the deposits. Tungsten, platinum, carbon, etc. are some of the common materials that can be deposited by FIB.

There are many similarities between FIB-induced deposition and etching process. Deposition and etching gases both can adsorb/desorb without reacting with the sample surface. However, some etching gases (for example, chlorine on silicon) are also able to react with the sample surface. For etching, sputtering due to FIB will add to the material removal. While in case of deposition, sputtering will complete the process. The basic requirements for precursor gas are same for both deposition and etching. The gas should stick to the surface for activation/dissociation for etching/deposition (Yao 2007).

2.5 Focused Ion Beam Limitations

Main issues/limitations in FIB milling for 3D micro/nanofabrication are as follows (Fu et al. 2000; Lehrer et al. 2001):

- *Overlap effect*—To fabricate a smooth surface, the spacing between the pixels should be less than the beam size. However, this results in more milling than intended and consequently depth milled is higher.
- *Beam tail effect*—This effect is due to the beam profile, which is generally Gaussian distribution. The tail of the beam results in undesired milling, thus making it difficult to control the depth and maintain the geometric accuracy.
- *Sputter yield*—In addition to other beam parameters, the sputter yield also depends on the angle of incidence of the ion beam. As the machining progresses, the surface geometry is changed, which in turn changes the effective angle at which the ion beam hitting the surface. Thus, the sputter yield changes with time, leading to the non-uniform material removal at different regions of the geometry.
- *Redeposition effect*—Based on the amount of materials removed and aspect ratio of the target geometry, some of the sputtered atoms redeposit on the sample surface itself, which results in fabrication inaccuracies.

In addition to the above-mentioned limitations, FIB also has scalability limitations. This makes FIB milling restricted to a smaller volume for material characterisation/removal applications; however, this issue is now being addressed by employing Xe-based plasma FIB (Burnett et al. 2016).

3 The Optimization Model

Focused ion beam process planning is important for fabrication of desired structures from accuracy point of view. Focused ion beam is a complex process and involves material interaction of ion beam during the process. An optimization model from the literature (Qin et al. 2013) is adapted here to optimize the dwell time for maintaining accuracy and fabrication of desired profile using FIB milling.

3.1 Scanning for FIB Milling

A charged beam of ions is scanned over the target substrate for machining in FIB milling. The substrate is divided into a grid of pixels for beam control and machining accuracy. The machining depth depends upon the dwell time, i.e. the time for which a particular pixel is scanned by FIB. However, when the beam is scanned over a certain pixel (say (x_j, y_j)), it results in milling at nearby pixel (say

(x_i, y_i)) due to the Gaussian distribution of beam. The effect of nearby pixels being milled due to FIB scanning at certain pixels is shown in Fig. 4. It is also clear from Fig. 4, if a particular pixel is scanned for a longer duration (more dwell time), it will result in more depth being milled. Thus, fabrication profile can be varied by controlling the dwell time. The standard Gaussian normal distribution is given by

$$G(x) = \frac{1}{\sqrt{2\pi\sigma^2}}\left[e^{-\frac{(x-\mu)}{2\sigma^2}}\right] \tag{5}$$

where σ is the standard deviation of the beam. The total depth being machined due to FIB milling can be written as

$$z_i = \sum_{j=1}^{M} Y_s t_j \left[e^{-\frac{[(x_i-x_j)^2+(y_i-y_j)^2]}{2\sigma^2}}\right] \tag{6}$$

where M is the number of pixels to be machined, z_i is the machined depth at ith pixel, Y_s is the erosion rate and t_j is the dwell time (which is decision variable) at jth pixel. The above equation can be written as

$$z_i = \sum_{j=1}^{M} a_{ij} t_j = \boldsymbol{A}_i \boldsymbol{t}, \quad \text{for } i = 1, 2, \ldots N \tag{7}$$

where N is the total number of pixels on substrate, a_{ij} is the milling coefficient and $\boldsymbol{A}$ is an $N \times M$ matrix to measure the milling coefficients.

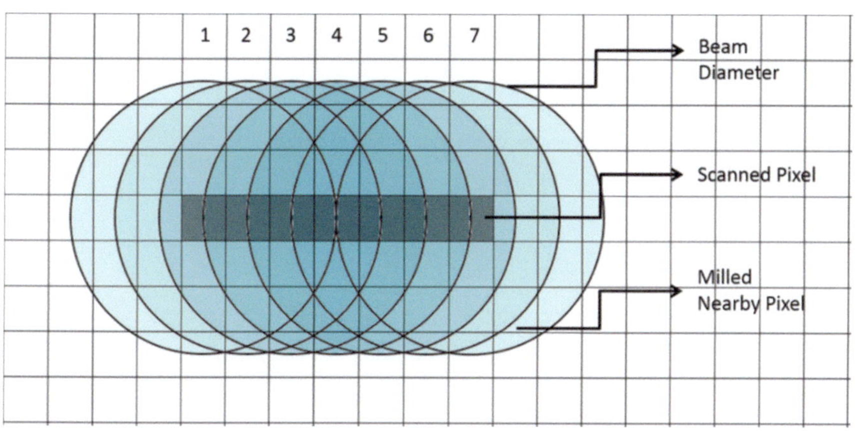

Fig. 4 Effect of beam shape and dwell time on fabrication depth at different pixels—colour intensity represents the fabrication depth

3.2 The Objective Function

The decision variable dwell time $\boldsymbol{t}$ should be optimized such that it results in material removal $\boldsymbol{At}$ (i.e. fabricated profile) close to the desired geometry profile z^{o}. A quadratic optimization problem is set up to minimize the error between $\boldsymbol{At}$ and z^{o} in the following form:

$$\begin{aligned} &\text{minimize } F(\boldsymbol{t}) = \|\boldsymbol{At} - z^{o}\|_2^2 \\ &\text{subject to, } \boldsymbol{t} \geq 0 \end{aligned} \tag{8}$$

The objective function in $F(\mathbf{t})$ is rewritten in the following form to represent the following standard quadratic programming problem:

$$\begin{aligned} &\text{minimize } F(\boldsymbol{t}) = \boldsymbol{t}^T \boldsymbol{Q}\boldsymbol{t} + 2\boldsymbol{p}^T \boldsymbol{t} + r \\ &\text{subject to, } \boldsymbol{t} \geq 0 \end{aligned} \tag{9}$$

where

$$\begin{aligned} \boldsymbol{Q} &= \boldsymbol{A}^T \boldsymbol{A}, \\ \boldsymbol{p} &= -\boldsymbol{A}^T z^{o} \\ r &= (z^{o})^T z^{o} \end{aligned} \tag{10}$$

3.3 Solution

A multiplicative updates algorithm from the literature (Yang and Laaksonen 2007) is used for solution of the above non-negative quadratic programming problem. If $\boldsymbol{t}^0$ is an initial value of dwell time $\boldsymbol{t}$, and $\boldsymbol{t}^k$ ($k > 0$) is the value from the kth update, algorithm can be written as

$$t_j^{k+1} \leftarrow \frac{\max\left[-p_j, 0\right]}{\left(\boldsymbol{Q}\boldsymbol{t}^k\right)_j} t_j^k \tag{11}$$

4 Simulation Results

The optimization algorithm from the previous section is used to simulate a spherical cavity using MATLAB. The beam current is taken as 20 pA, $X = Y = 600$ nm, and the pixel size is 3 nm. Results are shown in Fig. 5 on a 200 $\times$ 200 grid (since pixel size is 3 nm). The beam diameter at this current is 17 nm (Zeiss Auriga dual beam

FIB-SEM system), and step size is taken as 3 nm. The plot in Fig. 5a is the desired spherical shape with z^o = 60 nm. Figure 5b shows the simulation results obtained by FIB scanning. The difference between $\boldsymbol{At}$ (simulated geometry) and z^o (desired geometry), i.e. error is shown in Fig. 5c. It can be seen from the figure that the error is significant only at the boundaries. This is because no actual machining takes place at boundary pixels, and the depth obtained is only because of the milling at nearby pixels. Thus, the optimization model is capable of producing accurate results. Several other simulations have also been run using the optimization model for different profiles, in order to check the validity of the optimization algorithm, and no significant error was observed between desired and simulated profile.

Further, optimization of dwell time for desired and accurate 3D optical elements by FIB can be obtained using the developed algorithm. Fabrication of 3D optical elements such as lenses on optical fibre for waveguide coupling (Schiappelli et al. 2004; Melkonyan et al. 2017), Fresnel zone plates (FZPs) for X-rays focusing, etc. forms an important field of study due to the unique capability of these elements. Fresnel zone plates are very popular optical components for applications in X-ray microscopy (Quiney et al. 2006) with the capability of focusing X-rays. X-rays can have a high penetration depths and resolution, allowing the use of X-rays for

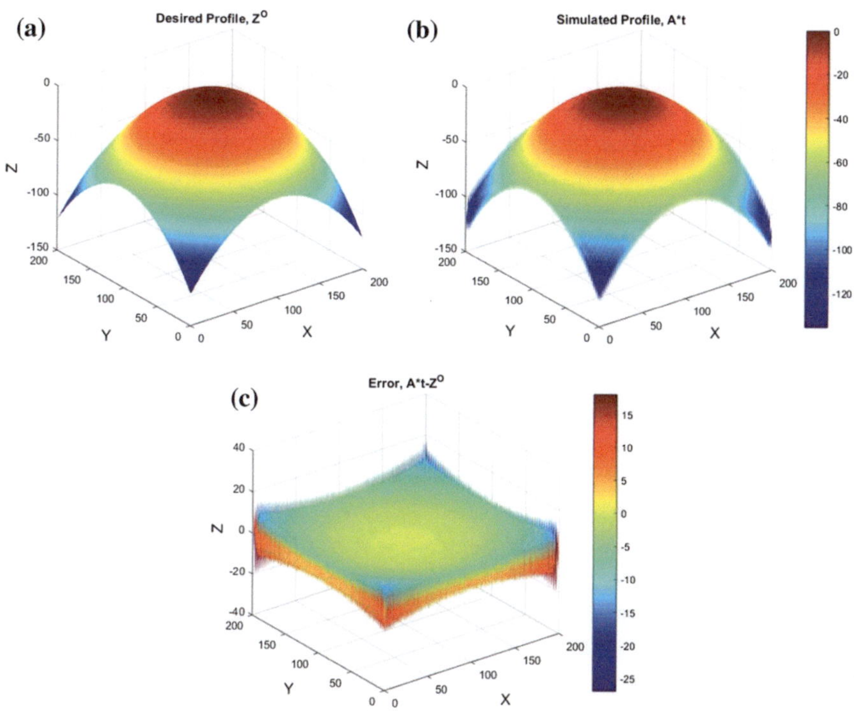

Fig. 5 Simulation results for a spherical profile obtained from optimization algorithm at a beam current of 20 pA and pixel size of 3 nm: **a** desired spherical profile, **b** simulated spherical profile and **c** error between the desired and simulated profile

material characterization applications (Attwood 1999). As a result, FZPs become important optical elements for applications in microscopic imaging and nanotechnology. Researchers all over the world have presented a variety of FIB fabrication methods for FZPs (Keskinbora et al. 2013a, b; Vijayakumar et al. 2014; Vijayakumar and Bhattacharya 2015; Sanli et al. 2015; Rodrigues Ribeiro et al. 2016). Experimental demonstrations of FZPs (Mote et al. 2010) as well as sub-wavelength focusing technique using metallic coatings also exist (Mote et al. 2008). However, there is still a need to explore this method for better control and improvement of fabrication accuracy. Here, we design and optimize FZP fabrication by FIB using the developed optimization algorithm as an example.

FZPs are diffractive optical elements and these are made of alternate opaque and transparent zones. The FZPs are designed according to

$$r_n = \sqrt{n\lambda f + n^2\lambda^2/4} \tag{12}$$

where r_n is the radius of nth zone, λ is the wavelength and f is the focal distance (Attwood 1999). The resolution of FZP depends on the outermost zone and is given by

$$\Delta r_{\text{Rayleigh}} = 1.22\Delta r \tag{13}$$

where $\Delta r_{\text{Rayleigh}}$ is the Rayleigh resolution limit and Δr is the width of outermost FZP zone (Attwood 1999).

Figure 6 show the simulation results for FZP. The simulation results shown here are for a binary FZP of diameter 100 μm, focal length of 150 μm and for a wavelength of 660 nm. The beam current for the simulation is taken as 500 pA, $X = Y = 100$ μm, and a large value of pixel size 200 nm is chosen. Results are shown in Fig. 6 on a 500 × 500 grid (since pixel size is 200 nm). The beam diameter at this current is 50 nm for Zeiss Auriga dual beam FIB-SEM system. The reason for selecting large pixel size for FZP simulation is to reduce the simulation time maintaining accuracy, as the size of FZP is large (100 μm). The results from Fig. 6c indicate a very small value of error. It is due to the binary nature of simulated FZP, as it did not involve any curved features. Next, an attempt is made to validate the optimization results with experiments.

5 Results of Focused Ion Beam Fabrication

Experiments have been carried out on Zeiss Auriga dual beam FIB-SEM system using Ga ion beam. This system allows smallest spot size of 8 nm at a beam current of 1 pA. It allows milling, deposition and imaging using ion and electron beam. Focused ion beam milling is a very powerful tool and it enables micro/nano-fabrication with great accuracy. Simulation profile of FZP in the previous section is fabricated using FIB milling for demonstration. The system allows

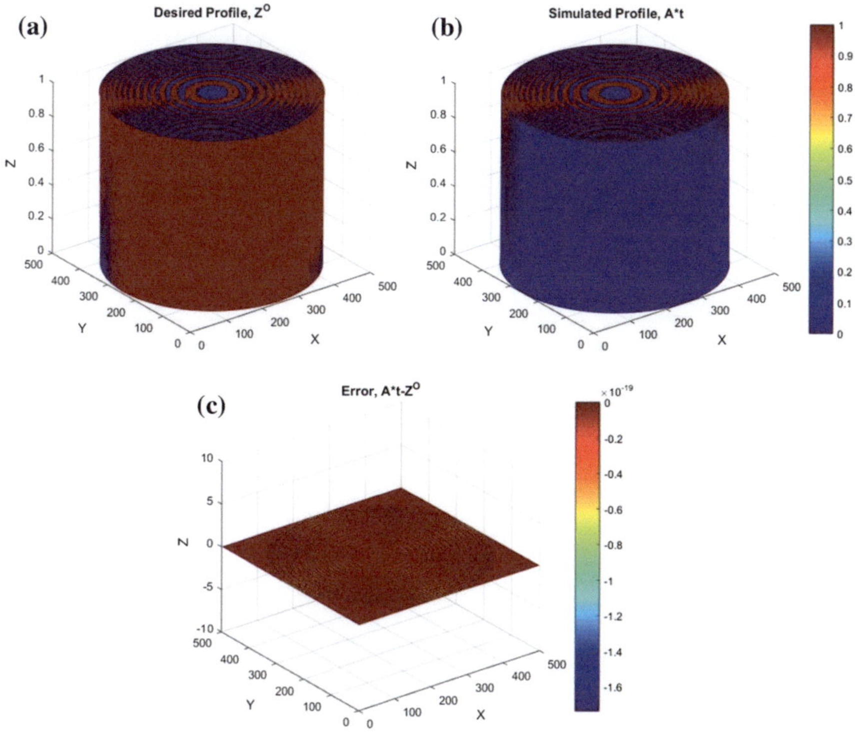

Fig. 6 Simulation results for fabrication of Fresnel zone plate obtained from optimization algorithm at a beam current of 500 pA and pixel size of 200 nm: **a** desired profile, **b** simulated profile and **c** error between the desired and simulated profile

feature-based milling, and bitmap images consisting of designed pattern can be milled by choosing desired beam current, dwell time, acceleration voltage, etc. A bitmap image consisting of ring patterns based on the FZP design is created using MATLAB. Figure 7a shows the bitmap image used for a binary FZP of diameter 100 μm, focal length of 150 μm for a wavelength of 660 nm. This bitmap image has been used on FIB-SEM system for fabrication on silicon. Focused ion beam is scanned over the black areas in the bitmap image, and milling takes place at these areas only. Scanning electron microscopic image of FZP fabricated on silicon using FIB milling is shown in Fig. 7b. The fabrication has been done using Gallium ion beam at a beam current of 500 pA and acceleration voltage of 30 kV. Next, atomic force microscopic (AFM) characterization of fabricated FZP is done. Only a portion of FZP is scanned due to the scanning limitations of AFM. Figure 8 shows the AFM characterization results of fabricated FZP. 2D and 3D profiles of FIB fabricated FZP in Fig. 8a, b, respectively, demonstrate the fine accuracy and capability of the process. Figure 8c represents the surface profile of fabricated FZP.

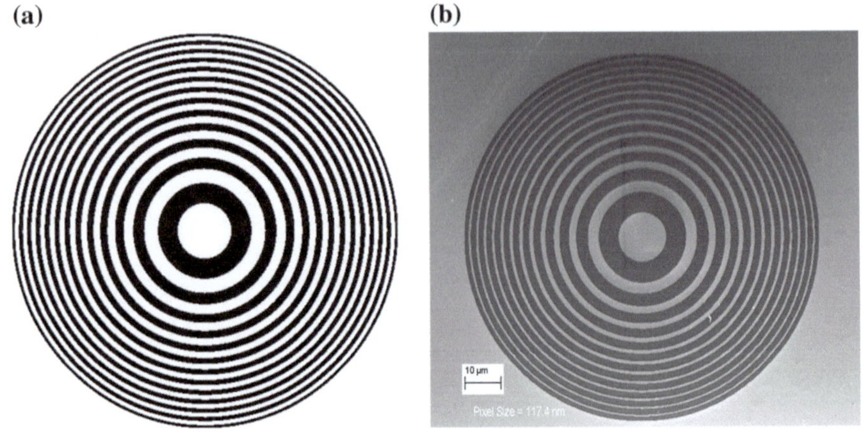

Fig. 7 Binary Fresnel zone plate fabrication by focused ion beam milling: **a** bitmap image used in the Zeiss Auriga dual beam FIB-SEM system, and **b** scanning electron microscopic image of fabricated FZP

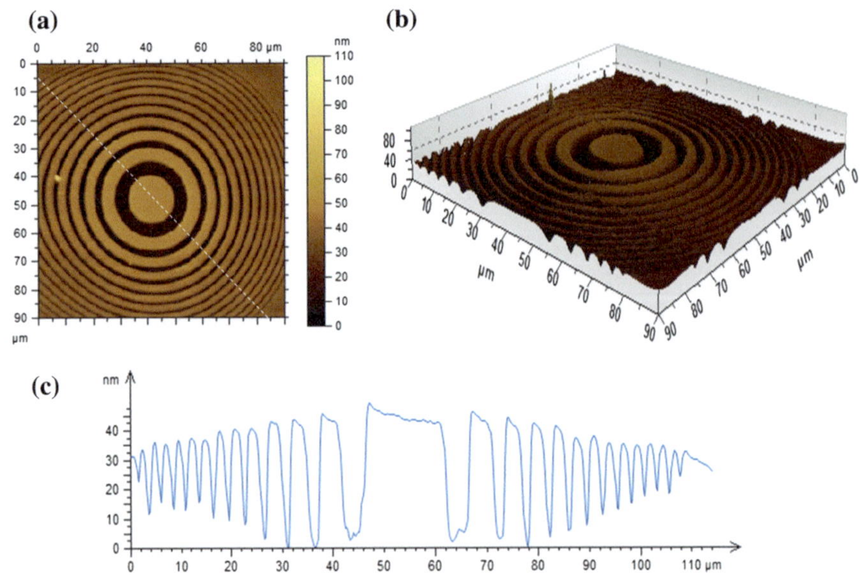

Fig. 8 Atomic force microscopic (AFM) results of focused ion beam milled Fresnel zone plate: **a** 2D AFM image, **b** 3D AFM image and **c** variation of surface profile along white line indicated in (**a**)

6 Conclusions

This work presented focused ion beam milling process development and optimization for fabrication of optical applications. An optimization model has been developed and used to simulate the FIB milling process for fabrication of desired profiles. Ion beam dwell time is an important parameter for the accurate fabrication. Beam dwell time is optimized for the required accuracy and an example has been presented for a spherical profile. Simulation results have been shown to predict the fabrication geometry fairly well. The amount of error between the desired geometry and the simulated geometry is quite low, indicating good results and validity of the optimization algorithm. Further, simulation results for optimization of FZP fabrication are presented, which have been able to predict the required fabrication geometry well at the given beam parameters. The error between the desired geometry and simulated FZP geometry is insignificant (<0.1%) due to its binary nature. The FZP structure is fabricated using FIB process on silicon and characterized using AFM. This approach can be further extended to fabrication of 3D optical elements and complex micro/nano-structures by FIB milling. To summarize,

- A framework for optimal solution of FIB dwell time through an iterative approach.
- Dwell time was optimized by formulation and solution of a quadratic programming problem, minimizing the error between desired and simulated geometry. Simulation examples were demonstrated for a spherical profile and FZP. Experiments were carried out for fabrication of a binary FZP by FIB.
- The approach can be extended further for FIB fabrication of 3D nanostructures and optical elements maintaining high accuracy.

Acknowledgements The authors acknowledge the financial support by IRCC, IIT Bombay via Seed Grant: Spons/ME/I14079-1/2014. The first author acknowledges the Ph.D. scholarship support by IITB Monash Research Academy.

References

Adams, D.P., and M.J. Vasile. 2006. Accurate focused ion beam sculpting of silicon using a variable pixel dwell time approach. *Journal of Vacuum Science and Technology B* 24: 836–844.

Adineh, V.R., B. Liu, R. Rajan, W. Yan, and J. Fu. 2015. Multidimensional characterisation of biomechanical structures by combining atomic force microscopy and focused ion beam: A study of the rat whisker. *Acta Biomaterialia* 21: 132–141.

Al-Abboodi, A., J. Fu, P.M. Doran, and P.P.Y. Chan. 2013. Three-dimensional nanocharacterization of porous hydrogel with ion and electron beams. *Biotechnology and Bioengineering* 110: 318–326.

Attwood, D. 1999. *Soft X-rays and extreme ultraviolet radiation, soft X-rays and extreme ultraviolet radiation: Principles and applications*. Cambridge University Press.

Burnett, T.L., R. Kelley, B. Winiarski, L. Contreras, M. Daly, A. Gholinia, M.G. Burke, and P. J. Withers. 2016. Large volume serial section tomography by Xe Plasma FIB dual beam microscopy. *Ultramicroscopy* 161: 119–129.

Fu, J., and S. Joshi. 2010. Optimization based geometric modeling of nano/micro scale ion milling of organic materials for multidimensional bioimaging. *Journal of Nanotechnology in Engineering and Medicine* 1: 031003.

Fu, Y.Q., N.K.A. Bryan, O.N. Shing, and N.P. Hung. 2000. Influence of the redeposition effect for focused ion beam 3D micromachining in silicon. *International Journal of Advanced Manufacturing Technology* 16: 877–880.

Glass, G.A., J.F. Dias, A.D. Dymnikov, and B. Rout. 2008. 900 keV gold ion sputter etching of silicon and metals. *Nuclear Instruments and Methods in Physics Research Section B: Beam Interactions with Materials and Atoms* 266: 3330–3331.

Keskinbora, K., C. Grévent, M. Bechtel, M. Weigand, E. Goering, A. Nadzeyka, L. Peto, S. Rehbein, G. Schneider, R. Follath, J. Vila-Comamala, H. Yan, and G. Schütz. 2013a. Ion beam lithography for Fresnel zone plates in X-ray microscopy. *Optics Express* 21: 11747–11756.

Keskinbora, K., C. Grévent, and U. Eigenthaler. 2013b. Rapid prototyping of Fresnel zone plates via direct Ga^+ ion beam lithography for high-resolution X-ray imaging. *ACS Nano* 7: 9788–9797.

Kim, C.S., S.H. Ahn, and D.Y. Jang. 2012. Review: Developments in micro/nanoscale fabrication by focused ion beams. *Vacuum* 86: 1014–1035.

Lehrer, C., L. Frey, S. Petersen, and H. Ryssel. 2001. Limitations of focused ion beam nanomachining. *Journal of Vacuum Science & Technology B: Microelectronics and Nanometer Structures Processing, Measurement, and Phenomena* 19: 2533.

Melkonyan, H., K. Sloyan, K. Twayana, P. Moreira, and M. Dahlem. 2017. Efficient fiber-to-waveguide edge coupling using an optical fiber axicon lens fabricated by focused ion beam. *IEEE Photonics Journal* 9: 1–9.

Melngailis, J. 1987. Focused ion beam technology and applications. *Journal of Vacuum Science and Technology B* 5: 469–495.

Mote, R.G., and L. Xiaomin. 2017. *Focused ion beam (FIB) nanofinishing for ultra-thin TEM sample preparation*, 155–176. Nanofinishing science and technology, micro and nanomanufacturing series. CRC Press.

Mote, R.G., S.F. Yu, B.K. Ng, W. Zhou, and S.P. Lau. 2008. Near-field focusing properties of zone plates in visible regime–new insights. *Optics Express* 16: 9554–9564.

Mote, R.G., S.F. Yu, A. Kumar, W. Zhou, and X.F. Li. 2010. Experimental demonstration of near-field focusing of a phase micro-Fresnel zone plate (FZP) under linearly polarized illumination. *Applied Physics B* 102: 95–100.

Orloff, J. 1993. High-resolution focused ion beams. *Review of Scientific Instruments* 64: 1105–1130.

Qin, R., J. Fu, Z. Yin, and C. Zheng. 2013. Large-scale process optimization for focused ion beam 3-D nanofabrication. *International Journal of Advanced Manufacturing Technology* 64: 587–600.

Quiney, H.M., A.G. Peele, Z. Cai, D. Paterson, and K.A. Nugent. 2006. Diffractive imaging of highly focused X-ray fields. *Nature Physics* 2: 101–104.

Reyntjens, S., and R. Puers. 2001. A review of focused ion beam applications in microsystem technology. *Journal of Micromechanics and Microengineering* 11: 287–300.

Rodrigues Ribeiro, R.S., P. Dahal, A. Guerreiro, P.A.S. Jorge, and J. Viegas. 2016. Compact solutions for optical fiber tweezers using Fresnel zone and phase lenses fabricated using FIB milling. In *Proceedings of SPIE*.

Sanli, U.T., K. Keskinbora, K. Gregorczyk, J. Leister, N. Teeny, C. Grévent, M. Knez, and G. Schütz. 2015. High-resolution high-efficiency multilayer Fresnel zone plates for soft and hard x-rays. In: *Proceedings of SPIE*.

Schiappelli, F., R. Kumar, M. Prasciolu, D. Cojoc, S. Cabrini, M. De Vittorio, G. Visimberga, A. Gerardino, V. Degiorgio, and E. Di Fabrizio. 2004. Efficient fiber-to-waveguide coupling by a lens on the end of the optical fiber fabricated by focused ion beam milling. *Microelectronic Engineering, Micro and Nano Engineering* 73: 397–404.

Tseng, A.A. 2005. Recent developments in nanofabrication using focused ion beams. *Small* 1: 924–939.

Utke, I., P. Hoffmann, and J. Melngailis. 2008. Gas-assisted focused electron beam and ion beam processing and fabrication. *Journal of Vacuum Science and Technology B* 26: 1197–1276.

Vijayakumar, A., and S. Bhattacharya. 2015. Design of multifunctional diffractive optical elements. *Optical Engineering* 54: 24104.

Vijayakumar, A., U. Eigenthaler, K. Keskinbora, G.M. Sridharan, V. Pramitha, M. Hirscher, J. P. Spatz, and S. Bhattacharya. 2014. Optimizing the fabrication of diffractive optical elements using a focused ion beam system. *Micro-Optics* 2014 (9130): 1–8.

Yang, Z., and J. Laaksonen. 2007. Multiplicative updates for non-negative projections. *Neurocomputing* 71: 363–373.

Yao, N. 2007. *Focused ion beam systems: Basics and applications*. Cambridge University Press.

Ziegler, J.F., and J.P. Biersack. 1985. The stopping and range of ions in matter. In *Treatise on heavy-ion science*, ed. D.A. Bromley, 93–129. US: Springer.

Machining Guidelines for Fabricating Microgrooves of Varied Cross Sections by Electrochemical Micromachining

V. Rathod, B. Doloi and B. Bhattacharyya

Abstract Increased surface area at the cross section of varied cross-sectional microgroove improves the product performance; hence, complex microgrooves of different cross sections are machined on various micro-products. Fabrication of straight microgroove on metallic surfaces is simple, whereas machining of varied cross-sectional microgroove on metallic surfaces becomes the challenging task from the aspects like fabrication of complex shaped microtool and suitable microgroove manufacturing process. Very few methods are capable to machine such complex shaped microgrooves. This chapter explores the potential of machining varied cross-sectional microgrooves on metallic surfaces by electrochemical micromachining. Important process parameters such as applied voltage, duty ratio and machining time have been controlled along the depth while machining the microgrooves. Influences of these parameters on width have been investigated, and results have been applied to devise the machining guidelines to fabricate varied cross-sectional microgrooves. The guidelines for machining straight, reverse tapered, barrel-shaped, double stepped, spherical based and microgroove with internal pocket have been developed. Finally, fabrications of these microgrooves have been demonstrated successfully by following developed guidelines. The developed guidelines can be applied for machining complex microgrooves for micro-coolers, micro-reactors and micro-mixers that need definite shape and size for their working.

Keywords EMM · Complex microgroove · Guidelines · Cross sections
Reverse taper · Barrel shape · Double step · Spherical base · Internal pocket

---

V. Rathod (✉)
Mechanical Engineering Department, Government Polytechnic,
Mumbai 400051, Maharashtra, India
e-mail: vurathodju@gmail.com

B. Doloi · B. Bhattacharyya
Production Engineering Department, Jadavpur University, Kolkata 700032, India
e-mail: bdoloionline@rediffmail.com

B. Bhattacharyya
e-mail: bb13@rediffmail.com

S. S. Pande and U. S. Dixit (eds.), *Precision Product-Process Design and Optimization*, Lecture Notes on Multidisciplinary Industrial Engineering,
https://doi.org/10.1007/978-981-10-8767-7_9

Nomenclatures

h	Inter-electrode gap
ρ_s	Specific electrolyte resistivity
τ	Charging time constant
c	Specific capacitance of double layer
V_m	Volume of material removed
C	Electrochemical constant of a material
E	Applied voltage
A	Active microelectrode surface area
t	Machining time
t_{on}	Pulse on-time
T	Total machining time
n	Total number of scans
L	Microgroove length
V_x	Scanning speed
σ	Standard deviation

1 Introduction

Micro-devices such as micro-actuators, micro-pumps, micro-mixers and micro-dies need microgrooves as one of the essential micro-features fabricated on it to fulfil their functional requirement. Recently, the demand for microgrooves with internal features has been increased, specifically in the applications of microthermal devices such as micro-coolers, micro-heat exchangers and micro-reactors. Specific cross-sectional shape, i.e. geometry of microgrooves, improves product functionality as well as product performance because of augmented inner surface area, which improves heat transfer rate well as rate of chemical reaction. The micro-products like microthermal devices and micro-reactors demand high mechanical strength at elevated temperature and higher thermal conductivity for effective heat transfer. Corrosion resistant is the main criterion in biomedical applications. Hence, properties like wear resistance, corrosion resistance, thermal conductivity and mechanical strength at elevated temperature are the important aspects in material selection of the micro-products. Advance engineering metals such as superalloys, titanium, nickel base alloys and stainless steel are most suitable for fabrication of the micro-products to fulfil such requirements. Microgrooves of various geometries are to be machined on various micro-products to fulfil their functional requirements. Hence, precise manufacturing of microgrooves with internal features forming different cross sections of the microgroove on metallic surfaces has become research issue that requires additional developments.

1.1 Literature Survey

Fabrication of plain microgrooves is comparatively easy and can be achieved by various traditional as well as non-traditional micromachining methods. Whereas the machining of varied cross-sectional microgroove becomes the challenging task from the aspects like development of complex shaped microtool and suitable microgroove manufacturing process. Chern et al. (2007) and Yan et al. (2009) fabricated various microgrooves on metal surfaces by traditional machining methods such as micro-end milling, shaping, fly cutting and diamond turning to explore the burr formation mechanism during machining. Authors observed microtool wear and burr formation as the significant issue in during machining of microgrooves. Nageswara and Kunzru (2007) applied photolithography or wet chemical etching for manufacturing of fine grooves on silicon. Authors find the various issues in controlling the depth of microgroove of different shapes, as well as the requirements of special equipment. Oh et al. (2006) applied laser beam micromachining for machining of high aspect ratio microgrooves on metallic surfaces. During the investigation, authors reported an issue of unavoidable heat affected zone and the challenges in regulating uniform cross section along the depth of microgrooves. Presently, very few micromachining methods such as micro electro-discharge machining (μ-EDM) are capable to machine microgrooves of different cross sections on metallic surfaces. Yan et al. (2010) developed microtools of different end shapes by wire-EDM and applied these microtools to fabricate the varied cross-sectional microgrooves on metallic surfaces by μ-EDM. Authors focused the requirement of individual end shaped microtools for each cross-sectional shape of the microgroove and microtool wear as the significant problem in this process, which restricts the use of developed microtool for machining of multiple micro-features.

Therefore, the potential of fabricating varied cross-sectional microgrooves by electrochemical micromachining (EMM) has been explored in this chapter. Important process parameters such as applied voltage, duty ratio and machining time are regulated while fabricating microgrooves using enveloped EMM setup. Influences of these parameters on width overcut have been investigated, and investigational results have been applied to devise the machining guidelines to fabricate varied cross-sectional microgrooves. The guidelines for machining straight, reverse tapered, barrel-shaped, double stepped, spherical based and microgroove with internal pocket have been developed. Finally, fabrications of these microgrooves have been demonstrated successfully by following developed guidelines. The developed guidelines can be applied for fabricating complex microgrooves suitable for micro-coolers, micro-reactors and micro-mixers that require specific geometry for product working.

2 Mathematical Analysis for Generation of Different Internal Features by Electrochemical Micromachining (EMM)

To machine the microgrooves of various cross-sectional shapes on metallic surfaces, principle of EMM has been studied. Material removal mechanism of the process has been analysed, so that material removal can be controlled during machining of microgroove.

2.1 EMM Principle

EMM works on the principle of anodic material dissolution, in which microtool and workpiece are immersed in an electrolyte with very small inter-electrode gap as cathode and anode, respectively. To improve the machining accuracy, ultrashort pulse power supply with minimum IEG is preferred. Stagnant electrolytes are used during machining because of micron-sized tools. During machining of micro-features, microtool moves with constant tool feed rate towards the workpiece, following the predetermined path (Kirchner et al. 2001).

In EMM, as the potential is applied between the microtool electrode and workpiece immersed in an electrolyte with small IEG. At the interface between electrode and electrolyte, a very thin layer exists called as double layer. The potential profile in the double layer becomes similar to that of an equivalent circuit which consists of capacitors and resistors. Electrolyte resistance is the product of the inter-electrode gap (h) and the specific electrolyte resistivity (ρ_s). Therefore, charging time constant (τ) of the double layer having specific capacitance (c) is

$$\tau = \rho_s ch \tag{1}$$

When IEG is small, the time constant for double layer charging is small. When applied pulse duration is longer than the charging time constant, the double layer becomes charged high enough for dissolution. However, at other places where the charging time constant is larger than the pulse duration, the double layer is not charged sufficiently for material dissolution. Hence, material dissolution occurs in areas where electrolyte resistance is lower during each pulse.

The volume of material dissolved in EMM can be determined by combining Faraday's first law and Ohm's law as (McGeough 1974)

$$V_{\mathrm{m}} = \frac{CEA}{\rho_s h} t \tag{2}$$

where V_{m} is the volume of material dissolved, C is the electrochemical constant of a material, E is applied voltage, A is active microelectrode surface area which takes part in machining, t is the machining time, ρ_s is specific electrolyte resistivity and h is inter-electrode gap. From above equation, it can be said that volume of material removed by EMM depends upon applied voltage, electrode size, type of electrolyte and its concentration, inter-electrode gap and machining time.

In EMM, double layer is charged and discharged frequently during every pulse at both the electrodes. Considering that, material dissolution takes place during pulse on-time only, and pulse off-time is kept long enough for heat dissipation and to flush away the dissolved materials from the machining zone. Therefore, Eq. (2) can be modified as

$$V_{\mathrm{m}} = \int_{0}^{t_{\mathrm{on}}} \frac{CEA}{\rho_s h}\,\mathrm{d}t \tag{3}$$

In fabrication of micro-features using sidewall insulated microtool of suitable size with specific parameter settings, the values of electrochemical constant, active electrode surface area, specific electrolyte resistivity and inter-electrode gap can be assumed as constants, and then, Eq. (3) can be further modified as

$$V_{\mathrm{m}} = \frac{CA}{\rho_s h} \int_{0}^{t_{\mathrm{on}}} E\mathrm{d}t \tag{4}$$

From Eq. (4), it can be concluded that during electrochemical micromachining of micro-features, volume of material removed, i.e. machining gap across front and side of the microtool, can be regulated by applied voltage, pulse on-time and machining time, which can be used in machining of complex micro-features.

For machining microgroove of desired depth microtool has to scan number of times for complete scan length, followed by down feed at the end of each scan which takes total machining time T, which can be given as

$$T = \frac{nL}{V_x} \tag{5}$$

where n is the total number of scans, L is the microgroove length and V_x is scanning speed. Hence, volume of material removed in total machining time T can be given as

$$V_{\mathrm{m}} = \left[\frac{CA}{\rho_s h} \int_{0}^{t_{\mathrm{on}}} E\mathrm{d}t\right] \frac{nL}{V_x} \tag{6}$$

From Eq. (6), it can also be concluded that scanning speed of the microtool also has a significant role in regulating volume of material removed and, in turn, regulates the front and side machining gap in layer-by-layer machining of microgrooves. Increased scanning speed reduces the total machining time, decreasing the total volume of material removed, that results in reduced machining gap at front and side, even though numbers of scans are same. Similarly, the reduced scanning speed requires longer machining time, increasing total volume of material removed and generating higher machining gap at front and side.

3 Experimental Details

Experiments have been performed utilizing in-house developed EMM setup which consisted of various subsystems. The same setup has been utilized for the development of disc microtool, required for the further machining of complex microgrooves.

3.1 Experimental Setup

Figure 1 depicts the photographic view of the EMM setup that has been utilized for the development of disc shape microtool, as well as in machining microgrooves of varying cross sections. Developed EMM setup consists of various subsystems integrated together such as mechanical machining unit, controller unit, desktop computer, piezoelectric transducer (PZT), pulse power supply, digital storage oscilloscope and machining chamber with work holding arrangement. The basic function of the mechanical machining unit is to provide the movements to microtool in X, Y and Z directions by combining three separate units of long travel linear stages arranged in gantry type structure, and also to hold microtool and workpiece. Each of the linear stage has a stepper motor of 0.3125 μm resolution and is controlled through machine controller unit through a desktop computer. PZT has been mounted on vertical stage, and microtool tool holder has been mounted on PZT, so that longitudinal vibrations can be transferred to the microtool. Amplitude of vibration and vibrational frequency of PZT has been regulated through power supply of PZT. Pulse generator provides the pulsed power supply with different parameters such as voltage amplitude, pulse frequency and duty ratio to the microtool and workpiece. Digital storage oscilloscope has been utilized for observing the nature of pulse being supplied, for online monitoring and during IEG setting. Microgrooves have been machined on stainless steel, measured and analysed utilizing measuring microscope and SEM micrographs, respectively.

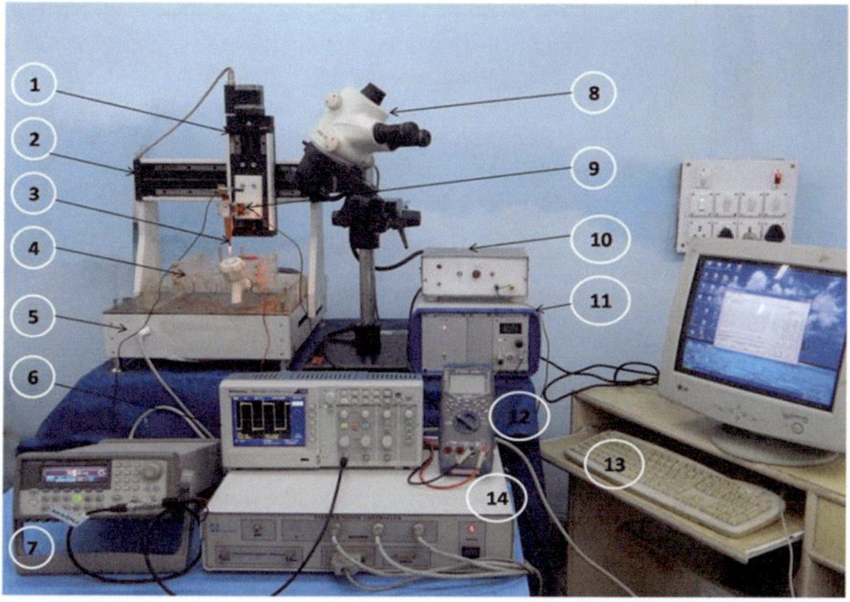

Fig. 1 Photographic view of developed EMM setup. *1* Long travel linear stage, *2* gantry bridge, *3* microtool holder, *4* machining chamber, *5* mechanical machine unit, *6* digital storage oscilloscope, *7* pulse generator, *8* steoreozoom microscope, *9* PZT, *10* PZT power supply, *11* amplifier, *12* digital multimeter, *13* desktop computer and *14* machine controller unit

3.2 Planning of Experiments

In-house developed sidewall insulated disc microtool has been utilized for machining of complex microgrooves by EMM. EMM characteristics such as side machining gap and linearity of microgroove have been studied to develop the strategy for machining the various cross-sectional shapes of the microgrooves. Varied cross-sectional microgrooves have been machined by layer-by-layer machining by scheming the important process parameters while machining the microgroove. Microtool vibration system has also been utilized during machining to ensure the supply of fresh electrolyte at IEG.

To study the influence of process parameters such as applied voltage, duty ratio and machining time over side machining gap and linearity, microgrooves have been machined on stainless steel using sidewall insulated disc microtool by layer-by-layer machining. Scanning type machining also takes care of supply of fresh electrolyte at narrow IEG during each scan. Therefore, layered machining strategy for microtool movement has been followed, in which microtool moves with horizontal tool feed of 109 μm/s for complete scan length and down feed of 0.625 μm at the end of each scan. Fabricated microgrooves have been analysed for the measurements and shape accuracy. Widths of the microgroove have been measured at different locations along its length, and side machining gap has been determined. Standard deviation (σ) for width of microgroove has been analysed to appraise the uniformity of microgroove along its length. Side machining gap characteristics have been analysed and applied for machining of different micro-features inside the microgrooves by controlling the process parameters during machining of microgrooves.

While machining deep microgrooves by EMM, continuous supply of fresh electrolyte at machining zone and taking away the process by-products from narrow IEG are the main challenges. Longitudinal vibrations of the microtool with few microns of amplitude of vibration help to improve the supply of fresh electrolyte by removing sludges from machining region. Vibration frequency of 88 Hz and 4–5 μm amplitude of vibration result in best machining performance, which has been decided through trial experiments (Bhattacharyya et al. 2007). Finally, SEM micrographs of the machined microgrooves with different internal micro-features have been analysed for measurements and for further analysis.

3.3 Layer-by-Layer Machining Strategy

In layer-by-layer machining, microtool moves for complete machining length with full scanning speed and feeds vertically downward by short distance at the end. Hence, for fabricating the microgrooves of desired length and depth, microtool has to scan to and fro along the full length of microgroove for '*n*' number of times to complete total depth, with vertical down feed by small distance, i.e. scan depth, each time.

3.4 Disc Shape Microtool and Its Sidewall Insulation

Microtool is the vital element in EMM. In EMM, microtools with finished surface quality are preferred because the characteristics of microtool such as shape, size and surface finish are conveyed to the workpiece. For microtool fabrication, tungsten is one of the best suitable materials, because of its various important properties like hardness and corrosion resistance. Figure 2 shows the microscopic image of fabricated disc shape microtool of diameter 200 μm and height 70 μm, developed from tungsten micro-rod specimen of diameter 200 μm by reverse EMM (Rathod et al. 2013). Machining conditions during fabrication of disc shape microtool are given in Table 1.

In EMM, use of disc microtool confines the dissolution of workpiece to disc region and reduces the taper formation at the sidewalls of micro-features as shown in Fig. 3 (Rathod et al. 2014). To confine the dissolution of workpiece material along the disc region, the radial difference between the disc radius and shank radius of the microtool electrode (*d*) must be greater than the difference between the

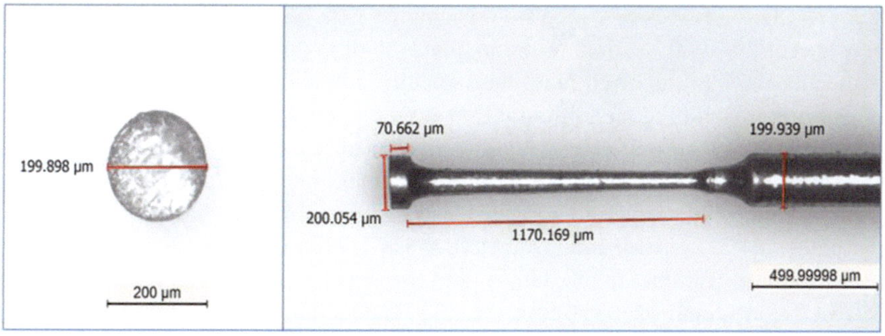

Fig. 2 Disc shape microtool fabricated by EMM

Table 1 Machining parameters in fabrication of disc microtool

Parameters	Details
Microtool (anode)	Tungsten specimen Ø200 μm
Workpiece (cathode)	SS sheet with central hole of Ø2.5 mm
Electrolyte	1 M NaOH
Applied voltage	1 V
Pulse frequency	1 MHz
Duty ratio	80%
Machining time	15 min

Fig. 3 Influence of disc shape on sidewall profile of micro-feature (Rathod et al. 2017)

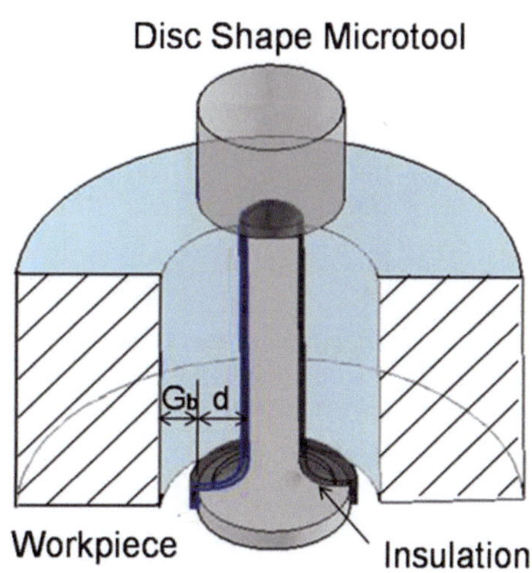

machining gap, i.e. $d > G_b$ (Kim et al. 2005). Machining gap depends upon machining parameters; ultimately, the desired radial difference between the disc and shank of the electrode also depends on the machining parameters. Fabrication of complex microgrooves of varied side machining gaps along its depth needs various disc microtools with various combinations of disc and shank diameters that is practically impossible, or utilizing the same disc microtool with sidewall insulation by varying machining conditions. Therefore, to avoid the over dissolution, to reduce the stray current effects and to minimize the taper formation along the sidewalls of microgroove while machining complex micro-features with varied side machining gaps, disc microtool sidewalls are insulated. Sidewall insulation of disc microtool makes the EMM non-sensitive to the microtool length during machining micro-features of higher depth; hence, uniform side machining gap can be obtained irrespective of depth of micro-features. Use of disc microtool assures the presence of fresh electrolyte at machining zone along with easy removal of EMM by-products from very small machining zone as compared with straight cylindrical microtool due to smaller shank diameter and smaller disc height that enhances the overall machining performance.

4 Results and Discussions

Machining of perfect micro-features of required dimension by EMM desires to know the accurate amount of side machining gaps, which may generate during machining. Therefore, to investigate the side machining gap characteristics, microgrooves have been machined utilizing developed EMM system setup, and influence of the major process parameters such as applied voltage, duty ratio and machining time on side machining gap and uniformity of microgrooves has been analysed. Experimental outcomes have been graphed to reveal the effect of various factors on side machining gap.

4.1 Influence of Applied Voltage on Side Machining Gap and Linearity

To study the effect of applied voltage on side machining gap and uniformity of the microgroove along its length, microgrooves have been fabricated with applied voltage 2.8–4.0 V, pulse frequency 5 MHz, 30% duty ratio and 0.15 M H_2SO_4 as an electrolyte. Figure 4 depicts the microscopic images of microgrooves machined at different applied voltages. Widths of the machined microgroove have been measured at various points of the microgroove as shown in the figure. Microgroove widths obtained at various voltages of 2.8, 3.2, 3.6 and 4.0 V are 305, 344, 356 and 425 μm, respectively. The increase in length of the machined microgrooves can also be seen as 2269, 2281, 2356 and 2332 μm, at various applied voltages.

Figure 5 plots the influence of applied voltage on side machining gap and standard deviation of machined microgrooves. From the figure, it can be observed

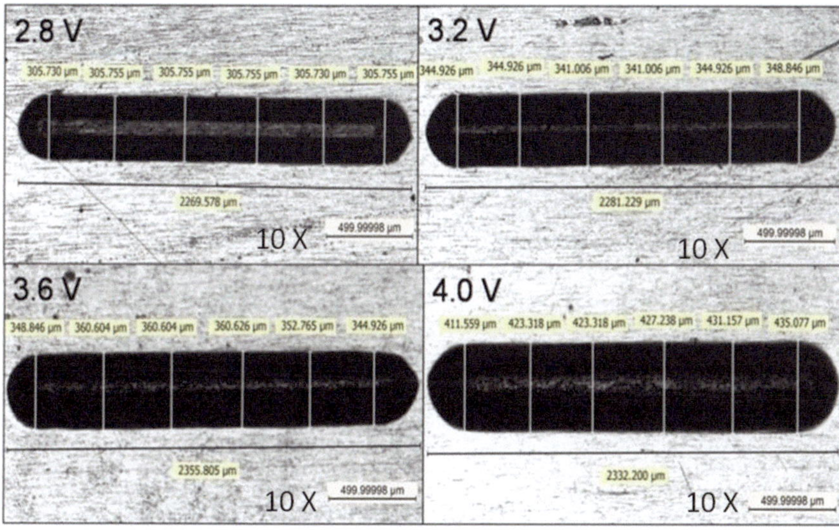

Fig. 4 Microgrooves machined at various applied voltages (Rathod et al. 2017)

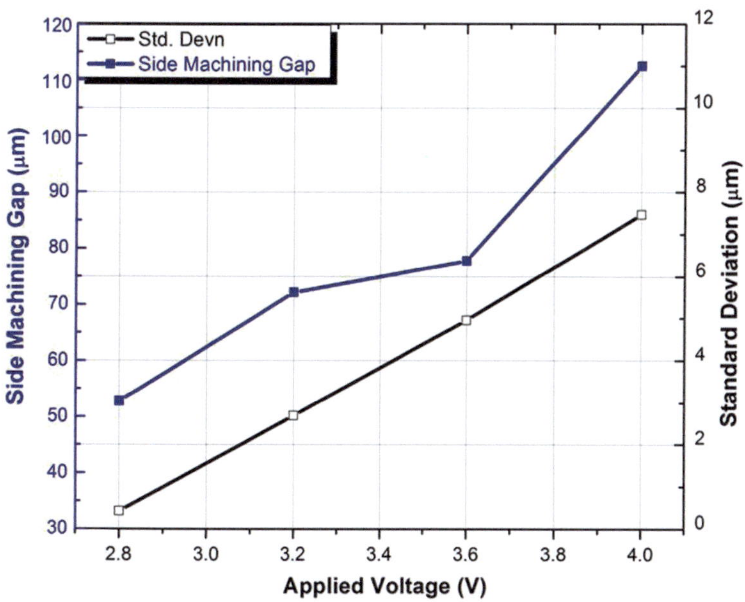

Fig. 5 Variation in machining gap and standard deviation with applied voltage (Rathod et al. 2017)

that with increase in applied voltage, side machining gap increases and linearity of microgroove decreases. The reason behind this is increased machining current at higher applied voltage that increases the current density and in turn reduces the machining localization. Finally, side machining gap increases with increase in applied voltage due to increased material dissolution as per Eq. (6). From above experiments, it can be said that applied voltage has a key role in regulating the side machining gap in EMM, and this factor can be utilized for fabricating microgrooves of varied side machining gaps.

4.2 Influence of Duty Ratio on Side Machining Gap and Linearity

The ratio of pulse 'on' time to the total pulse period is the duty ratio, and it implies the percentage of time for which pulse remains 'on'. Material dissolution occurs during the pulse 'on' time only; hence, pulse 'on' time is very important parameter in EMM. To examine the influence of pulse 'on-time' over side machining gap and uniformity of machined microgrooves, duty ratio has been varied from 30 to 60%, i.e. pulse 'on' time from 60 to 120 ns, with pulse frequency 5 MHz, applied voltage 2.8 V and 0.15 M H_2SO_4 as an electrolyte. Figure 6 shows the microscopic images of machined microgrooves at different duty ratios. Microgroove widths obtained at different duty ratios of 30, 40, 50 and 60% are 297, 367, 418 and 583 μm, respectively. The length of machined microgroove can also be observed as 2230, 2277, 2332 and 2371 μm, at different applied voltages. Figure 7 demonstrates the change in side machining gap and standard deviation of microgrooves machined at different duty ratios. Increase in side machining gap and reduction in linearity of microgroove with increase in duty ratio can be seen from Figs. 6 and 7. This can be explained with the help of nature of pulse waveforms observed during machining.

Figure 8 shows the digital images of pulse waveforms obtained by digital storage oscilloscope. Pulse parameters as can be seen from Fig. 8a are applied voltage 5 V, pulse frequency 5 MHz, pulse period 200 ns, duty ratio 30% and pulse width, i.e. pulse on-time 60 ns. Single pulse consists of pulse on-time and pulse off-time, and pulse on-time increases with increase in duty ratio. Dissolution of workpiece occurs during pulse 'on' time, and process by-products like sludge and heat are taken out from machining region during pulse 'off' time mainly. Increased pulse on-time with increases in duty ratio increases the volume of material removed as per Eq. (6). Increased machining current at higher duty ratio reduces machining localization, which increases material dissolution that increases machining gap finally. From these experiments, it is observed that duty ratio also has an important role in regulating side machining gap in EMM and can be used for machining of micro-features having different machining gaps.

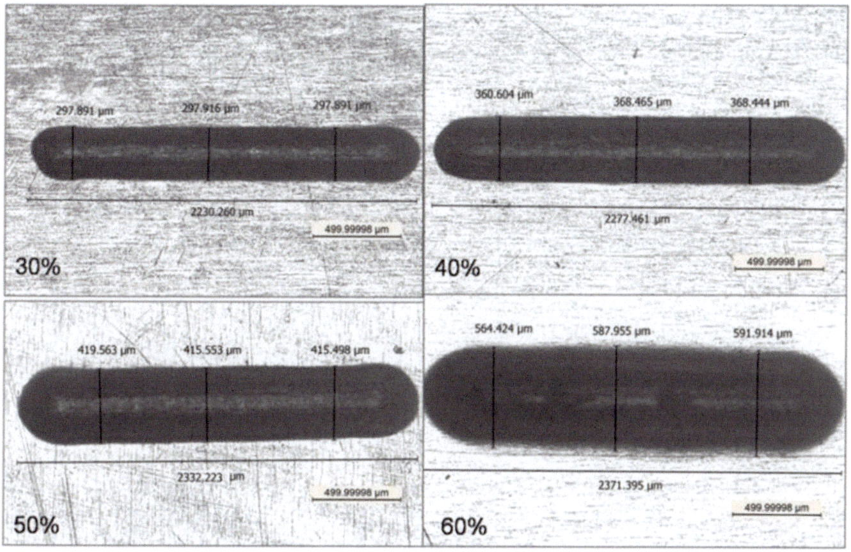

Fig. 6 Machined microgrooves at different duty ratios (Rathod et al. 2017)

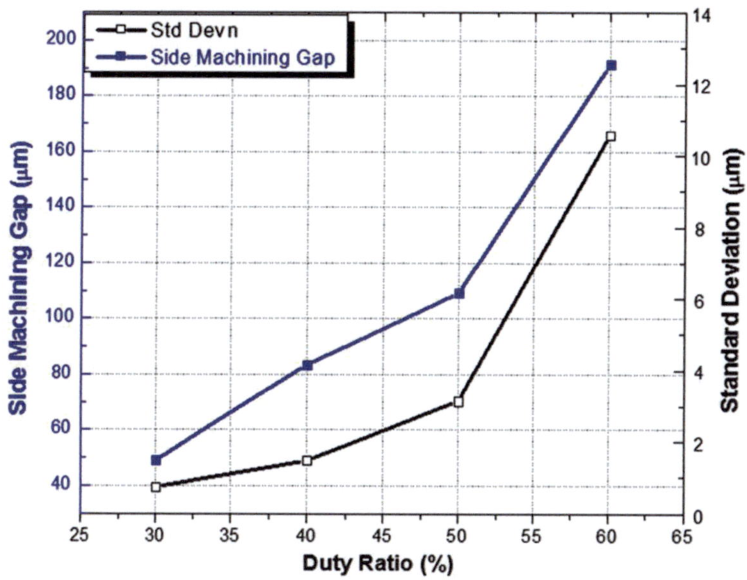

Fig. 7 Variation in machining gap and standard deviation with duty ratio (Rathod et al. 2017)

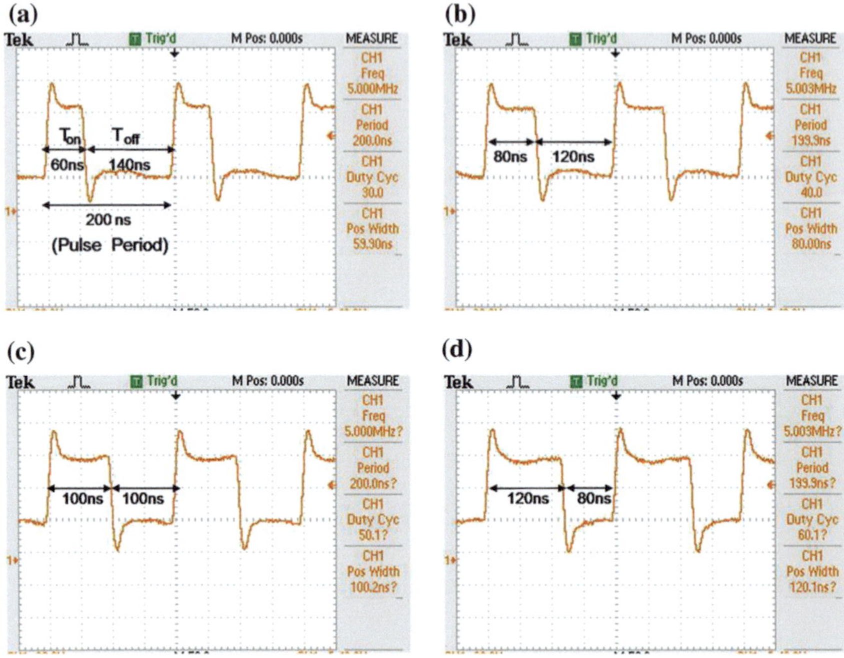

Fig. 8 Pulse waveforms at: **a** 30%, **b** 40%, **c** 50% and **d** 60% duty ratio (Rathod et al. 2017)

4.3 Influence of Machining Time on Side Machining Gap and Linearity

To study the effect of total machining time over side machining gap and linearity, microgrooves have been machined to various machining times from 3 to 12 min, with applied voltage 2.8 V, pulse frequency 5 MHz, duty ratio 30% and 0.15 M H_2SO_4 as an electrolyte with inter-electrode gap of 10 μm with no vertical microtool feed. Figure 9 depicts the microscopic images of the machined microgrooves at various machining times. Microgroove widths obtained at different machining time of 3, 6, 12 and 15 min are 260, 272, 292 and 313 μm, respectively. Similarly, the lengths of machined microgroove obtained are 2054, 2114, 2198 and 2223 μm, respectively. Figure 10 shows the variation in side machining gap and linearity of the microgroove machined at various machining times. From the figure, it can be observed that side machining gap rises with machining time, and linearity of microgroove enhances initially with increase in machining time, and afterwards, it diminishes. This is due to the continuous dissolution of workpiece with increased machining time, which ultimately increases the side machining gap. Inter-electrode gap increases with progress of time that diminishes the machining localization, and finally reduces the linearity and increases the side machining gap of the

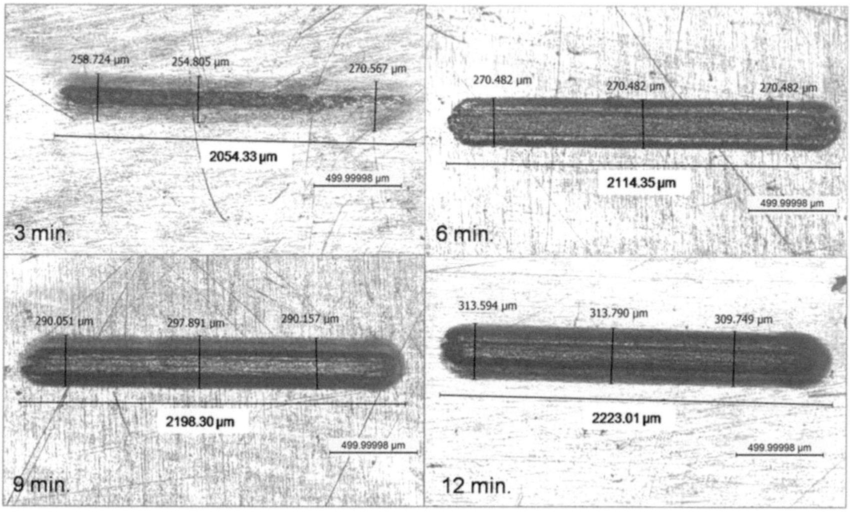

Fig. 9 Machined microgrooves at various machining times (Rathod et al. 2017)

microgroove. From these experiments, it can be said that machining gap varies also with machining time and can be used for machining of micro-features with different side machining gaps.

5 Machining of Varied Cross-Sectional Microgrooves

Influence of various process parameters on side machining gap has been analysed through the experiments. Variation in machining gap with process parameters offers the possibility of machining different micro-features inside the microgrooves to generate various cross-sectional microgrooves. Utilizing sidewall insulated disc microtool with longitudinal vibrations, adopting layer-by-layer machining strategy and by varying the process parameters during machining of microgroove, fabrication of various cross-sectional microgrooves has been demonstrated (Rathod et al. 2017).

5.1 Machining of Reverse Tapered Microgroove

Microgrooves with reverse taper are very much difficult to be machined by various present micromachining techniques, whereas such microgrooves can be machined without any difficulty by increasing duty ratio gradually along the depth of microgroove during machining. Figure 11 illustrates the planned profile of reverse

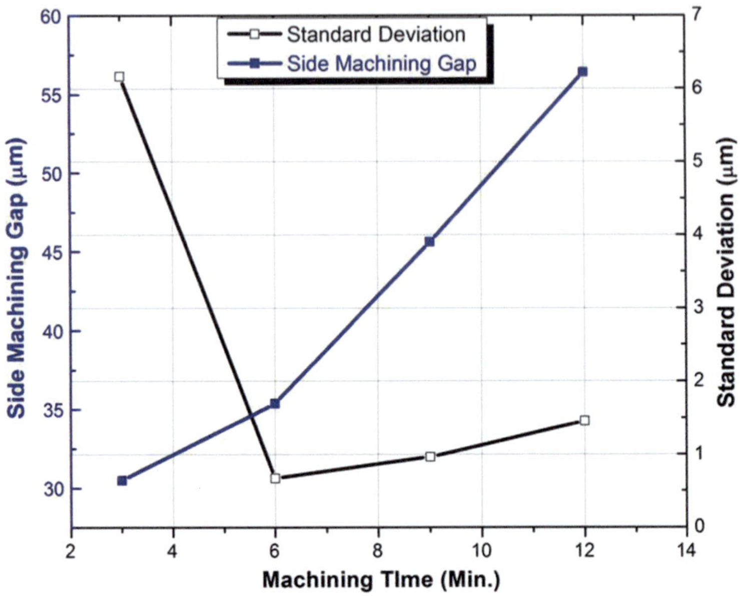

Fig. 10 Deviation in machining gap and standard deviation with machining time (Rathod et al. 2017)

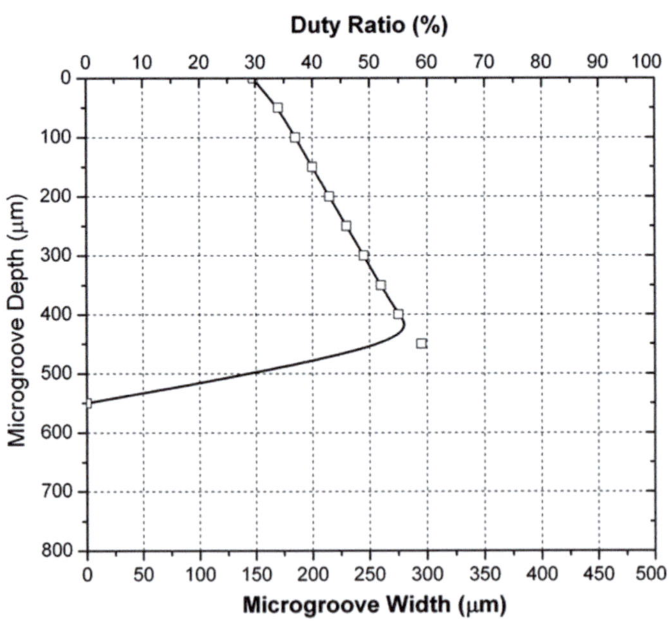

Fig. 11 Planned profile for reverse tapered microgroove (Rathod et al. 2017)

tapered microgroove as per the experimental findings, with machining parameters as applied voltage 3.2 V, pulse frequency 5 MHz, 0.15 M H_2SO_4 as an electrolyte and scanning speed of 109 μm/s with tool feed of 0.625 μm at the end of each scan by varying duty ratio gradually from 30 to 60% along the depth during machining.

Increase in duty ratio during machining of the microgroove leads to the more material dissolution, i.e. higher width of microgroove that gives the reverse tapered microgroove. Application of insulated disc-shaped microtool controls over dissolution of workpiece material along sidewalls of the microgroove, as well as stray current effects even at higher duty ratio. Thus, when duty ratio increases with respect to the depth during machining of microgroove, the width increases only at front end of the microtool and top face width, i.e. entry width of microgroove remains unchanged. Hence, microgroove having gradual increase in width towards the bottom can be fabricated by this method. It is also possible to fabricate the microgrooves of various reverse taper angles by controlling the duty ratio during machining of microgroove. Figure 12 depicts the SEM images of machined reverse tapered microgroove. From the figure, entry width 354.14 μm and bottom width 553.20 μm with reverse taper angle of 55° for total depth of 534.94 μm can be seen clearly. Curved surface is generated at the bottom of the microgroove, due to the nature of the current distribution. In disc shape microtool, more current flows through the core of the disc, in comparison with the periphery because of shorter path of current flow at core. Higher current at core dissolves more material in central portion, i.e. bottom of the microgroove as compared to periphery, leading the curved surface at the bottom of the microgroove.

5.2 *Machining of Barrel-Shaped Microgroove*

Applied voltage directly influences the material removal in EMM. Hence, varying cross-sectional microgrooves can also be obtained by regulating the applied voltage during microgroove machining. Barrel-shaped microgroove that has increased

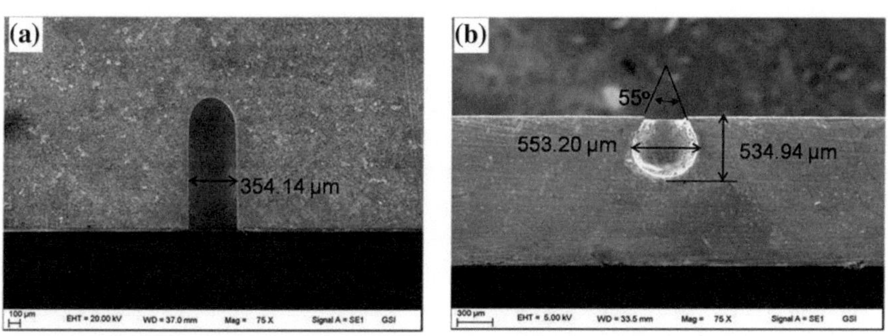

Fig. 12 SEM micrograph of machined microgroove (Rathod et al. 2017)

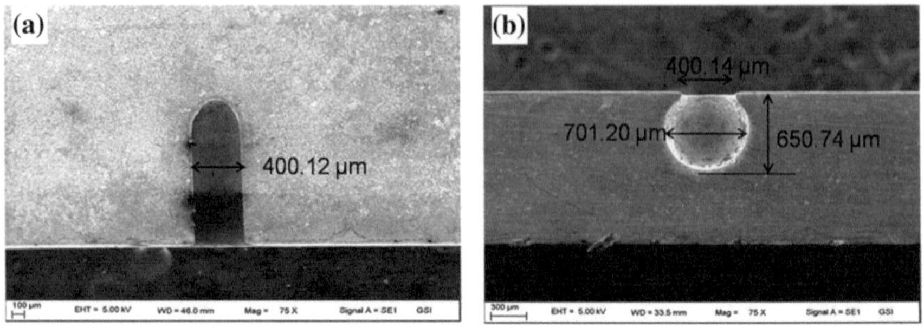

Fig. 13 SEM micrograph of barrel shape microgroove (Rathod et al. 2017)

width up to mid and reduced width for remaining depth can be machined by varying applied voltage from 3.2 to 3.8 V up to midsection of microgroove and again decreased from 3.8 to 3.2 V for remaining depth with pulse frequency 5 MHz, duty ratio 30% and 0.15 M H_2SO_4. Figure 13 shows the SEM micrographs of machined microgroove with entry width 400 μm with increasing width. It can also be seen that maximum diameter of 701 μm at the mid of microgroove, with total depth of 650.74 μm.

Increase in applied voltage enlarges side machining gap at front face of the microtool resulting in increased width of microgroove; further continuation of machining with reduced applied voltage leads to the reduced machining gap. Therefore, a microgroove of increasing width up to desired depth and reducing width for remaining depth can be machined. Microgrooves of different opening widths as well as different internal widths can be fabricated by combination of applied voltage and other parameters during machining of microgroove. Reduced scanning speed enlarges the opening width that also supports the fabrication of microgrooves of required dimension.

5.3 *Machining of Spherical-Based Microgroove*

In EMM, machining gap can be regulated not only by changing the pulse parameters but also by changing the machining time. Microgrooves with uniform width up to required depth having spherical base are challenging to machine by the present techniques. Such microgrooves can be fabricated easily by machining microgrooves up to desired depth with constant parameters and further prolongation of machining at higher duty ratio, with no microtool down feed for generation of spherical base. Figure 14 shows the SEM micrograph of microgroove machined with applied voltage 3.2 V, pulse frequency 5 MHz, duty ratio 30%, 0.15 M H_2SO_4 as an electrolyte and scanning speed of 109 μm/s with 0.625 μm vertical feed at the end of scan up to 250 μm depth. After that, microgroove has been machined with increased duty ratio of 50% for 10 min with no vertical down feed.

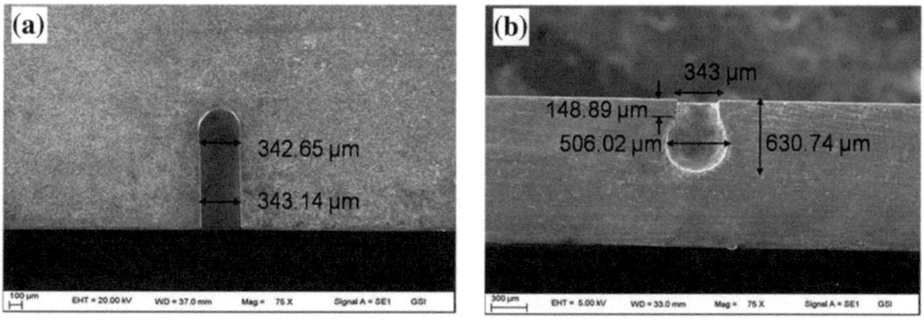

Fig. 14 SEM micrograph of microgroove with spherical base (Rathod et al. 2017)

The figure shows the entry width of 343 μm up to 149 μm depth with having spherical base of 506 μm diameter, machined for the total depth of 630.74 μm.

Sidewall insulated disc microtool limits over dissolution of material even at higher machining time and higher duty ratio. Therefore, at increased machining time and duty ratio, width of microgroove, i.e. machining gap, increases at front end of microtool only, and width at an entrance of microgroove remains unchanged. Hence, a microgroove with constant width up to required depth with spherical base of higher diameter can be machined. Because of the nature of current distribution at the front end of the disc shape microtool, spherical base is generated at base of straight microgroove. Increased duty ratio reduces machining localization, which results in material dissolution from wider area and generates sphere of higher diameter at bottom of machined microgroove, since there is no vertical tool feed. Microgrooves of different widths with spherical base of various diameters can be generated on metallic surfaces by regulating different process parameters during machining.

5.4 *Machining of Stepped Microgroove*

Microgrooves of very complex profiles can be machined by devising process parameter along the depth during machining of microgrooves. Schematics of the profile planned for stepped microgroove are as shown in Fig. 15. Machining of stepped microgroove has been planned into two steps. Microgroove of desired depth 'Z_1' has been planned with higher duty ratio of 60% for depth of 300 μm, and microtool travels for distance 'Z_2', i.e. 100 μm without machining to overcome the depth overcut generated at higher duty ratio. Afterwards, microgroove at the base has been planned with lower duty ratio of 30% for required depth of 'Z_3', i.e. 200 μm. Other process parameters have been kept identical during machining for both steps of the microgrooves as applied voltage 3.2 V, pulse frequency 5 MHz, 0.15 M H_2SO_4 as an electrolyte and scanning speed of 109.375 μm/s.

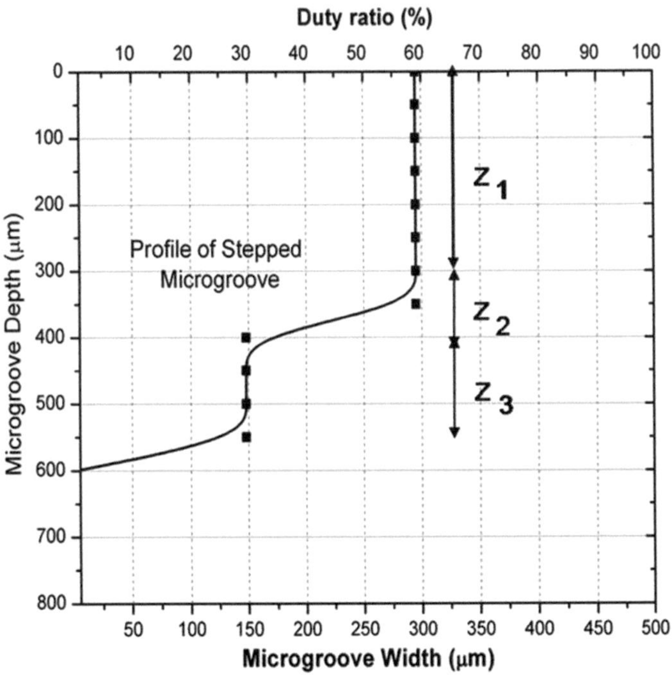

Fig. 15 Profile planned for stepped microgroove

Figure 16 depicts the SEM image of double stepped microgroove fabricated by EMM using disc microtool. The figure shows the average microgroove width of 561 μm at the top face and inner microgroove of 323 μm width. Total depth of the microgroove is 602 μm, with inner groove depth of 241 μm. Machining with increased duty ratio results in higher machining gap due to increased pulse 'on' time, in first step. Afterwards, material dissolution with reduced duty ratio results in smaller side machining gap. From Figs. 15 and 16, it can be seen that, by adopting scanning type strategy of microtool movement during machining, microgrooves with multiple steps of desired shape and size can be fabricated by EMM utilizing the same disc shape microtool by controlling the EMM parameters. Rounded surface is formed at the base of fabricated microgroove in comparison with the planned flat base of the microgroove. This is due to the nature of the current distribution through the microtool and poor machining localization at the front face of the microtool. However, flat surfaces can also be obtained by improving the process parameters, i.e. with lower applied voltage, and ultrashort pulses.

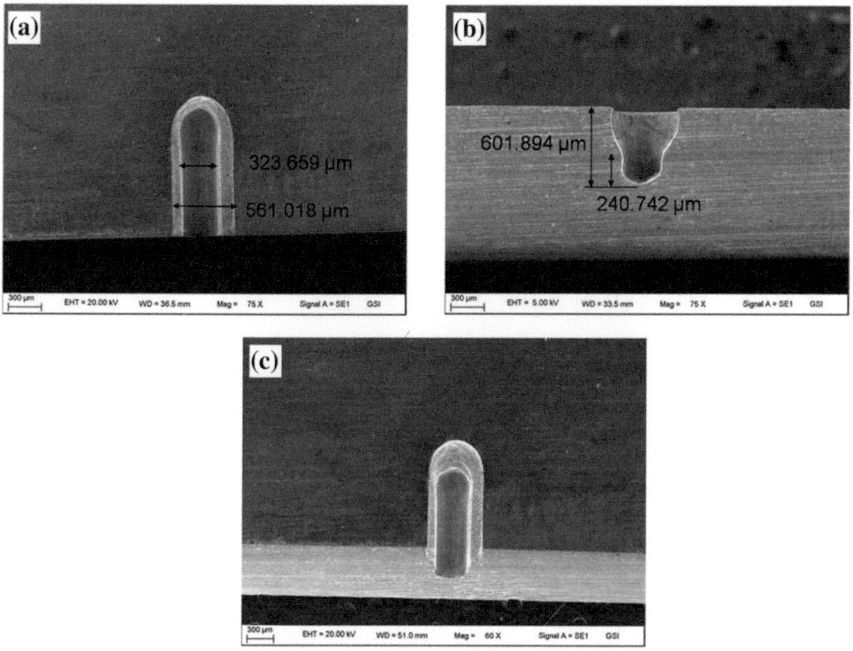

Fig. 16 SEM micrograph of stepped microgroove (Rathod et al. 2017)

5.5 *Machining of Microgroove with Internal Pocket*

Figure 17 shows the planned profile for machining the microgroove having internal pocket. Microgroove with narrow opening is intended to be fabricated with smaller duty ratio for desired depth as indicated by 'Z_1', followed by the fabrication with increased duty ratio for internal pocket of desired dimension as shown by 'Z_2' with other identical parameters.

Figure 18 depicts the SEM images of fabricated microgroove with internal pocket using disc microtool with applied voltage 3.2 V, pulse frequency 5 MHz, 30% duty ratio for 200 μm depth and 60% duty ratio for 400 μm depth, 0.15 M H_2SO_4 as an electrolyte and scanning speed of 109.375 μm/s with 0.625 μm vertical feed to the microtool at the end of each scan. Average opening width of 325 μm with the depth 121 μm. The internal pocket of 527 μm × 671 μm can be seen for the total depth of 794 μm. From the figures, it can be seen that by following this scheme of machining, microgrooves with varied opening width and internal pocket of desired dimensions, i.e. width and depth, can be fabricated utilizing the same disc shape microtool.

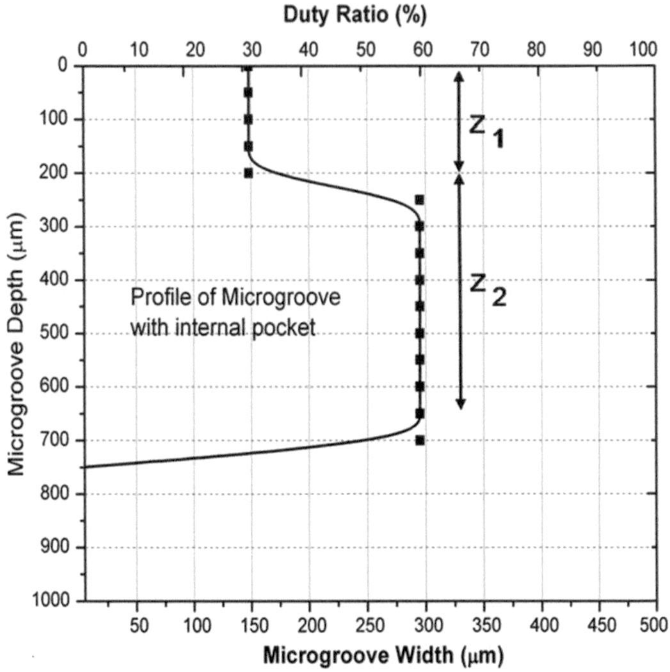

Fig. 17 Planned profile of microgroove with internal pocket

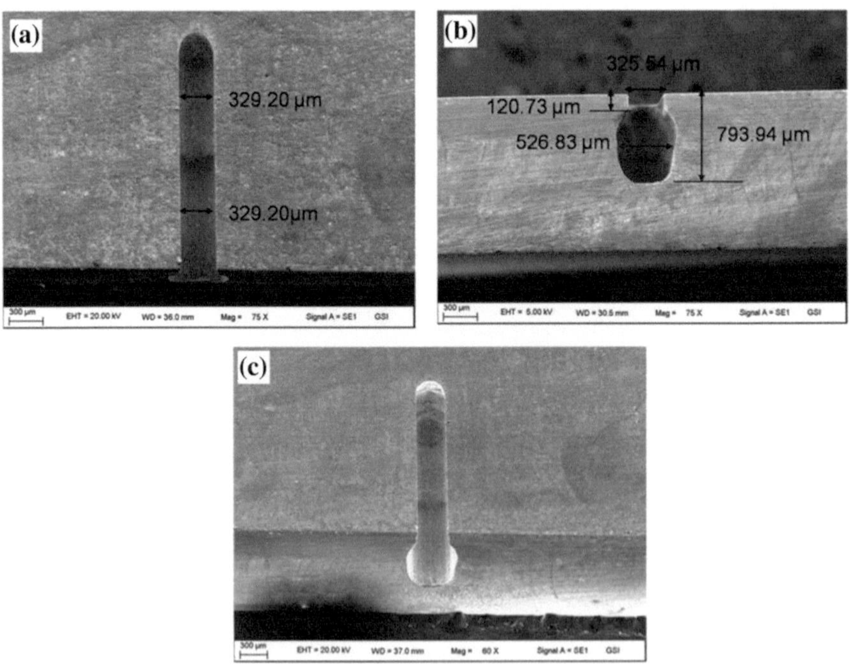

Fig. 18 SEM micrograph of stepped microgroove

6 Guidelines for Machining Varied Cross-Sectional Microgrooves

Investigational outcomes have been used for planning the desired cross section of the microgroove. Tables 2, 3, 4, 5, 6 and 7 provide the guidelines for machining microgrooves with different internal features forming different cross-sectional microgrooves, machined at applied voltage 3.2 V, pulse frequency 5 MHz, 0.15 M H_2SO_4 as an electrolyte and scanning speed of 109 µm/s with tool feed of 0.625 µm at every scan by varying duty ratio gradually from 30 to 60% during machining of microgroove. Use of sidewall insulated disc microtool with microtool vibrations has been considered in process. Tabulated data will be helpful to plan and machine different features inside the microgroove.

Table 2 provides the guidelines for controlling the duty ratio during fabrication of straight microgrooves of required depth. The total depth of microgroove considered is 700 µm. For machining straight microgroove of required depth, duty ratio is maintained constant throughout the machining depth. When the duty ratio is kept 30%, i.e. DR_1, with other constant parameters as mentioned above, the width of microgroove predicted from the experimental results is W_1, i.e. 296 µm. Similarly, by varying duty ratio to different values ranging from 40, 50 and 60%, i.e. DR_2, DR_3 and DR_4, microgrooves of different widths 368, 418 and 582 µm can be machined for required depth.

Table 3 presents the guidelines of varying the duty ratio in fabrication of reverse tapered microgrooves of required taper angle and depth. Total depth of the microgroove considered is 700 µm. Machining is started with 30% duty ratio, and it is increased gradually by 1° for every 20 µm depth, i.e. with vertical tool feed, to get the reverse tapered microgroove. In this way, microgrooves with different reverse taper angles can also be machined by changing the duty ratio at the start of machining of microgroove, as well as by varying the rate of change of duty ratio along the depth.

Table 4 gives the guidelines of varying the duty ratio along the depth of microgroove to machine barrel shape microgrooves of required dimensions. Total depth of the microgroove considered is 700 µm. Duty ratio is increased up to the mid-depth of microgroove and is reduced gradually after the mid-depth, to get the barrel shape of microgroove. Machining is started with 30% duty ratio and increased gradually by 1° for every 10 µm of vertical tool feed up to 400 µm, and duty ratio is gradually reduced with the same rate for remaining depth. Likewise, microgrooves of various barrel dimensions can also be fabricated by changing the duty ratio at the start of microgroove machining, as well as by varying the rate of change of duty ratio along the depth.

Table 5 illustrates the guidelines of varying the duty ratio during fabrication of V-shaped microgrooves. Total depth of the microgroove considered is 700 µm. Duty ratio is reduced gradually up to the required depth to achieve the *V* shape of microgroove. Machining is started with 60% duty ratio and is gradually reduced by 1° for every 20 µm of vertical tool feed required depth. Likewise, microgrooves of

Table 2 Guidelines for machining straight microgroove

Pt.	Microgroove depth	Duty ratio				Microgroove width				Microgroove shape
		DR_1	DR_2	DR_3	DR_4	W_1	W_2	W_3	W_4	
a	00	30	40	50	60	296	368	418	582	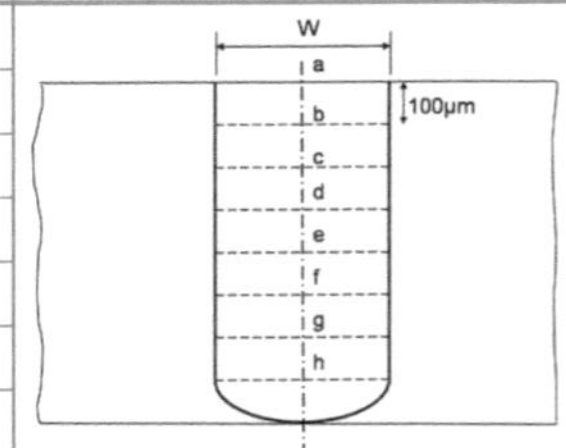
b	100	30	40	50	60	296	368	418	582	
c	200	30	40	50	60	296	368	418	582	
d	300	30	40	50	60	296	368	418	582	
e	400	30	40	50	60	296	368	418	582	
f	500	30	40	50	60	296	368	418	582	
g	600	30	40	50	60	296	368	418	582	
h	700	30	40	50	60	296	368	418	582	

Table 3 Guidelines for machining reverse taper microgroove

Pt.	Microgroove depth	Duty ratio	Microgroove width	Microgroove shape
a	00	30	296	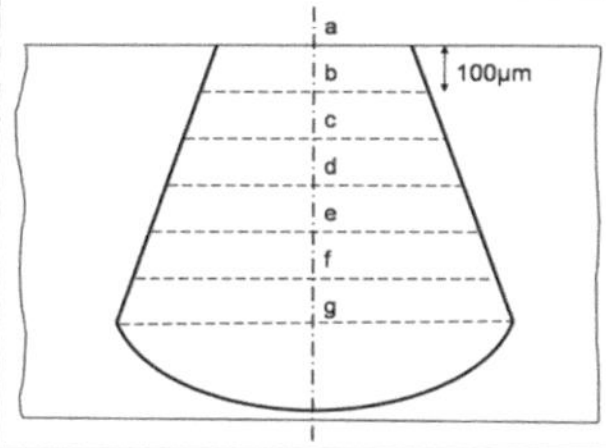
b	100	35	332	
c	200	40	368	
d	300	45	393	
e	400	50	418	
f	500	55	500	
g	600	60	582	

Table 4 Guidelines for machining barrel shape microgroove

Pt.	Microgroove depth	Duty ratio	Microgroove width	Microgroove shape
a	00	30	296	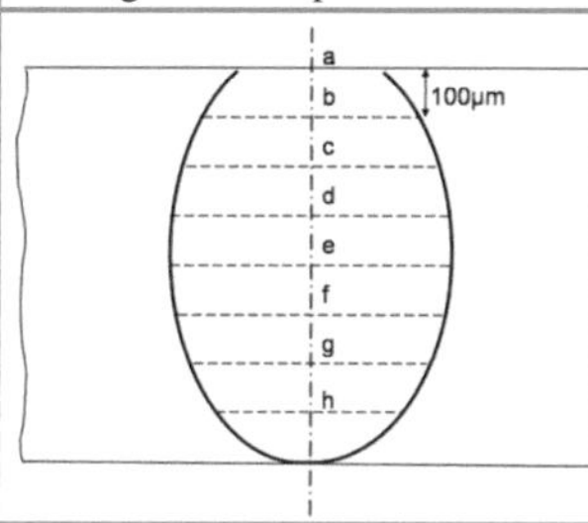
b	100	40	368	
c	200	50	418	
d	300	60	582	
e	400	50	418	
f	500	40	368	
g	600	30	296	

Table 5 Guidelines for machining *V* shape microgroove

Pt.	Microgroove depth	Duty ratio	Microgroove width	Microgroove shape
a	00	60	582	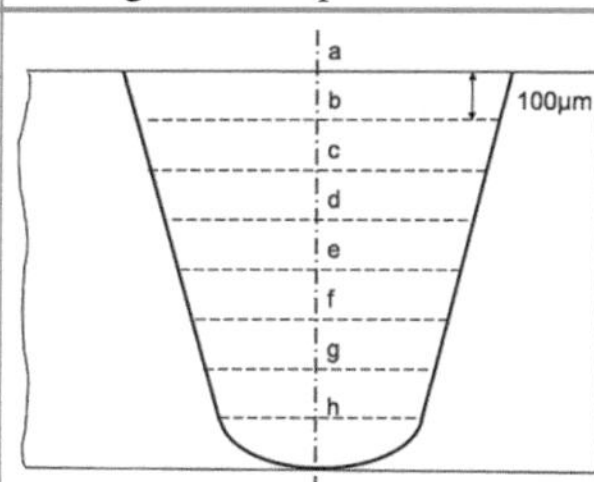
b	100	55	500	
c	200	50	418	
d	300	45	393	
e	400	40	368	
f	500	35	332	
g	600	30	296	

various *V* angles can also be fabricated by changing the duty ratio at the start of microgroove machining, as well as by varying the rate of change of duty ratio along the depth.

Table 6 offers the guidelines of varying the duty ratio during fabrication of straight microgrooves with spherical base. Microgroove is machined with fixed duty ratio up to certain depth to generate straight microgroove, and afterwards, duty ratio is increased to higher value and machined for predetermined time without

Table 6 Guidelines for machining spherical base microgroove

Pt.	Microgroove depth	Duty ratio	Microgroove width	Microgroove shape
a	00	30	296	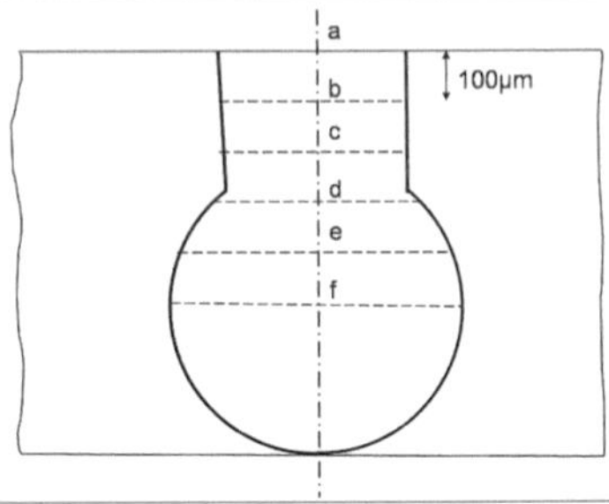
b	100	30	296	
c	200	30	296	
d	300	30	296	
e	400	30	368	
f	500	50[a]	520	

[a]Machining for 10 min without any vertical feed

Table 7 Guidelines for machining double microgroove

Pt.	Microgroove depth	Duty ratio	Microgroove width	Microgroove shape
a	00	60	582	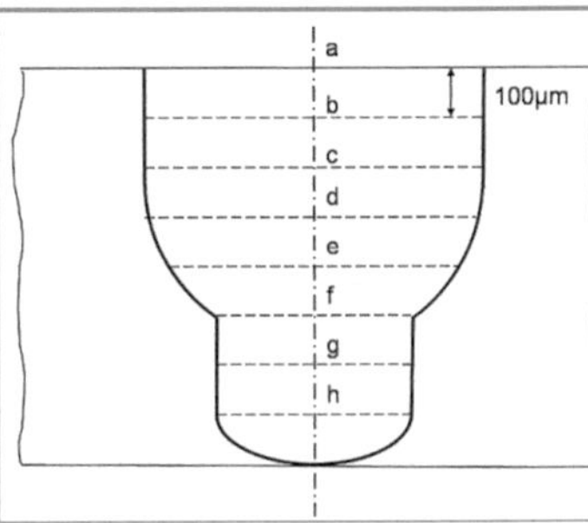
b	100	60	582	
c	200	60	582	
d	300	60	582	
e	400	00	430	
f	500	30	296	
g	600	30	296	
h	700	30	296	

vertical tool feed, to achieve the spherical base. Microgroove is machined with 30% duty ratio up to 400 μm depth, and at the end, duty ratio is increased to 50% and machined for 10 min with no vertical tool feed to generate the spherical base. Likewise, microgrooves of various base diameters can also be fabricated by regulating the duty ratio at the start of microgroove machining, as well as by changing the duty ratio and machining time at the end.

Table 7 presents the guidelines of varying the duty ratio during fabrication of double stepped microgrooves. Outer microgroove has been fabricated with higher duty ratio up to desired depth, microtool is fed down for short distance without machining to clear depth overcut and inner microgroove is machined with reduced duty ratio for the remaining depth to achieve double stepped microgroove. Outer microgroove is machined with 60% duty ratio up to 300 μm depth, and then, microtool is fed down for 100 μm without any material dissolution to clear depth overcut, and finally, inner microgroove is machined with 30% duty ratio for the remaining depth to generate double stepped microgroove. Similarly, microgrooves having number of steps with different dimensions of outer and inner microgrooves can be machined by changing the duty ratio at appropriate depth, during machining.

7 Conclusions

An application of sidewall insulated disc microtool with longitudinal vibration has been proposed in fabrication of various cross-sectional microgrooves by EMM. Machining characteristics, viz. side machining gap and linearity of microgroove, have been analysed through experiments. From experimental results, it has been concluded that

(i) Side machining gap, i.e. width of microgroove, varies with process parameters such as applied voltage, duty ratio and machining time. By regulating these process parameters along the depth of microgroove, it is possible to plan and fabricate different internal micro-features inside the microgrooves.
(ii) To develop the machining guidelines for various cross sections of the microgrooves, duty ratio has been regulated from 30 to 60%, with other process parameters as applied voltage 3.2 V, pulse frequency 5 MHz, 0.15 M H_2SO_4 as an electrolyte and scanning speed 109 μm/s, with tool feed of 0.625 μm at every scan. Guidelines have been developed for straight, reverse tapered, barrel shape, *V* shape, spherical base and double stepped microgrooves.
(iii) Insulation of lateral surface of disc microtool prevents the over dissolution of workpiece during fabrication of complex microgrooves, minimizes the stray current effect and also makes EMM process non-sensitive to the microtool depth.
(iv) Layer-by-layer machining in addition to the microtool vibration assists in removing slush and process by-products from IEG ensuring the presence of fresh electrolyte at machining zone and enhances the machining performance while machining complex microgrooves.
(v) Finally, fabrication of reverse tapered microgroove having reverse taper angle 55° for total depth of 534.94 μm, barrel shape microgroove of 400 μm entry width, 701 μm wide at mid, for total depth 650.74 μm, spherical base microgroove having width 343 μm for 149 μm depth, and spherical base of 506 μm dia. for total depth 630.74 μm, stepped microgroove with double steps having 561 μm width at top for 361 μm depth and inner groove 323 μm wide for 241 μm depth, microgroove with internal pocket with straight opening of 325 μm for 121 μm depth and internal pocket of 527 μm × 671 μm for the total depth of 794 μm has been demonstrated successfully.

Designed guidelines can be used for practicing engineers and researchers in this field for planning and machining varied cross-sectional microgrooves as per the requirement. Machining guidelines have been designed considering variation of duty ratio along the depth to obtain required cross-sectional shaped microgroove. However, other parameters such as applied voltage and machining time can also be considered for designing machining guidelines, which may help to combine the effect of variation of more than one parameter to minimize the total machining time.

References

Bhattacharyya, B., M. Malapati, J. Munda, and A. Sarkar. 2007. Influence of tool vibration on machining performance in electrochemical micro-machining of copper. *International Journal of Machine Tools and Manufacture* 47: 335–342.

Chern, G.L., Y.J. Engin Wu, J.C. Cheng, and J.C. Yao. 2007. Study on burr formation in micro-machining using micro-tools fabricated by Micro-EDM. *Journal of Precision Engineering* 31 (2): 122–129.

Kim, B.H., S.H. Ryu, D.K. Choi, and C.N. Chu. 2005. Micro electrochemical milling. *Journal of Micromechanics and Micro-engineering* 15: 124–129.

Kirchner, Viola, Laurent Cagnon, Rolf Schuster, and Gerhard Ertl. 2001. Electrochemical machining of stainless steel microelements with ultra-short voltage pulses. *Applied Physics Letters* 79 (11): 1721–1723.

McGeough, J.A. 1974. *Principles of electrochemical machining*. London, UK: Chapman and Hall.

Nageswara, Rao P., and Deepak Kunzru. 2007. Fabrication of micro channels on stainless steel by wet chemical etching. *Journal of micromechanics and microengineering* 17: N99–N106.

Oh, Kwang H., M.K. Lee, and S.H. Jeong. 2006. Laser micromachining of high aspect-ratio-metallic groove for application to micro thermal devices. *Journal of Micromechanics and Microengineering* 16: 1958.

Rathod, V., B. Doloi, and B. Bhattacharyya. 2013. Parametric investigation into the fabrication of disk microelectrodes by electrochemical micromachining. *Journal of Micro and Nano-Manufacturing-ASME Transactions* 1: 041005-1–041005-11.

Rathod, V., B. Doloi, and B. Bhattacharyya. 2014. Sidewall insulation of microtool for electrochemical micromachining to enhance the machining accuracy. *International Journal of Materials and Manufacturing Processes* 29 (3): 305–313.

Rathod, V., B. Doloi, and B. Bhattacharyya. 2017. Fabrication of microgrooves with varied cross-sections by electrochemical micromachining. *International Journal of Advance Manufacturing Technology* 92: 505–518.

Yan, J., K. Uchida, N. Yoshihara, and T. Kuriyagawa. 2009. Fabrication of micro end mills by wire EDM and some micro cutting tests. *Journal of Micromechanics and Microengineering* 19 (2): 025004-1–025004-9.

Yan, Jiwang, Takuya Kaneko, Kazunori Uchida, Nobuhito Yoshihara, and Tsunemoto Kuriyagawa. 2010. Fabricating microgrooves with varied cross-sections by electro-discharge machining. *The International Journal of Advanced Manufacturing Technology* 50: 991–1002.

Optimization of Process Parameters in Powder-Mixed EDM

S. Tripathy and D. K. Tripathy

Abstract Electrical discharge machining (EDM) finds wide application all around the world for producing complex contour on very hard metals and alloys. It finds numerous applications in automobile, die making, aerospace, and electronic industries. Low volumetric material removal rate and reduced surface quality are few limitations of the process. Powder-mixed electro-discharge machining (PMEDM) is a novel advancement to EDM that has enhanced the machining capabilities of the process. The present chapter deals with the investigations on the effect of process variables like powder concentration (C_p), peak current (I_p), pulse-on-time (T_{on}), duty cycle (DC), and gap voltage (V_g) on output responses like material removal rate (MRR), tool wear rate (TWR), electrode wear ratio (EWR), surface roughness (SR), recast layer thickness (RLT), and microhardness (HVN) for PMEDM of H-11 hot work tool steel. The work material possesses excellent mechanical characteristics and finds huge application in various industries. Taguchi's L_{27} orthogonal array has been implemented to conduct the experiments using graphite powder-mixed dielectric with the copper tool. Single-objective optimization has been performed to determine the optimum set of input parameters. Analysis of variance (ANOVA) has been implemented to determine the effect of each process parameter and to establish the optimal setting of process parameters. Multi-objective optimization using technique for an order of preference by similarity to ideal solution (TOPSIS) was used to determine the optimum set of process variables to obtain maximum MRR and HVN with minimum TWR, EWR, SR and RLT simultaneously. Predicted results on verification with confirmation tests improve the preference values by 0.1021 with TOPSIS. The recommended settings of process parameters is found to be $C_p = 6$ g/l, $I_p = 3$ A, $T_{on} = 200$ µs, DC = 80% and $V_g = 50$ V from TOPSIS. The microstructures were examined with

S. Tripathy
Mechanical Engineering Department, Siksha 'O' Anusandhan (Deemed to be University), Bhubaneswar 751030, India

D. K. Tripathy (✉)
Indian Institute of Technology Kharagpur, Kharagpur, India
e-mail: dkt1946@yahoo.co.in

S. S. Pande and U. S. Dixit (eds.), *Precision Product-Process Design and Optimization*, Lecture Notes on Multidisciplinary Industrial Engineering,
https://doi.org/10.1007/978-981-10-8767-7_10

scanning electron microscope (SEM) to find the presence of surface defects and identify modifications on the surface.

Keywords PMEDM · Taguchi · TOPSIS · Material removal rate
Tool wear rate · Electrode wear ratio · Surface roughness · Recast layer thickness
Microhardness

1 Introduction

Novel manufacturing theory makes use of non-conventional sources of energy like light, sound, chemical, electrical, mechanical, ions and electrons. The technical and industrial advancement has led to the growth of very hard materials which are difficult to machine but are widely used in nuclear, aerospace and other industries. With the progress in the field of material science, advancement of latest metallic, ceramic materials and composite materials has been witnessed which possess excellent mechanical properties, thermal characteristics and electrical conductivity. Spark erosion machining techniques or non-traditional machining processes are used for machining such exotic materials. Non-traditional machining processes do not employ any conventional tools for material removal. The intricacy of the contour, size, requirement of product accurateness and high surface quality can be overcome by implementing non-traditional methods. Presently, non-conventional processes acquire infinite capabilities, but exhibit poor material removal rates. Enormous developments have taken place in the past few years for the improvement of MRR. With increase in removal rate, the cost efficiency of the process gets maximized, leading to greater application of non-traditional techniques for machining. Electrical discharge machining (EDM) is widely used for making tools, dies and parts with higher accuracy. Modern electric discharge machines have been established globally as a benchmark in manufacturing. They are proficient of machining geometrically complex components such as composites, superalloys, ceramics, heat-treated tool steels, heat-resistant steels, carbides etc. which find application in mould making industries, aeronautics, aerospace and nuclear industries. EDM has reached the recent fields of sports, optics, medicinal and surgical instruments, automotive and R&D areas.

The process of EDM came into existence in 1943, with its foundation in Moscow by the Russian scientists Boris and Natalya Lazarenko. The researchers reported that by immersing the electrodes in oil, steady sparks were generated than air. The phenomenon was inverted and controlled sparks were used for erosion. EDM machine was first developed during war by Lazarenkos. It was very useful for the erosion of metals like tungsten or tungsten carbide. Advancement for understanding the erosion phenomenon evolved in the 1950s. In 1960s, the growth in semiconductor industry endorsed substantial development in EDM machines with increased reliability that produced surfaces with superior quality. The development then led to the design of generators, automation, servo-control and robotics. During

1980s micro-machining using EDM gained a good deal of interest. The EDM process became globalized during this period. Subsequently, advancements to EDM emerged in the 1990s using neural networks, response surface methodology, fuzzy control, central composite design, Taguchi optimization etc. The research in this field is still in progress with innovative ideas of adding additives to the dielectric fluid like conductive powders, nano-sized particles, carbon nanotubes etc.

In EDM process the tool and workpiece material are separated by a small gap in the presence of a dielectric medium. High-frequency-controlled pulses are generated which creates a plasma channel due to the continuous movement of electrons and ions. The temperature in the discharge gap rises to a range of 8000–12,000 °C which causes melting and vaporization. PMEDM improves the process capabilities of EDM by producing surfaces with superior finish and less cracks. Adding fine powders to the dielectric decreases its insulating strength by increasing the inter-electrode gap. The removal of debris becomes easier in the presence of powders. On applying a voltage of 80–320 V an electric field is formed in the range of 10^5–10^7 V/m generating positive and negative charges on powder particles. This energizes powder particles and they move in zigzag manner forming clusters in the sparking area. Bridging occurs under the sparking area creating several discharges in a single pulse. The rapid sparking and erosion from the workpiece surface improve the machining rate. Widening of plasma channel produces stable sparks which form craters with improved surface finish. Material is removed from both the electrodes which combine with the powder particles and modifies the surface properties of the machined surface. Consequently, the MRR increases, TWR reduces and uniform sparks produce corrosion-resistant surfaces. The presence of abrasive powder changes the sparking pattern and improves the surface properties, increases the microhardness gets increased and micro-cracks get minimized. Added powders may be aluminium, chromium, graphite, silicon, titanium etc.

Pecas and Henriques (2008) reported that PMEDM process performance depends upon powder type, concentration, grain size, electrode area, constituents of the tool and workpiece material. Kumar et al. (2009) reviewed the outcome of mixing the dielectric fluid with various powders and additives. Assarzadeh and Ghoreishi (2013) implemented response surface methodology with desirability technique to model and optimize the process parameters during EDM of CK-45 die steel using Al_2O_3 powder-mixed dielectric to improve the MRR. Singh et al. (2012) investigated the effect of process parameters on SR for machining H-11 with the copper tool in the presence of Al powder in the dielectric. Negative polarity of the tool electrode reduced the SR. Talla et al. (2015) conducted multi-objective optimization of PMEDM using Taguchi, GRA and Principal Component Analysis (PCA) to control the process parameters. Lal (2015) performed multi-objective optimization with Taguchi-based GRA for wire EDM. Sidhu et al. (2014) reported the optimal process setting for machining of three types of MMCs using PMEDM. MRR, TWR, SR and surface integrity were examined to determine the significant process parameters. The responses were jointly optimized using technique for order of preference by similarity to ideal solution (TOPSIS) and optimal process conditions were recognized. Singh et al. (2015) examined the outcome of adding graphite powder to the dielectric on the surface properties of superalloy Super Co 605. The results showed that an optimization between microhardness and surface finish can be achieved by this method of machining. Batish and Bhattacharya (2012) studied the material migration occurring between electrode and powder-suspended dielectric fluid for enhancing the surface properties in terms of microhardness for H-11 and H-13 steels. Aluminium, copper, graphite and tungsten powders were added to kerosene, EDM oil, and refined mineral oil as dielectric. Sidhu et al. (2014) optimized MRR, TWR, SR and surface integrity for three different metal matrix composites using TOPSIS and the ranking was done as per the severity of surface defects. Gadakh (2012) applied TOPSIS method for solving multiple criteria optimization problem in wire electrical discharge machining (WEDM) and obtained the most suitable set of process parameters.

From the available literature it is evident that several researchers have reported results using different powders mixed during EDM, but performance characteristics of H-11 during PMEDM needs investigation. The quality of the machined component is defined by various output characteristics such as MRR, TWR, SR, recast layer thickness, microhardness obtained on the machining surface etc. Thus, investigating the significance of the process variables in relation to the output performance characteristics becomes vital. Therefore the problem of PMEDM can be considered as a multi-objective optimization problem. The aim of this study is to obtain a single-optimal setting of various input parameters to obtain a single-output characteristic as a whole. Multi-attribute decision-making techniques like TOPSIS have not yet been used to find the optimal setting during PMEDM of H-11. The present work is a stride in this direction. Taguchi design of experiments is used to conduct the experiments using an L_{27} orthogonal array. An effort has been made to find an optimal set of process variables by using multi-objective optimization using TOPSIS to get maximum MRR and minimum TWR, EWR, SR, RLT and maximum HVN by adding graphite powder to the dielectric in different concentrations. ANOVA has been used to create a relationship among the significant input parameters on the output responses. A comparative study for the EDM and PMEDM surface characteristics has been done using scanning electron microscope. The optimal parameter setting obtained from Taguchi and TOPSIS can be used for quality improvement in industrial applications involving PMEDM.

2 Materials and Methods

The machines, materials and design of experiment technique adopted for the estimation of output responses are highlighted in this section. The procedure used for optimization has also been presented in this section.

2.1 EDM Machine Set-up

The electric discharge machine, model ELECTRONICA-ELECTRAPULS PS 50ZNC has been used for the experiments. Commercial-grade EDM oil has been used as a dielectric fluid. To facilitate the generation of a rectangular form of current pulses for discharging, dielectric fluid was energized by "Current Pulse Generator" and associate controller. The current and voltage waveforms were recorded on a "Digital Storage Oscilloscope". Figure 1 shows the machine used in the present study.

Working tank of the machine had a capacity of 300 L for the circulation of dielectric fluid. The powder particles were required to be added in different

Fig. 1 EDM machine used for the experiment

Fig. 2 PMEDM setup (Tripathy and Tripathy 2017b, c)

concentrations for which changing the entire dielectric fluid and removing the powder particles from the circulating system would have been difficult. The existing circulation system might have choked due to the presence of powders and debris. To minimize the cost, avoid the wastage of dielectric and for effective use of powder particles, a separate machining tank has been designed with a capacity of 20 L. It consists of a machining tank to perform the operation placed in the working tank of EDM. A workpiece fixture assembly was placed in it to hold the workpiece. The machining tank was filled up with the dielectric fluid. A pump was installed to ensure proper distribution of powder in dielectric fluid. To avoid the powder from settling down, a stirring arrangement was installed. Each run was carried out for a time duration of 15 min. The setup of the tank is shown in Fig. 2.

2.2 *Selection of Materials*

The mechanical properties and composition of the workpiece and tool material used for the present experiment have been discussed in this section.

2.2.1 Workpiece Material

H-11 hot work tool steel is the workpiece material. This steel possesses very high strength, abrasion resistance, wear resistance, compressive strength, hardenability, toughness and it is not susceptible to hot cracking. The presence of chromium in the H-11 steel resists oxidization whereas molybdenum prevents corrosion in

Table 1 Mechanical properties of H-11 steel

Properties	H-11
Density	7.81 g/cm^3
Modulus of elasticity	210 GPa
Poisson's ratio	0.30
Yield strength	1650 MPa
Ultimate tensile strength	1990 MPa
Specific gravity	7.8
Melting temperature	1427 °C
Thermal conductivity	42.2 W/mK
Hardness	55 HRC

Table 2 Chemical composition of H-11 steel

Constituent	C	Si	Mn	P	S	Cr	Mo	Co	Cu	V	Fe
% Composition	0.39	1	0.5	0.03	0.02	4.75	1.1	0.01	0.01	0.5	Balance

non-oxidizing environments. The mechanical properties of the H-11 steel are given in Table 1. Applications of H-11 is found in aircraft components, structural use, die casting dies, extrusion tooling, forging dies, piercing tools, hot work punches, tool holders, ejector pins etc. The properties cause a great challenge during machining by conventional methods. The chemical composition of this material as obtained by glow discharge-optical emission spectrometer is given in Table 2. Each surface of the workpiece and tool were machined using CNC-milling machine to get smooth mirror-like surfaces. The electrodes were subjected to surface grinding for proper contact and alignment of surfaces during machining. The dimension of the workpiece used for this EDM operation was 120 × 60 × 25 mm.

2.2.2 Tool Material

Electrolytic copper electrode (99.9%) has been used as the tool electrode material. A square tool of dimension 20 × 20 × 60 mm has been used to perform the machining operation. The mechanical properties of the tool material are given in Table 3. The work material was mounted on the T-slot table and positioned at the desired place and clamped. The electrode was clamped and its alignment was

Table 3 Properties of tool electrode material

Electrode material	Density (g/cc)	Specific heat (J/kg/K)	Thermal conductivity (W/mK)	Electrode resistivity (μΩ)	Hardness BHN
Electrolytic copper	8.9	386	388	1.69	48

checked. The machining was performed for time duration of 15 min. Finally the essential power switches were switched 'ON' for operating the desired machine settings.

2.3 Process Parameters

The process parameters chosen for the present research work are powder concentration (C_p), peak current (I_p), pulse on time (T_{on}), duty cycle (DC) and gap voltage (V_g) to study their effect on output parameters e.g. material removal rate (MRR), tool wear rate (TWR), electrode wear ratio (EWR), surface roughness (SR), recast layer thickness (RLT) and microhardness (HVN) based upon the significant effect on the EDM and PMEDM process and the extensive literature review presented. The methodology to assess the performance characteristics are discussed below.

2.3.1 Material Removal Rate

High material removal rate is the most desirable output response for any machining process which leads to increased productivity. After each machining operation, the workpiece material was taken out and weighed to find out the weight loss. Weight of the workpiece before and after the experiment was measured using an electronic balance with a least count of 0.001 g. The time duration of each experimental run was recorded using a digital stop watch. From the weight loss obtained, the material removal rate was calculated for different experimental runs. MRR is calculated using the volume loss from the workpiece material as cubic millimetre per minute (mm^3/min). The MRR is expressed as:

$$\begin{aligned} \text{MRR}\left(\text{mm}^3/\text{min}\right) &= \frac{\text{Wear weight of the work piece}}{\rho \times \text{time}} \\ \text{MRR} &= w_i - w_f/\rho * T \end{aligned} \tag{1}$$

where w_i and w_f are initial and final weights of the workpiece before and after the machining process, ρ is density of the workpiece material and T is the machining time in minutes.

2.3.2 Tool Wear Rate

The tool material for machining is selected based upon the principle that the material should have low resistance to electricity and high melting point. The tool electrode was taken out and weighed after each machining operation to find out the weight loss. Weight of the tool before and after the experiment was measured to determine the tool wear rate. The TWR is expressed as:

$$\text{TWR}\ \left(\text{mm}^3/\text{min}\right) = \frac{\text{Wear weight of the tool}}{\rho \times \text{time}}$$
$$\text{TWR} = t_i - t_f/\rho * T \qquad (2)$$

where t_i and t_f are initial and final weights of the tool before and after the machining process, ρ is density of the tool and T is the machining time in minutes.

2.3.3 Electrode Wear Ratio

The electrode wear ratio is dependent on the material removal rate and the tool wear rate. Lower EWR is desirable to enhance the productivity of the process. EWR can be defined as "the ratio of weight of the electrode wear to the weight of the workpiece wear after machining" and is expressed as:

$$\text{EWR}\ (\%) = \frac{\text{“Wear weight of the tool”}}{\text{“Wear weight of the work piece”}} \times 100$$
$$\text{EWR} = w_t/w_w * 100 \qquad (3)$$

where w_t and w_w are the wear weights of the tool and workpiece material measured after the machining operation is carried out in relation to the TWR and MRR.

2.3.4 Surface Roughness

Larger is the vertical deviation, rougher is the surface. The surface roughness is measured based upon various statistical descriptors out of which centre line average method is mostly used. SR is the arithmetic mean of the deviations from the mean line. The expression for R_a is given as:

$$R_a = \left(\frac{1}{L}\right) \int_0^L |y(x)| \mathrm{d}x \qquad (4)$$

where, L is the sampling length, y is the profile curve sampled by the set of N points and x is the profile direction. The roughness of the surface was measured using a surface roughness tester (Talysurf, Rank Taylor Hobson, England, Model-Surtronic S-100 series).

2.3.5 Recast Layer Thickness

During the machining process, a small amount of material gets re-solidified after being melted due to the refrigeration effect of the dielectric fluid. This layer is known as the recast layer. Material transfer also takes place from the powder suspended in the dielectric fluid and also from the electrode to the machined surface. Beyond this layer lies the heat-affected zone and the base material. In order to find out the structural features present below the machined surface and the distribution of cracks in the recast layer, specimens were cut from the machined surface in a traverse direction and were then mounted for metallographic studies. The recast layer thickness was measured using a scanning electron microscope (FESEM, model: Supra 55, Zeiss, Germany) for all the experiments by taking three sets of readings for a particular experiment and considering the average of the three values as the average recast layer thickness.

2.3.6 Microhardness

After the metallographic analysis, the samples were measured for hardness in a microhardness tester (LM 247AT of LECO) under a load of 10 mgf. The purpose of obtaining the microhardness of the material before and after machining was to examine the change in hardness and its effect on the machining surface due to the addition of the powder particles during machining. The microhardness was measured at three different locations and the average value was considered to be the microhardness of the machined specimen.

2.4 Design of Experiments

Taguchi's technique uses a philosophy and methodology to improve the quality of the process and minimizes the cost involved to carry out the process by optimizing the product design using statistical concepts. The effect of various machining process parameters such as concentration of powder (C_p), peak current (I_p), pulse on time (T_{on}), duty cycle (DC) and gap voltage (V_g) on various output responses like material removal rate (MRR), tool wear rate (TWR), electrode wear ratio (EWR), surface roughness (SR), recast layer thickness (RLT) and microhardness (HVN) on H-11 hot work tool steel was investigated using Taguchi's parameter design and Analysis of variance (ANOVA) helps to determine the statistically significant parameters affecting the responses. Predicted results obtained by Taguchi's technique were verified through confirmatory tests for validation and minimization of errors. Taguchi prescribes the use of orthogonal arrays (OA) for experiments. The design of experiments involves the assignment of important and influencing parameters to appropriate columns in the array with the use of linear graphs or triangular tables as suggested by Taguchi. The use of array in the design provides almost identical experimental runs. The most important stage in the DOE lies in the selection of control factors and their levels. The results are further analysed to establish the optimal condition for a product or process, estimation of the contribution of individual parameters affecting the response, and to determine the optimum response under the best condition. The best condition may be determined by analysing the behaviour of minimum effects of each of the parameters which provides the trend of each parameter and its influence on the process. ANOVA is applied to the results which help to determine the percentage contribution of each parameter for a stated level of confidence. The ANOVA table suggests which parameters need to be controlled. Taguchi suggests two ways to carry out the complete analysis. In the first case, the results of a single run or the average of repetitive runs are analysed through main effects plot and ANOVA (raw data analysis). The second approach uses signal-to-noise (S/N) ratio to the previous steps. The S/N ratio is a concurrent quality metric linked to the loss function. As S/N ratio maximizes, the loss associated gets minimized. S/N ratio provides the robust set of operating conditions for the process. S/N ratio is defined as "the ratio of the mean of the signal to the standard deviation of the noise". It is denoted by 'η' and it has the unit of dB. The present analysis involves the use of Taguchi's orthogonal array to conduct the experiments and the optimum setting is obtained by analyzing the main effect plot aided by the raw data analysis aided by S/N data analysis followed by ANOVA. Based upon the trial experiments and extensive literature survey, the significant machining parameters taken into consideration are concentration of powder (C_p), peak current (I_p), pulse on time (T_{on}), duty cycle (DC) and gap voltage (V_g) and their effect on the output responses has been investigated. Three sets of experimental runs have been performed at each condition. The control factors were selected based on the literature survey and some preliminary investigations. L_{27} Taguchi's orthogonal array was used for the experiments as shown in Table 4.

2.4.1 Signal-to-Noise Ratio

Establishment of the loss function with its appropriate k value and to use it as a figure of merit is not easy and cost-effective. Thus, the loss function is transformed

Table 4 Selection of levels for the factors

Factors with symbol and units	Levels		
	Level 1	Level 2	Level 3
Concentration of graphite powder 'C_p' (g/L)	0	3	6
Peak current 'I_p' (A)	3	6	9
Pulse on time 'T_{on}' (μs)	100	150	200
Duty cycle 'DC' (%)	7	8	9
Gap voltage 'V_g' (V)	30	40	50

to S/N ratio. S/N ratio being a concurrent statistics uses two or more characteristics of distribution and converts them into a single figure of merit. A higher value of S/N ratio implies that signal is much higher than the random effect of noise factors. The equations for calculating S/N ratios for "lower-the-better (LB), higher-the-better (HB) and nominal-the-best (NB)" type of characteristics are:

$$\text{LB: S/N Ratio} = -10\log_{10}\left[\frac{1}{n}\sum_{i=1}^{n} y_i^2\right] \tag{5}$$

$$\text{HB: S/N Ratio} = -10\log_{10}\left[\frac{1}{n}\sum_{i=1}^{n} y_i^{-2}\right] \tag{6}$$

$$\text{NB: S/N Ratio} = -10\log_{10}\left[\frac{1}{n}\sum_{i=1}^{n} (y_i - y_0)^2\right] \tag{7}$$

where

y sample mean
z standard deviation for the number of observations in each trial.
y_o nominal value of the characteristic

2.4.2 Selection of Orthogonal Array and Parameter Assignment

Selection of orthogonal array depends on the number of controllable parameters, their interactions with each other, and the number of levels to be selected. The number of controllable parameters with their levels considered in the present work is shown in Table 4. The minimum DOF required for the experiment are the sum of all the degrees of freedom of the factors. In the present experimental investigation, five three-level factors are considered for the study. As per Taguchi's experimental design philosophy, a set of three levels assigned to each parameter has two degrees of freedom. Thus, total DOF of the experiment becomes 10. The OA to be used should have more than 10 DOF. Hence, an L_{27} OA was chosen which has 27 trials and 26 DOF, assuming the interaction effect of process parameters negligible as shown in Table 5. The additional DOF may be used to measure the random error. Twenty-seven experiments each were conducted for three different powders mixed with the dielectric fluid using Taguchi's experimental design methodology, repeating each experiment three times. The designs, plots, and analysis have been carried out with MINITAB 15 software.

Table 5 Taguchi's L_{27} standard orthogonal array

Column no.	1	2	3	4	5
Run	C_p	I_p	T_{on}	DC	V_g
1	1	1	1	1	1
2	1	1	1	1	2
3	1	1	1	1	3
4	1	2	2	2	1
5	1	2	2	2	2
6	1	2	2	2	3
7	1	3	3	3	1
8	1	3	3	3	2
9	1	3	3	3	3
10	2	1	2	3	1
11	2	1	2	3	2
12	2	1	2	3	3
13	2	2	3	1	1
14	2	2	3	1	2
15	2	2	3	1	3
16	2	3	1	2	1
17	2	3	1	2	2
18	2	3	1	2	3
19	3	1	3	2	1
20	3	1	3	2	2
21	3	1	3	2	3
22	3	2	1	3	1
23	3	2	1	3	2
24	3	2	1	3	3
25	3	3	2	1	1
26	3	3	2	1	2
27	3	3	2	1	3

2.4.3 Analysis of Variance

Analysis of variance (ANOVA) was performed to determine the significant effect of process parameters on performance characteristics. The main effect plots for factors show the trend of the influence of factors toward the process. The importance of process variables with respect to output responses helps to determine the optimum set of parameters using ANOVA from Minitab15 software. Various steps involved in the analysis are:

Step 1: Total of all results (T):

$$T = \sum_{i=1}^{n} \sum_{j=1}^{R} y_{ij}, \quad (8)$$

where, y_{ij} is the value of the characteristic in the ith trial and jth repetition.

Step 2: Correction Factor (C.F.):

$$\text{C.F.} = T^2/N, \text{ where } N \text{ is the total number of experiments i.e. } 3 \times 27 = 81. \quad (9)$$

Step 3: Total sum of squares (SS_T):

$$SS_T = \sum_{i=1}^{n} \sum_{j=1}^{R} y_{ij} - \text{C.F.} \quad (10)$$

Step 4: Sum of squares of parameter A (SS_A):

$$SS_A = \left[\frac{A(1)^2}{N_{A1}} + \frac{A(2)^2}{N_{A2}} + \frac{A(3)^2}{N_{A3}}\right] - \text{C.F.} \quad (11)$$

where, N_{A1}, N_{A2} and N_{A3} are the number of experiments with parameter A at levels 1, 2 and 3, respectively. The sums of squares for all the factors are calculated similarly.

Step 5: Error sum of squares:

$$SS_e = SS_T - (\text{Summation of sum of squares of all the parameters}) \quad (12)$$

where, e stands for the error.

Step 6: Degrees of freedom:

$$\text{Total DOF} = (\text{total number of trials} - 1) = (R \times n - 1) = 80 \quad (13)$$

DOF of each parameters = (Number of levels of each parameter − 1) = 2
The DOF for all parameters are calculated in the similar way.

Step 7: Mean square of variance (V):

$$V_A = \text{Variance due to parameter } A = \frac{SS_A}{\text{fA}} \quad (14)$$

Variance for the other parameters is obtained in the similar manner.

Step 8: Percentage contribution (P):
P_A = Percentage contribution of parameter A towards mean of the response

$$P_A = \left(\frac{SS_A}{SS_T}\right) \times 100 \quad (15)$$

Similarly, the percentage contribution of all other parameters is calculated.

Step 9: F-ratios:
The F-ratio is defined as the ratio of variance due to a parameter and due to its error.

$$F_A = \frac{V_A}{V_e} \quad (16)$$

The F-ratio is calculated for all the parameters in the similar manner.

2.5 *Multi-objective Optimization*

Taguchi's experimental philosophy is focused on optimizing the process parameters in the perspective of a single quality criterion which does not give sufficient idea about the influence on other performance characteristics involved. The performance of the product is evaluated by various response parameters. Taguchi technique cannot solve a multi-response optimization problem. Hence, multi-objective optimization techniques are implemented wherein the quality characteristics are optimized and the results for the best levels are obtained. Taguchi technique is often combined with multi-objective optimization techniques to switch a multi-decision-making technique to a single-objective optimization problem. The decision maker assigns different priority weights to the responses basing upon their relative importance.

2.5.1 Technique for Order of Preference by Similarity to Ideal Solution (TOPSIS)

The present study focuses on finding the most suitable set of process variables for PMEDM using TOPSIS to obtain maximum MRR and HVN along with minimum TWR, EWR SR, and RLT with graphite powder-mixed dielectric. The optimum set of input parameters should be identified to improve the machining process for performance characteristics and surface quality and improve the machining characteristics of H-11 tool steel. Analysis of variance (ANOVA) helps to identify the statistically significant input parameters that affect the performance parameters.

TOPSIS is a capable multi-objective decision-making tool for solving complex decision-making problems in manufacturing domain where number of criteria, alternatives and their interactions play a significant role and their simultaneous effect influences the process to a great deal. TOPSIS uses preference grades with ranking order to find the most suitable set of input variables for the performance measures. TOPSIS can be used suitably for solving any type of decision-making problems. TOPSIS is a multi-optimization technique used to find the best alternative from a finite set. The technique is based upon the principle that the chosen criteria should have the shortest distance from the positive ideal solution and greatest distance from the negative ideal solution, the best solution being the closest to the ideal solution. The steps involved in carrying out the procedure of TOPSIS are:

Step 1 The decision matrix is the first step of TOPSIS which consist of 'n' attributes and 'm' alternatives and is represented as (Tripathy and Tripathy 2017b, c):

$$D_m = \begin{bmatrix} x_{11} & x_{12} & x_{13} & \cdots & \cdots & x_{1n} \\ x_{21} & x_{22} & x_{23} & \cdots & \cdots & x_{2n} \\ x_{31} & x_{32} & x_{33} & \cdots & \cdots & x_{3n} \\ \vdots & \vdots & \vdots & \ddots & \ddots & \vdots \\ \vdots & \vdots & \vdots & \ddots & \ddots & \vdots \\ x_{m1} & x_{m2} & x_{m3} & \cdots & \cdots & x_{mn} \end{bmatrix} \tag{17}$$

where x_{ij} is the performance of ith alternative with respect to jth attribute.

Step 2 Normalized matrix is obtained by the following expression:

$$r_{ij} = \frac{x_{ij}}{\sqrt{\sum_{i=1}^{m} x_{ij}^2}} \quad j = 1, 2, \ldots, n. \tag{18}$$

Step 3 The weight of each attribute was assumed to be w_j (j = 1, 2, …, n). Weighted normalized decision matrix $V = [v_{ij}]$ may be obtained as:

$$V = w_j r_{ij} \tag{19}$$

where, $\sum_{j=1}^{n} w_j = 1$.

Step 4 The most suitable and least suitable solutions may be obtained from the following expressions:

$$\begin{aligned} V^+ &= \left\{ \left(\sum_i^{\max} v_{ij} | j \in J \right), \left(\sum_i^{\min} | j \in J | i = 1, 2, \ldots m \right) \right\} \\ &= \{ v_1^+, v_2^+, v_3^+, \ldots, v_n^+ \} \end{aligned} \tag{20}$$

$$V^- = \left\{ \left(\sum_i^{\max} v_{ij} | j \in J \right), \left(\sum_i^{\min} | j \in J | i = 1, 2, \ldots m \right) \right\} = \left\{ v_1^-, v_2^-, v_3^-, \ldots, v_n^- \right\} \tag{21}$$

Step 5 The separation of alternatives from positive ideal solution is given by:

$$S_i^+ = \sqrt{\sum_{j=1}^n \left(v_{ij} - v_j^+ \right)^2}, \quad i = 1, 2, \ldots m \tag{22}$$

The separation of alternatives from negative-ideal solution is given by:

$$S_i^- = \sqrt{\sum_{j=1}^n \left(v_{ij} - v_j^- \right)^2}, \quad i = 1, 2, \ldots m \tag{23}$$

Step 6 The relative closeness of the alternative to the ideal solution is calculated and expressed as:

$$P_i = \frac{S_i^-}{S_i^+ + S_i^-} \quad i = 1, 2, \ldots m \tag{24}$$

Step 7 The P_i value was ranked in descending order to identify the set of alternatives having the preferred solutions.

Preference value of alternatives can be calculated from their nearness to ideal solution. The ratio of negative ideal separation measure divided by the sum of negative ideal separation measure and the positive ideal separation measure gives the relative closeness. The normalized matrix, weighted normalized decision matrix, separation of alternatives from positive and negative ideal solutions and preference values for TOPSIS for the runs with the ordered rankings are presented in Tables 25, 26, 27 and 28 respectively. The most suitable value of performance measure is the one which is close to the ideal solution of the performance measure and has the maximum preference value with the highest rank.

3 Results and Discussions

Experimental results obtained by varying the chosen set of input parameters for the selected output responses have been presented in this section. The discussion of the obtained results and the variations in their behaviour has also been detailed in the present section.

3.1 Experimental Results

Experiments were conducted using Taguchi's L_{27} orthogonal array. The results are represented in Table 6.

3.2 Effect of Process Parameters on Response Characteristics

The influence of input parameters on response characteristics has been highlighted in the present section. The variation of the output response with each input parameter has been discussed with the help of respective graphs.

Table 6 L_{27} experimental design with response variables

Run	C_p	I_p	T_{on}	DC	V_g	Avg MRR	Avg TWR	Avg EWR	Avg SR	Avg RLT	Avg HVN
1	0	3	100	7	30	2.564	0.0172	0.6718	3.8	13.8	784
2	0	3	100	7	40	2.649	0.0194	0.735	4.1	14.4	778
3	0	3	100	7	50	2.735	0.0224	0.8216	4.5	15.2	786
4	0	6	150	8	30	4.529	0.0277	0.6118	4.87	18.76	795
5	0	6	150	8	40	5.470	0.0307	0.5614	5.45	19.24	804
6	0	6	150	8	50	6.666	0.0367	0.5505	5.86	19.76	798
7	0	9	200	9	30	9.401	0.3895	4.1430	6.5	20.58	809
8	0	9	200	9	40	10.256	0.4868	4.7471	7.47	21.6	802
9	0	9	200	9	50	10.940	0.5243	4.7928	9.2	22.24	811
10	3	3	150	9	30	2.87	0.0123	0.429	2.48	22.6	833
11	3	3	150	9	40	3.54	0.0134	0.380	2.97	25.67	862
12	3	3	150	9	50	3.97	0.0152	0.383	3.42	25.31	841
13	3	6	200	7	30	4.42	0.0171	0.387	4.16	28.86	884
14	3	6	200	7	40	6.638	0.0186	0.2813	4.37	26.74	898
15	3	6	200	7	50	8.54	0.0192	0.224	4.82	29.2	876
16	3	9	100	8	30	9.26	0.0235	0.254	5.74	31.19	905
17	3	9	100	8	40	9.82	0.0261	0.266	6.41	34.42	929
18	3	9	100	8	50	11.04	0.0291	0.2639	6.72	37.64	976
19	6	3	200	8	30	5.56	0.0349	0.628	1.64	24.36	845
20	6	3	200	8	40	6.41	0.0378	0.5897	1.92	26.63	878
21	6	3	200	8	50	7.23	0.0398	0.5506	2.23	27.59	859
22	6	6	100	9	30	8.37	0.0448	0.5363	2.62	29.21	893
23	6	6	100	9	40	8.858	0.0463	0.523	2.94	32.49	904
24	6	6	100	9	50	11.23	0.0496	0.441	3.36	35.93	916
25	6	9	150	7	30	11.86	0.0586	0.494	3.82	36.2	941
26	6	9	150	7	40	12.07	0.0753	0.624	4.48	37.56	966
27	6	9	150	7	50	12.54	0.0880	0.7018	5.24	38.92	952

3.2.1 Effect of Process Parameters on Material Removal Rate

The goal of machining is high material removal and low tool wear. The MRR is the volume of material being removed per minute during machining. Experimental results show that machining without powder increases the MRR as I_p and T_{on} increase whereas the TWR, SR and RLT increase hampering the surface texture. Addition of powder in varying concentration modifies the surface properties. It causes a decrease in insulating strength of the dielectric fluid while increasing the inter-electrode gap making the removal of debris easier. Rapid sparks are generated due to the bridging effect which erodes material faster from the workpiece and increases the machining rate. Widening of plasma channel generates consistent sparks creating shallow craters on the machined region with superior surface finish. When graphite powder is added in C_p of 0 g/L, the MRR varies in a range of 2.564–10.940 mm^3/min. When C_p increases to 3 g/L, the MRR shows slight increase from 2.87–11.04 mm^3/min. Further increase of C_p to 6 g/L improves the MRR in a range of 5.56–12.54 mm^3/min. The variation of MRR with the input parameters is shown in Fig. 3a–e. As the I_p increases, the MRR increases significantly as more thermal energy is produced in the discharge channel as the electrical power increases. As T_{on} increases, the MRR slightly decreases and then tends to increase with the increase in discharge energy and heat transferred to the section. The decrease in MRR is due to the constant flushing pressure which is not sufficient to remove the molten material causing redeposition on the surfaces. The abrasive property of the powder promotes makes the removal of debris easier. The discharge gap distance increases with V_g, which consequently increases the material deposition on the surface which may be minimized under proper flushing conditions.

3.2.2 Effect of Process Parameters on Tool Wear Rate

Tool wear rate is determined as the volumetric material removed per minute during machining from the tool surface. The main aim of machining should be less tool wear for more amount of material removed from the workpiece surface. When no powder is added to the dielectric, the tool wear varies within a range of 0.0172–0.5243 mm^3/min. With the increase in C_p, the tool wear rate shows reduction. The TWR shows an increase with further increase in the concentration of powder to compensate for the more amount of MRR. With more MRR, TWR also shows an increase irrespective of the concentration of powder added to the dielectric as shown in Fig. 4a. More amount of MRR is observed while the I_p is more. Thus, more TWR occurs when more MRR is achieved with increased I_p. With the addition of 0 g/L of graphite powder, the maximum TWR is 0.5243 mm^3/min which reduces to a maximum value of 0.1517 mm^3/min when 3 g/L of graphite powder. On addition of 6 g/L of graphite powder, the maximum TWR reaches to a value of 0.2341 mm^3/min which is higher than that achieved with a C_p of 3 g/L. When graphite powder is added to the dielectric, more amount of carbon gets deposited on both the tool and workpiece surface. The resolidified layer on the

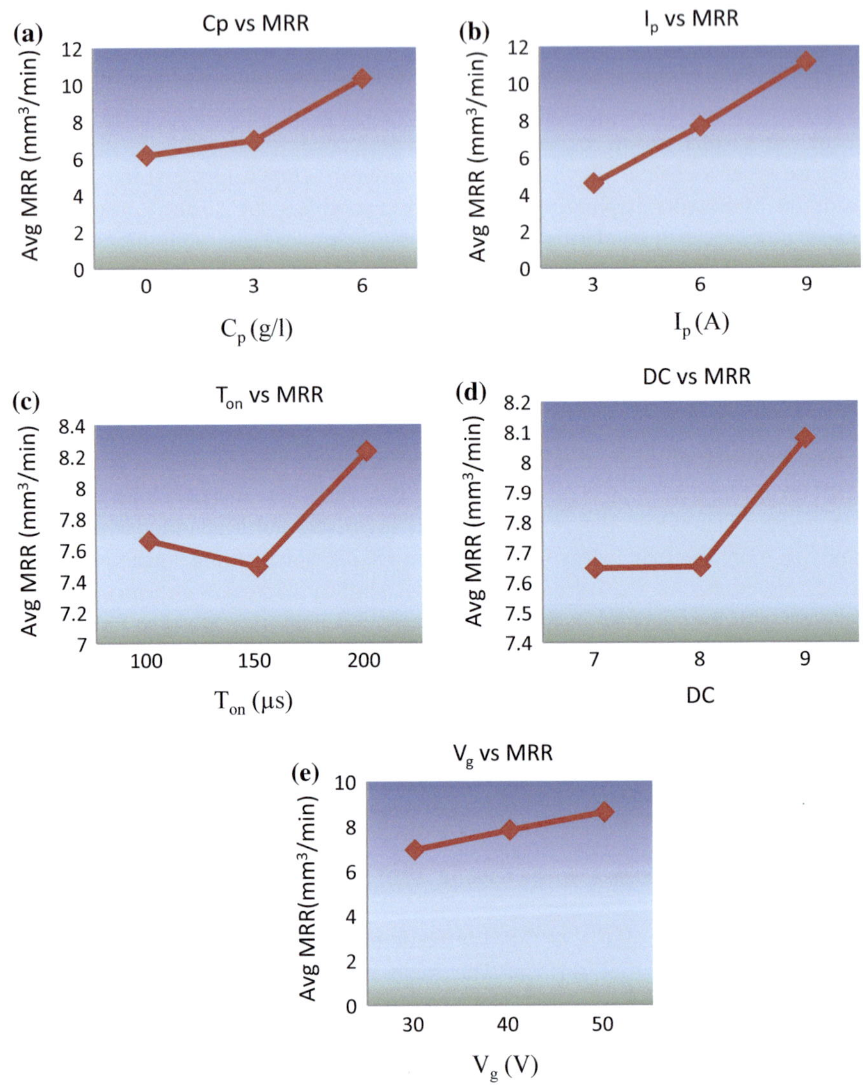

Fig. 3 Variation of MRR with process parameters (Tripathy and Tripathy 2017b, c)

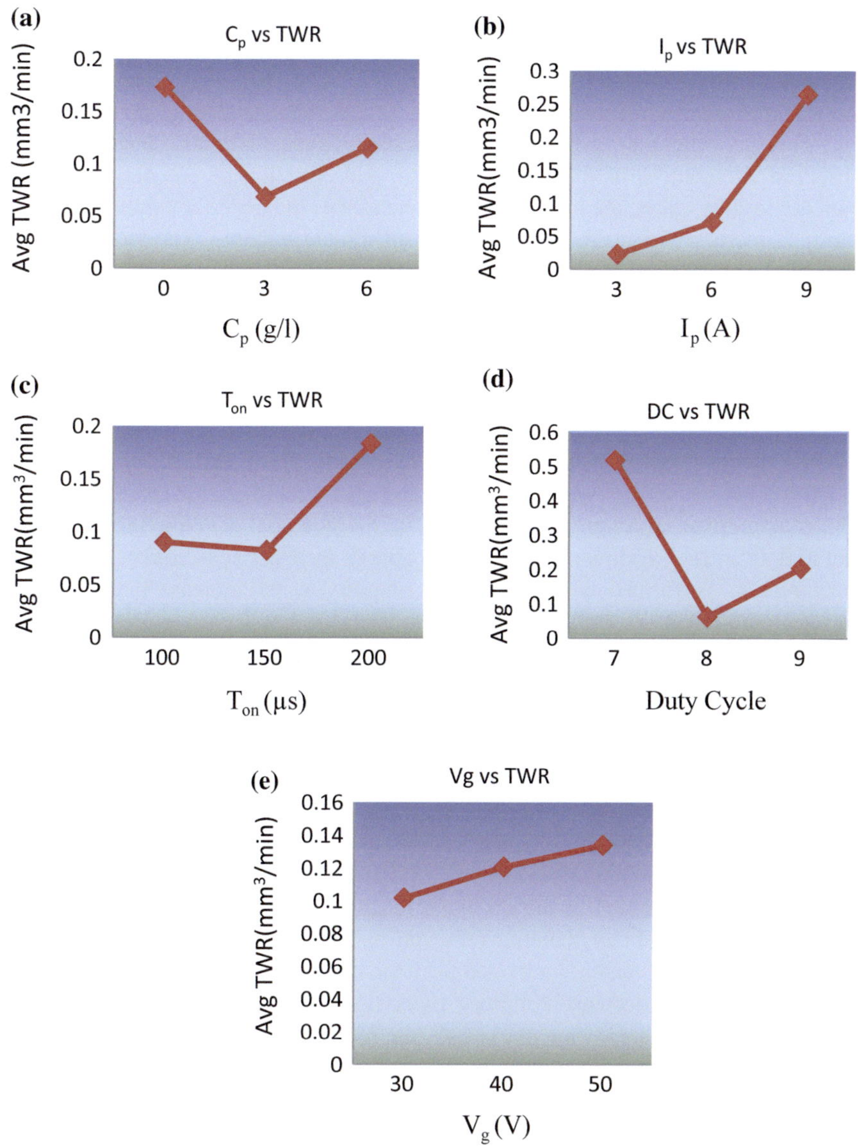

Fig. 4 Variation of TWR with process parameters (Tripathy and Tripathy 2017b, c)

workpiece leads to the recast layer formation and causes damage to the workpiece surface. Increased carbon content can be removed with proper flushing, which if insufficient would lead to increased TWR. Figure 4b–e show that with the increase in current and pulse on time the MRR and the TWR consequently increase due to the increased thermal energy in the discharge channel. This phenomenon is observed irrespective of the increase in the concentration of powders. With the increase in duty cycle, the tool wear rate increases as the material removal rate also show an increase due to the rapid ejection of molten material. The tool wear rate also shows an increase with the increase in gap voltage as the increased discharge gap distance increases the deposition of material on the machined surface and to perform the erosion of material from this deposited layer, the TWR shows an increase. With proper flushing conditions, higher machining rates can be achieved.

3.2.3 Effect of Process Parameters on Electrode Wear Ratio

The characteristic of a perfect tool should be the potential of removing maximum material from the workpiece with the capability to resist self-erosion. When the machining is performed without addition of powder, the increase in I_p and T_{on}, increases the EWR. With the addition of powder, the EWR decreases which is the result of less tool wear with more material removal from the workpiece. Figure 5a–e show that with the increase in concentration of powder-mixed to the dielectric, the EWR tends to decrease initially up to C_p of 3 g/L of powder but on increasing the C_p to 6 g/L, the EWR increases to a value much lower than that achieved with a C_p of 0 g/L. The nature of the variation of EWR can be related to the machining rates achieved by the addition of powder to the dielectric. The maximum value of EWR at 0, 3 and 6 g/L is 4.7928, 1.2777 and 1.711% respectively. With the increase in I_p, the EWR increases for graphite powder-mixed dielectric. This is because of the more thermal energy generated in the discharge channel. The EWR shows variation in relation to the amount of material removed and the corresponding TWR for the maximum amount of material removed from the workpiece. With the increase in T_{on}, the EWR shows a decrease followed by a significant increase with the further increase of T_{on}. With the increase in DC, the EWR tends to decrease initially followed by a significant rise with the increase in DC. This is because of the fact that for a C_p of 3 g/L, the amount of material removed from workpiece is less than that with a C_p of 6 g/L, with a subsequent increase in TWR. Hence, the EWR initially decreases, then increases due to the rapid ejection of molten material with the increase in DC. With the increase in V_g, the EWR increases due to the increase in discharge gap distance which leads to more amount of material deposition on the top surface of the workpiece leading to the increase in TWR for the required amount of MRR in presence of proper flushing.

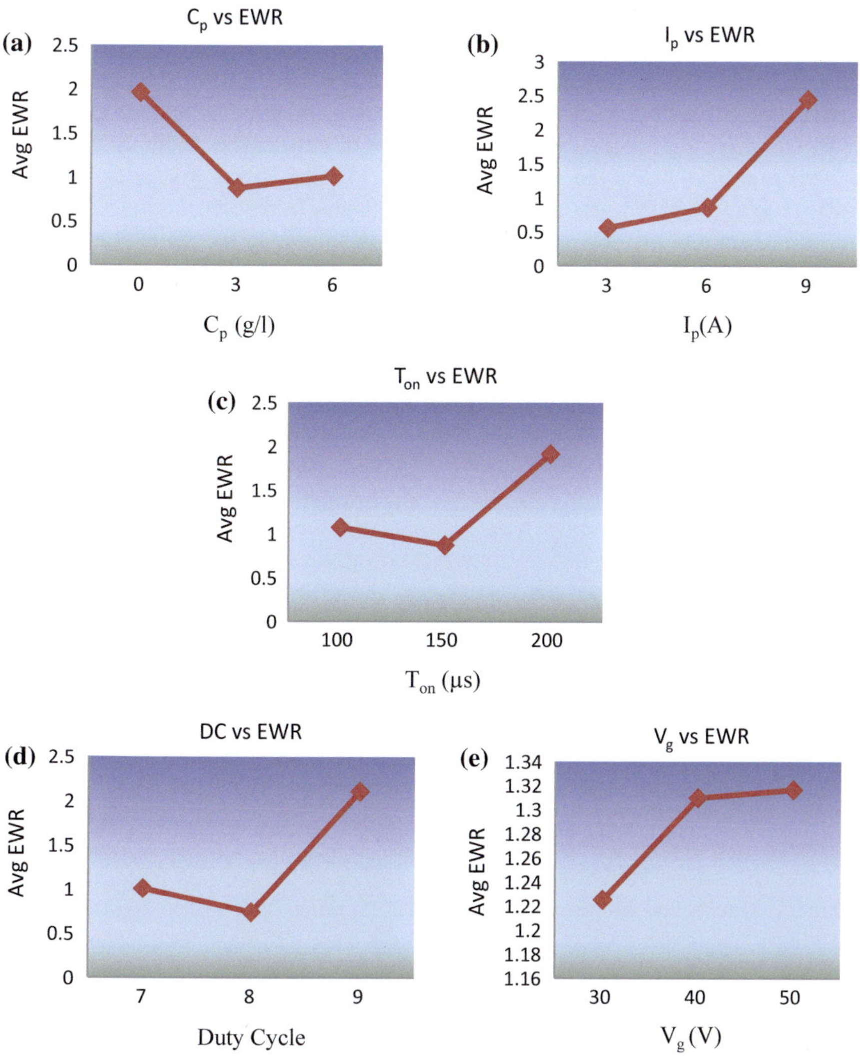

Fig. 5 Variation of EWR with process parameters

3.2.4 Effect of Process Parameters on Surface Roughness

The SR during PMEDM depends on size of the crater and variation of recast layer. From the experimental results, it can be seen that the roughness of the surface varies within a range of 3.8–6.19 μm without addition of powder. On addition of graphite powder in a C_p of 3 g/L, the SR lies between 1.94 and 5.65 μm. As the C_p increases to 6 g/L, the SR varies in a range of 2.41–6.19 μm. Influence of input parameters on SR is shown in Fig. 6a–e. With the rise in pulse current, the SR increases as the large dispersive energy cause impulsive forces and violent sparks which result in formation of large craters leading to the increase of SR as shown in Fig. 6b. Complete flushing does not occur during the pulse—off-time which increases the SR due to resolidification. Adding foreign particles in proper size and quantities reduce the SR during machining. With the increase in T_{on}, more heat is transferred to the section causing more material removal and increased SR. During PMEDM, the plasma flushing efficiency increases which improves the surface texture. Material deposition and carbide formation increases the SR. SR increases with increase in V_g as increased discharge gap distance minimizes the effect of induced energy at the workpiece surface thus increasing the deposition on the machined surface producing increased SR.

3.2.5 Effect of Process Parameters on Recast Layer Thickness

RLT increases with the increase in machining rate. It can be seen from Fig. 7a–e that increase in I_p increases the RLT. The temperature on the surface thus rises to the melting temperature consequently increasing the amount of material removed. Since the dielectric fluid does not get sufficient time to remove the molten material, the RLT increases. Increasing the C_p also increases the RLT as more amount of material is removed in presence of powder particles and redeposition occurs if flushing is insufficient which tends to increase the RLT. With the increase in T_{on}, the amount of heat transferred to the section increases. Adding graphite powder while machining and the breakdown of the dielectric fluid increases the carbon content. During PMEDM, as the T_{on} increases, the RLT tends to increase followed by reduction as the process becomes stable with the increase in discharge rate. Increase in DC, increases the RLT by a subsequent decrease which is due to the conduction of heat into the workpiece with a decrease in discharge duration. The abrasive effect of the powder particles decreases the RLT with the increase in DC. Increase in V_g increases the RLT due to the increase in discharge gap distance, which reduces the effect of induced energy at the workpiece and increases material deposition.

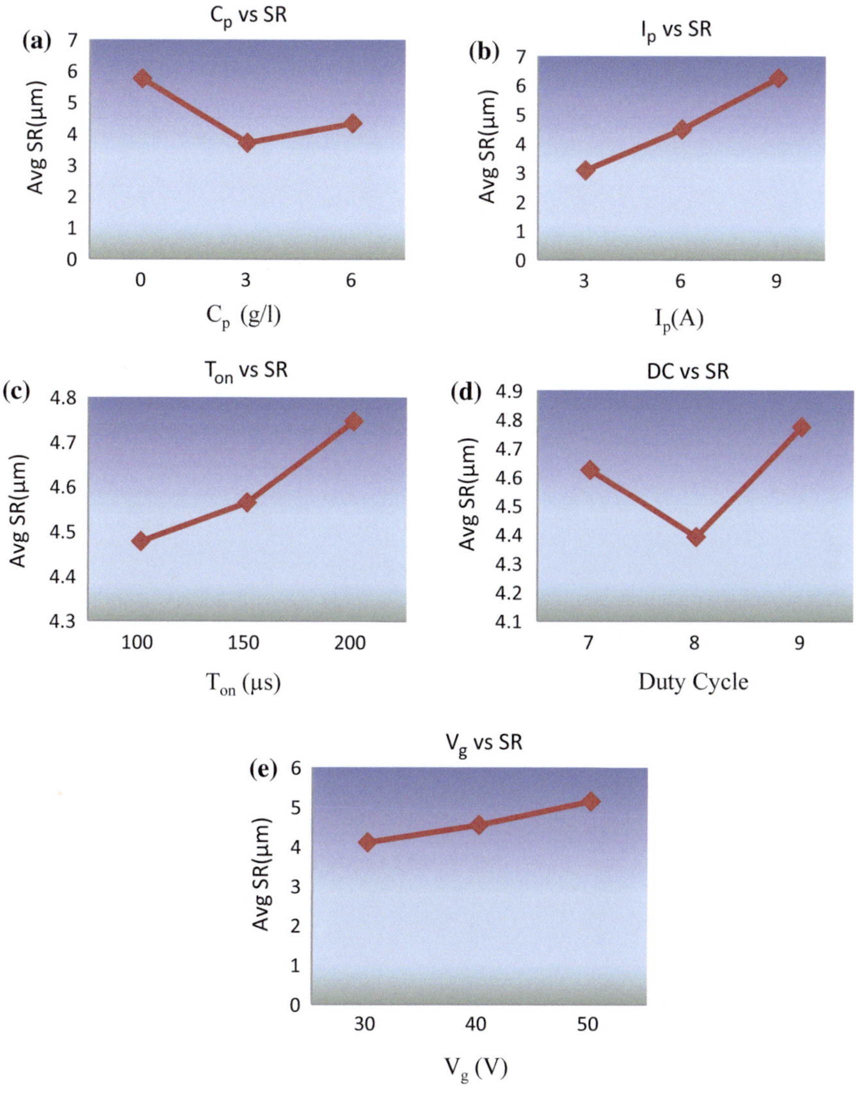

Fig. 6 Variation of SR with process parameters (Tripathy and Tripathy 2017b, c)

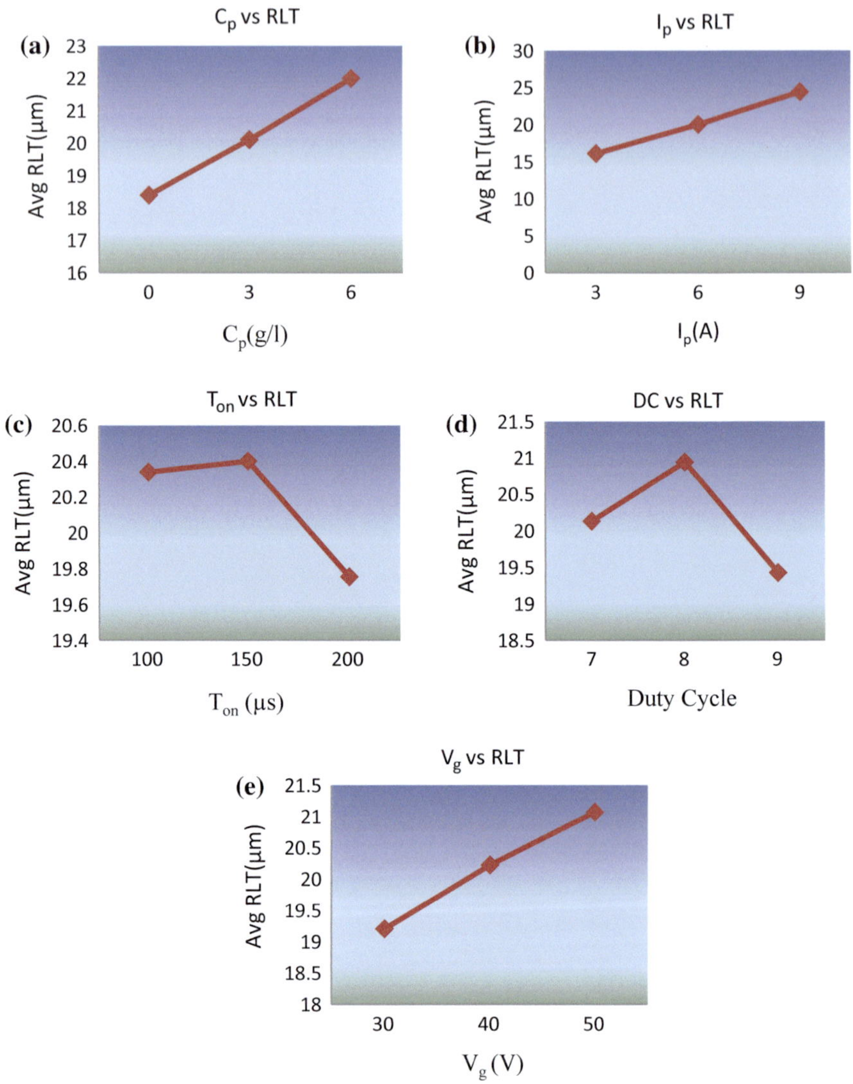

Fig. 7 Variation of RLT with process parameters (Tripathy and Tripathy 2017b, c)

3.2.6 Effect of Process Parameters on Microhardness

The microhardness of the parent material before machining has been found to be 621. The microhardness of the material after PMEDM was nearly double of the value before machining. It can be observed from Fig. 8a–e that the increase in C_p and I_p, increase the HVN values appreciably. Melting and deposition phenomenon increase the HVN. It may be observed that the surface quality is increased with the increase in HVN. The HVN increases while machining without the addition of powder with the increase in the values of I_p and T_{on} as high current increases the pulse energy. As the discharge column expands at higher pulse on time more heating of the surface occurs which releases the stresses and lowers the HVN at higher pulse on time. Increase in DC decreases the HVN as a consequence of the decrease in HVN with the variation of T_{on}. As V_g increases HVN also increases due to improper flushing of the debris during the no spark condition.

3.3 *ANOVA Results for the Output Responses*

The ANOVA results of each output response have been discussed in this section. The optimum set of process parameters has also been identified using the main effect plots for S/N ratio and means.

3.3.1 Analysis of Variance for Material Removal Rate

From the main effect plots shown in Fig. 9, the influence of different factors on the process can be visualized. The significance of the process variables in relation to MRR was investigated and the optimum combination of parameters was determined from the ANOVA analysis presented in Tables 7 and 8 respectively. MRR being higher-the-better type of quality characteristic, from the response curves as shown in Fig. 9a, it can be revealed that the third level of parameters of C_p, third level of I_p, third level of T_{on}, third level of DC and third level of V_g may offer maximum MRR. It may be noticed that the parameters C_p, I_p, T_{on} and V_g show similar trend of variation in both the main effect plots whereas DC shows the best value in the third level from the main effect plot of means and second level as a best value from the main effect plot for S/N ratio. In order to arrive at the final optimal setting, the relative contribution of the mean and S/N values has been considered from the ANOVA table.

The relative contribution of DC towards mean value of MRR (0.36%) is lower than that of S/N ratio (2.02%). Therefore, considering the level corresponding to the higher relative contribution, the optimum combination of input parameters for best MRR is the third level of parameters of C_p, third level of I_p, third level of T_{on}, second level of DC and third level of V_g which has been obtained at the main effect plot of S/N ratio.

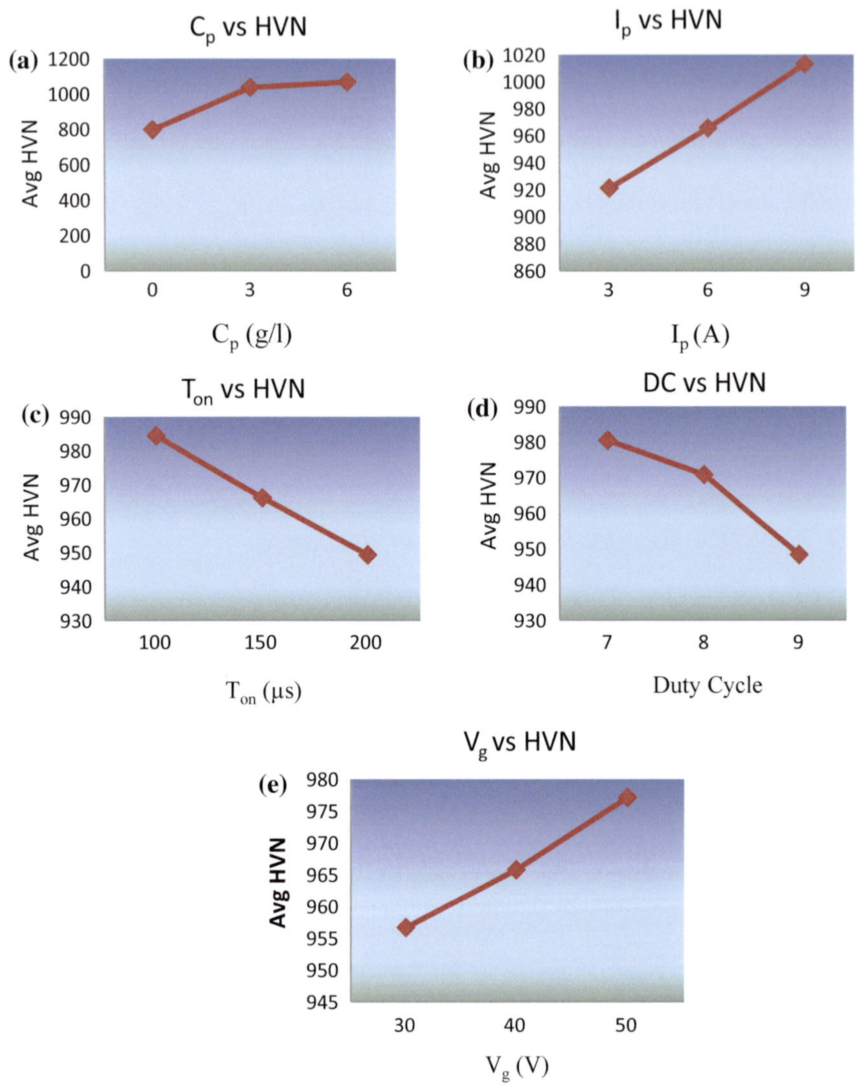

Fig. 8 Variation of HVN with process parameters

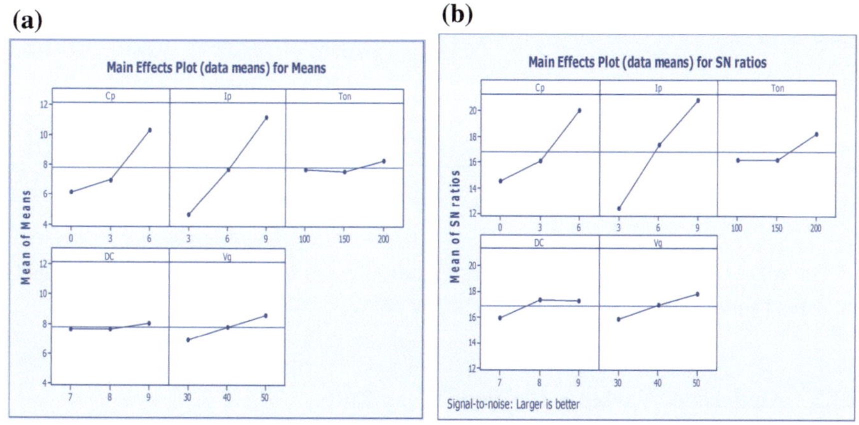

Fig. 9 Main effect plot for means and S/N ratio of material removal rate. **a** Main effect plot for means of MRR (Tripathy and Tripathy 2017a). **b** Main effect plot for S/N ratio of MRR

Table 7 ANOVA table for means of MRR

Source	DF	Seq SS	Adj SS	Adj MS	F	P	P (%)
C_p	2	87.870	87.870	43.9349	119.53	0.000	28.97
I_p	2	193.092	193.092	96.5458	262.66	0.000	63.68
T_{on}	2	2.711	2.711	1.3554	3.69	0.048	0.89
DC	2	1.099	1.099	0.5496	1.5	0.254	0.36
V_g	2	12.570	12.57	6.285	17.1	0.000	4.14
Error	16	5.881	5.881	0.3676			
Total	26	303.222					

$S = 0.6063$, R-Sq = 98.1%, R-Sq(adj) = 96.8%

Table 8 ANOVA table for S/N ratio of MRR

Source	DF	Seq SS	Adj SS	Adj MS	F	P	P (%)
C_p	2	145.504	145.504	72.752	191.26	0.000	27.64
I_p	2	323.339	323.339	161.67	425.02	0.000	61.42
T_{on}	2	23.662	23.662	11.831	31.1	0.000	4.49
DC	2	10.635	10.635	5.317	13.98	0.000	2.02
V_g	2	17.180	17.180	8.59	22.58	0.000	3.26
Error	16	6.086	6.086	0.38			
Total	26	526.406					

$S = 0.6167$, R-Sq = 98.8%, R-Sq(adj) = 98.1%

Table 9 Response table for means of MRR

Level	C_p	I_p	T_{on}	DC	V_g
1	6.134	4.586	7.655	7.646	6.941
2	6.938	7.654	7.489	7.650	7.820
3	10.299	11.132	8.228	8.076	8.611
Delta	4.165	6.546	0.740	0.430	1.671
Rank	2	1	4	5	3

The response table for all the process parameters is shown in Table 9. The ranks are based upon delta statistics that evaluate relative magnitude of effects.

3.3.2 Analysis of Variance for Tool Wear Rate

Figure 10 represents the influence of different factors on the process. The significance of the process variables in relation to TWR was investigated and the optimum combination of parameters was determined from the ANOVA analysis presented in Tables 10 and 11 respectively.

TWR being lower-the-better type of quality characteristic, the tendency of deviation of the response curves as shown in Fig. 10 demonstrate that the optimal set of parameters for TWR is the second level of parameters of C_p, first level of I_p, second level of T_{on}, second level of DC, and first level of V_g. The response table for all the process parameters is shown in Table 12.

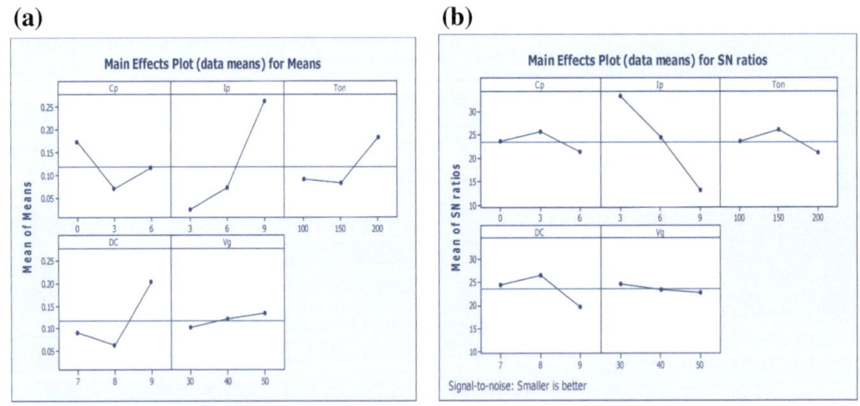

Fig. 10 Main effect plot for means and S/N ratio of tool wear rate. **a** Main effect plot for means of TWR (Tripathy and Tripathy 2017a). **b** Main effect plot for S/N ratio of TWR

Table 10 ANOVA table for means of TWR

Source	DF	Seq SS	Adj SS	Adj MS	*F*	*P*	*P* (%)
C_p	2	0.0493	0.0493	0.0247	35.68	0.000	9.57
I_p	2	0.2923	0.2923	0.1461	211.4	0.000	56.74
T_{on}	2	0.0571	0.0571	0.0286	41.34	0.000	11.09
DC	2	0.1005	0.1005	0.0503	72.72	0.000	19.51
V_g	2	0.0047	0.0047	0.0024	3.42	0.058	0.919
Error	16	0.01106	0.01106	0.00069			
Total	26	0.515157					

S = 0.02629, R-Sq = 97.9%, R-Sq(adj) = 96.5%

3.3.3 Analysis of Variance for Electrode Wear Ratio

Figure 11 shows the influence of different factors on the process. The significance of the process variables in relation to EWR was investigated and the optimum combination of parameters was determined from the ANOVA analysis presented in Tables 13 and 14 respectively.

EWR being lower-the-better type of quality characteristic, the trend of deviation of the response curves as shown in Fig. 11 demonstrate that the optimal set of parameters for EWR is the second level of parameters of C_p, first level of I_p, second level of T_{on}, second level of DC and first level of V_g. The response table for all the process parameters is shown in Table 15.

3.3.4 Analysis of Variance for Surface Roughness

Figure 12 demonstrates the influence of different factors on the process. The significance of the process variables in relation to SR has been investigated and the optimum combination of parameters was determined from the ANOVA analysis presented in Tables 16 and 17 respectively.

Table 11 ANOVA table for S/N ratio of TWR

Source	DF	Seq SS	Adj SS	Adj MS	*F*	*P*	*P* (%)
C_p	2	85.29	85.29	42.645	26.54	0.000	3.699
I_p	2	1840.29	1840.29	920.145	572.73	0.000	79.83
T_{on}	2	116.89	116.89	58.444	36.38	0.000	49.68
DC	2	220.35	220.35	110.175	68.58	0.000	9.558
V_g	2	16.74	16.74	8.371	5.21	0.018	7.11
Error	16	25.71	25.71	1.607			
Total	26	2305.26					

S = 1.268, R-Sq = 98.9%, R-Sq(adj) = 98.2%

Table 12 Response table for means of TWR

Level	C_p	I_p	T_{on}	DC	V_g
1	0.1727	0.0223	0.0901	0.0906	0.1016
2	0.0682	0.0706	0.0823	0.0621	0.1206
3	0.1151	0.2632	0.1836	0.2034	0.1338
Delta	0.1045	0.2408	0.1012	0.1413	0.0322
Rank	3	1	4	2	5

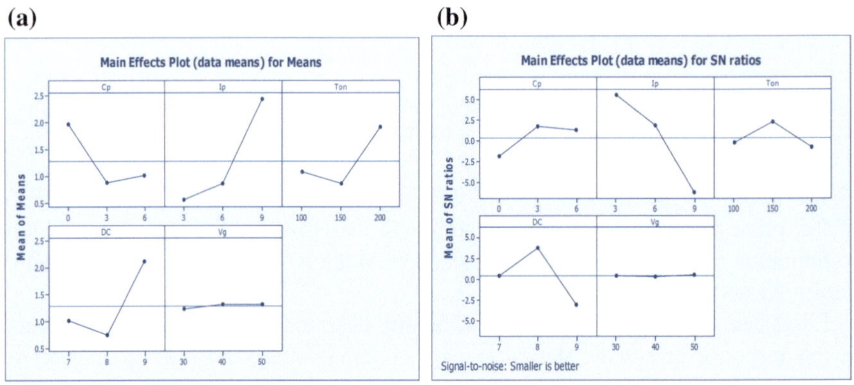

Fig. 11 Main effect plot for means and S/N ratio of electrode wear ratio. **a** Main effect plot for means of EWR. **b** Main effect plot for S/N ratio of EWR

Table 13 ANOVA table for means of EWR

Source	DF	Seq SS	Adj SS	Adj MS	F	P	P (%)
C_p	2	6.2531	6.253	3.1265	86.21	0.000	15.57
I_p	2	18.3438	18.343	9.1719	252.91	0.000	45.66
T_{on}	2	5.5277	5.5277	2.7638	76.21	0.000	13.76
DC	2	9.4095	9.4095	4.7047	129.73	0.000	2.34
V_g	2	0.0462	0.0462	0.0231	0.64	0.542	0.115
Error	16	0.5803	0.5803	0.0362			
Total	26	40.1606					

$S = 0.1904$, R-Sq = 98.6%, R-Sq(adj) = 97.7%

SR being lower-the-better type of quality characteristic, the tendency of deviation of the response curves demonstrate that the optimal set of parameters for SR is the second level of parameters of C_p, first level of I_p, first level of T_{on}, second level of DC and first level of V_g as shown in Fig. 12. The response table for all the process parameters is shown in Table 18.

Table 14 ANOVA table for S/N ratio of EWR

Source	DF	Seq SS	Adj SS	Adj MS	F	P	P (%)
C_p	2	71.56	71.56	35.78	19.15	0.000	7.08
I_p	2	644.33	644.33	322.167	172.41	0.000	63.81
T_{on}	2	49.41	49.41	24.704	13.22	0.000	4.89
DC	2	214.38	214.38	107.190	57.36	0.000	21.23
V_g	2	0.16	0.16	0.081	0.04	0.958	0.015
Error	16	29.90	29.90	1.869			
Total	26	1009.74					

$S = 1.367$, R-Sq = 97%, R-Sq(adj) = 95.2%

Table 15 Response table for means of EWR

Level	C_p	I_p	T_{on}	DC	V_g
1	1.9594	0.5549	1.0704	1.0086	1.2252
2	0.8766	0.8599	0.8676	0.7383	1.3094
3	1.0145	2.4358	1.9126	2.1037	1.3161
Delta	1.0828	1.8809	1.0450	1.3654	0.0909
Rank	3	1	4	2	5

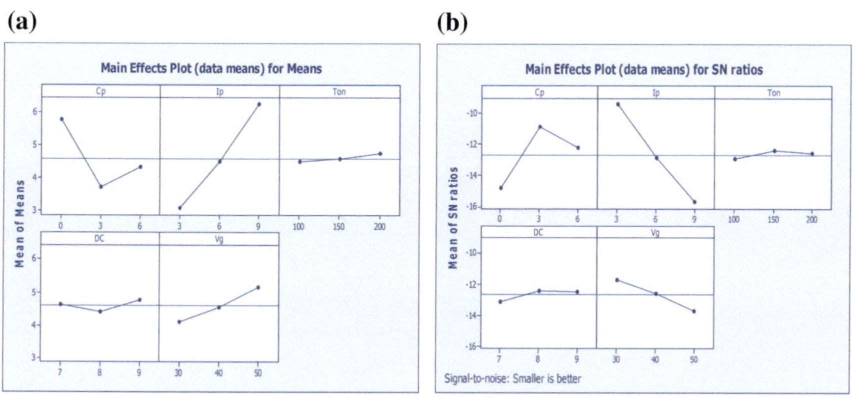

Fig. 12 Main effect plot for means and S/N ratio of surface roughness. **a** Main effect plot for means of SR (Tripathy and Tripathy 2017a). **b** Main effect plot for S/N ratio of SR

3.3.5 Analysis of Variance for Recast Layer Thickness

Figure 13 shows the influence of different factors on the process. The significance of the process variables in relation to RLT has been investigated and the optimum combination of parameters was determined from the ANOVA analysis presented in Tables 19 and 20 respectively.

Table 16 ANOVA table for means of SR

Source	DF	Seq SS	Adj SS	Adj MS	F	P	P (%)
C_p	2	19.7174	19.7174	9.8687	79.15	0.000	26.96
I_p	2	45.4767	45.4767	22.7383	182.55	0.000	62.2
T_{on}	2	0.3391	0.3391	0.1695	1.36	0.285	0.463
DC	2	0.6681	0.6681	0.3341	2.68	0.099	0.913
V_g	2	4.9155	4.9155	2.4577	19.73	0.000	6.723
Error	16	1.9929	1.9929	0.1246			
Total	26	73.1096					

$S = 0.3529$, R-Sq = 97.3%, R-Sq(adj) = 95.6%

Table 17 ANOVA table for S/N ratio of SR

Source	DF	Seq SS	Adj SS	Adj MS	F	P	P (%)
C_p	2	72.595	72.595	36.2973	187.75	0.000	26.31
I_p	2	178.867	178.867	89.4333	462.59	0.000	64.84
T_{on}	2	1.125	1.125	0.5626	2.91	0.084	0.407
DC	2	2.449	2.449	1.2245	6.33	0.009	0.887
V_g	2	17.713	17.713	8.8564	45.81	0.000	6.421
Error	16	3.093	3.093	0.1933			
Total	26	275.842					

$S = 0.4397$, R-Sq = 98.9%, R-Sq(adj) = 98.2%

Table 18 Response table for means of SR

Level	C_p	I_p	T_{on}	DC	V_g
1	5.750	3.070	4.478	4.624	4.102
2	3.708	4.477	4.564	4.391	4.543
3	4.331	6.242	4.747	4.773	5.143
Delta	2.042	3.172	0.269	0.382	1.041
Rank	2	1	5	4	3

RLT being lower-the-better type of quality characteristic, the tendency of deviation of the response curves as shown in Fig. 13, demonstrate that the optimal set of parameters for RLT is the first level of parameters of C_p, first level of I_p, third level of T_{on}, third level of DC and first level of V_g. The response table for all the process parameters is shown in Table 21.

3.3.6 Analysis of Variance for Microhardness

From the main effect plots shown in Fig. 14, the influence of different factors on the process can be visualized. The significance of the process variables in relation to

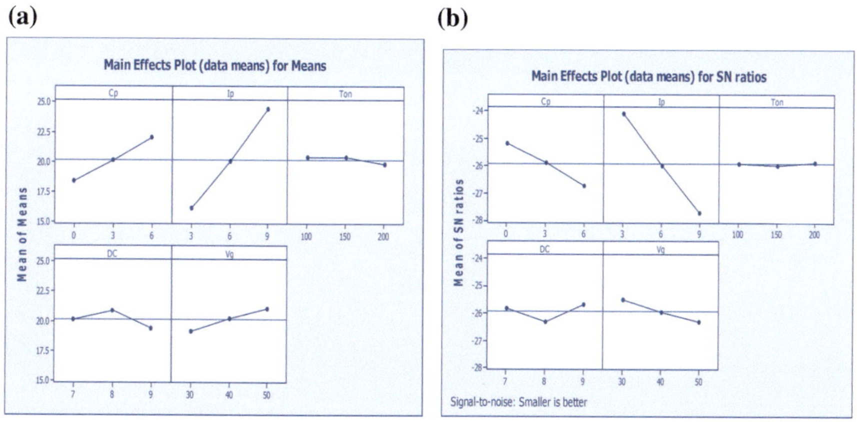

Fig. 13 Main effect plot for means and S/N ratio of recast layer thickness. **a** Main effect plot for means of RLT (Tripathy and Tripathy 2017a). **b** Main effect plot for S/N ratio of RLT

Table 19 ANOVA table for means of RLT

Source	DF	Seq SS	Adj SS	Adj MS	*F*	*P*	*P* (%)
C_p	2	58.191	58.191	29.095	66.13	0.000	14.42
I_p	2	309.912	309.912	154.956	352.20	0.000	76.81
T_{on}	2	2.286	2.286	1.143	2.60	0.105	5.26
DC	2	10.323	10.323	5.161	11.73	0.001	2.55
V_g	2	15.708	15.708	7.854	17.85	0.000	3.89
Error	16	7.039	7.039	0.440			
Total	26	403.459					

$S = 0.6633$, R-Sq = 98.3%, R-Sq(adj) = 97.2%

Table 20 ANOVA table for S/N ratio of RLT

Source	DF	Seq SS	Adj SS	Adj MS	*F*	*P*	*P* (%)
C_p	2	10.8879	10.8879	5.4439	61.61	0.000	14.19
I_p	2	59.2309	59.2309	29.6154	335.16	0.000	77.28
T_{on}	2	0.0459	0.0459	0.0229	0.26	0.774	0.059
DC	2	2.1335	2.1335	1.0668	12.07	0.001	2.78
V_g	2	2.9230	2.9230	1.4615	16.54	0.000	3.81
Error	16	1.4138	1.4138	0.0884			
Total	26	76.6349					

$S = 0.2973$, R-Sq = 98.2%, R-Sq(adj) = 97.0%

Table 21 Response table for means of RLT

Level	C_p	I_p	T_{on}	DC	V_g
1	18.40	16.09	20.34	20.13	19.20
2	20.10	20.01	20.40	20.94	20.22
3	21.99	24.39	19.75	19.43	21.07
Delta	3.59	8.29	0.65	1.51	1.87
Rank	2	1	5	4	3

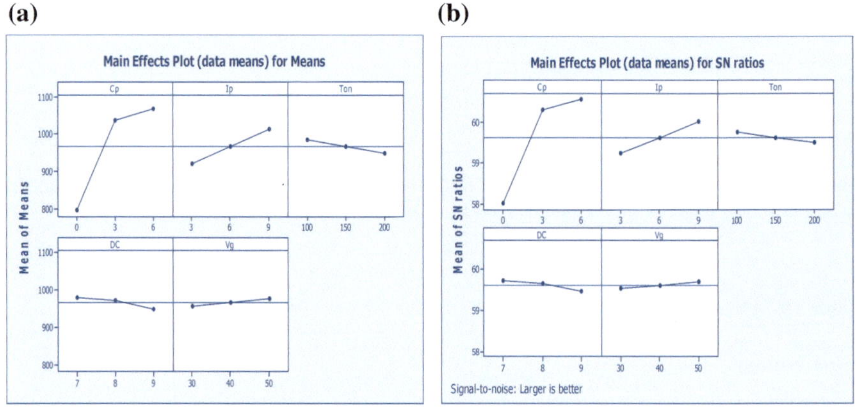

Fig. 14 Main effect plot for means and S/N ratio of microhardness. **a** Main effect plot for means of HVN. **b** Main effect plot for S/N ratio of HVN

HVN has been investigated and the optimum combination of parameters was determined from the ANOVA analysis presented in Tables 22 and 23 respectively.

HVN being higher-the-better type of quality characteristic, the tendency of deviation of the response curves as shown in Fig. 14, demonstrate that the optimal setting of parameters for HVN is the third level of parameters of C_p, third level of I_p, first level of T_{on}, first level of DC and third level of V_g. The response table for all the process parameters is shown in Table 24.

3.4 Multi-objective Optimization Using TOPSIS

The normalized matrix, weighted normalized decision matrix, separation of alternatives from positive and negative ideal solutions and preference values for TOPSIS obtained for experimental runs with ranks are represented in Tables 25, 26, 27 and 28 respectively. The weights given to different parameters are MRR, TWR and SR = 0.2, EWR = 0.1, RLT and HVN = 0.15 (Sum = 1).

It may be observed that the experimental run #21 has the most suitable multiple performance characteristics having the highest preference order followed by #20

Table 22 ANOVA table for means of HVN

Source	DF	Seq SS	Adj SS	Adj MS	*F*	*P*	*P* (%)
C_p	2	395,513	395,513	197,756	636.42	0.000	87.73
I_p	2	38,012	38,012	19,006	61.16	0.000	8.431
T_{on}	2	5551	5551	2776	8.93	0.002	12.11
DC	2	4885	4885	2442	7.86	0.004	1.08
V_g	2	1888	1888	944	3.04	0.076	0.418
Error	16	4972	4972	311			
Total	26	450,821					

$S = 17.63$, R-Sq = 98.9%, R-Sq(adj) = 98.2%

Table 23 ANOVA table for S/N ratio of HVN

Source	DF	Seq SS	Adj SS	Adj MS	*F*	*P*	*P* (%)
C_p	2	34.9657	34.9657	17.4829	825.92	0.000	90.13
I_p	2	2.7475	2.7475	1.3737	64.90	0.000	7.082
T_{on}	2	0.3203	0.3203	0.1601	7.56	0.005	0.825
DC	2	0.2922	0.2922	0.1461	6.90	0.007	0.753
V_g	2	0.1285	0.1285	0.0642	3.03	0.076	0.331
Error	16	0.3387	0.3387	0.0212			
Total	26	38.7929					

$S = 0.1455$, R-Sq = 99.1%, R-Sq(adj) = 98.6%

Table 24 Response table for means of HVN

Level	C_p	I_p	T_{on}	DC	V_g
1	796.3	921.1	984.3	980.4	956.7
2	1035.8	965.4	966.0	970.8	965.8
3	1067.4	1013.0	949.2	948.3	977.1
Delta	271.1	91.9	35.1	32.1	20.4
Rank	1	2	3	4	5

and #19. The higher preference values are considered as optimum, therefore, considering the preference values as higher-the-better type of quality characteristic, the third level of parameters of C_p, first level of I_p, third level of T_{on}, second level of DC and third level of V_g offer maximum grades and are considered to be the optimum set of process parameters. The optimal parametric combination is $C_{p3}I_{p1}T_{\text{on}3}\text{DC}_2V_{g3}$.

Table 25 Normalized matrix

Run	MRR	TWR	EWR	SR	RLT	HVN
1	0.0581	0.0182	0.0730	0.1498	0.1293	0.1547
2	0.0601	0.0205	0.0798	0.1616	0.1349	0.1535
3	0.0620	0.0237	0.0893	0.1773	0.1424	0.1551
4	0.1027	0.0292	0.0664	0.1919	0.1758	0.1569
5	0.1241	0.0324	0.0610	0.2148	0.1803	0.1586
6	0.1512	0.0387	0.0598	0.2310	0.1852	0.1574
7	0.2133	0.4115	0.4502	0.2562	0.1929	0.1596
8	0.2327	0.5144	0.5159	0.2944	0.2024	0.1582
9	0.2482	0.5539	0.5209	0.3626	0.2084	0.1600
10	0.0708	0.0182	0.0600	0.0764	0.1475	0.1896
11	0.0836	0.0260	0.0728	0.0890	0.1424	0.1951
12	0.0984	0.0199	0.0472	0.1095	0.1466	0.1904
13	0.1466	0.0652	0.1038	0.1430	0.1712	0.2003
14	0.1643	0.0481	0.0684	0.1383	0.1861	0.2076
15	0.1719	0.0670	0.0910	0.1639	0.191	0.2026
16	0.1744	0.1002	0.1341	0.1742	0.2268	0.2121
17	0.2371	0.1435	0.1413	0.1978	0.2359	0.2182
18	0.2693	0.1602	0.1388	0.2227	0.248	0.2234
19	0.1470	0.0280	0.0445	0.095	0.1537	0.1945
20	0.1736	0.0296	0.0399	0.1088	0.1773	0.1997
21	0.1825	0.0282	0.0360	0.1214	0.1831	0.203
22	0.2141	0.0984	0.1073	0.1470	0.1804	0.2088
23	0.2234	0.1580	0.1650	0.1659	0.1988	0.204
24	0.2644	0.1339	0.1182	0.1919	0.2189	0.2182
25	0.2900	0.1972	0.1587	0.2215	0.2420	0.2224
26	0.2978	0.1742	0.1365	0.2408	0.2475	0.22
27	0.3103	0.2473	0.186	0.244	0.2532	0.2249

3.4.1 Confirmatory Experiment for TOPSIS

After the evaluation of optimal parameter setting, prediction and confirmation for the enhancement of quality characteristic using the optimal setting has been examined. The results have been presented in Table 29.

Table 29 shows that the optimum set of parameters obtained from TOPSIS gives an increased MRR with a rise in the value from 2.564 to 9.6 mm^3/min. For this subsequent rise in MRR, the TWR exhibits an increase from a value of 0.0172–0.0395 mm^3/min. The EWR value decreases from 0.6718 to 0.4115%. The surface roughness increases from 3.8 to 4 μm as more amount of material removal leads to the formation of rough surfaces. The recast layer thickness reduces from 13.8 to 13.2 μm and the microhardness value shows an increase from 784 to 1105

Table 26 Weighted normalized decision matrix

Run	MRR	TWR	EWR	SR	RLT	HVN
1	0.0116	0.0036	0.0073	0.0299	0.0194	0.0232
2	0.012	0.0041	0.0079	0.0323	0.0202	0.023
3	0.0124	0.0047	0.0089	0.0354	0.0213	0.0232
4	0.0205	0.0058	0.0066	0.0383	0.0263	0.0235
5	0.0248	0.0064	0.0061	0.0429	0.0270	0.0238
6	0.0302	0.0077	0.0059	0.0462	0.0277	0.0236
7	0.0426	0.0823	0.04502	0.0512	0.0289	0.0239
8	0.0465	0.1028	0.0515	0.0588	0.0303	0.0237
9	0.0496	0.1107	0.05209	0.0725	0.0312	0.024
10	0.0141	0.0036	0.0060	0.0152	0.0221	0.0284
11	0.0167	0.0052	0.0073	0.0178	0.0213	0.02927
12	0.0196	0.0039	0.00472	0.0219	0.022	0.02856
13	0.0293	0.013	0.0104	0.0286	0.0256	0.03004
14	0.0328	0.0096	0.0068	0.0276	0.0279	0.0311
15	0.0343	0.0134	0.0091	0.0327	0.0286	0.0304
16	0.0348	0.02005	0.01341	0.0348	0.034	0.0318
17	0.0474	0.0287	0.01413	0.0395	0.0353	0.0327
18	0.0538	0.03205	0.0138	0.0445	0.0372	0.0335
19	0.0294	0.0056	0.0044	0.0190	0.023	0.0292
20	0.0347	0.0059	0.0039	0.0217	0.0266	0.0299
21	0.0365	0.0056	0.0036	0.0242	0.0274	0.0305
22	0.0428	0.0196	0.0107	0.0294	0.027	0.0313
23	0.0446	0.0316	0.0165	0.0332	0.0298	0.0306
24	0.0528	0.0267	0.0118	0.0383	0.0328	0.0327
25	0.058	0.0394	0.0158	0.0443	0.0363	0.0334
26	0.0595	0.0348	0.0136	0.0481	0.037133	0.033
27	0.062	0.0494	0.0186	0.0488	0.03799	0.03375

by adding graphite powder. The concentration of graphite powder causing more amount of material removal is 6 g/L. The improvement in preference value for ideal solution = 0.1021.

3.4.2 ANOVA for TOPSIS

The influence of process parameters on performance characteristics may be determined by ANOVA. The result for preference solution using ANOVA is given in Table 30. The results of factor responses are considered by using higher-the-better criteria by means of MINITAB software. Table 31 indicates that C_p, I_p, T_{on} and DC are parameters which have a significant contribution towards improvement in the value of preference solution while the role of V_g is insignificant.

Table 27 Separation of alternatives from positive and negative ideal solutions

Run	S^{+}	S^{-}
1	0.0537	0.1251
2	0.0541	0.1235
3	0.0549	0.1215
4	0.04923	0.1202
5	0.04821	0.119
6	0.0465	0.1179
7	0.0988	0.0486
8	0.1205	0.0391
9	0.1323	0.0386
10	0.0484	0.1310
11	0.04584	0.1284
12	0.04329	0.1287
13	0.0379	0.1172
14	0.0336	0.1221
15	0.0359	0.1165
16	0.0413	0.1086
17	0.0424	0.1032
18	0.0464	0.1017
19	0.0334	0.1295
20	0.0293	0.1288
21	0.0285	0.1285
22	0.0307	0.1141
23	0.0411	0.1015
24	0.0375	0.1080
25	0.0507	0.0972
26	0.0498	0.1011
27	0.0616	0.0900

The response table for all the process parameters is shown in Table 31. The table demonstrates the ranks based upon the delta statistics which compare the relative magnitude of effects.

3.5 *Microstructure Analysis*

The SR has been observed to be dependent on the recast layer distribution. The existing thermal conditions damage the surface and make it irregular. The mechanism of melting and mixing of powder in proper concentrations vary the surface properties of the material resulting in its modification. The presence of foreign particles if added in appropriate quantities reduces the SR of the machined parts. Figure 15a, b represent the microstructures obtained at the optimum set of process

Table 28 Estimation of preference value with rank order

Run	Preference value	Order
1	0.6997	18
2	0.6954	19
3	0.6887	20
4	0.7096	16
5	0.7117	14
6	0.7171	15
7	0.3297	25
8	0.2449	26
9	0.2260	27
10	0.7306	11
11	0.737	10
12	0.7483	8
13	0.7557	7
14	0.7841	5
15	0.764	6
16	0.7245	12
17	0.7086	17
18	0.6867	21
19	0.7948	3
20	0.8145	2
21	0.81835	1
22	0.7881	4
23	0.7118	13
24	0.7425	9
25	0.6570	23
26	0.6704	22
27	0.5938	24

parameters with the addition of graphite powder to the dielectric fluid. The surface quality of the machined sample tends to improve as the recast layer thickness and microcracks formation reduce in comparison to machining without the addition of powder. The increase in plasma flushing efficiency during PMEDM results in the ejection of molten material and resolidification. Under constant flushing pressure, with the increase in C_p the SR increases. This is due to the formation of carbide layers resulting from the increased level of carbon from the graphite powder particles. The experimental findings depict that adding powder reduces the SR to a huge extent, but multi-objective optimization helps to determine the most suitable set of parameters with simultaneous optimization of chosen parameters to resulting in improved surface properties of the machined surface.

Table 29 Results of confirmatory experiment

Initial factor setting		Optimal condition
		Experimental
Level	$C_{p1}I_{p1}T_{on1}DC_1V_{g1}$	$C_{p3}I_{p1}T_{on3}DC_2V_{g3}$
Concentration of graphite powder (g/L)	0	6
Peak current (A)	3	3
Pulse on time (μs)	100	200
Duty cycle (%)	7	8
Gap voltage (V)	30	50
MRR (mm^3/min)	2.564	9.6
TWR(mm^3/min)	0.0172	0.0395
EWR (%)	0.6718	0.4115
SR (μm)	3.8	4
RLT(μm)	13.8	13.2
HVN	784	1105
Value of preferred solution	0.732	0.8341

Improvement in preference value for ideal solution = 0.1021

Table 30 ANOVA table for preference solution

Source	DF	Seq SS	Adj SS	Adj MS	F	P	P (%)
C_p	2	0.188	0.1880	0.0940	127.55	0.000	29.67
I_p	2	0.2575	0.2575	0.1287	174.71	0.000	40.64
T_{on}	2	0.0524	0.0524	0.0262	35.60	0.000	8.28
DC	2	0.1215	0.1215	0.0607	82.43	0.000	19.17
V_g	2	0.0023	0.0023	0.0012	1.58	0.237	0.366
Residual error	16	0.0117	0.0118	0.0007			
Total	26	0.6337					

$S = 0.02715$, R-Sq = 98.1%, R-Sq(adj) = 97.0%

Table 31 Response table for means of preference value

Level	C_p	I_p	T_{on}	DC	V_g
1	0.5581	0.7475	0.7162	0.7009	0.6877
2	0.7377	0.7427	0.6973	0.7429	0.6753
3	0.7323	0.5379	0.6146	0.5843	0.6650
Delta	0.1797	0.2095	0.1016	0.1586	0.0227
Rank	2	1	4	3	5

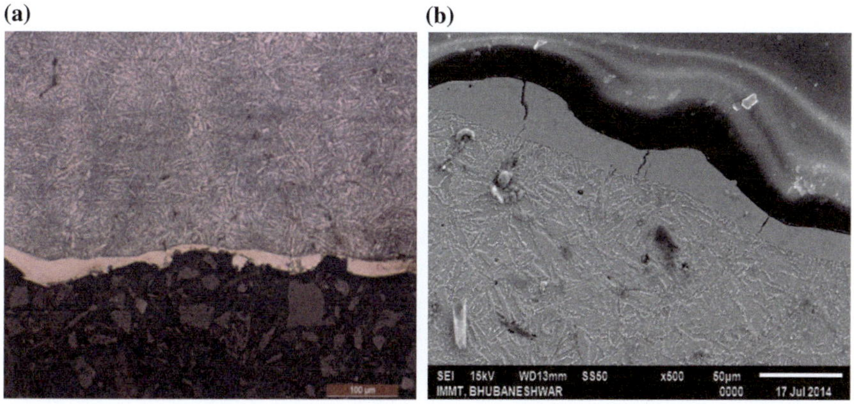

Fig. 15 SEM of surfaces and sub-surfaces for the machining with graphite powder-mixed dielectric. **a** Optical microscopy image of recast layer. **b** SEM micrograph of recast layer

4 Conclusion

The present investigation aims to determine the optimal setting for the process variables to increase the MRR and HVN and minimize the TWR, EWR, SR and RLT simultaneously for H-11 hot work tool steel by adding graphite powder to the dielectric fluid. Taguchi's technique has been implemented to perform the experiments by altering C_p, I_p, T_{on}, DC and V_g. Single-objective optimization has been carried out and an optimum set of process parameters have been identified for the response parameters. Further, multi-objective optimization has been performed using TOPSIS to identify the optimum set of input parameters that improve the process performance. The findings from the present work are as follows:

1. Maximum MRR can be achieved at $\boldsymbol{C_{p3}I_{p3}T_{\text{on}3}\text{DC}_2V_{g3}}$ when machined using PMEDM. I_p, powder concentration, V_g and T_{on} are the parameters which have significant contribution toward improvement in MRR while the role of DC is insignificant.
2. Minimum TWR can be obtained at $\boldsymbol{C_{p2}I_{p1}T_{\text{on}2}\text{DC}_2V_{g1}}$. Parameters which have significant contribution toward improvement in TWR are C_p, I_p, T_{on} and DC are while the role of V_g is insignificant.
3. $\boldsymbol{C_{p2}I_{p1}T_{\text{on}2}\text{DC}_2V_{g1}}$ offer minimum EWR when machined using PMEDM. C_p, I_p, T_{on} and DC are parameters which have significant contribution toward improvement in EWR while the role of V_g is insignificant.
4. Minimum SR is obtained at $\boldsymbol{C_{p2}I_{p1}T_{\text{on}1}\text{DC}_2V_{g1}}$. Parameters which have a significant contribution towards improvement in SR are C_p, I_p and V_g are while the role of T_{on} and DC is insignificant.

5. Minimum RLT is obtained at $\boldsymbol{C_{p1}I_{p1}T_{\mathrm{on}3}\mathrm{DC}_3V_{g1}}$. Parameters which have significant contribution towards improvement in RLT is C_p, I_p and V_g are while the role of T_{on} is insignificant.
6. Maximum HVN can be achieved at $\boldsymbol{C_{p3}I_{p3}T_{\mathrm{on}1}\mathrm{DC}_1V_{g3}}$ when machined using PMEDM. I_p, powder concentration, DC and T_{on} are the parameters which have significant contribution toward improvement in HVN while the role of V_g is insignificant.
7. The multi-objective optimization results show that C_p of 6 g/L, I_p of 3 A, T_{on} of 200 μs, DC of 80% and V_g of 50 V i.e. $\boldsymbol{C_{p3}I_{p1}T_{\mathrm{on}3}\mathrm{DC}_2V_{g3}}$ is the optimal setting using TOPSIS. The optimal setting obtained can develop the performance of the quality characteristics under consideration.
8. Confirmatory test shows improvement of 0.1021 in the preferred values for the optimum set using TOPSIS as compared to the initial setting, which is satisfactory.
9. The significant machining parameters affecting the process characteristics at 95% confidence interval were determined using ANOVA. The adjusted R^2 value was found to be 98.1% which means that 98.1% of the response variables fit the linear model.
10. The microstructure analysis was done for the optimal setting which shows improved properties due to less crack formation, lower roughness values and small thickness of recast layer.
11. The model is appropriate for use to identify the most suitable set of input parameters for the required performance characteristics. The outcome of the present research work will be a substantial aid to the industries concerned with the use of materials processed through PMEDM.
12. Adding powder particles to the dielectric widens the gap, improves the flushing and makes the process stable. However, powders should be added in appropriate concentrations as they tend to settle down in the tank and cause difficulty in stirring.

References

Assarzadeh, S., and M. Ghoreishi. 2013. A dual response surface-desirability approach to process modeling and optimization of Al_2O_3 powder-mixed electric discharge machining (PMEDM) parameters. *International Journal of Advanced Manufacturing Technology* 64: 1459–1477.

Batish, A., and A. Bhattacharya. 2012. Mechanism of material deposition from powder, electrode and dielectric for surface modification of H-11 and H-13 die steels in EDM process. *Materials Science Forum* 701: 61–75.

Gadakh, V.S. 2012. Parametric optimization of wire electric discharge machining using TOPSIS method. *Advances in Production Engineering & Management* 7 (4): 157–164.

Kumar, S., R. Singh, T.P. Singh, and B.L. Sethi. 2009. Surface modification by electrical discharge machining: A review. *Journal of Materials Processing Technology* 209: 3675–3687.

Lal, Shyam, Sudhir Kumar, Zahid A. Khan, and Arshad N. Siddiquee. 2015. Multi-response optimization of wire electrical discharge machining process parameters for Al7075/Al2O_3/SiC hybrid composite using Taguchi-based grey relational analysis. *Journal of Engineering Manufacture* 1–9.

Pecas, P., and E. Henriques. 2008. Electrical discharge machining using simple and powder mixed dielectric: The effect of the electrode area on the surface roughness and topography. *Journal of Materials Processing Technology* 200: 250–258.

Sidhu, Sarabjeet Singh, Ajay Batish, and Sanjeev Kumar. 2014. ED Machining of Particulate Reinforced MMC's. *International Journal of Mechanical, Aerospace, Industrial, Mechatronics and Manufacturing Engineering* 8(3):509–515.

Singh, Baljinder, Paramjit Singh, Gaurav Tejpal, and Singh Gurtej. 2012. An experimental study of surface roughness of H-11 in EDM process using copper tool electrode. *International Journal of Advanced Manufacturing Technology* 3 (4): 30–33.

Singh, A.K., S. Kumar, and V.P. Singh. 2015. Effect of the addition of conductive powder in dielectric on the surface properties of superalloy Super Co 605 by EDM process. *International Journal of Advanced Manufacturing Technology* 77: 99–106.

Talla, Gangadharudu, Deepak Kumar Sahoo, S. Gangopadhyay, and C. K. Biswas. 2015. Modeling and multi-objective optimization of powder mixed electric discharge machining process of aluminium/alumina metal matrix composite. *Engineering Science and Technology* 1–5.

Tripathy, S., and D.K. Tripathy. 2016. Multi-attribute optimization of machining process parameters in powder mixed electro-discharge machining using TOPSIS and grey relational analysis. *Engineering Science and Technology: an International Journal* 19 (1): 62–70.

Tripathy, S., and D.K. Tripathy. 2017a. Multi-response optimization of machining process parameters for powder mixed electro-discharge machining using grey relational analysis and topsis. *Machining Science and Technology* 21 (3): 362–384.

Tripathy, S., and D.K. Tripathy. 2017b. Optimization of process parameters and investigation on surface characteristics during EDM and powder mixed EDM.' In *Innovative design and development practices in Aerospace and Automotive Engineering*, eds. R. Bajpai, and U. Chandrasekhar, Lecture notes in Mechanical Engineering, pp. 385–391. Singapore: Springer.

Tripathy, S., and D.K. Tripathy. 2017c. Grey Relational Analysis and its application on surface properties during EDM and Powder Mixed EDM. *Journal of Engineering Science and Technology* 12 (9): 2374–2392.

Optimization of Single Pitch Error and MRR in a WEDM Gear Cutting Process

Kasinath Das Mohapatra and Susanta Kumar Sahoo

Abstract Gears are the key elements to the manufacturing processes. Gears having teeth or wheels mesh together with other gears to transmit torque. Accurate motion transfer and minimum running noise are the two most important features for these gears which depend on the amount of errors, i.e. the errors present in the pitch of the gears. Gears can be manufactured by different manufacturing processes such as shaping, forming, hobbing, milling and broaching. Gears are one of the crucial components of the highly accurate miniaturized devices such as pumps and motors, electronics, business machines, home appliances, automotive parts, measuring instruments, timing devices, MEMS, etc. It is also used in the scientific, industrial and domestics areas. Gear cutting by WEDM finds its application in many industrial areas of aeronautical and electrical industries that requires precise finishing and appropriate accuracy. Geared devices can change the torque, speed and direction of the power source. Copper, bronze, brass, stainless steel and aluminium are the most frequently used materials for these gears. In the present investigation, miniature copper gears of 2 mm thickness are cut by WEDM and the machine parameters are optimized.

Keywords Addendum · Dedendum · Electron dispersive spectroscopy
Gears · Pulse on time · Pulse off time · Single pitch error · Wire tension
Wire feed rate

K. D. Mohapatra (✉) · S. K. Sahoo
Department of Mechanical Engineering, National Institute of Technology Rourkela, Rourkela, India
e-mail: kitu.kasinath1@gmail.com

S. K. Sahoo
e-mail: sks@nitrkl.ac.in

S. S. Pande and U. S. Dixit (eds.), *Precision Product-Process Design and Optimization*, Lecture Notes on Multidisciplinary Industrial Engineering,
https://doi.org/10.1007/978-981-10-8767-7_11

Nomenclature

D	Pitch circle diameter
F	Fisher's value
F_{pt}	Single pitch error
F_p	Theoretical pitch
h	Workpiece thickness
K	Diameter of the wire
l	Length of cut
m	Alternatives
N	Number of teeth
n	Objectives
P	Probability value
P_d	Diametric pitch
R^2	Regression coefficient
T_{on}	Pulse on time
T_{off}	Pulse off time
t	Time between the cutting length
V_C	Cutting speed
Wf	Wire feed rate
W_i	Weight of the criterion
Wt	Wire tension
X_{ij}	Response
Y_i	Modified coefficient ratio

1 Introduction

Wire EDM or Wire Electric Discharge Machining process is an unconventional machining process used for cutting of different complex shapes of a conductive material. Till today, manufacturing industries face difficulties to design and cut intricate shapes using other conventional processes. Wire EDM makes it easier to cut different types of conductive materials making it much faster using altered parameter settings. The mechanism of the wire EDM is the involvement of vaporization and melting in which the materials are removed by thermoelectric erosion process by an electric spark created between a very thin wire (usually brass of diameter 0.25 mm) and an electrically conductive workpiece material. The wire in the spool is continuously fed to the workpiece to cut the material from the workpiece and the process occurs in an optimized speed such that there is a less chance of breakage in the wire. A high-frequency DC pulse power is used in the spark gap between the wire and the workpiece. This causes the existence of tiny sparks and the energy contained in it removes a fraction of workpiece material. The machine uses distilled water as a dielectric as it acts also as a coolant to the material and wire. Wire EDM is used for cutting many complex geometries and 3D shapes which are difficult to machine by other conventional and non-conventional processes.

The present work deals with the cutting of small miniature gears made of copper with thickness 3 mm. The machining is carried out using brass wire as the tool electrode having a diameter of 0.25 mm and distilled water as the dielectric fluid. The experiment was carried out using four input parameters at two levels each repeated twice. Pulse on time, wire feed rate, pulse off time, and wire tension are taken as the process parameters and material removal rate and single pitch error were taken as the output parameters for the experiments. Totally 32 teeth have been obtained by conducting 32 sets of experiments, each parameter being machined considering two gear teeth. Totally four gears have been obtained each having 16 numbers of teeth and the responses, i.e. material removal rate and pitch error have been obtained and assessed from these experiments. A Multi-objective optimization technique using MOORA (Multi-objective Optimization using Ratio Analysis) method has been performed to know the optimum settings out of the 32 combinations obtained for the experiment. ANOVA (Analysis of Variance) table and response table has been obtained to know the significant process parameters affecting the output responses. The effects of response parameters on different process parameters have been analysed. Microstructural analysis has been carried out using SEM (Scanning Electron Microscope) for the wire (brass) and the workpiece (copper) to study and investigate the various defects, microstructures and surface characteristics of the cutting surface of the machined gear. Electron Dispersive Spectroscopy (EDS) was carried out to know the particular compound and composition present in the material. The optimum settings obtained for the gear can be used for manufacturing and production of large quality miniature gears.

2 Literature Review

Many works on WEDM, in general, gear cutting by WEDM in particular, have been carried out by the different researchers. On surface integrity of miniature spur gears by WEDM was carried out by Gupta and Jain (2014). They found that the combination of low discharge energy parameters leads to better accuracy and surface finish of the miniature gears. Gupta and Jain (2014) analysed and optimized the micro-geometry of miniature spur gears by wire EDM. Menz (2003) developed a 3D micro-structuring of ceramics and materials for high aspect ratio in micro-wire EDM process. The micro-geometry of miniature spur gears by wire EDM process were studied by Taylor et al. (2013). They suggested that in order to manufacture high-quality miniature gears, the use of pulse on time and voltage should be low. Yeh et al. (2013) studied the surface characteristics of polycrystalline silicon using phosphorous dielectric on wire electric discharge machining process. A multi-objective optimization was carried out for a 3D surface topography by Ming et al. (2014) in machining of YG 15 in WEDM. The spark erosion machining of miniature gears was studied by Gupta et al. (2015). Mandal et al. (2015)

modelled and optimized C-263 superalloy using multi-cut strategy in wire EDM. They used the desirability function for the prediction of the optimal settings. Habib and Okada (2015) studied the movement of the wire electrode in WEDM process. They found the backward deflection of the wire in fine WEDM process. Wentai et al. (2015) investigated the change in the wire tension in high-speed wire electric discharge machining. The wire tension was found to be capable enough to remain stable during the machining process. Kuriakose et al. (2003) studied the data mining applied to the wire EDM process. They conducted the experiments collected from the conducted data and they also tested it on additional data. They arrived at a conclusion that the model that was built using data mining provides good results with desired accuracy. With the improvement in the surface roughness, the material removal rate was a major concern for the wire EDM process. A development of the cylindrical wire EDM process, surface integrity and roundness was studied by Qu et al. (2002). The accuracy improvement in the wire EDM by real-time wire tension control was examined by Yan and Huang (2004). They attached dynamic absorbers to the idle rollers of wire transportation mechanism so that the vibration of wire tension is reduced during wire feeding. They concluded that with the attachment of the dynamic absorbers, small steady-state error and fast transient response can be obtained with geometrical contour error of corner cutting reduced to approximately 50%. Miller et al. (2004) investigated the spark cycle on material removal rate in wire EDM of advanced materials. They used metal bond diamond grinding wheels and porous metal foams in their experiment. They identified five types of constraints developed during machining, i.e. wire breakage, machine slide speed limit, MRR due to short circuit and pulse on time upper and lower limits. Further the capability of WEDM process was also established by them to machine different advanced materials. In the meantime, several researchers tested the wire EDM parameters. Liao et al. (2004) conducted Analysis of Variance (ANOVA) and *Fischer* test (*F*-test) in order to obtain a fine surface finish in wire EDM. They found that low conductivity of dielectric should be amalgamated for the discharge spark to take place with the best surface finish of 0.22 μm. They further concluded that the traditional circuit should be modified using low power for ignition in order to achieve a good surface roughness during machining. The cutting parameters were still a major concern when Kanlayasiri and Boonmung (2007) investigated the effects of wire EDM machining parameters on surface roughness of newly developed DC53 die steel. They investigated on the machining parameters such as pulse on time, pulse off time, wire tension and pulse peak current. Further they employed quantitative testing methods in place of qualitative testing techniques on residual analysis. Their findings concludes that based on ANOVA, pulse peak current and pulse on time were the significant variables to affect the surface roughness of wire-EDMed DC53 die steel. As the wire deforms resulting in the deviations in the inclination angle of machined parts, Plaza et al. (2009) developed some original models for the prediction of angular error in WEDM taper cutting. The two original models were developed by them for the prediction of angular error to decrease the experimental load and contribute a general approach to the problem. Their findings showed that the taper angle and the part thickness were the most influencing variables in the problem. The angular error in wire EDM taper cutting was studied by Sanchez et al. (2008) when he designed a new approach to the prediction of angular error in taper cutting. They derived a quadratic equation for the prediction of angular error in taper cutting. They reported that the influence of the angular error is mostly due to the angle and the part thickness determining wire's mechanical behaviour. Soon with the advancement of the taper cutting, the corner cutting were also given importance and hence the corner cutting accuracy of thin parts using wire electric discharge machining was carried out by Dodun et al. (2009). Their results disclosed that the outside corners can be obtained with a small thickness and corner angle having the shape of post-yield bending when accompanied by the machining error. Yan and Liu (2009) designed, analysed and experimented the study of a high-frequency power supply for finish cut of wire EDM. They developed a high-frequency power supply for the improvement of surface quality in wire EDM. Their findings revealed that the electrolytic effect of tungsten carbide can be reduced by pulse duration ratio in a high discharge frequency of more than 500 kHz.

From the literature review, it was observed that very few works have been carried out in gear cutting of copper by wire EDM process. In view of it, the objective of the present work is to focus on producing spur gears made of copper, to maximize the MRR and minimize the error present in the gears by optimizing the responses and to obtain a best combination from the present settings so that the time can be reduced. In addition to it, the machining of copper gear by wire EDM process at the obtained optimized setting will be an advantage in producing copper gears by achieving faster machining rate.

As the machining of gears by copper have an application in various transportation oils such as diesel, petroleum, semiconductor industries, lubricant oils, clocks, radiators, aerospace, printing, papermaking, mining industries, electric motors, trucks, air brakes, etc., the obtained optimized setting can be an advantage over other combinational settings to produce high-quality miniature spur gears made of copper.

3 Experimental Setup

3.1 Machines, Materials and Specifications

The experimental work is carried out using an ELECTRONICA EPULSE 15® machine. As copper gears have important applications in many industries mentioned above, so copper was chosen as the workpiece material of thickness 3 mm. A brass wire of diameter 0.25 mm is taken as the tool electrode for the machining operation. Distilled water is used as the dielectric fluid for carrying out the present work. Figure 1 displays the schematic diagram of wire EDM process. The composition of the brass wire is a pure brass of 65% copper and 35% of zinc. The machine uses a high-voltage DC servo stabilizer of 400 V.

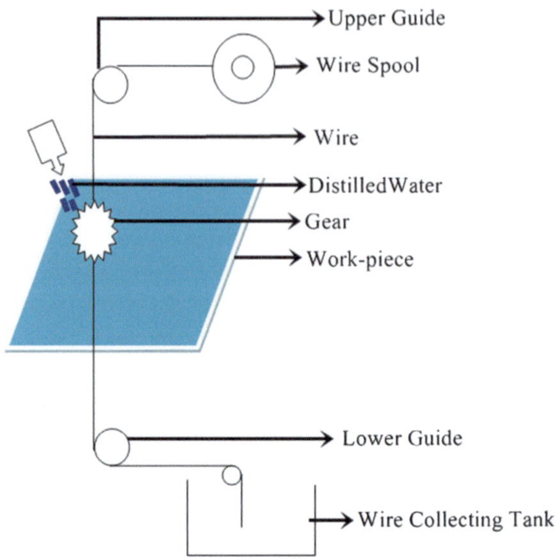

Fig. 1 Schematic diagram of wire EDM process

3.2 Parameter Settings and Gear Specifications

The machine parameters are chosen according to the machine restraints and past literature survey. To conduct the experiment, four input parameters were chosen at two different levels repeated twice. The combinations were obtained from Minitab 16® software with 32 numbers of setting combinations. Pulse on time (T_{on}), pulse off time (T_{off}), wire tension (Wt) and wire feed rate (Wf) were taken as the process parameters.

The other wire EDM parameters such as peak current, servo voltage, peak voltage, water pressure and servo feed rate were kept constant for the experiment. The response parameters such as Material removal rate and single pitch error (SPE) were obtained after the machining operation.

The model of the gear is constructed using ELCAM® software giving necessary geometric parameters. The gear model was designed giving required pressure angle of 20°, pitch circle diameter (D) of 10 mm and number of teeth (N) as 16. Table 1 depicts the parameters and range considered for conducting the experiment.

Table 1 Parameters and range considered for the experiment

Input parameters	Units	Symbol	Levels	
			I	II
Pulse on time	μs	Ton	105	110
Pulse off time	μs	Toff	53	58
Wire feed rate	m/min	Wf	3	5
Wire tension	Kg-f	Wt	5	7

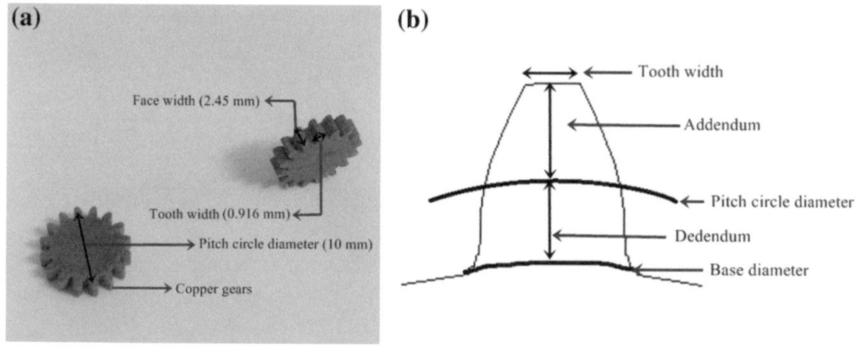

Fig. 2 Gears obtained after the machining operations **a** specifications of the gear and **b** gear terminologies

Figure 2 shows the different gears obtained after the machining operation, its specifications and terminologies. The experiment was carried out for all the 32 sets of the combinations considering each two teeth's as one set. So an all total of 4 gears were obtained having sixteen numbers of teeth on one gear. So in one gear eight sets of combinations were carried out. The gears obtained were measured for calculating the single pitch error produced in the teeth. Table 2 depicts the specification of gears obtained for the machining process.

3.3 *Calculation of the Response Parameters*

The response parameters obtained were measured for the experimental investigation and analysis. The material removal rate is obtained by taking the cutting speed, wire diameter and workpiece thickness into consideration. Cutting speed is calculated by taking length and time into consideration. Higher the pulse on time, more is the removal of material. The *material removal rate* (mm^3/min) is calculated by the following formula:

$$\mathrm{MRR} = V_C \times h \times k, \tag{1}$$

Table 2 Gear specifications obtained for the machining operation

Material	Copper
Profile	Involute
Pressure angle	20°
Number of teeth	16
Pitch circle diameter	10 mm
Face width	2.45 mm
Tooth width	0.916 mm

where V_c is the cutting speed in mm/min, K is the diameter of the wire (0.25 mm) and h is the workpiece thickness (3 mm).

Similarly, the *cutting speed* is calculated by the following formula:

$$V_C = 60 \times l/t \text{ mm/min}, \tag{2}$$

where l is the cut length of two teeth's of the gear in mm and t is the time between the cutting length.

The machining time was calculated by using stopwatch from the starting point of the one tooth to the ending point of the same tooth for one gear tooth.

During the machining of the gears by wire EDM, some deviations occur while cutting of the gear tooth. These are nothing but the type of errors produced in the gear tooth during the machining operation. These errors arise due to noise and certain environmental condition during machining. Single pitch error is one of the pitch deviation error occurs during the machining of the gear tooth.

Single pitch error is calculated by taking two teeth into consideration. It is the difference between the theoretical and actual measured values of a pitch for a given pair of teeth or the deviation between the actual measured pitch values between any two adjacent tooth surfaces and the theoretical circular pitch.

$$\text{Single pitch error } (F_{\text{Pt}}) = \text{Theoretical Circular Pitch} - \text{Actual measured pitch values} \tag{3}$$

The actual measured values of pitch were obtained after the machining by taking the average of the two left flanks and right flanks distance into consideration.

The *theoretical circular pitch* (F_P) is calculated by the following formula:

$$F_p = \pi/P_d = \pi D/N, \tag{4}$$

where D is the pitch circle diameter in mm (10 mm), N is the number of teeth and P_d is the diametric Pitch.

Figure 3 depicts the image scale legend for 20× magnification zoom, where magnification 20× corresponds to 14.537 μm for 0.5 mm length. Similarly, the measurement of single pitch error for one combination (L_{21}) out of 27 combinations is depicted in Fig. 4. The measurement of single pitch error in Fig. 4 is measured in microns. The micron is then converted to millimetre according to the 20× scale.

The readings for the single pitch error were measured by using suitable software Caliper pro 4.2. From the Fig. 4a, A to B signifies the measurement of left flank and right flank for one gear tooth. A to a signifies the measurement of left flank measured from the software. Similarly, b to B signifies the measurement of right flank measured from the software. The readings were assessed and averaged thrice for the measurement and the average of the left flank and right flank were calculated for the measurement of the pitch error.

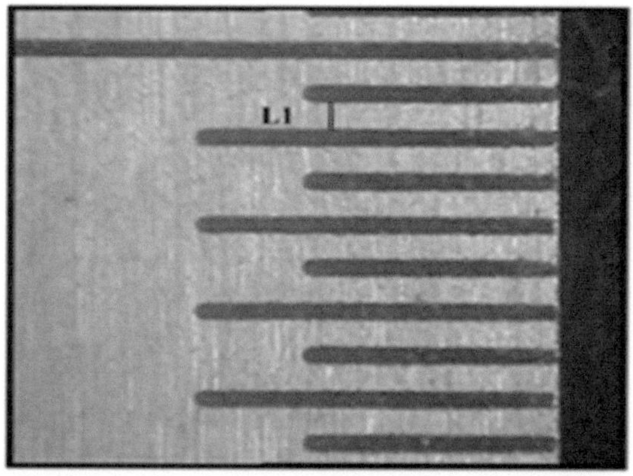

Measurement Name	Description	Results (Micron)	Calibration (Micron/Pixel)
L1	Length	14.537 Micron	X: 100/264 Y: 100/264

Fig. 3 Image scale legend while measuring the single pitch error at 20× magnification zoom

4 Multi-objective Optimization Using MOORA

The experiment was conducted for all the 32 sets of combinations. The output parameters, i.e. MRR and single pitch error were optimized using ratio analysis and is called multi-objective optimization using ratio analysis (MOORA). This method was first introduced by Brauer (2007), in order to solve various conflicting decision-making problems. The multi-objective response parameters are converted to single-objective response parameters by this method. In the present work, the MRR should be maximized and the single pitch error should be minimized. The decision matrix showing the performance of different alternatives can be written in the form:

$$X = \begin{bmatrix} X_{11} & \cdots & X_{12} & \cdots & X_{1n} \\ X_{21} & \cdots & X_{22} & \cdots & X_{2n} \\ \cdots & \cdots & \cdots & \cdots & \cdots \\ X_{m1} & \cdots & X_{m2} & \cdots & X_{mn} \end{bmatrix} \tag{5}$$

The method starts with a matrix of responses of different alternatives to different objectives:

$$\left(X_{ij}\right)_{\text{MRR}} \text{ and } \left(X_{ij}\right)F_{\text{pt}} \tag{6}$$

where X_{ij} is the response of the ith objective to the jth alternative, $i = 1, 2, \ldots, n$ are the objectives, $j = 1, 2, \ldots, m$ are the alternatives.

To achieve the problem, normalization is done so that the values transforms into a dimensionless number [0, 1].

(a)

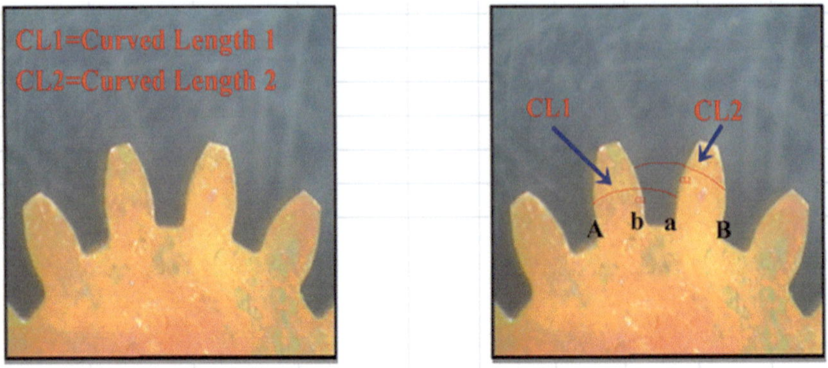

Measurement Name	Description	Results (Micron)	Calibration (Micron/Pixel)
CL1	Curved Length	154.474 Micron	X: 100/123 Y: 100/123
CL2	Curved Length	172.899 Micron	X: 100/123 Y: 100/123

(b)

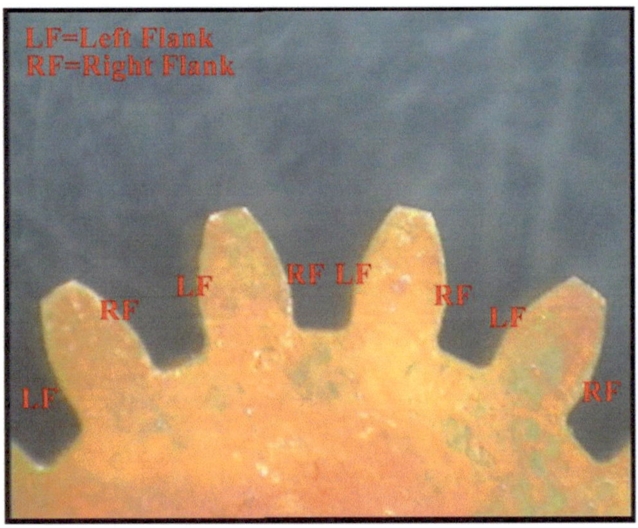

Fig. 4 Measurement of single pitch error at 20× magnification zoom

The summation for MRR and F_{pt} were calculated as follows:

$$\left(X_{ij}\right)_{\text{MRR}} = \sqrt{\sum_{i=1}^{m} X_{ij}^{2}} \text{ and } \left(X_{ij}\right) F_{\text{pt}} = \sqrt{\sum_{i=1}^{n} X_{ij}^{2}}. \quad (7)$$

Step 1: The normalized decision matrix by vector method is calculated by the formula:

$$X_{ij,\mathrm{MRR},F\mathrm{pt}} = \frac{X_{ij}}{\sqrt{\sum_{i=1}^{m} X_{ij}^2}}, \quad i = 1,\ldots,n; j = 1,\ldots,m. \tag{8}$$

The MOORA method has two different components, ratio system and reference point approach. The responses are added in case maximization (MRR) and subtracted in case of minimization (single pitch error) for optimization.

Step 2: The ranking value according to ratio system is as follows:

$$Y_i = \sum_{i=1}^{g} X_{ij}^* - \sum_{i=g+1}^{n} X_{ij}^*, \tag{9}$$

where $\sum_{i=1}^{g} X_{ij}^*$ and $\sum_{i=g+1}^{g} X_{ij}^*$ are for the benefits and non-benefits criteria respectively.

i = 1, 2, …, g are the objectives to be maximized and i = g+1, g + 2, …, n are the objectives to be minimized and Y_i is the normal assessment of the alternative j with respect to all objectives.

If there are some attributes more important than others then the Ranking value becomes

$$Y_i = \sum_{i=1}^{g} W_i X_{ij}^* - \sum_{i=g+1}^{n} W_i X_{ij}^* \tag{10}$$

where W_i is the weight of the ith criterion.

Step 3: The Y_i values are ranked and the alternative with the highest value is chosen as the best alternative.

Step 4: The modified coefficient ratio is given by the following formula:

$$Y_i = \sum_{i=1}^{g} X_{ij}^* / \sum_{i=g+1}^{n} X_{ij}^*. \tag{11}$$

Table 3 depicts the experimental design values of both input and output parameters obtained by using L_{32} orthogonal array and the optimization was carried out by combining the output responses using Ratio analysis. The multi-objective responses, i.e. MRR and single pitch error were optimized to single-objective response MCR (Modified Coefficient Ratio) using ratio analysis. From the above table, the highest coefficient rank corresponds to the 24th setting of the combinations which is considered as the best alternative. So T_{on} 110-T_{off} 53-Wf 5-Wt 7 is chosen the best combination of the gear made of copper to be manufactured in the field of wire EDM within these settings.

Table 3 MOORA Optimization of L32 orthogonal array

S. No.	Input parameters				Output parameter		MOORA	
	T_{on}	T_{off}	Wf	Wt	MRR	F_{Pt}	MOORA coefficient	rank
1	105	53	3	5	1.26	1.2	0.6841	24
2	105	53	3	5	1.24	1.08	0.7510	18
3	105	53	3	7	1.31	1.14	0.7498	19
4	105	53	3	7	1.30	1.18	0.7205	21
5	105	53	5	5	1.34	1.22	0.7182	22
6	105	53	5	5	1.37	1.22	0.7343	20
7	105	53	5	7	1.35	1.13	0.7783	17
8	105	53	5	7	1.23	1.16	0.6947	23
9	105	58	3	5	1.04	1.09	0.6228	25
10	105	58	3	5	1.08	1.25	0.5629	32
11	105	58	3	7	1.06	1.16	0.5981	30
12	105	58	3	7	1.10	1.14	0.6298	26
13	105	58	5	5	1.04	1.12	0.6018	28
14	105	58	5	5	1.03	1.12	0.6004	27
15	105	58	5	7	1.05	1.15	0.5948	29
16	105	58	5	7	1.02	1.14	0.5788	31
17	110	53	3	5	2.25	1.07	1.3700	3
18	110	53	3	5	2.10	1.06	1.2914	8
19	110	53	3	7	2.13	1.02	1.3650	5
20	110	53	3	7	2.23	1.08	1.3452	4
21	110	53	5	5	2.35	1.11	1.3793	2
22	110	53	5	5	2.26	1.21	1.2125	12
23	110	53	5	7	2.06	1.14	1.1784	15
24	*110*	*53*	*5*	*7*	*2.18*	*1.03*	*1.3802*	*1*
25	110	58	3	5	2.09	1.04	1.2993	7
26	110	58	3	5	2.03	1.06	1.2489	10
27	110	58	3	7	2.08	1.11	1.2172	14
28	110	58	3	7	2.02	1.05	1.2493	11
29	110	58	5	5	1.87	1.09	1.1125	16
30	110	58	5	5	2.28	1.15	1.2839	6
31	110	58	5	7	2.09	1.08	1.2620	9
32	110	58	5	7	1.98	1.04	1.2341	13

5 Analysis of the Experiment

5.1 Microstructural Study of Material and the Wire

The objective of carrying out the material analysis is to investigate and analyse the various surface defects of the copper plate material at the optimized settings obtained from the wire EDM parameters. The surface metallography needs to be investigated as wire EDM is a thermal dominant process and the very high temperature has a significant impact on the process-induced surface integrity including microstructure change, surface characteristics and grain structure.

The experiment was conducted and the output parameters, i.e. MRR and single pitch error were optimized using MOORA method to get the best alternative. The material and the wire with the best alternative T_{on} 110 µs, T_{off} 53 µs, Wf 5 m/min and Wt 7 kg-f are now analysed for the various qualitative and microstructural analyses of the study.

5.1.1 Study of Grain Structure

The optimized setting of the teeth was first seen under the optical microscope and then the image is viewed under scanning electron microscope to know the presence of grains after etching. In order to see the grain structure, proper etchant should be applied to the machined gear tooth. The etchant used for copper is distilled water (50 ml) and Nitric acid (50 ml).

Figure 5a, b depicts the optical image of the gear visualized at the surface at 400× zoom and Fig. 5c depicts the SEM image of the gear at 300× zoom after the etchant has been applied. From the figures, it can be observed that some grains like structures or grain boundaries are formed on the machined material surface after etching. The formations of grains are due to the structure, composition or phase of the metal that creates electrochemical potentials when exposed to an etchant.

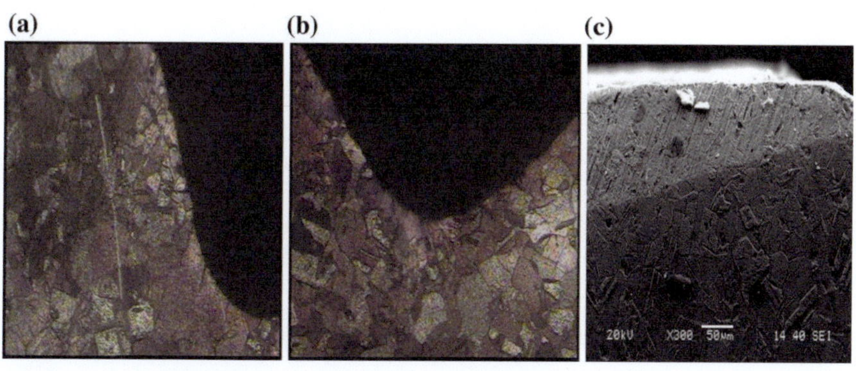

Fig. 5 Optical and SEM image of the copper at 400× and 300× zoom

5.1.2 Study of Surface Characteristics Using Scanning Electron Microscope (SEM)

The gear tooth is placed under scanning electron microscope and the micrographs were studied for both the material and the brass wire. Figure 6 describes the SEM image of the workpiece material at 10,000× zoom. It can be observed that there are some voids and cracks formation on the surface of the workpiece material. Some dark regions spots such as pores were clearly visible on the work material. High-speed spark energy density and wire feed rate were the reasons for the occurrence of such voids and cracks due to the sparks generated from the wire. It is clear that the surface characteristic of the WED machined surface depends upon the applied discharge energy.

Figure 7 depicts the SEM analysis of the copper material at 5000× and 50,000× magnification zoom. From the Fig. 7a, it can be observed that some holes and heat-affected zone (HAZ) were formed on the material surface. The heat-affected zone is associated with grain growth, high tensile residual stress, porosity, and dielectric fluid. The formation of the hole is due to the high discharge energy of the spark with a high spark on time. The heat-affected zone formation is due to the change in temperature, rapid heating, continuous high sparks and quenching in the WEDM process. The bull eye was also marked as shown in Fig. 7a and the occurrence of bull eye is due to the presence of oxygen-rich layer. A lump of debris was also identified at high discharge energy settings as depicted in Fig. 7b. The dielectric carries out some of the molten material at high discharge energy settings. The leftover molten material re-solidifies to form lumps of debris.

Figure 8 depicts the SEM analysis of the copper material at 10,000× magnification zoom. Pores and holes were formed on the material surface as shown in Fig. 8a. The pores are formed due to the frequent melting expulsion on the wire

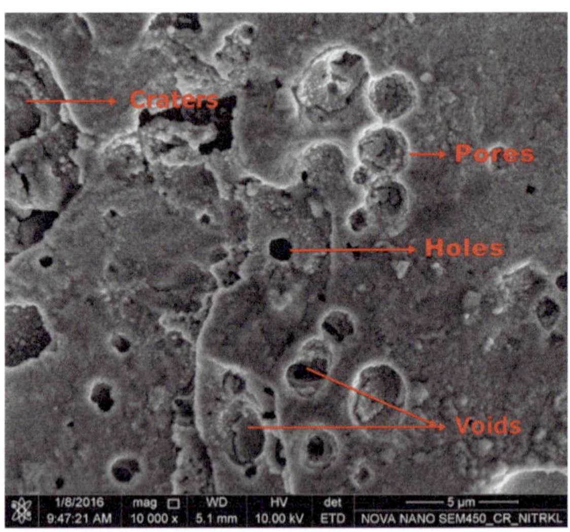

Fig. 6 Micrographs of the copper material at 10,000× zoom

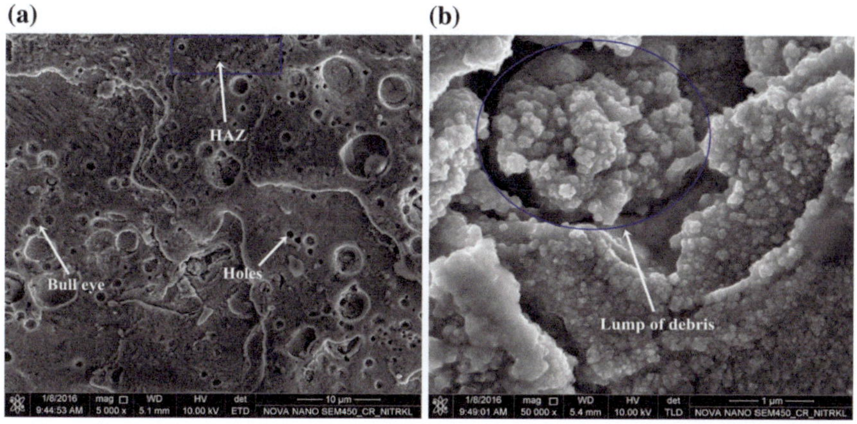

Fig. 7 SEM analysis of the copper material at **a** 5000× and **b** 50,000× magnification zoom

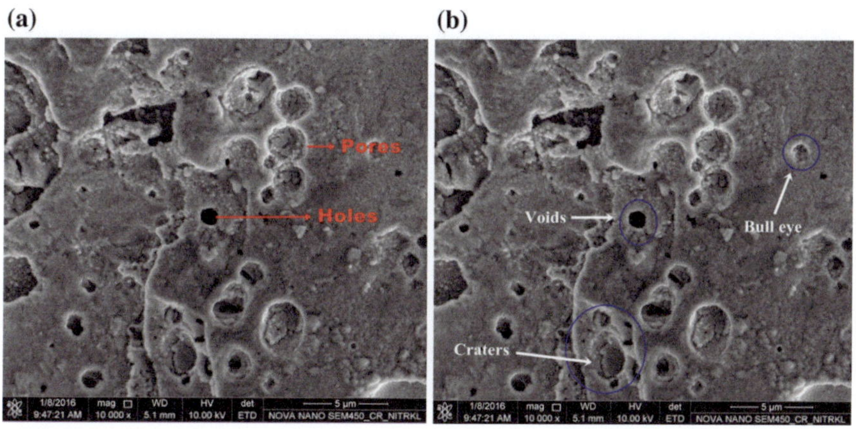

Fig. 8 SEM analyses of the copper material at 10,000× magnification zoom

surface due to high pulse on time. High spark energy density at high wire feed rate causes the material to form surface holes during machining operation. Similarly, Fig. 8b shows voids and craters due to high spark energy density at high pulse on time causing the sparks to create such formations on the material surface.

Figure 9 depicts the SEM analysis of the copper material at 20,000× magnification zoom. Craters and voids were visible on the material surface as displayed in Fig. 9. The violent sparks with high-energy density and the various forces occurring in the wire lag lead to the formation of voids and craters on the material surface. Similarly, the melted debris was identified on the material surface as shown in Fig. 9b due to the deposition of the melted material formed due to the solidification of the spark at high discharge energy settings. The high discharge energy of the

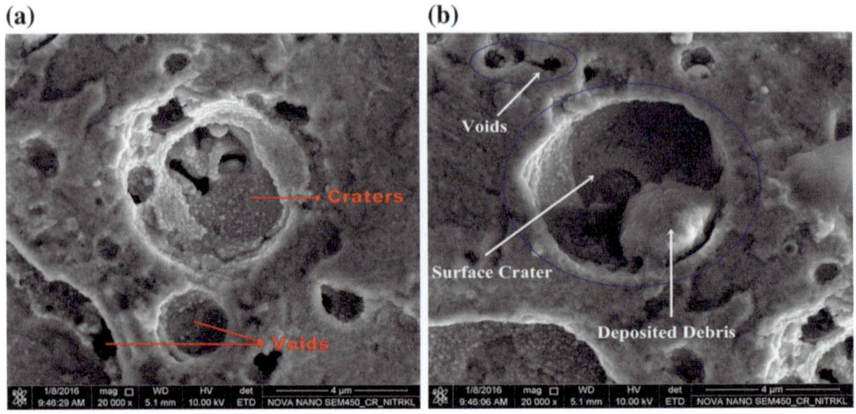

Fig. 9 SEM analyses of the copper material at 20,000× magnification zoom

plasma channel results in melting of the material but it is not sufficient to produce a high exploding pressure which can spray all the molten metal away from the machined surface. The gas bubbles banish from the molten material when the remaining molten material solidifies on the surface resulting in micro-voids.

Figure 10 depicts the microstructural analysis carried out for the wire at the optimized setting of the response T_{on} 110 µs, T_{off} 53 µs, Wf 5 m/min, and Wt 7 kg-f. Figure 10a displays the affected damaged zone indicating that the wire is being affected by the sparks produced from the wire at high discharge rate. Some cracks and melted deposits were also marked over the wire surface as shown in Fig. 10b. High discharge energy at high servo voltage and wire feed rate results in the formation of such cracks. Moreover high peak current and induced stresses exceeding the wires ultimate tensile strength are the reasons for the formation of cracks.

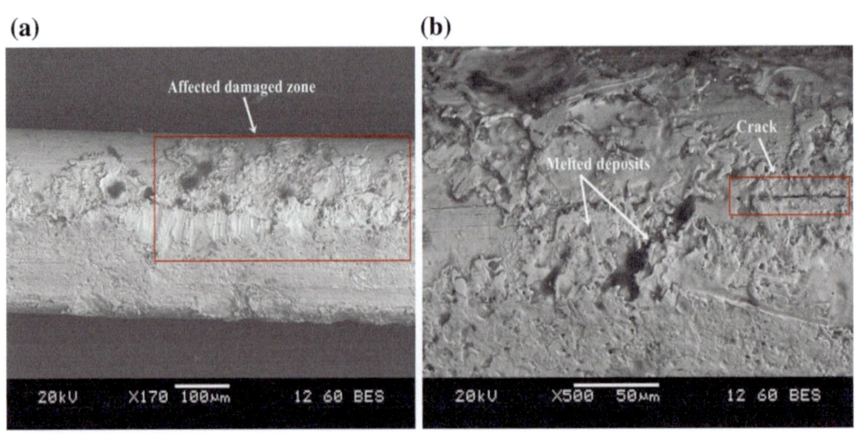

Fig. 10 Microstructural analysis of the wire

5.2 *Elemental Analysis*

The very high temperatures of the sparks (up to 45,000 K) subsequently melt and vaporize the surface of the workpiece material during the discharge process. As a result complex chemical reactions take place between wire material, workpiece material, and the dielectric. The analysis is measured by using EDS so that the element composition of the machined surfaces and wires can be analysed.

5.2.1 EDS of Copper Material

EDS or Energy-Dispersive X-ray Spectroscopy is used for analysing chemical characteristics of the material.

The EDS of the copper plate is shown in Table 4.

The elements present on the workpiece material are indicated by the peaks corresponding to their energy levels. The weight percentage of the carbon was found to be 20.68 and that of the copper was found to be 79.32% in the *K*th shell. The EDS of the copper material is shown in Fig. 11. The presence of carbon and oxygen are negligible due to significantly reduced discharged energy.

5.2.2 EDS of Brass Wire

The EDS analysis of the brass wire electrode at the optimized settings (T_{on} 110 μs, T_{off} 53 μs, Wf 5 m/min and Wt 7 kg-f) was exposed at high discharge energy settings to know the presence of foreign elements such as Cu, O, and Zn on the WED machined surface migrated in considerable quantities. Figure 12 depicts the EDS analysis of brass wire electrode.

The section of the machined wire surface scanned was subjected to EDS and the weight % and atomic % of the elements formed was analysed. Table 5 shows the EDS analysis of brass wire electrode. It was concluded that the elements present in the brass wire after machining were Cu and Zn with the weight % of Cu as 54.66% in the *L*th shell and Zn as 45.34% in the *L*th shell. The presence of copper was detected due to the diffusion of the material between the brass wire electrode to the workpiece material.

Table 4 EDS of the copper material

Element	Weight %	Atomic %
C K	20.68	57.99
Cu K	79.32	42.01
Total	100	

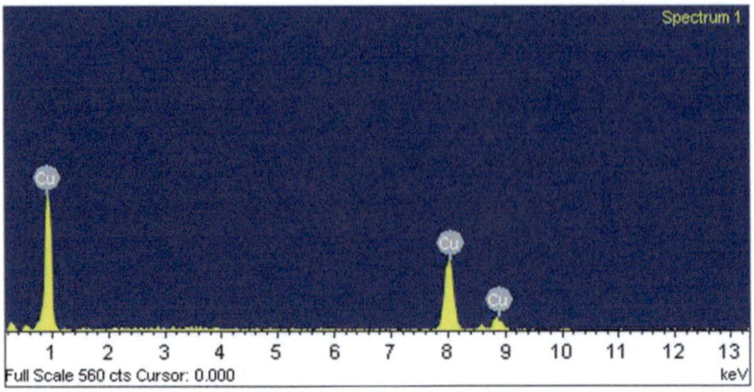

Fig. 11 EDS of the copper material

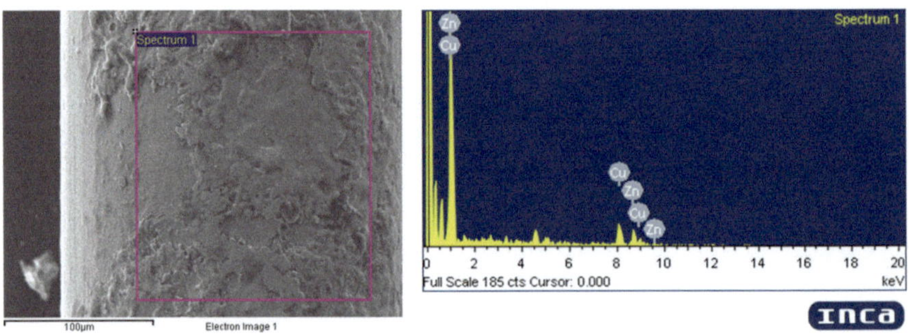

Fig. 12 EDS of brass wire electrode

Table 5 EDS analysis of brass wire electrode

Element	Weight %	Atomic %
Cu L	54.66	55.36
Zn L	45.34	44.64
Total	100	

6 Results and Discussion

6.1 ANOVA Table for Means

Table 6 shows ANOVA for the means. The factors affecting the response can be determined by the ANOVA table.

If the probability (*P*) value <0.05, the factor is said to be significant. From Table 6, it was observed that pulse on time and pulse off time are the most crucial factors contributing the most in the prediction of the output responses. The R-Square also corresponds to 97.52% of the total value for the experiment.

Table 6 ANOVA table for means

Source	DF	SS	MS	F (Fisher's)	*P*
T_{on}	1	3.006	3.006	1031.5	0.000
T_{off}	1	0.085	0.085	29.41	0.000
Wf	1	0.004	0.004	1.40	0.248
Wt	1	0.0003	0.0003	0.11	0.739
Error	27	0.078	0.0029		
Total	31	3.175			

Table 7 Response table for means

Level	T_{on}	T_{off}	Wf	Wt
1	0.6638	1.0221	0.9816	0.9671
2	1.2769	0.9186	0.9591	0.9736
Delta	0.6131	0.1035	0.0226	0.0064
Rank	1	2	3	4

6.2 Response Table for Means

The rank of the process parameters are determined by the response table. The response table of the process parameters is shown in Table 7.

From the response table, it was analysed that the pulse on time is the most determining parameter followed by pulse off time, wire feed rate and wire tension.

6.3 Statistical Models for MRR and Single Pitch Error

Kanlayasiri and Boonmung (2007) suggested some multiple linear regression models that are suitable in predicting various performance measures in wire EDM process. The regression equations of the statistical models of the responses were obtained by using suitable software by selecting the input and output responses of the model. The experimental results were obtained after the gear has been machined by wire EDM process. Regression coefficient (R^2) 0.975 and 0.942 in MRR and pitch error, respectively, indicates a good sign of statistical model values with the experimental. The statistical regression models for MRR and single Pitch error can be given by the following equations:

$$\text{MRR} = -16.5 + 0.190 * T_{on} - 0.0389 * T_{off} + 0.0056 * \text{Wf} - 0.0136 * \text{Wt} \quad (12)$$

and $$F_{Pt} = 2.79 - 0.0141 * T_{on} - 0.00258 * T_{off} + 0.0127 * \text{Wf} - 0.0111 * \text{Wt}. \quad (13)$$

Figure 13 displays the comparison between statistical and experimental model values of MRR. Similarly, the comparison between the statistical and the experimental values for single pitch error is displayed in Fig. 14. Both the graph confirms a good agreement of the results between the statistical and the experimental model values.

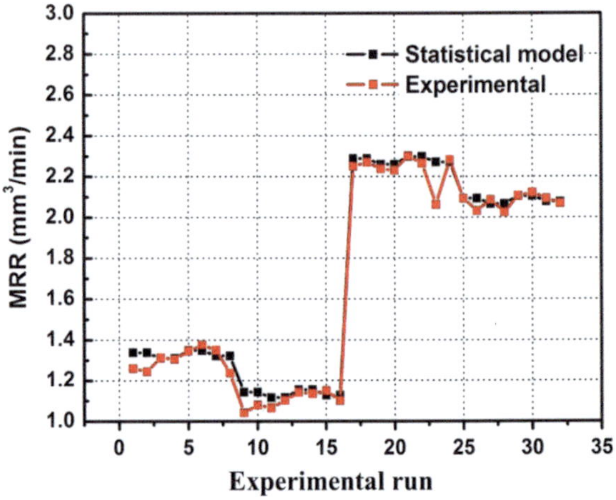

Fig. 13 Comparison between statistical and experimental values of MRR

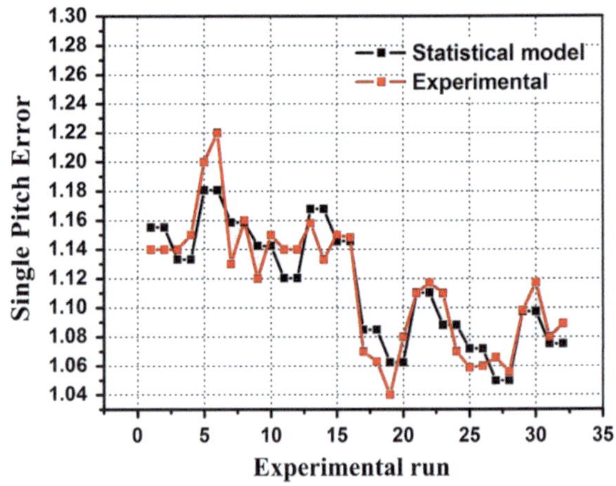

Fig. 14 Comparison between statistical and experimental values of single pitch error

6.4 *Various Plots Obtained During Analysis*

6.4.1 Main Effect Plot

Figure 15 depicts the main effect plot diagram for the modified coefficient ratio (MCR). Figure 15 depicts the main effect plot diagram for the modified coefficient ratio (MCR). The MRR should be maximized so that the cutting speed and the machining rate should be higher while machining of the materials. Similarly, the errors in the pitch should be lower or in other words, the single pitch error should be minimized. The obtained MCR after optimizing the multi-objective responses should be considered as highest the better and the combinations for the best alternative can be evaluated by ordering their ranks. The mean for each input factors with respect to the MCR was analysed and is depicted in Fig. 15. From the Fig. 15a, the MCR increases with the increase in the pulse on time. The modified coefficient ratio has the highest impact on pulse on time 110 μs rather than pulse on time 105 μs. High values of pulse on time results in high discharge energy of the spark which further results in higher values of MCR.

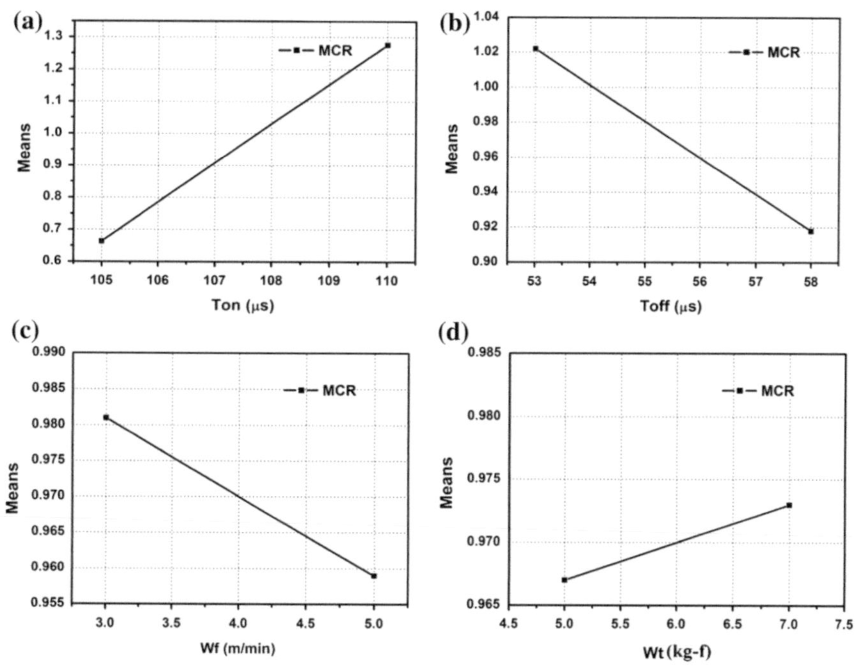

Fig. 15 Main effect plot for modified coefficient ratio

Similarly from the Fig. 15b, the MCR decreases with the increase in the pulse off time. With the increase in the wire feed rate, the MCR decreases as shown in Fig. 15c. Similarly the MCR increases with the increase in the wire tension as the discharge energy drawn from the wire is more at low wire feed rate and is shown in Fig. 15d.

6.4.2 Surface and Contour Plots

Surface plots and contour plots are designed to express the potential relationship between the input and output variables. The contour plot is a substitute of 3D surface plot. The contour plot and surface plots can hold two continuous factors at a time. The contour plots can be represented in terms of area, contour lines and area and contour lines.

Figure 16 depicts the 3D surface plot analysed for the MCR with respect to the input variables. The region with the darkest colour indicates highest MCR while the region with the lightest colour indicates lowest MCR. Figure 16a depicts the 3D surface plot of MCR versus T_{on} and T_{off}. From Fig. 16a, it can be concluded that the highest MCR corresponds to T_{on} 107 μs and T_{off} 53 μs. Figure 16b depicts the 3D surface plot of MCR versus T_{on} and Wt. Similarly from Fig. 16b, it can be resolved that the highest MCR corresponds to Wt 5 kg-f and T_{on} 110 μs.

Figure 17 depicts the contour of MCR versus T_{on} and T_{off} of areas only. The contour plot displays a stationary ridge topographical map with the rows increasing in a uniform manner. The darkest region corresponds to highest MCR (>1.35) which reveals that the MCR is highest in the region of T_{on} 110 μs and T_{off} 53 μs. With the increase in the pulse off time, the MCR goes on decreasing in between the range of T_{on} 109–110 μs. This indicates that a higher material removal rate and lower single pitch error can be achieved at low values of pulse off time and high values of pulse on time.

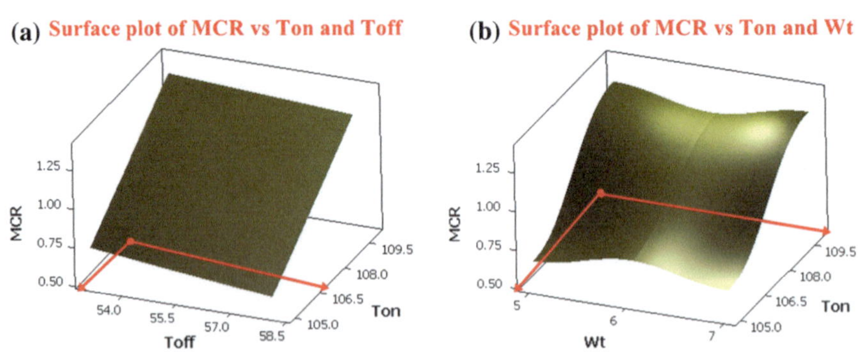

Fig. 16 3D surface plot for modified coefficient ratio

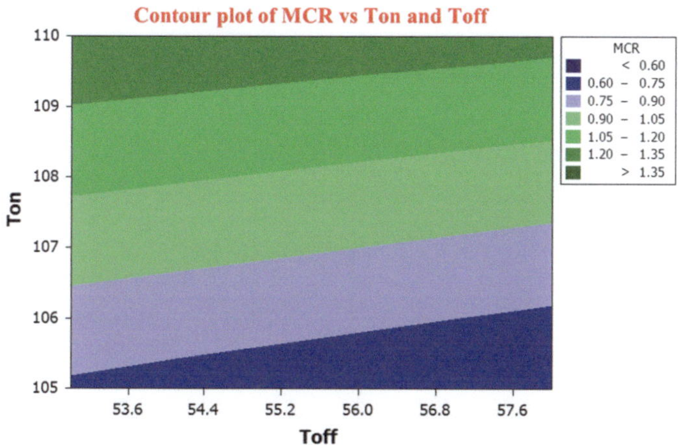

Fig. 17 Contour plot for modified coefficient ratio

6.5 Effects of Response Parameters on Process Parameters

6.5.1 Effect of Material Removal Rate

Figure 18 depicts the effect of MRR on process parameters. Figure 18a depicts the effect of MRR at increasing pulse on time and pulse off time at constant wire feed rate and wire tension. With the increase in the pulse on time at increased pulse off time, the MRR tends to increase. With the further increase in the wire tension at constant wire feed rate, the MRR increases as shown in Fig. 18b. The increase in the MRR is due to the high pulse on time at high discharge energy settings causing the spark intensity to increase thereby increasing the cutting speed and machining rate. With the further increase in the wire feed rate, the MRR also increases as shown in Fig. 18c. This is due to the fact that at high pulse on time and due to higher wire feed rate, the cutting time is reduced with the occurrence of high sparks thus increasing the material removal rate. With the further increase in the wire tension (Fig. 18d), the MRR increases due to the high reaction forces cause the spark to create more heat and temperature thereby increasing the machining rate and cutting speed.

6.5.2 Effect of Single Pitch Error

Figure 19 depicts the effect of single pitch error on process parameters. Figure 19a depicts the effect of single pitch error at increasing pulse on time and pulse off time at constant wire feed rate and wire tension. There is a decrease in the single pitch error as depicted from Fig. 19a. High discharge energy at low wire feed rate and wire tension result in uniform machining of the gear tooth without any deviation.

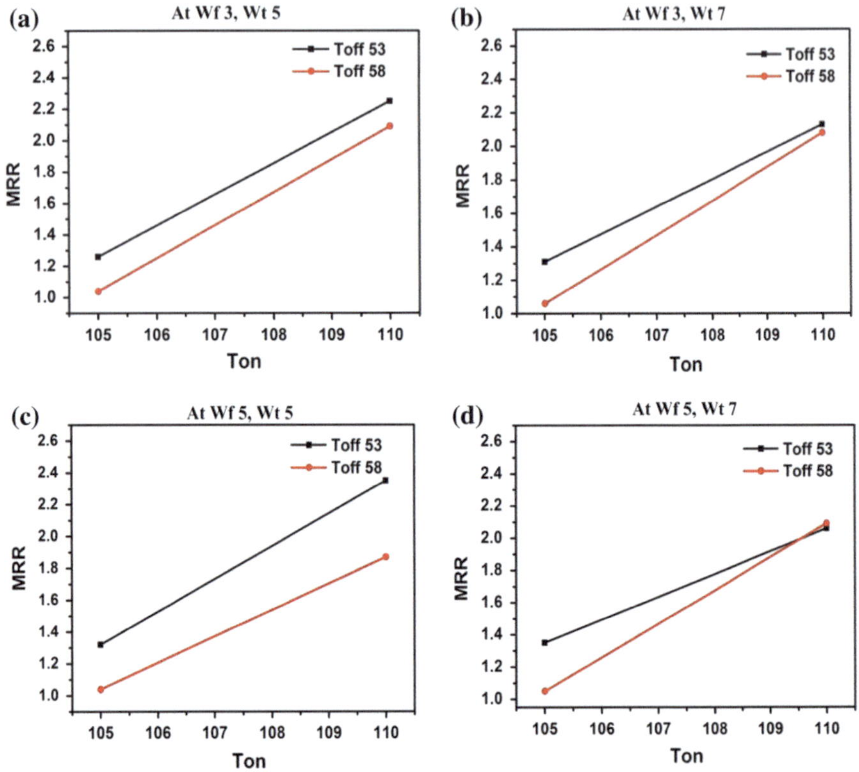

Fig. 18 Effect of MRR on process parameters

With the further increase in the wire tension at low wire feed rate, the wire is stretched more tightly in between the upper and lower guide providing uniform cutting of the gear tooth with little/no deviations. Hence, the error tends to minimize or decrease (1.17) even if the pulse on time is increased as displayed in Fig. 19b. The sparks produced from the wire remains unchanged at high discharge energy and high pulse off time at low wire feed rate. With the increase in the wire feed rate (Fig. 19c) and at low pulse off time (53 μs), the error slightly increases (1.21) due to the rapid movement of the wire resulting in the irregular cutting of the gear profile at high discharge settings. With the further increase in the wire tension (Fig. 19d), at high discharge energy and low pulse off time, the pitch error tends to increases. The increase in the pitch error is due to the fact that at high discharge energy, wire feed rate and wire tension the occurrence of the spark is more but due to the low pulse off time (53 μs), the spark termination time and voltage for the gap is reduced for the period and this results in the plasma channel to draw more discharge energy from the wire resulting in an increase in the spark efficiency thus leading to irregularity of the gear tooth and deviations in the profile.

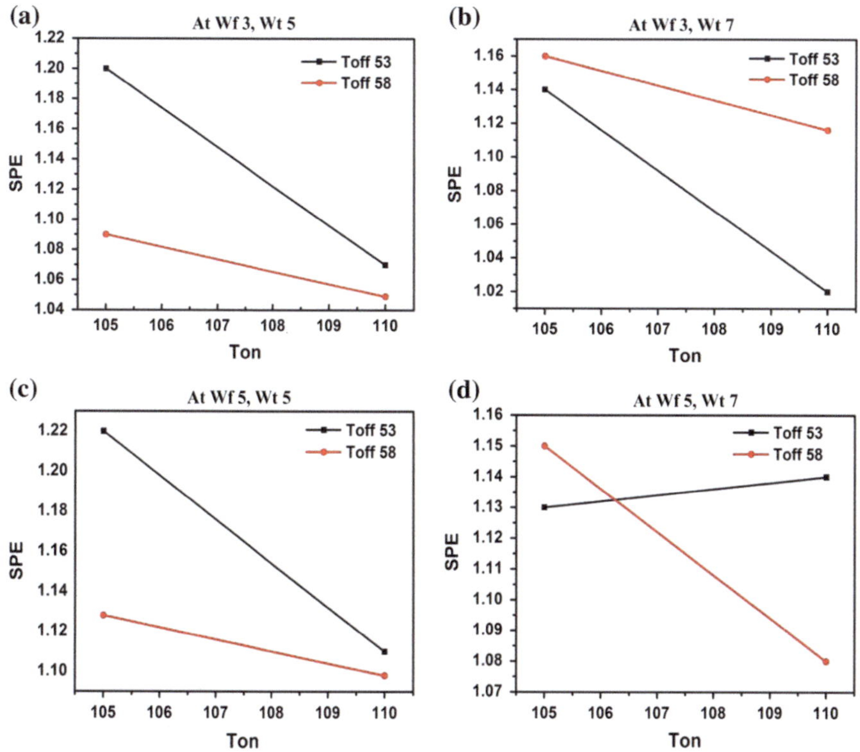

Fig. 19 Effect of single pitch error on process parameters

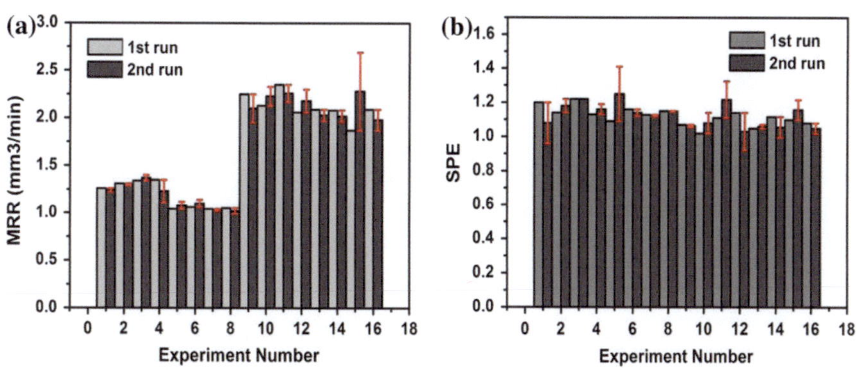

Fig. 20 Error graphs obtained for MRR and single pitch error

Figure 20 depicts the error graphs obtained for both MRR and single pitch error in machining of the gear profile. As the replication is made twice, the error graphs were obtained considering the output responses of the first and second run and it was observed from both the figures that the error variation for both the responses differs by a very small amount satisfying the stability and accuracy of the model.

7 Conclusions

The study focused on the effect of machining parameters on the output responses of the gear cut copper material in the main cut operation and analysis of surface characteristics by the wire EDM process. Key findings can be summarized as follows.

- The experiment was conducted for the gear cutting process by wire EDM. The response parameters were optimized using MOORA method and a combination was obtained for the best alternative.
- Microscopic study of the material was carried out at the optimized setting of the response and it was found that some grain like structures were formed on the workpiece material after the material has been etched by the etchant, i.e. distilled water and nitric acid.
- Microstructural study also reveals that voids and craters were formed on the surface of the workpiece as viewed under SEM at higher magnification zoom. Micro-voids were formed due to the solidification of the molten material on the material surface. Furthermore holes, debris and cracks were also identified due to high discharge energy settings resulting in high intensity of the spark. Some cracks and melted deposits were also marked over the wire surface at high discharge energy, servo voltage and wire feed rate due to the increase in the induced stresses and peak current exceeding the wires tensile strength.
- EDS characterization was done to know the elemental composition of the machined surfaces and wires. The percentage of copper was found to be 79.32 in the workpiece material. Similarly, the EDS characterization was also performed for the brass wire depicting the weight percentage of copper to be 54.66% and Zznc to be 45.34%.
- ANOVA table and response table were depicted showing that pulse on time and pulse off time were the most significant factors affecting the responses. The regression equations were obtained and the experimental model was validated with the statistical model values illustrating a good sign of stability of the model.
- Various plots like main effect plot for means, surface plots and contour plots were analysed to describe the potential relationship between the variables. The effects of different responses on individual process parameters were plotted and investigated. The graph reveals that the MRR increases and single pitch error decreases with the increase in the pulse on time. The MCR decreases with the increase in the wire feed rate and increases with the increase in the wire tension due to low wire feed rate and high discharge energy drawn from the wire.
- The error graphs were obtained for both MRR and single pitch error and it was observed that the error variation for both the responses differs by a very small

amount satisfying the stability and accuracy of the model. The optimized setting or combination obtained in the present work can be further used in machining and manufacturing of high-quality copper gears.

References

Brauers, W.K.M. 2007. *Optimization methods for a stakeholder society: A revolution in economic thinking by multi objective optimization*. Boston: Kluwer Academic.

Dodun, O., A.M.G. Coelho, L. Statineanu, and G. Nagit. 2009. Using wire electrical discharge machining for improved corner cutting accuracy of thin parts. *Interntional Journal of Advanced Maufacturing Technology* 41: 858–864.

Gupta, K., and N.K. Jain. 2014a. On surface integrity of miniature spur gear manufactured by wire electrical discharge machining. *International Journal of Advanced Manufacturing Technology* 72: 1735–1745.

Gupta, K., and N.K. Jain. 2014b. Analysis and optimization of micro-geometry of miniature spur gears manufactured by wire electric discharge. machining. *Precision Engineering* 38 (4): 728–737.

Gupta, K., N.K. Jain, and R.F. Laubscher. 2015. Spark erosion machining of miniature gears: a critical review. *International Journal of Advanced Manufacturing Technology* 80: 1863–1877.

Habib, S., and A. Okada. 2015. Study on the movement of wire electrode during fine wire electrical discharge machining process. *Journal of Materials Processing Technology* 227: 147–152.

Kanlayasiri, K., and S. Boonmung. 2007. An investigation on effects of wire-EDM machining parameters on surface roughness of newly developed DC53 die steel. *Journal of Materials Processing Technology* 187–188: 26–29.

Kuriakose, S., K. Mohan, and M.S. Shunmugam. 2003. Data mining applied to wire-EDM process. *Journal of Materials Processing Technology* 142: 182–189.

Liao, Y.S., J.T. Huang, and Y.H. Chen. 2004. A study to achieve a fine surface finish in Wire-EDM. *Journal of Materials Processing Technology* 149: 165–171.

Mandal, A., A.R. Dixit, A.K. Das, and N. Mandal. 2015. Modeling and optimization of machining nimonic C-263 super alloy using multi-cut strategy in WEDM. *Materials and Manufacturing Processes* 31 (7): 37–41.

Menz, W. 2003. Micro wire EDM for high aspect ratio 3D microstructuring of ceramics and metals. *Microsystem Technologies* 11: 250–253.

Miller, S.F., A.J. Shih, and J. Qu. 2004. Investigation of the spark cycle on material removal rate in wire electrical discharge machining of advanced materials. *International Journal of Machine Tools and Manufacture* 44: 391–400.

Ming, W., Z. Zhang, G. Zhang, Y. Huang, J. Guo, and Y. Chen. 2014. Multi-objective optimization of 3D-surface topography of machining YG15 in WEDM. *Materials and Manufacturing Processes* 29(5): 37–41.

Plaza, S., N. Ortega, J.A. Sanchez, I. Pombo, and A. Mendikute. 2009. Original models for the prediction of angular error in wire-EDM taper-cutting. *Interntional Journal of Advanced Maufacturing Technology* 44: 529–538.

Qu, J., A.J. Shih, and R.O. Scattergood. 2002. Development of cylindrical wire electrical discharge machining process, part 2: surface integrity and roundness. *Journal of Manufacturing Science and Engineering* 124: 708–714.

Sanchez, J.A., S. Plaza, N. Ortega, M. Marcos, and J. Albizuri. 2008. Experimental and numerical study of angular error in wire-EDM. *International Journal of Machine Tools and Manufacture* 48: 1420–1428.

Taylor, P., K. Gupta, and N.K. Jain. 2013. On micro-geometry of miniature gears manufactured by wire electrical discharge machining. *Materials and Manufacturing Processes* 28: 37–41.

Wentai, S., L. Zhidong, and Q. Mingbong. 2015. Wire tension in high-speed wire electrical discharge machining. *International Journal of Advanced Manufacturing Technology* 82: 379–389.

Yan, M.T., and P.H. Huang. 2004. Accuracy improvement of wire-EDM by real-time wire tension control. *International Journal of Machine Tools and Manufacture* 44: 807–814.

Yan, M.T., and Y.T. Liu. 2009. Design, analysis and experimental study of a high-frequency power supply for finish cut of wire-EDM. *International Journal of Machine Tools and Manufacture* 49 (10): 793–796.

Yeh, C., K. Wu, and J. Lee. 2013. Study on surface characteristics using phosphorous dielectric on wire electrical discharge machining of polycrystalline silicon. *International Journal of Advanced Manufacturing Technology* 69: 71–80.

Novel Finishing Process Development for Precision Complex-Shaped Hemispherical Shell by Bulk Plasma Processing

D. Sam Dayala Dev, Enni Krishna and Manas Das

Abstract It is not uncommon that inertial sensor technology demands 'zero' surface and subsurface defects on sensing element microstructures, which are polished to very fine surface finish. Non-contact type unconventional finishing techniques are being developed to augment or replace chemo-mechanical polishing (CMP) technique for finishing such microstructures to remove subsurface damage. The aim of this study is to develop an atomistic material removal mechanism by chemical vaporization process. The novelty of this process is combining the merits of low-pressure plasma etching by ion such as isotropic material removal on all the surfaces simultaneously and merits of atmospheric plasma process such as chemical vaporization rather than physical bombardment by ions. This achieves defect-free extremely fine-polished surfaces. A finite element based Comsol® software package is used to model dielectric barrier excited RF discharge for helium and oxygen gas as processing gas and reactive gas, respectively. The gas composition, pressure, electrode configurations and power of RF excitation are studied with respect to oxygen radical formation and their uniformity of distribution in the chamber. Accordingly, plasma chamber is designed and built with Zerodur material with an optical window to achieve a deterministic process, which is capable of simultaneously polishing entire complex 3D surfaces including cavities where no tool or beam can reach. Plasma is established with helium as processing gas and oxygen and SF_6 are used as reactive gases at medium pressure of 20 mbar. The atomic emission spectroscopy is used to monitor the various oxidation states of silica and established correlation with respect to material removal rate and oxidation states of

D. S. D. Dev · E. Krishna
ISRO Inertial Systems Unit, Indian Space Research Organisation,
Thiruvananthapuram 695013, India
e-mail: samddev@gmail.com

E. Krishna
e-mail: kris.chem99@gmail.com

M. Das (✉)
Department of Mechanical Engineering, Indian Institute of Technology Guwahati,
Guwahati 781039, Assam, India
e-mail: manasdas@iitg.ernet.in

S. S. Pande and U. S. Dixit (eds.), *Precision Product-Process Design and Optimization*, Lecture Notes on Multidisciplinary Industrial Engineering,
https://doi.org/10.1007/978-981-10-8767-7_12

silica Si II and Si III. Fine-tuning of these parameters is done while polishing hemispherical shell based on atomic emission spectroscopy observations and established process with material removal rate of 0.008 mm^3/min. After repeated polishing cycles with cumulative 48 h of material removal, surface roughness (Ra) of 3.6 nm is achieved from as-machined shell of Ra = 903 nm.

Keywords Nanofinishing · Plasma polishing · Atomic emission spectroscopy Fused silica polishing

1 Introduction

The state-of-the-art inertial sensors demand on sensing microstructures, ultra-fine-surface finish with 'zero' surface or subsurface damage (SSD). Mostly, these microstructures of complex surfaces are realized in silica or fused silica materials. Fused silica has been used heavily in laser systems and optical systems which demand very high precision and complex shape. The most challenging task in optical component fabrication is achieving surface accuracy and surface finish. The need for shape accuracy and extreme surface finish is to be accomplished at the same time which demands innovative process for finishing optical glasses. The material processing technology has advanced to such an extent that 3D homogeneous fused silica is readily available with impurities limited to tens of parts per billion (ppb). Hence, the bulk material is not limiting the performance of the microstructures but the surface does. When the surface becomes thinner and thinner, zero surface and subsurface damage can totally modify the performance of these components. Therefore, the major challenge is to regain the bulk material properties on the surface to atomistic level while ensuring zero defects on the surface and subsurface. The normal line of manufacture of these brittle materials is through machining using diamond-coated milling cutter with/without ultrasonic assistance and further wet etching of the surface to eliminate all cracks and defects.

Wet etching invariably degrades the topography significantly while removing microcracks. Microcracks/bond strain gets removed but surface topography changes as shown in Fig. 1. As surface topography and surface finish get poor and subsurface defects are not completely removed, precision grinding is adopted to correct the topography and surface finish before resorting to conventional polishing technique such as chemo-mechanical polishing (CMP). Optical industries are able to achieve sub-nanometer surface finish with variance of CMP process.

In CMP process, apart from the chemical effect of the slurry on the surface, each abrasive makes a brittle machining for material removal. By the very nature of the

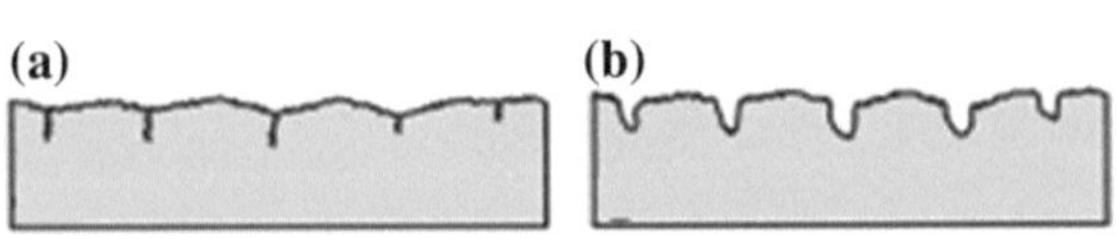

Fig. 1 Schematic of **a** machined surface, and **b** wet-etched surface

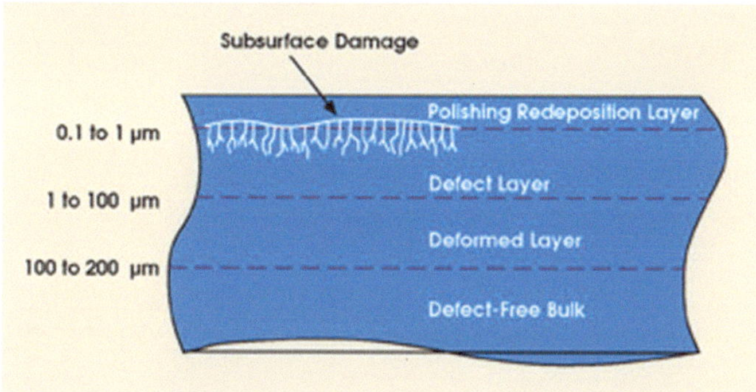

Fig. 2 Typical surface and subsurface defects occur in CMP process. With permission from Shena et al. (2005), copyright (2005) Elsevier

CMP process, the surface defects are submerged and redeposited layer forms which is chemical and mechanical property wise distinctly different from the bulk material. Also, these processes are not deterministic resulting in poor and inconsistent yield hiking the cost of these components. Typical surface and subsurface defects occur in CMP process as shown in Fig. 2. Therefore, processes such as magnetorheological finishing (MRF), elastic emission machining (EEM), etc., are developed as alternate finishing methods with least contact forces. Even though MRF is able to achieve very high surface finish with low SSD, it does not fully eliminate. Their applicability in finishing complex shape is very limited. Therefore, a non-contact type alternate line of finishing for glass or brittle components like plasma polishing is conceived in the present work, which is capable of atom-by-atom material removal from the surface at different material removal rates to replace wet etching and fine polishing processes.

1.1 Plasma Polishing

The plasma is known as a wide variety of macroscopically neutral substance having many interacting free electrons and ionized atom or molecules. In other words, plasma, which is called the fourth state of matter, is a partially ionized gas having same number of +Ve and −Ve particles. Ion is having positive charge and it is a molecule or atom of gas with a removed electron. Radical is a neutral particle (atom or molecule) that exists in a state of incomplete chemical bonding, and therefore it is chemically reactive. It is generated due to the fracture of a gas molecule by a high-energy electron collision. Equation (1) shows the oxygen radical ($\dot{o}$) formation.

Table 1 Plasma species for neutral molecules of density $10 \times 10^{16}/cm^3$

Radicals	$10 \times 10^{14}/cm^3$
Electrons	$10 \times 10^{8}/cm^3$
Positive ions	$10 \times 10^{8}/cm^3$

Table 2 Plasma material removal mechanism spectrum

Decreasing pressure	Plasma characteristics	Increasing energy
<0.1 mbar	Physical (sputtering) momentum transfer directional etching	High energy and surface damage is high
0.2 mbar	Reactive ion etching physical and chemical variable anisotropy	Low energy
0.5 mbar	Chemical plasma etching, fast and isotropic material removal	
1–100 mbar	*Cold chemical plasma, bulk acting and isotropic material removal*	*Low-energy cold plasma model assumes zero plasma temperature*
1 bar	Atmospheric pressure plasma, cold chemical isotropic jet plasma (small aperture)	Very low energy, high surface integrity

$$O_2 + e^- \longrightarrow 2\dot{O} \tag{1}$$

There can be different forms of O_2 radicals such as O_2a1d, O_2b1s, O1d and O1s depending on the orbital position of the colliding electron. A typical plasma for a given neutral molecule of density $10 \times 10^{16}/cm^3$ contains typically the following species as shown in Table 1.

Therefore, more number of radicals (million times) than electrons or ions are available in the plasma. Radicals are formed more easily and also their lifetime is higher than electrons or ions. Ions do not etch or remove material. Ions affect the process by energetic (physical) bombardment on the surface, influencing the chemical process of atom-by-atom material removal. Radicals take part in dry etching or material removal process. Another property of radicals is their chemical activeness. Also, radicals react with the workpiece surface and volatile products are generated. Table 2 shows the wide spectrum of plasma material removal mechanisms for wide range of plasma gas pressure. The medium pressure of 1–100 mbar has the unique feature of being cold chemical plasma with a capability to act on the bulk of the specimen resulting in isotropic material removal. Table 3 lists the merits and demerits of existing low pressure and atmospheric plasma processing techniques with respect to the proposed medium-pressure plasma process. The generation of uniformly distributed plasma in a suitably designed chamber is possible with medium-pressure plasma process.

Also, in situ passivation and cleaning can be carried out on the specimen. All surfaces of 3D complex objects can be simultaneously and uniformly treated

Table 3 Plasma pressure versus polishing application

Applications and properties	Low-pressure plasma		Medium-pressure plasma		Atmospheric plasma	
	Advantages	Disadvantages	Advantages	Disadvantages	Advantages	Disadvantages
Plasma generation	Uniformly distributed plasma inside the chamber	Complex vacuum technology is required	*Plasma is distributed uniformly within plasma chamber*	*Complex vacuum and flow control is required*	Plasma treatment is possible directly even at the conveyor belt. No vacuum is necessary	The treatable area is limited to approximately 8–12 mm at a given time
3D objects	All items in the plasma chamber are treated simultaneously. Also, cavities can be treated from inside	None known	*All surfaces can be treated uniformly and simultaneously. Also, cavities can be treated from inside wide aperture*	*Establishing isotropic chemically reactive plasma is a challenge. Custom designed cavity may be required*	Local surface treatment is possible	Complex robotic technology is necessary. Treatment of surfaces with deep grooves is limited. Small aperture process
Surface integrity and surface finish of optics	Limited application	Due to anisotropy polishing cannot be done	*Very good as it is non-contact chemical reactive plasma*	*Only incremental surface finish improvement*	Surface integrity and surface finish improved	Very small area only can be covered

provided the shape, size and configuration of the plasma chamber and electrodes are suitably designed.

Plasma can be generated by either high DC voltage excitation or capacitive coupled radio frequency (CCRFD) excitation. CCRFD consists of two typically plane parallel electrodes of either same or different surface areas. The electrodes are placed inside the vacuum chamber in direct contact with the plasma or isolated from the plasma by a dielectric medium. If the electrode is behind a dielectric barrier, then it is called 'Dielectric barrier RF excitation'. Generally, the chamber is grounded. One electrode is joined to an RF generator via an impedance matching network and another electrode is connected to an RF high voltage waveform.

Due to difference between electron mass (m_e) and ion mass (m_i) ($m_i/m_e \gg 1$), the ion inertia is much higher than the electron inertia. Therefore, electrons have much higher energies than ions, and electrons will be lost in the walls quickly, if the plasma gets in contact with the boundary wall. As a consequence, a sheath of positive space charge will develop adjacent to the wall. In the sheath, an electric field accelerates ions towards the wall and repels electrons confining them in the discharge. In the sheath, quasi-neutrality is violated. In the quasi-neutral bulk, an ambipolar field coupled ion and electron diffusion. Depending on the choice of electrode surface area (chamber geometry) electrode gap, applied voltage waveforms, gas mixture, pressure and power, a variety of capacitively coupled discharges can be generated. Each type provides unique features useful for particular application. The gas mixture and pressure play the major role in defining the plasma characteristics. The choice of processing gas is essential for chemical processes in the plasma itself and at surfaces in contact with the plasma. Depending on the gas, the discharge might be electropositive or electronegative. Capacitive coupled dielectric barrier RF discharges can be operated at low pressure to less than 10^{-2} mbar for an isotropic etching process or at a high pressure, e.g. atmospheric pressure plasma jets. High-pressure micro-discharges can provide high radical density which is essential for material removal.

In the present study, a low power medium-pressure 'cold' plasma atomistic finishing process is conceived to achieve isotropic polishing of complex 3D surfaces concurrently including inside surfaces where no tool or beam can reach.

1.2 Literature Survey

The literate survey on plasma-based polishing methods done by different researchers is given below. Optical media is polished by the Gerhard et al. using dielectric barrier discharge gas plasma (DBDG) at atmospheric pressure (Gerhard et al. 2013). They achieved approximately 20% reduction in surface roughness along with 80% improvement in waviness with increased surface energy on optical materials including fused silica. The material removal is explained by ion bombardment and de-excitation of argon species. It is explained that the plasma discharge causes high electric field strength at roughness peaks. Hence, the waviness improves distinctly. Wang et al. studied the surface changes in terms of surface modulus and hardness using atmospheric pressure plasma polishing (APPP) and they demonstrated improved surface mechanical properties (Wang et al. 2011). They reported that there is a decrease in surface residual stresses as the deformed layer is removed atom by atom from the surface. Jin et al. investigated the impacting factors on the surface roughness of Zerodur material by atmospheric pressure plasma jet (APPJ) polishing using He, O_2 and SF_6 (Jin 2010). Yao et al. (2010) reported the chemical machining of Zerodur material using APPJ. Wang et al. (2011) reported a novel non-thermal APPP for polishing SiC optics. Reactive gas such as CF_4 was introduced into the plasma area to react with the molecules at the surface of the SiC optics. Wang et al. (2009) developed highly stable SF_6 and Ar/O_2 capacitive coupled RF discharged method. Atmospheric pressure plasma with different working gases has been successfully developed to achieve sub-nanometer surface finish without surface damage but on a small aperture/area on the workpiece. However, this process is not suitable for bulk polishing of complex-shaped or freeform surfaces and inaccessible areas in precision microstructures. Inside surfaces, where the plasma torch cannot reach, atmospheric plasma fails to be useful.

Liu et al. (2009) reported the low-pressure plasma polishing process in the range of 10^{-2}–0.5 mbar and it is capable of reducing/removing the subsurface damage on complex freeform surface. However, there is no appreciable improvement in the surface finish but surface integrity is enhanced in terms of surface residual stresses. Enough literature is available with respect to the low-pressure plasma etching with available custom-built industrial equipment and machines.

Wang et al. (2006) developed a novel non-thermal atmospheric pressure plasma for polishing SiC optics. Reactive gas such as CF_4 was introduced into the plasma area to react with the molecules at the surface of the SiC optics. As the interaction is a pure chemical process, the material removal is at the atomic and no surface damage is caused.

Dev et al. (2016) targeted to develop an atomistic finishing process combining the best of low-pressure plasma processing capabilities such as removal of surface and subsurface damaged layer simultaneously on all the surfaces of the component and atmospheric pressure plasma polishing capabilities such as fine polishing and isotropic material removal. The process is novel from the fact that it is capable of polishing simultaneously entire complex 3D surfaces including cavities where no tool or beam can reach.

There is numerous application of freeform optics in reflective, refractive and diffractive optical system which demand finishing multiple surfaces on a single microstructure or inside surfaces which cannot be accessed by any tool or jets (either abrasive jets or plasma jets). Freeform optics are designed to improve the optical performance by removing the optical aberrations, improving the performance level many folds. Nanofinishing of such freeform surfaces in optics and sensor microstructure realized from glass or glass-like materials is thought to be not achievable. However, newer and innovative processes are being developed for nanofinishing such components.

Low-pressure plasma is known for the high MRR, and anisotropic plasma machining has been effectively used for such application. As the pressure of the plasma is increased, the isotropic property is gained and the energy level keeps coming down. The atmospheric pressure plasma is a low-energy isotropic plasma polisher. Inside surfaces or where the plasma torch cannot reach, atmospheric plasma fails to be useful.

Enough literature is available with respect to low-pressure plasma etching/polishing and custom-built industrial equipment and machine are also being available. Also, a very active research is being carried out in atmospheric pressure plasma polishing for wider applications. However, there is no literature available in medium-pressure isotropic cold plasma polishing.

1.3 Motivation of the Present Work

The navigation grades inertial sensors such as gyros and accelerometers have sensing microstructures which demand high order of precision and surface finishing to enhance the surface integrity which has direct impact on the performance of the sensors. The motivation behind the present work is to develop a finishing process to enhance the surface integrity of the polished surface up to 2–3 Å. Variants of conventional chemo-mechanical polishing are adopted to achieve such surface finish but the yield has been poor as the processes are not deterministic to achieve 'Zero defects' in critical locations. Most of such optics passes through stringent inspection with white light. However, when inspected in laser light, the subsurface damage (SSD) shows up as spots of light scattering resulting in rejection. The typical yield in laser light inspection is of the order of 10–12% and is primarily due to the contact mode of polishing which results in molecular level distortions/strained bonds. The laser gyro components and hemispherical resonator gyro (HRG) components are shown in Figs. 3 and 4, respectively.

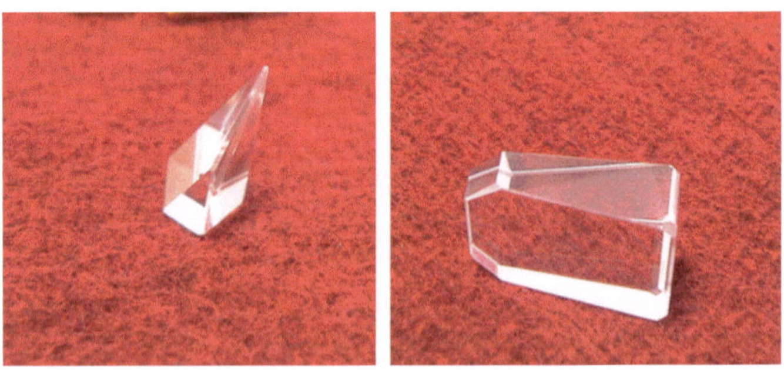

Fig. 3 ISRO laser gyro components

Fig. 4 Hemispherical resonator gyro (HRG) components

Hence, a non-contact type surface finishing method in which the surface integrity is enhanced while maintaining or improving the surface finish is studied. The expected surface material removal is limited to 10–1000 nm. Another application is where the microstructures display complex and/or freeform surfaces which needs to be polished to sub-nanometer finish from as-machined or ground workpiece while maintaining very high surface integrity level with respect to subsurface damage and residual stresses. Added to above requirements, the process shall not damage/degrade the topography and geometrical tolerances. The scope of present activity is focussed towards the development of a non-conventional non-contact type surface finishing process applicable to glass, fused silica and similar brittle materials.

2 Comsol Simulation

The primary aim is to identify the electron density and uniformity of distribution and O_2 radical's distribution with respect to the shape of the chamber, position and shape of the electrodes. Study the electron temperature to confirm the electron temperature is benign and 'cold' plasma conditions are maintained. Also, the Comsol model is validated with experimental atomic spectroscopy results and analytical model. The following objectives are identified:

- Investigate the effect of mixture of helium and oxygen gas on the SiO_2 specimen inside the plasma chamber by varying the volume concentration and pressure.
- Study the effect of shape, size and position of specimen inside the plasma chamber.

Table 4 The main reactions occurring inside the plasma chamber

Electron impact reactions	Heavy species transport reactions	Surface reactions
e + He => e + He	$He^* + O_2$ => O_2^++ He + e	He^* => He
e + He => e + He^*	$He^+ + O_2$ => O^+ + O + He	He^+ => He
e + He => 2e + He^+	$He^+ + O_2$ => O_2^+ + He	O_2^+ => O_2
e + O_2 => 2e + O_2^+		$2O^+$ => O_2
e + O_2 => e + O_2a1d		
e + O_2a1d => e + O_2		
e + O_2 => e + O_2b1s		
e + O_2b1s => e + O_2		
e + O_2 => e + O + O		
e + O_2 => e + O + O1d		
e + O_2 => e + O + O1s		

- Also, design the shape, size and position of electrode outside the plasma chamber to get homogeneous distribution of reactive species and ions.

In the present case, a two-dimensional frequency transient study under microwave plasma (MWP) is used for simulation. Electric field is setup using out-of-plane vector considering the signal in transverse electric mode. Reduced electron transport properties are considered to account for the heavy species transport. The pressure and temperature of the plasma are taken as 25 mbar and 300 K, respectively, and are considered to be constant throughout the operation of the plasma. Plasma setup is made considering the plasma to be static. The different species used for current simulation are, in molecular state: He and O_2, ions: He^+, e, O, O^+, O_2^+, and radicals: O_2a1d, O_2b1s, O1d, O1s, He^*. To study the plasma chemistry, the main reactions occurring inside the plasma chamber are shown in Table 4.

For each of the heavy species, the molecular weight, potential characteristic length and potential energy minimum are required to compute the correct diffusivity and mobility. The initial number density for electron is considered as $1 \times 10^{17}/m^3$. Considering the plasma to be neutral consisting of ions and electrons, the initial number density and initial mole fractions of different species (ions and electrons) are to be equal. Since the radicals (excited species) are formed at a later part during reaction, the initial mole fraction of the radicals is considered very small (negligible), i.e. $1–5 \times 10^{-8}$.

2.1 Computational Domain and Boundary Conditions

The geometry of the two-dimensional computational domain as shown in Fig. 5 is prepared similar to the proposed experimental setup. The cross-sectional area of the setup is prepared into three domains, i.e. dielectric, plasma and electrode (Fig. 5).

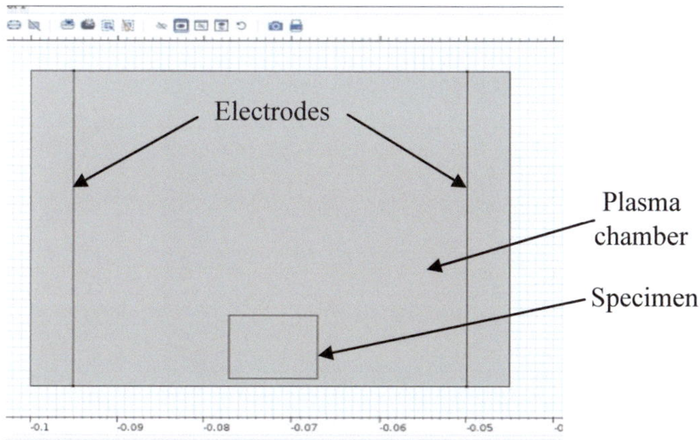

Fig. 5 2D computational domain of the plasma chamber

The electrodes are placed at the extreme boundaries on both left and right sides of the setup. The setup dimensions considered are (55 × 40) mm^2, and the thickness of the dielectric chamber is 5 mm. A specimen of SiO_2 (quartz) is taken at the bottom of the chamber. The dimension of the specimen is (10 × 8) mm^2. A physics-controlled mesh of finer nature is used to carry out the simulation. The mesh at the boundaries is made finer to accommodate the collision of the charged particles. Free triangular mesh is considered.

The four boundaries of the plasma chamber are considered as the four walls in the model containing the plasma within the domain. The four boundaries of the chamber are grounded so as the surface reactions to take place at the boundaries. The plasma is excited using an RF frequency as a port applied to one electrode while another electrode opposite to the first electrode is grounded. Square-wave RF signal is used for exciting the plasma, and the power level considered is 40 W.

2.2 *Comsol Simulation Results and Discussion*

It is a transient analysis and steady state is reached before one second. The plasma condition at 1 s is taken for comparison study for all cases. The electron density and O_2 radials such as O1s and O_2b1s are studied. In Fig. 6a, b, the electron density is shown without and with the specimen, respectively, inside the plasma chamber. The electron density varies from 1×10^{16} to 6×10^{16} 1/m^3 for both the cases. However, the distribution profile is altered while the specimen is placed inside the chamber. Hence, it is understood that the shape and position of the fused silica influence the distribution of electron density inside the polishing chamber.

After that a cylindrical fused silica specimen of varying sizes with respect to the plasma chamber is placed inside the plasma chamber to understand the effect of

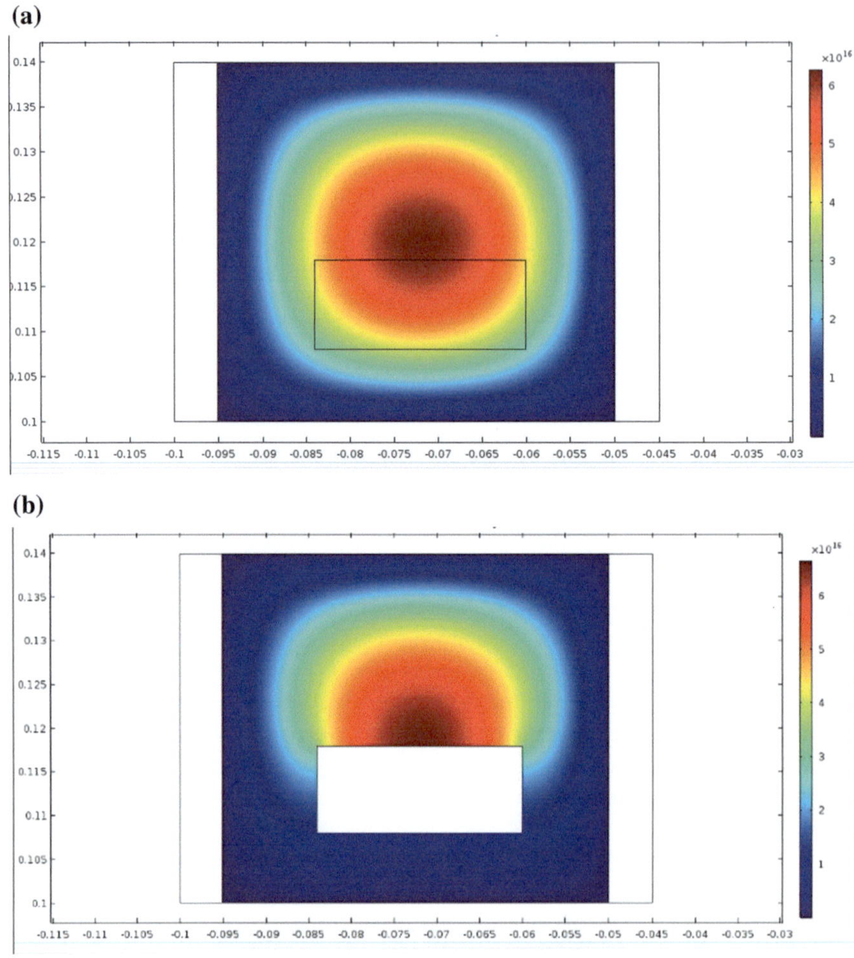

Fig. 6 Electron density **a** without, and **b** with specimen inside plasma chamber

specimen size inside the plasma chamber (Fig. 7). In Fig. 7a–c, the specimen sizes are approximately three-fourth, half and one-fourth of the chamber size, respectively. As shown in Fig. 7, the electron density distribution is not uniform when the specimen is large (Fig. 7a) compared to the cross section of the chamber. In other words, when the free volume is less, the electron density is lower and non-uniform. As the specimen size reduces, i.e. free volume of the chamber increases, the uniformity of the electron density distribution improves distinctly and electron density also increases. The same is true for the distribution of O_2 radials such as O1s and O_2b1s (results not shown here). However, there is no variation in the number density of radicals irrespective of the size of the specimen.

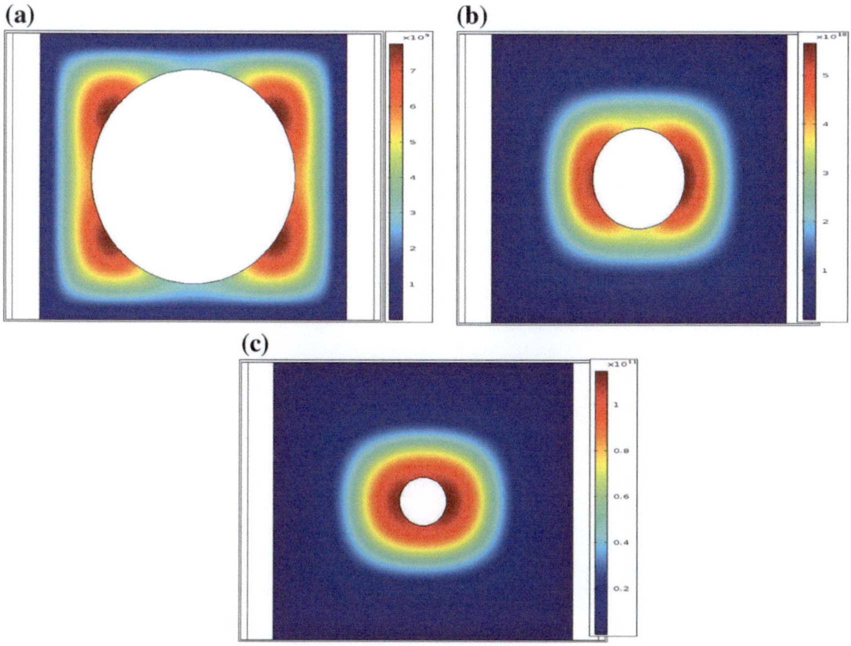

Fig. 7 Electron density when specimen sizes are **a** three-fourth, **b** half and **c** one-fourth of chamber size

From Fig. 8, it is observed that as the O_2% increases, the number density of O1s radicals increase. However, the distribution of O1s radical is unsymmetrical with respect to the specimen up to at 1% and symmetrical at 2% and it loses its symmetrical distribution on further increase of O_2%. Therefore, 2% O_2 composition may be optimum for processing fused silica components.

Based on the Comsol simulation results, the plasma chamber along with electrodes is modified to suit target specimen of hemispherical shell made of fused silica. Figure 9a shows the plasma chamber with two parallel electrodes with respect to the position of the HRG shell. Figure 9b shows the same plasma chamber with four-sided electrodes. The corresponding distribution of O_2b1s radicals is shown in Fig. 10. The number density of O_2b1s radicals is more by 18% for four-sided electrode configuration than electrode placed at top and bottom. Hence, more material removal is expected while polishing with four-sided electrodes.

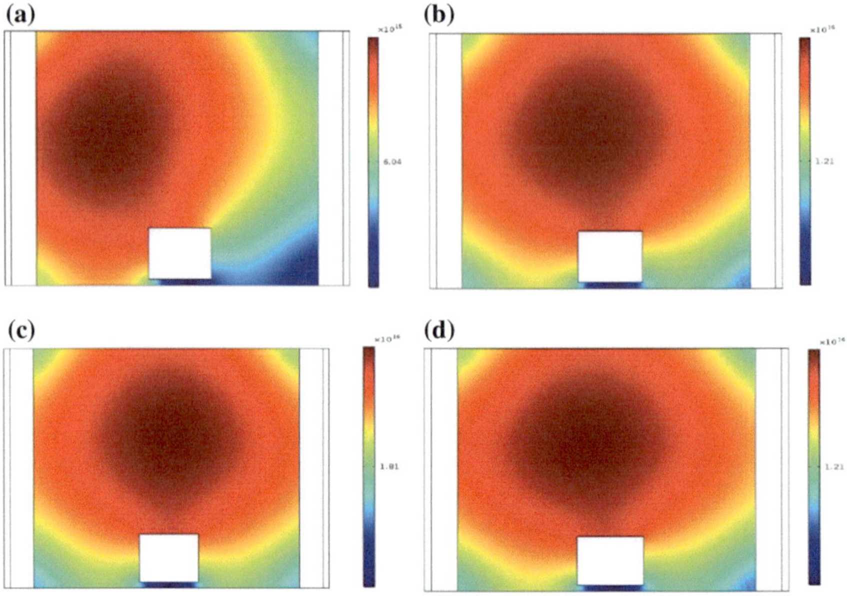

Fig. 8 Number density variation of O1s radical with increasing O_2 concentration (%) **a** 1%, **b** 1.5%, **c** 2%, **d** 5%

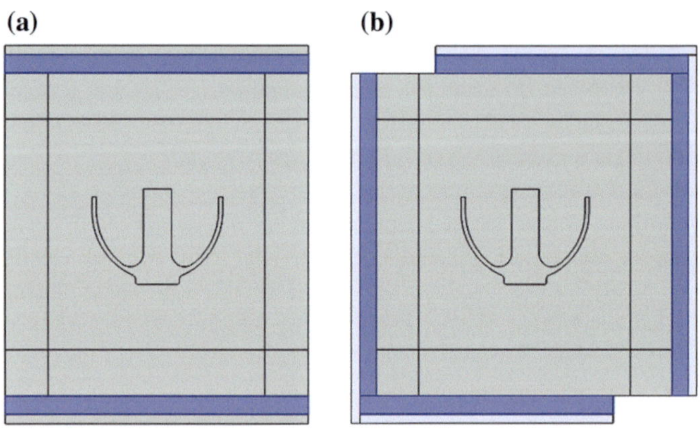

Fig. 9 2D computational domain of the plasma chamber with **a** two parallel electrodes at top and bottom, and **b** four-sided electrodes

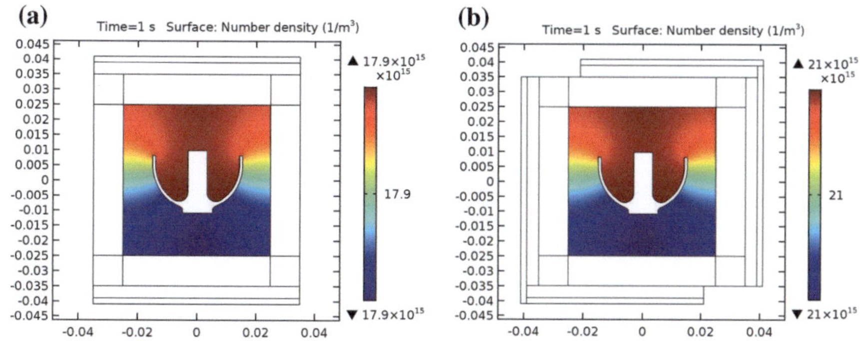

Fig. 10 Distribution of number density of O_2b1s radicals inside plasma chamber with **a** two parallel electrodes at top and bottom, and **b** four-sided electrodes

3 Design and Development of Experimental Setup

The configuration of the experimental setup developed based on simulation results of plasma analysis using finite element based software package Comsol®. The basic requirements of the present plasma polishing set up are as follows:

- The setup should have hermetically sealed chamber to maintain required vacuum/low pressure.
- The chamber should have interface to vacuum system and multiple gas feed lines for processing and reactive gases.
- The body of the plasma chamber shall act as a dielectric for RF excitations for the plasma with electrode placed outside the chamber.
- The chamber shall have optically transparent window for wavelengths 200–1200 nm in order to characterize the plasma by optical emission spectroscopy.

Based on above requirements, the plasma chamber is fabricated using Zerodur material body with fused silica cap as shown in Fig. 11a. The fused silica lid plate is optically transparent in the required wavelength. The chamber is connected to the vacuum system and gas feed lines through glass tubes. The glass tube feed lines are connected to the chambers by glass blowing technique. The schematic diagram of the experimental setup is shown in Fig. 11b.

3.1 *Optical Emission Spectroscopy*

It is required to interpret the mechanism of material removal, i.e. the generation of reactive species and generation reacted products with time during plasma polishing. The optical emission spectroscopy is a very useful tool to continuously monitor the relative spectroscopy densities of plasma species in the excited electronic states.

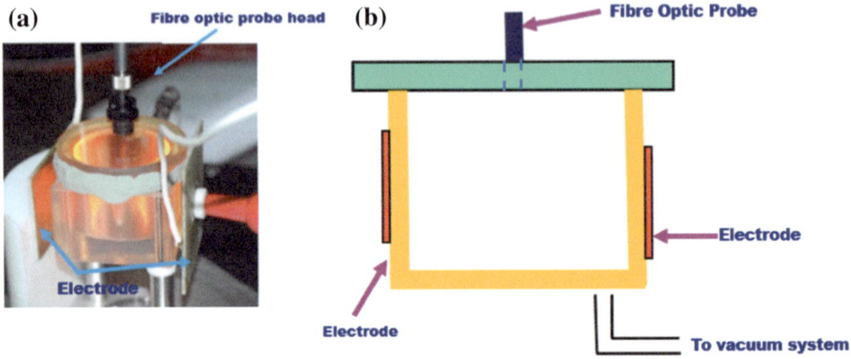

Fig. 11 **a** Actual photograph, and **b** schematic diagram of plasma polishing experimental setup

In a plasma, the excited atoms obtain enough energy, so that outer shell electron will transit to an excited state with higher energy. Since it is unstable in that state, it settles to a low energy level emitting electromagnetic radiation giving rise to an optical emission spectrum. The wavelength and energy change of the spectrum are related as

$$\lambda = hc/\Delta E \tag{2}$$

where ΔE is the energy difference between two levels of atomic state, λ is the wavelength of emission, h is the Plank's constant, and c is the speed of light. The transition and hence relative densities of the species are identified based on measured spectral line (λ) in the plasma and National Institute of Standards and Technology (NIST) atomic spectra database. The fibre optic probe head of atomic emission spectrometer (Ava spec ULS 2048 from Avantes Inc. Netherland) is placed at the fused silica cap of the experimental setup. The optical emission spectroscopy and its probe positioned at plasma chamber are shown in Fig. 12a, b, respectively.

The plasma chamber shall enable generation of reactive species uniformly covering the components to be polished. The preliminary experimental setup was

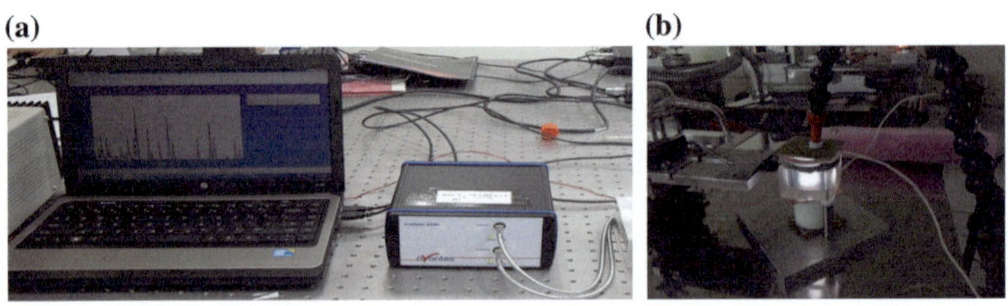

Fig. 12 **a** Optical emission spectroscopy set up, **b** its probe positioned at plasma chamber

Table 5 Specification of plasma processing setup

Discharge volume	Gap between the electrodes	Discharge frequency	Substrate used	Processing and reactive gases
30 cc (D = 40 mm, H = 25 mm)	50 mm	40.68 MHz	Fused silica	He, Ne and O_2 and SF_6

targeted to finish components of size ranging between 5 and 30 mm. The schematic diagram of experimental setup is shown in Fig. 12b. For a plasma polishing process to happen, generation of the reactive species and their adsorption at the component surface is essential. Then, the chemical reaction of the radicals (reactive species) and physical bombardment of the ions results in material removal atom by atom. The reacted volatile product diffuses away from the surface to be flushed out. Hence, it is very important of higher generation of the reactive species and their homogeneity over the component (all over the surfaces) for uniform material removal from complex freeform surfaces. The medium range pressure level is chosen to eliminate the damaging effects of ion impingement. In addition, RF excitation is used to limit electron heating while allowing chemical interaction of free radicals with surface atoms of workpiece. The RF excitation frequency of 40.68 MHz is chosen to minimize ion bombardment. The exact shape and size of the electrodes are configured based on information from simulation study using Comsol®. The specification of plasma processing setup is given in Table 5.

3.2 *Experimental Procedure*

The first step followed is to characterize the processing gas and reactive gas mixtures and fine-tune the gas mixture ratio and pressure. Preliminary experiments are conducted by characterizing the plasma with He, Ne and O_2 as processing gases, and O_2 and SF_6 as reactive gases at different pressures, different gas mixture ratios and RF excitation power level. Based on the observed experimental results by the optical emission spectroscopy, the selected gases are He as processing gas, oxygen as catalyst cum radical donor, and SF_6 as radical/ion donor (Liu 2009).

3.2.1 Characterization of Plasma for Reactive Gases

Reactive radicals such as fluorine are to be present for sustained material removal by plasma. The chemically active neutral substances (i.e. free radicals) of the F* which is dissociated from SF_6 will react with silica to form volatile products as per the following equation:

Table 6 Observed silica ion, fluorine and SiF transition lines

Element	Line (nm)	I-rel	Energy Ev Lower–Upper	Transition Lower–Upper	Quantum no Lower–Upper
II	450.7152	1996	33.04–35.79	3d′ 3P0–4f′ 2[5/2]	2–3
F III	449.5077	7763	55.12–7.88	4f 2 (Jin 2010) 0–5 g 2 (Wang et al. 2011)	9/2–11/2
SiF	449.58	600	0.00–2.84	X2Pi–A2Sig+	3–3
Si III	482.897	12303	25.99–28.55	4f 3F0–5 g 3G	4–5
Si IV	480.1437	132	39.08–41.67	6f 2F0–8d 2D	7/2–5/2
Si II	519.286	2512	16.36–18.75	4p′ 4D–4d′ 4F	5/2–7/2
Si II	518.525	1259	16.35–18.74	4p′ 4D–4d′ 4F	3/2–5/2
Si II	518.19	1000	16.34–18.73	4p′ 4D–4d′ 4F	1/2–3/2

$$SiO_2 + 4F^* \longrightarrow SiF_4 \uparrow + O_2 \uparrow \quad (3)$$

Another possible chemical reaction, i.e. the positive F^+ ion, reacts with the substrate to form volatile products as per the following equation:

$$4F^+ + SiO_2 + 4e^- \longrightarrow SiF_4 \uparrow + O_2 \uparrow \quad (4)$$

Accordingly, SF_6 gas is introduced as reactive gas. With this chemistry of gases, the silica and fluorine peaks are predominant between 440 and 520 nm as shown in Table 6 from NIST database.

Plasma processing is done at 20 mbar pressure and 40 W power, where uniform plasma distribution is obtained. Experiments are carried out to study the effect of SF_6/O_2 ratio on material removal rate (MRR) of the shell and it is shown in Fig. 13. MRR is the highest when the ratio of SF_6/O_2 is 2. The rate of material removal is 1.1 mg/h. The mean material removal rate of 0.008 mm^3/min is achieved for the optimized reactive gas mixture. The gas mixture selected after a series of experiments is He: 19.25 mbar, O_2: 0.25 mbar and SF_6: 0.5 mbar.

For further experiments, each cycle duration is fixed to 20 min and the chamber is evacuated to flush out the volatile products. After that the chamber is refilled for

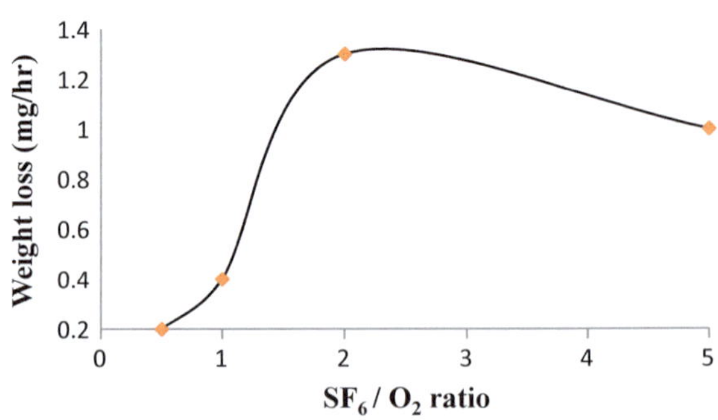

Fig. 13 Weight loss versus ratio of SF_6/O_2

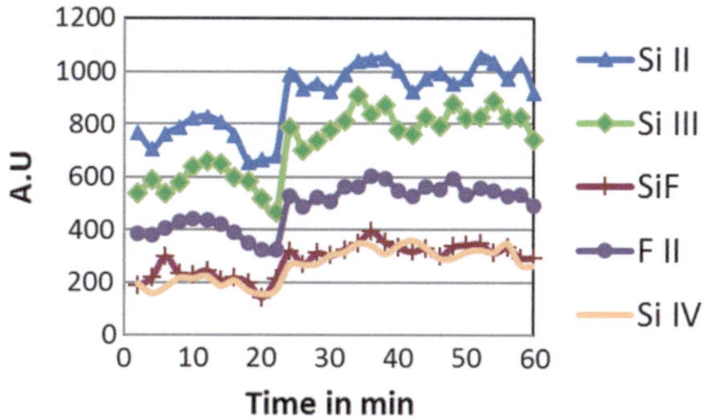

Fig. 14 Silicon ions and SiF transitions with time

next cycle of experiments. Multiple such cycles are performed and the silicon ion and SiF transitions are plotted with time (Fig. 14). For a weight loss of 1.1 mg/h for the given specimen, the mean material removal rate is 0.01 mm^3/min. Compared to the MRR for atmospheric pressure plasma reported in the literature, the present MRR is low.

3.3 *Initial Stage Polishing of Hemispherical Shell*

Based on the parameters optimized during preliminary experimental study, plasma processing of fused quartz hemispherical shell is carried out. Different views of the shell as compared with Indian one rupee coin are shown in Fig. 15. After one level of plasma processing with three cycles of 20 min duration, the surface roughness profile is recorded as shown in Fig. 16b from its initial as-machined condition (Fig. 16a). When the number of cycles is further repeated, the MRR is reduced drastically. It is suspected that certain polymers are forming on the specimen surface and due to which MRR is reduced for further cycles. The surface residual stress or molecular level strained bonds are expected to be better with plasma-processed

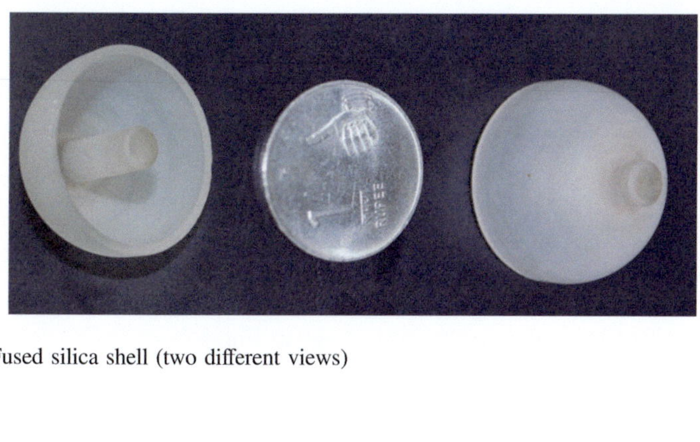

Fig. 15 Fused silica shell (two different views)

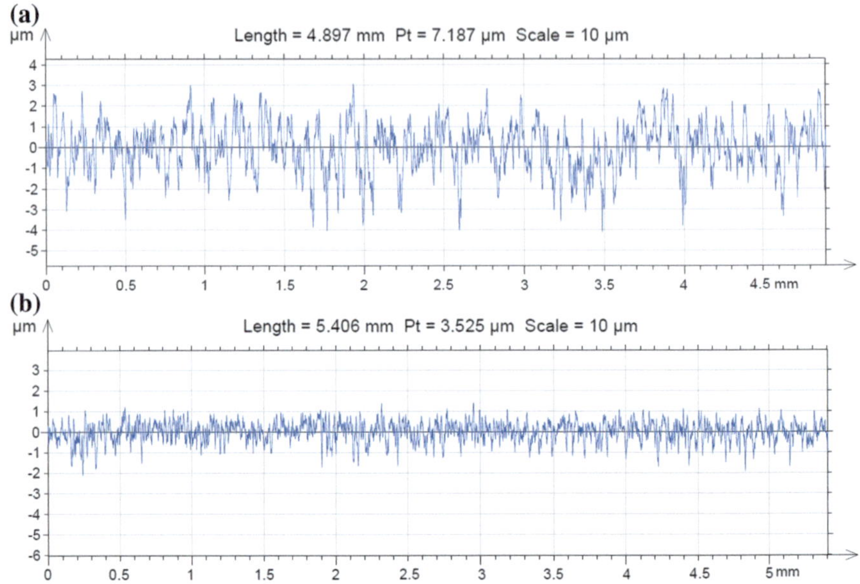

Fig. 16 Surface profiles **a** as machined (Ra = 0.90 μm, Rz = 6.44 μm), **b** after three cycles of 20 min duration plasma polishing (Ra = 0.38 μm, Rz = 2.81 μm)

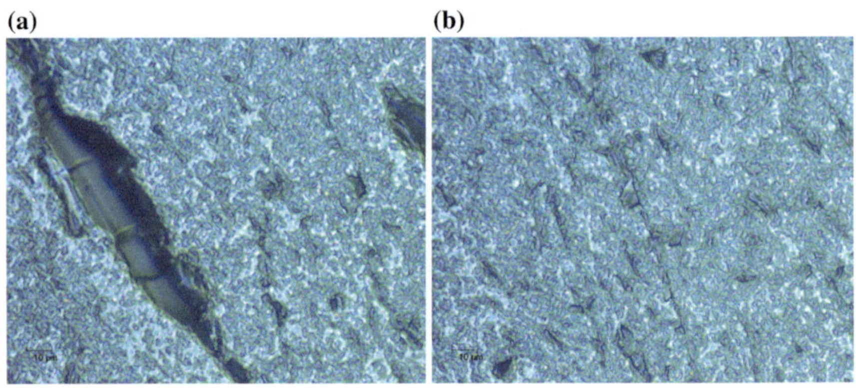

Fig. 17 AFM surface image **a** as machined, and **b** after three cycles of 20 min duration of plasma processing

surface. The surface finish is improved from 0.90 to 0.38 μm only. However, the surface cracks and damages are effectively removed from the plasma-processed surface as shown in atomic force micrograph (AFM) image (Fig. 17b) from the as-machined surface as shown in AFM image in Fig. 17a.

3.3.1 Fine Plasma Polishing of Hemispherical Shell

The shell is subjected to cumulative plasma polishing for 48 h and the product is shown in Fig. 18. After multiple sessions of plasma polishing cycles, the component is left with patches of opaque regions apart from transparent surface.

The surface finish is measured by non-contact Brooker contour GTK-A surface profilometer. The 3D surface roughness profile of the shell from areas having non-transparent patches and transparent region is shown in Fig. 19a, b, respectively. In Fig. 19b, the surface finish is improved to 3.6 nm from the shell's initial surface roughness value of 902 nm. It is also observed from Fig. 19a, i.e. from the areas of non-transparent patches, that the surface finish has improved to 32 nm and it is due to the presence of few fine digs, i.e. crater-like defects. These defects are primarily due to the ultrasonic-assisted brittle machining, i.e. milling operation adopted to shape the job from a cylindrical bar. Suitable process modification is required for further improvement of the shell surface. This gives confidence that medium-pressure plasma polishing is capable of polishing complex surface such as hemispherical shell.

Fig. 18 Hemispherical shell of fused silica after plasma polishing

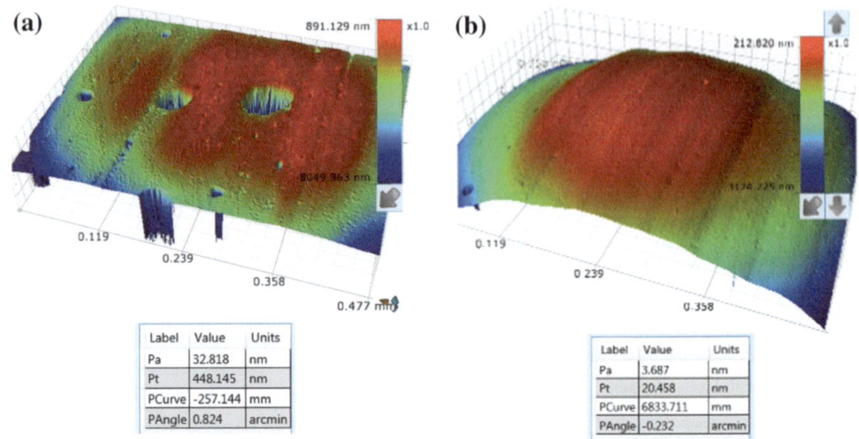

Fig. 19 3D contour surface profiles on curved shell (Fig. 18) surface **a** with defects (Pa = 0.0328 μm, Pt = 0.448 μm), and **b** on transparent surface (Pa = 0.0036 μm, Pz = 0.045 μm)

4 Conclusions

Comsol® finite element simulation of the plasma is used to analyse the distribution of the radicals inside the plasma chamber with and without specimens and for the ratio of free volume to the total available volume of the chamber for uniform distribution of radicals. The ratio of processing gas and reactive gas is identified for optimum distribution of radicals. In the presence of hemispherical shell component, the four-sided electrode provides higher density of radicals. The experimental setup is designed and built based on preliminary inputs from Comsol® analysis, and four-sided electrode is used for shell processing. The gas mixture is fine-tuned to get best MRR of 0.008 mm^3/min. The atomic emission spectroscopy is used to monitor the various oxidation states of silica and to establish correlation with respect to MRR and intensity of oxidation states of silica Si II and Si III. With initial plasma processing for one-hour duration, the surface finish is improved from Ra = 0.90 nm to Ra = 0.38 nm. The AFM images after pre- and post-plasma polishing confirms elimination of large cracks/indents. The plasma process could efficiently play the role of wet etching in removing the surface cracks and damages without degrading the surface topography. Cumulative plasma polishing of 48 h has resulted in surface finish of the order of 3.6 nm (Pa) in most of the locations. However, in the areas with patches, the surface roughness of 32.82 nm (Pa) is obtained which is attributed to the excessive damaged surface due to ultrasonic milling process adopted for shell shaping prior to polishing.

References

Dev, D.S.D., E. Krishna and M. Das. 2016. A Novel Plasma assisted atomistic surface finishing on freeform surfaces of fused silica. *International Journal of Precision Technology* 6: 262–276.

Gerhard, C., T. Weihs, A. Luca, and S. Wieneke. 2013. Polishing of optical media by dielectric barrier discharge inert gas plasma at atmospheric pressure. *Journal of the European Optical Society—Rapid Publications* 8: 13081–13085.

Jin, H.L., B. Wang and F.H. Zhang. 2010. Effect on surface roughness of zerodur material in atmospheric pressure plasma jet processing. *Proceedings of SPIE* 7655: 76552X 1–7.

Liu, W., D. Wang, M. Hu, Y. Wang, H. Liang and L. Hang. 2009. Roughness evaluation of fused silica plasma polishing. *Advanced optical Manufacturing Techniques-Proceeding of SPIE* 7282: 72822T 1–5.

Shena, J., S. Liuc, K. Yi, H. He, J. Shaoa, and Z. Fan. 2005. Subsurface damage in optical substrates. *Optik—International Journal for Light and Electron Optics* 116: 288–294.

Wang, B., J. Zhang, and S. Dong. 2006. Application of atmospheric pressure plasma in the ultrasmooth polishing of SiC optics. *Materials Science Forum* 532–533: 504–507.

Wang, B., J. Zhang, and S. Dong. 2009. New development of atmospheric pressure plasma polishing. *Chinese Optics Letters* 7: 537–538.

Wang, D., W. Liu, Y. Wu, L. Hang, H. Yu, and N. Jin. 2011. Material removal function of the capacitive coupled hollow cathode plasma source for plasma polishing. *Physics Procedia* 19: 408–411.

Yao, Y.X., B. Wang, J.H. Wang, H.L. Jin, Y.F. Zhang, and S. Dong. 2010. Chemical machining of Zerodur material with atmospheric pressure plasma jet. *CIRP Annals-Manufacturing Technology* 59: 337–340.

Ultrasonic Assisted Turning: A Comparative Study of Surface Integrity

Varun Sharma, Pulak M. Pandey, Uday S. Dixit, Anish Roy and Vadim V. Silberschmidt

Abstract The surface integrity of machined sample has detrimental effects on the product's life. The machining parameters have been found to significantly affect the surface integrity during the turning operation. Surface roughness, residual stresses and microhardness are the most commonly used parameters to study the surface integrity. In this chapter, an attempt has been made to present a comparative analysis of conventional turning (CT) and ultrasonically assisted turning (UAT) processes with plain and textured cutting tools. The effect of ultrasonic power (measured in terms of amplitude) on surface roughness has been studied during UAT process. An increase in the amplitude improved the surface finish of the machined specimen, significantly. The surface residual stresses generated in the machined part during UAT have been compared for plain and self-lubricating cutting inserts using XRD analysis. An attempt has been made to evaluate the residual stresses generated during the process by using commercially available finite element method package, ANSYS. The generation of a large compressive residual stresses during UAT process with self-lubricating cutting inserts signifies

V. Sharma
Department of Mechanical Engineering, BITS Pilani, Pilani, India
e-mail: varun.sharma@pilani.bits-pilani.ac.in

P. M. Pandey (✉)
Department of Mechanical Engineering, IIT Delhi, New Delhi, India
e-mail: pmpandey@mech.iitd.ac.in

U. S. Dixit
Department of Mechanical Engineering, IIT Guwahati, Guwahati, India
e-mail: uday@iitg.ernet.in

A. Roy · V. V. Silberschmidt
Wolfson School of Mechanical and Manufacturing Engineering, Loughborough University, Loughborough, UK
e-mail: A.Roy3@lboro.ac.uk

V. V. Silberschmidt
e-mail: v.silberschmidt@lboro.ac.uk

S. S. Pande and U. S. Dixit (eds.), *Precision Product-Process Design and Optimization*, Lecture Notes on Multidisciplinary Industrial Engineering,
https://doi.org/10.1007/978-981-10-8767-7_13

better fatigue life of the component. The microhardness measurements are used to demarcate the Machine Affected Zone (MAZ) for UAT and CT process. A comparative analysis between CT and UAT has also been presented in terms of surface integrity to demonstrate better machining regime found in UAT using self-lubricating cutting inserts.

Keywords Surface integrity · Vibration · Conventional turning
Ultrasonic assisted turning · Self-lubricating inserts

1 Introduction

High cutting forces are generated during machining of difficult-to-cut materials. These forces are said to have detrimental effects not only on cutting tool life but also on surface integrity of the machined specimen. Research in recent past has been focused on eliminating adverse impacts of the use of lubricants and coolants. Thus, in order to reduce the carbon footprints of machining, dry machining is preferred (Dixit et al. 2012; Sharma and Pandey 2016e). Researchers have also observed that application of high-intensity ultrasonic power causes a favourable reduction in machining forces. This is often attributed to reduction in material hardness due to acoustic softening (Siddiq and Sayed 2011; Maurotto et al. 2012). This effect of ultrasonic vibrations on hard-to-cut materials is useful in turning, drilling and milling operations.

The work reported in this chapter primarily focuses upon presenting a comparative analysis between CT and UAT process by primarily focussing on the surface integrity (surface roughness, microhardness and residual stresses) of the machined specimen. In the following sections, a brief introduction to UAT process has been presented followed by the research done so far, in this regard.

2 Ultrasonic Assisted Turning

From a review of the literature, ultrasonic-vibration assisted machining of metals first started in the late 1950s (Skelton 1969; Brehl and Dow 2008). Over the years, ultrasonically assisted turning (UAT) has been improved upon significantly. In UAT, high-frequency and low-amplitude vibrations are superimposed on the cutting tool. This vibratory tool movement, in turn, allows for a periodic separation between rotating workpiece and tool. The tool movement may be in one direction (1D-UAT; Fig. 1a), or in a combination of two directions (2D-UAT; Fig. 1b) to yield linear or elliptical motions of cutting edge, respectively (Sharma 2017).

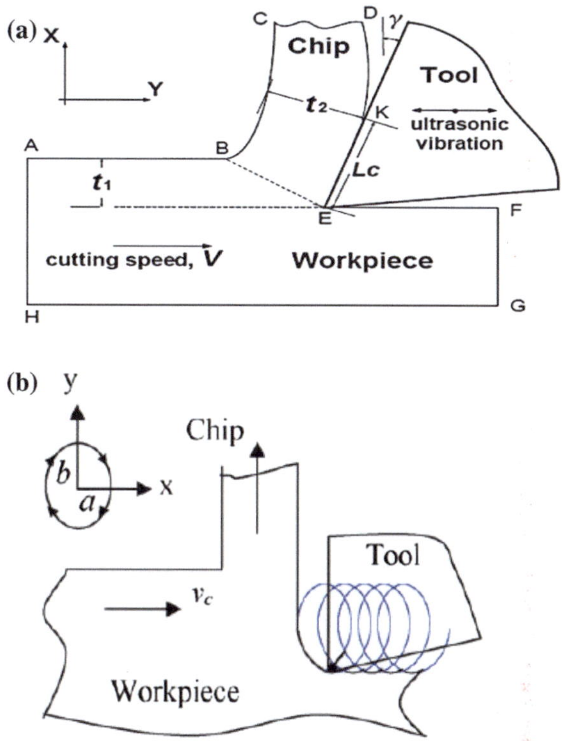

Fig. 1 Schematic representing relative movement of tool and workpiece for **a** 1D-UAT (Patil et al. 2014), **b** 2D-UAT (Nath et al. 2009)

The following paragraphs highlight the schematic of 1D-UAT and 2D-UAT setups developed by various researchers and the mechanism involved in UAT process, which makes it a better alternative to CT for machining of hard-to-cut materials.

2.1 One-Dimensional UAT

One-dimensional UAT (1D-UAT) was introduced in late 1950s and was introduced for machining materials that could not be easily turned using a diamond cutting tool. In the mid-1980s 1D-UAT was used to machine steel, glass and brittle ceramics. It was observed that 1D-UAT lead to better surface integrity. Research in 1D-UAT gained popularity with the advent of difficult-to-cut materials like Ni- and Ti-based alloys. 1D-UAT involves the use of piezoelectric or magnetostrictive actuator to create a reciprocating harmonic motion of high-frequency and low-amplitude vibrations. A suitably shaped acoustic booster and horn is generally used to amplify this ultrasonic motion (Sharma 2017). The cutting tool is then attached to the end of the horn as shown in Fig. 2. The amplitude at the horn end varies from 3 to 20 μm.

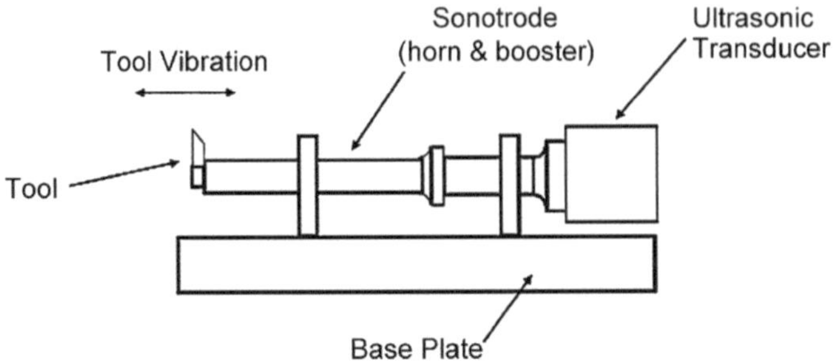

Fig. 2 One-dimensional ultrasonic assisted turning setup (Brehl and Dow 2008)

2.2 Two-Dimensional UAT

Two-dimensional UAT (2D-UAT) was introduced in the 1990s to achieve enhanced force reduction and improved surface finish beyond 1D-UAT. This process adds the vertical harmonic motion to the horizontal motion produced by 1D-UAT. The superimposition of two motions leads to an elliptical motion of the cutting insert. The vibrational amplitude in each axis may or may not be same and is defined as the amplitude of vibration in major axis (A) $\times$ amplitude of vibration along minor axis (B). It is represented as A μm $\times$ B μm (Sharma 2017).

2.3 Resonant Two-Dimensional UAT Systems

The resonant 2D-UAT system creates circular or elliptical motion by making the supporting structure to vibrate at a resonant frequency. By mounting the centre of mass of tool away from midpoint, a two-dimensional motion may be achieved. Brinksmeier and Gläbe (1999) found a bending vibration of supporting structure in direction of depth of cut. When this mode of vibration was combined with horizontal mode of vibration, an elliptical tool path was obtained. Moriwaki and Shamoto (1995) developed resonant 2D system as shown in Fig. 3. The piezoelectric actuators attached to beam side face produced a bending in horizontal and vertical directions on activation. The combination bending motions produced by vibrations resulted in an elliptical path of tool attached at the opposite end.

2.4 Non-resonant 2D-UAT Systems

In this variant of 2D-UAT, a sinusoidal voltage signal applied to piezo actuators resulted in an expansion and contraction of piezoelectric crystals. This motion was then converted to an elliptical motion of the tool with the help of mechanical

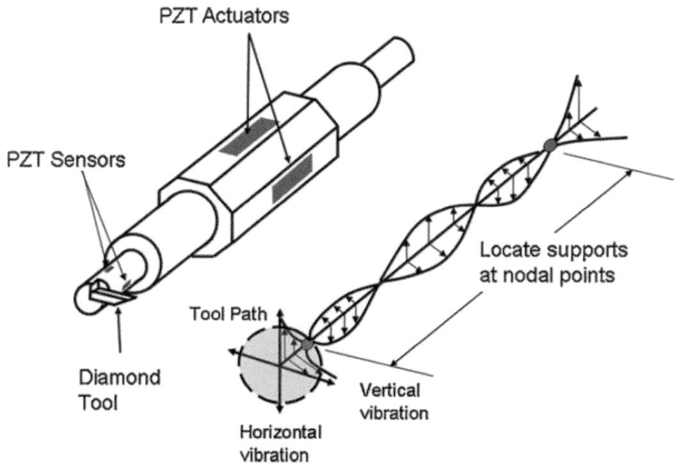

Fig. 3 Resonant 2D system (Brehl and Dow 2008)

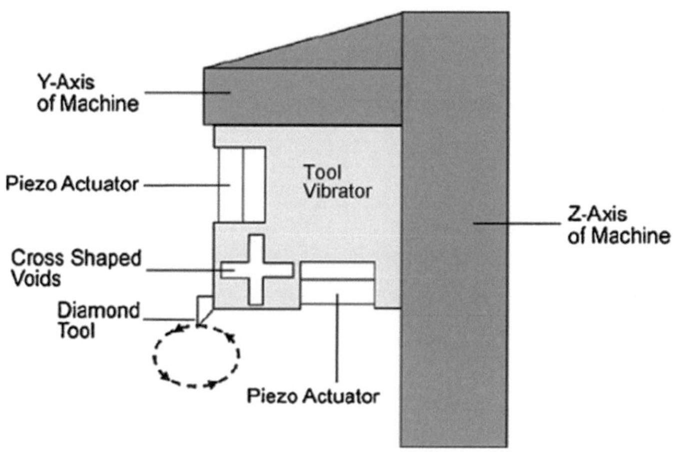

Fig. 4 Non-resonant 2D system (Brehl and Dow 2008)

linkages. The piezoelectric actuators were oriented at a right angle to each other. The tool was mounted on one side of the cross-shaped flexure to limit the crosstalk between two tool motions (Fig. 4). Crosstalk is the transfer of energy back and forth between two modes that results in a distorted tool path in resonant 2D-UAT systems. In order to reduce this effect, phase difference is provided by designing an internal cross-shape which in turn results in an elliptical tool path.

From the above illustrations, it can be deduced that in 1D-UAT, there occurs a linear motion of the cutting insert, whereas an elliptical motion of the cutting insert occurs in 2D-UAT process. The present research attempt has been focused on enhancing the efficacy of dry UAT process by combining the dual benefits of intermittent cutting of UAT and lubrication of textured cutting insert. The following subsection highlights the processing mechanism of 1D-UAT process.

3 Mechanism of 1D-UAT Process

The mechanism of UAT may be divided into four stages during a single cycle of vibration as shown in Fig. 5 (Patil et al. 2014). In stage 1, cutting tool approaches workpiece material, while in stage 2, the cutting tool makes contact with workpiece. The tool is in full contact with workpiece and starts penetrating into workpiece in stage 3.

When highest level of stress is achieved, it marks maximum penetration depth attained by cutting tool. This stage of UAT is similar to a conventional turning (CT) process as tool is in full contact with workpiece. In the last stage, i.e. in stage 4, the direction of cutting tool is reversed and the tool disengages from the workpiece. This stage of UAT results in a reduction of stress level attained in workpiece. Thus, the stress is variable in UAT, ranging from the maximum in stage 3 to quite less in other stages. As a result, a reduction in the average cutting force is obtained.

Vivekananda et al. (2014) highlighted that for intermittent cutting to occur the following conditions must satisfy:

$$v_C > V, \tag{1}$$

where v_C represents critical cutting speed and V denotes cutting speed. These parameters may be calculated using the following equations:

$$v_C = 2\pi af \tag{2}$$

$$V = \pi DN \tag{3}$$

where a is amplitude and f represents frequency of vibration, D is diameter and N denotes rotational speed of workpiece.

The following paragraphs discuss the research attempts by various researchers pertaining to the surface roughness and residual stresses during conventional and ultrasonic assisted turning processes (Sharma 2017).

3.1 Studies on Surface Roughness

Wang and Zhao (1987) were the pioneers in studying the surface roughness by providing ultrasonic vibration to cutting tool. It was reported that UAT results in an

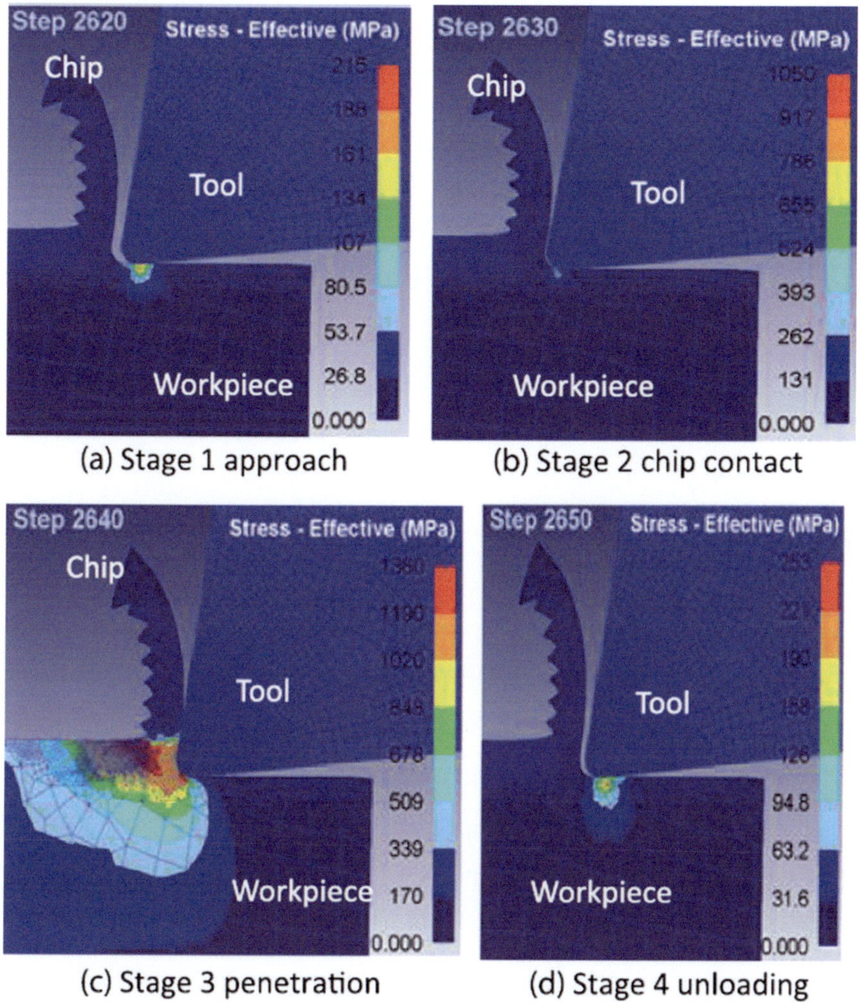

Fig. 5 The schematic representation of four-stage mechanism followed by UAT (Patil et al. 2014)

improvement of the surface finish over CT process. Silberschmidt et al. (2014) reported a twofold improvement in roughness value of Inconel 718 and Ti 15-3-3-3 by UAT when compared to CT. A reduction in chatter and a stable cutting mechanism were observed that resulted in a centre line average surface roughness, R_a, of 0.505 μm during machining Inconel 718 with UAT.

Muhammad et al. (2014) reported 50% improvement in R_a value of $(\alpha + \beta)$ Ti-based alloys. In another research attempt, Nath et al. (2007) found low feed rate to help in attaining better surface finish during UAT. Gao et al. (2002) reported tool tip to be located lower than rotating centre of workpiece by approximately three times the amplitude of vibration, in order to have less ripples formation. A surface

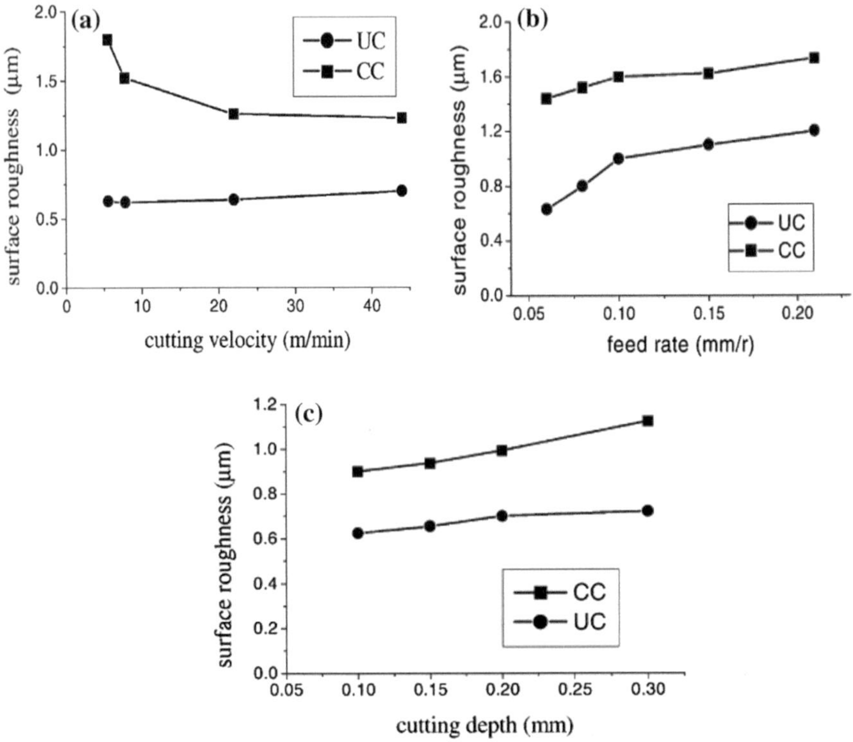

Fig. 6 Variation of surface finish with **a** cutting velocity; **b** feed rate; **c** cutting depth (Gao et al. 2002) (*CC* conventional cutting; *UC* ultrasonic cutting)

roughness of order of 30% less was found in machined specimen by providing ultrasonic vibration to cutting tool (refer Fig. 6).

In another research attempt, Lin and Zhong (2006) studied the effects of ultrasonic and cutting parameters on roughness value. The authors reported that high amplitude (3.7 μm) decreased surface roughness by 15% with reference to surface roughness obtained by UAT with an amplitude of 2.2 μm (Fig. 7).

Patil et al. (2014) related improvement in roughness with tool workpiece contact ratio (TWCR). The study indicates optimum cutting velocity, which provided best surface finish. At cutting velocity of 20 m/min, 40% improvement in surface finish could be achieved during UAT of Ti6Al4 V. However, average surface roughness was observed to increase with a further increase in cutting speed (refer Fig. 8).

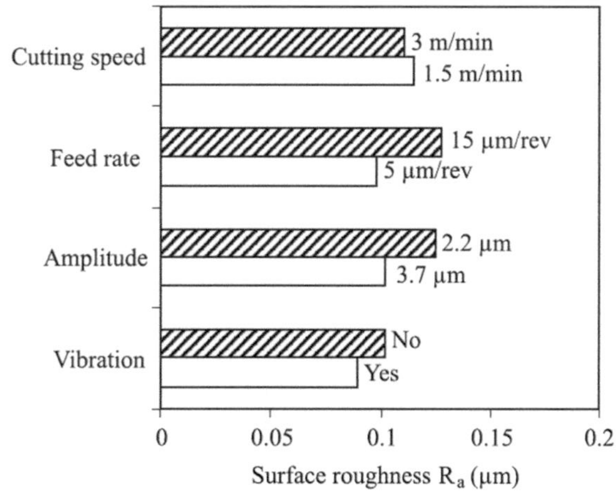

Fig. 7 Surface roughness variation for cutting and vibration parameters for aluminium-based composite material (Lin and Zhong 2006)

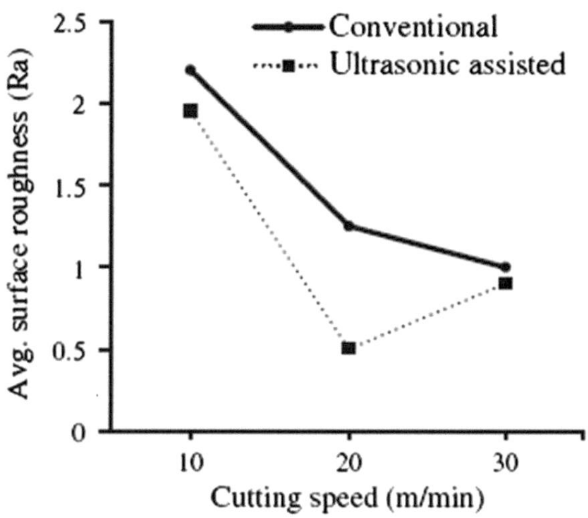

Fig. 8 Effect of cutting speed on surface roughness (Patil et al. 2014)

3.2 *Studies on Residual Stresses*

Ahmed et al. (2006) measured residual stresses and hardness using nanoindentation tests (Fig. 9). The machined specimens of CT and UAT were held in a plastic mould and a number of indentations were made along the depth of the machined specimen. In Fig. 9, the middle region indicated by 'plastic' implies the hardness of plastic mould and is to be ignored in the analysis. It was found that average hardness of the hardened layer in UAT was 60% less than that found during CT of machined part.

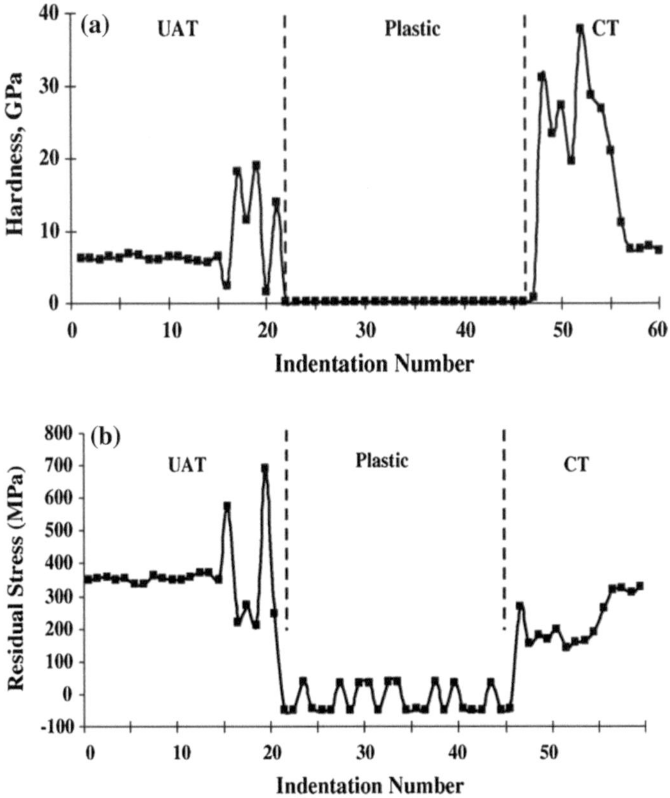

Fig. 9 Variation in **a** hardness and **b** residual stresses for UAT and CT process (Ahmed et al. 2006)

Numerical studies using the Johnson–Cook material model revealed that a high value of compressive residual stresses in UAT is generated as compared to CT (Fig. 9b).

In another research attempt, Silberschmidt et al. (2014) observed an increase in hardness of the surface layer with variation in depth of cut (Fig. 10). It was found that hardness of the CT machined part varied from the bulk (un-machined part). Besides this, since in UAT, the strain rate is expected to be high (when compared to CT), the hardness should have been more; however, Fig. 10 shows the reverse effect. This indicates that the strain rate was not the only factor that affected hardness of machined specimen.

Nestler and Schubert (2014) applied ultrasonic vibrations in radial direction to generate compressive residual stresses in machined part as compared to vibrations in cutting direction. Interestingly, ultrasonic vibration in the feed direction had no effect on residual stresses (Fig. 11).

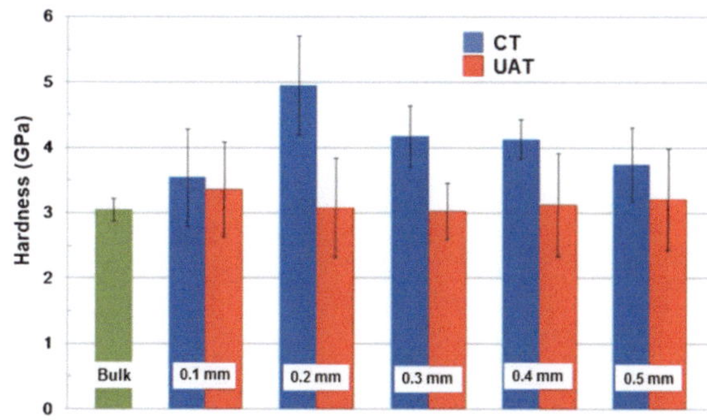

Fig. 10 Surface hardness variation in CT and UAT (Silberschmidt et al. 2014)

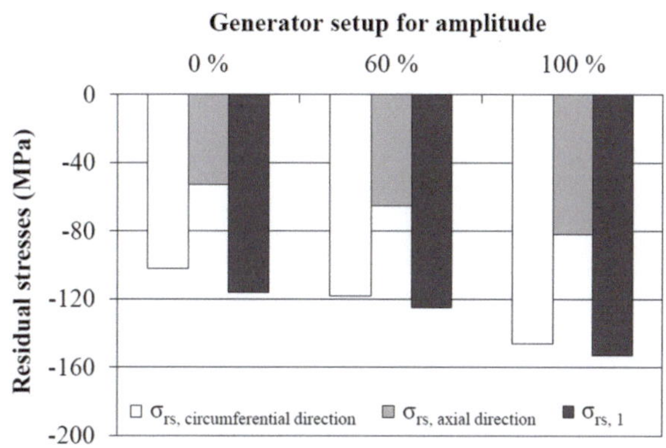

Fig. 11 Amplitude effect on generation of residual stresses in UAT (Nestler and Schubert 2014)

To further enhance efficacy of dry UAT process, the intended research attempts to use self-lubricating textured cutting inserts during UAT process. Sharma and Pandey (2016a, b, c), Sharma et al. (2016) discussed the benefits of self-lubricating cutting insert during UAT process. UAT with self-lubricating cutting inserts resulted in minimum cutting forces accompanied with better surface finish as compared to plane cutting insert. The most favourable method of enhancing the fatigue life is by inducing compressive residual stresses. The present chapter focuses on studying the surface integrity during conventional turning with plain tool (PCT) as well as textured tool (TCT). Surface roughness, residual stress and microhardness measurements at the machined specimen cross section have been taken as criteria for studying the surface integrity of machined specimen.

4 Fabrication of Ultrasonically Assisted Turning Setup

Figure 12a shows the schematic of UAT setup fabricated for the experimentation. The most crucial part in the setup fabrication has been the development of ultrasonic horn (Fig. 12b). The ultrasonic vibratory tool assembly comprised an ultrasonic generator, piezoelectric crystals, booster and a customized horn. The

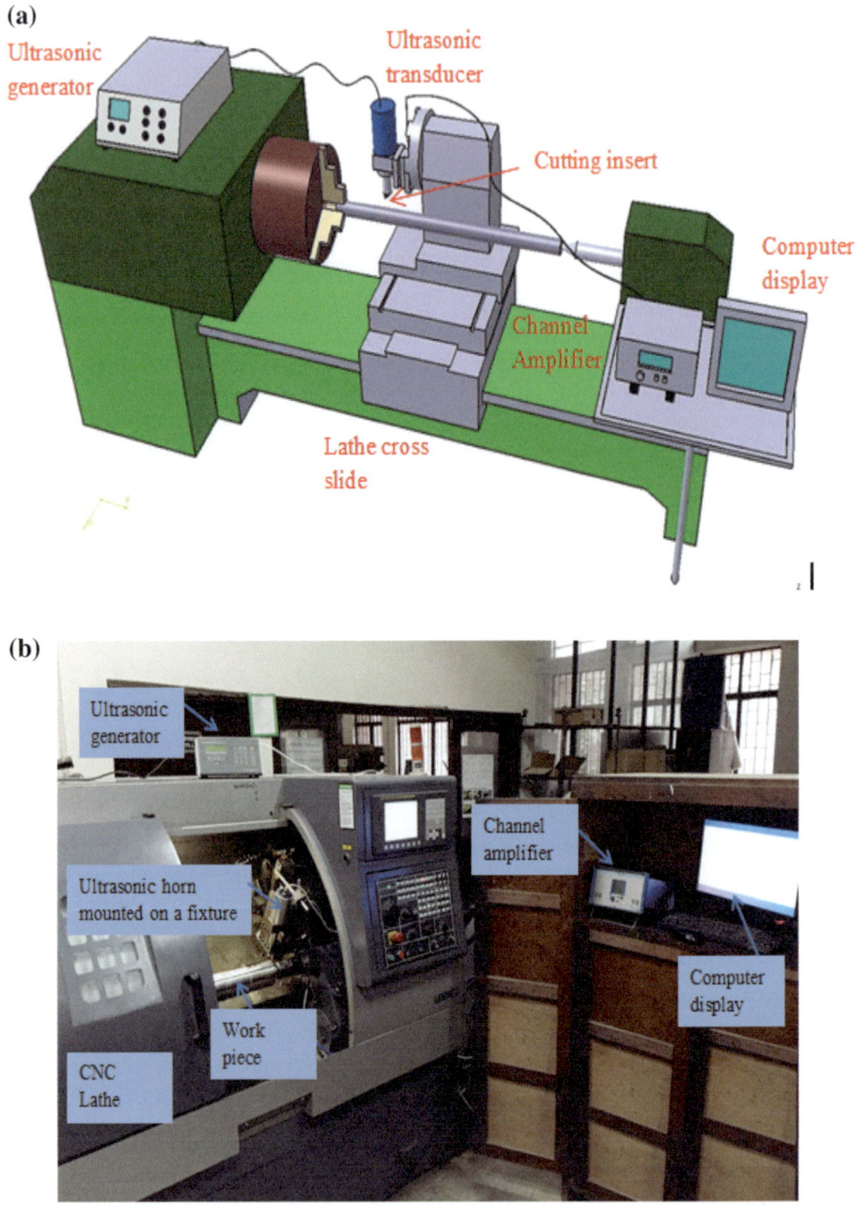

Fig. 12 **a** Schematic and **b** actual setup for UAT experimentation (Sharma and Pandey 2016d)

ultrasonic generator (make: Chromtech, Taiwan) generated electrical signals (1200 W, 20 kHz), which were then converted into mechanical vibrations through a stack of piezoelectric crystals. A customized stepped horn (Fig. 13a) was manufactured to boost the vibrations. The ultrasonic generator (Make: Chromtech,

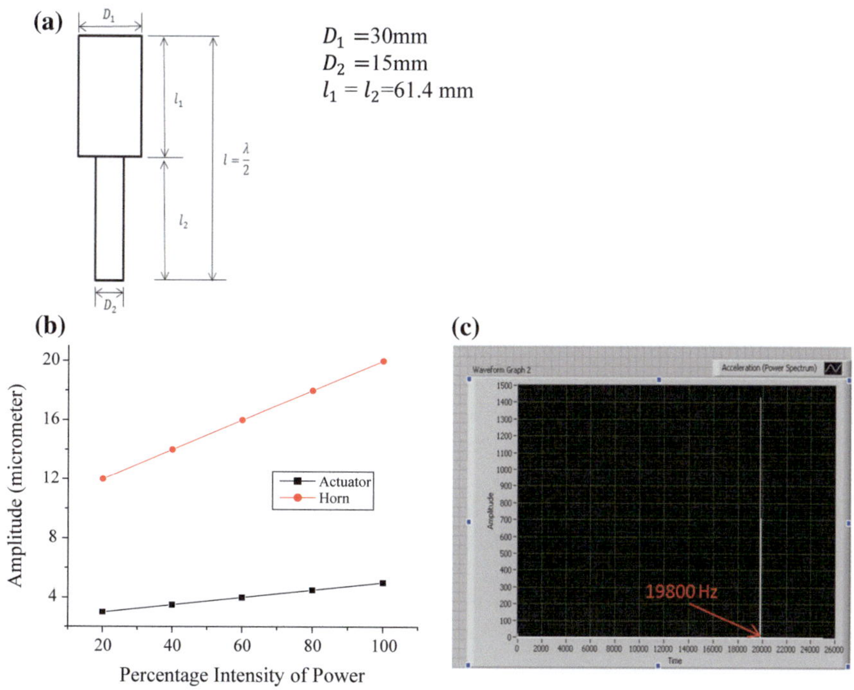

Fig. 13 **a** Schematic of stepped horn; **b** calibration of amplitude with percentage intensity of ultrasonic power; **c** natural frequency of stepped horn using accelerometer (Sharma 2017)

Taiwan) used for present experimentation operated at a frequency of 20 kHz and was with a variable power output source having maximum supply of 1200 W.

In order to analyze the modal and harmonic analysis of the stepped horn, a commercially available FE package, ANSYS was used. The modal analysis revealed that the 12th mode of vibration corresponded to the longitudinal mode with a natural frequency of 19.8 kHz.

Figure 13b shows the calibration graph of amplitude with the percentage intensity of ultrasonic power varied with the ultrasonic generator. Figure 13c presents the frequency of the horn measured using accelerometer. It could be inferred that the horn had a natural frequency of 19.8 kHz.

The developed horn was then mounted on the dynamometer with the help of a fixture. In order to prevent the resonance of fixture with the Kistler dynamometer, the natural frequency of the fixture was measured using hammer test. The maximum third natural frequency of the fixture was around 1000 Hz, which was far less than the natural frequency of the Kistler dynamometer, i.e. 3.2 kHz. Thus, the condition of resonance could be prevented during machining.

4.1 Work Material

4340 hardened steel of dimensions 40 mm in diameter and 400 mm in length has been used as the workpiece material. The elemental composition of the material as measured by EDX analysis is 0.4%C, 1.05%Ni, 1.05%Cr, 0.65%Mn, 0.21%Si and rest Fe. The material hardness was constant with a variation of ±1 HRC throughout the cross section of the cylindrical workpiece.

4.2 Cutting Tools

CNMA 120404 THM was used for performing the turning experiments. The selection of this cutting insert was made due to its lack of chip breaker or coating, which facilitated an easy manufacture of a textured pattern. In order to create texture pattern, femtosecond pulsed Nd-YAG laser (wavelength = 1064 nm, pulse energy = 25 MJ, repetition rate = 20 kHz) was used. After texturing, the cutting inserts were cleaned in an ultrasonic acetone bath for 20 min. Figure 14 shows the SEM image of developed textured pattern filled with CaF_2 solid lubricant under microscopic observation.

4.3 Sample Preparation

The machined workpiece was cut into sample size of 10 mm × 5 mm × 5 mm using wire-cut EDM. This method was used for the sample preparation, as it resulted in least heat affected zone (HAZ) and thus, facilitated accurate readings for microhardness profiles.

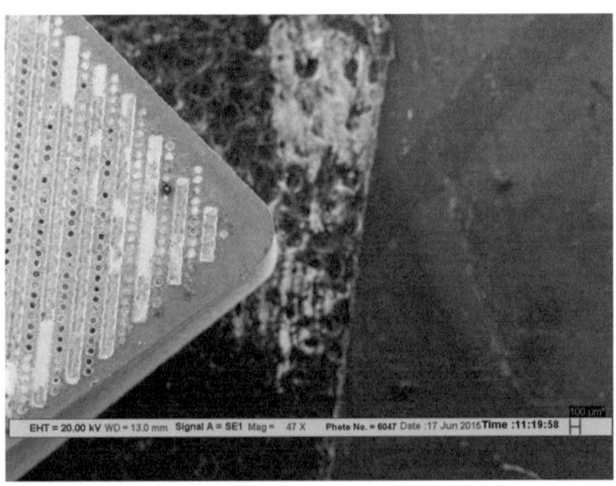

Fig. 14 SEM image of the textured cutting insert (47× and 100 µm scale) used for present experimentation

4.4 Microhardness Test

For microhardness measurement, cold mounting of samples was performed using a resin powder. The samples were then polished using waterproof SiC paper with grit size ranging from 200 to 1200. The microindentation was measured at a depth of 15, 30, 45, 60, 70, 90, 150 and 300 μm from machined surface to bulk of specimen.

4.5 Measurement of Residual Stress

Among the different techniques that are available for the measurement of residual stresses, XRD is preferred. This is due to the high spatial resolution of XRD accompanied with the measurement in all crystalline materials (Sharma and Pandey 2016d).

In the present work, Panalytical X'pert XRD unit has been used to evaluate the residual stresses with $\sin^2\Psi$ technique. The XRD system was calibrated with respect to the stress-free sample. CuKα was used as the target material with ⟨220⟩ plane. The other parameters for the measurement can be found elsewhere in other publications of present researchers (Sharma and Pandey 2016d).

The three-dimensional strain tensor is represented as follows:

$$\varepsilon_{ij} = \begin{bmatrix} \varepsilon_{11} & \varepsilon_{12} & \varepsilon_{13} \\ \varepsilon_{21} & \varepsilon_{22} & \varepsilon_{23} \\ \varepsilon_{31} & \varepsilon_{32} & \varepsilon_{33} \end{bmatrix}. \tag{4}$$

For calculating the stress tensor, Hooke's law is used, providing

$$\sigma_{ii} = \frac{E}{1+\mu}\varepsilon_{ii} + \frac{\mu E}{[(1+\mu)(1-2\mu)]}\Delta, \tag{5}$$

$$\sigma_{ij} = \frac{E}{1+\mu}\varepsilon_{ij}, \tag{6}$$

$$\Delta = \frac{\mathrm{d}V}{V} = \sum_{i=1}^{3}\varepsilon_{ii}. \tag{7}$$

As per the ASTM standard, the stress components are represented by the tensor σ_{ij} given as follows:

$$\sigma_{ij} = \begin{bmatrix} \sigma_{11} & \tau_{12} & \tau_{13} \\ \tau_{21} & \sigma_{22} & \tau_{23} \\ \tau_{31} & \tau_{32} & \sigma_{33} \end{bmatrix}. \tag{8}$$

The stress strain relationship in any direction of a component with respect to the azimuth phi (φ) and polar angle psi (Ψ) is defined as follows (Sharma and Pandey 2016d):

$$\begin{aligned} \varepsilon_{\varphi\Psi}^{hkl} &= \frac{1}{2}s_2^{hkl}\left[\sigma_{11}\cos^2\varphi\sin^2\Psi + \sigma_{22}\sin^2\varphi\sin^2\Psi + \sigma_{33}\cos^2\Psi\right] \\ &+ \frac{1}{2}s_2^{hkl}\left[\tau_{12}\sin(2\varphi)\sin^2\Psi + \tau_{13}\cos\varphi\sin(2\Psi) + \tau_{23}\sin\varphi\sin(2\Psi)\right] \cdot \\ &+ s_1^{hkl}[\sigma_{11} + \sigma_{22} + \sigma_{33}] \end{aligned} \tag{9}$$

For measuring the residual stress, sample (5 mm × 5 mm × 7 mm) was removed by wire EDM from the cutting length. The samples were then electropolished using 80:20 proportion of perchloric acid and ethanol as the electrolyte.

5 Results and Discussion

The following section highlights the trends in the variation of surface roughness, residual stress and microhardness variation for PCT (CT: Plain tool), TCT (CT: Textured tool), PUAT (UAT: Plain tool) and TUAT (UAT: Textured tool) to analyze the surface integrity.

5.1 *Surface Roughness*

In order to quantify the surface characteristics produced by CT and UAT, an optical profiler was used. Figure 15 shows the images from the optical profilometer of CT

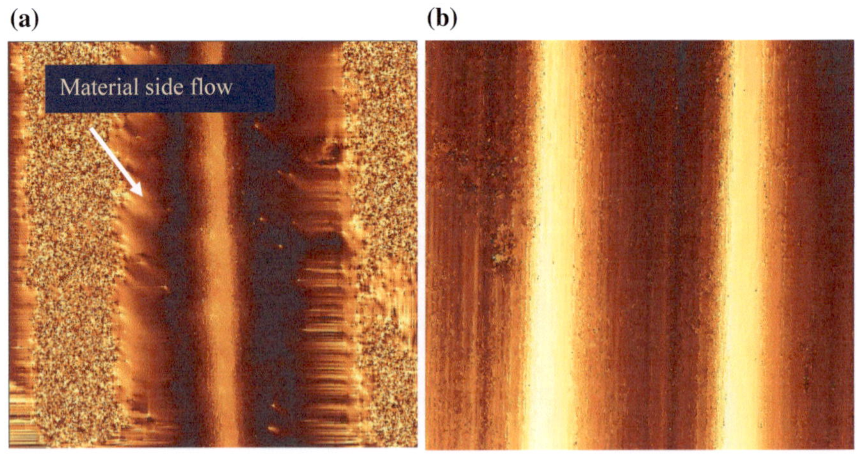

Fig. 15 Images of the machined surface using optical profilometer for **a** PCT; **b** TUAT surface

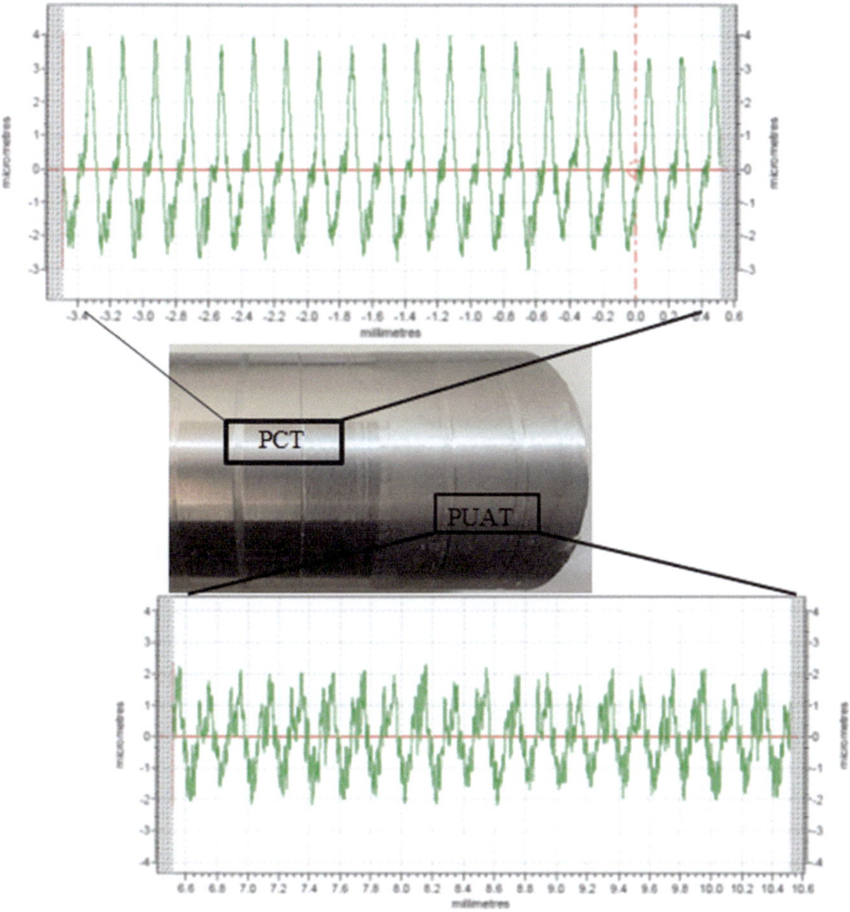

Fig. 16 Machined surface analysis for conventional and ultrasonic assisted turned specimens

and UAT machined specimen. It may be inferred that the material side flow is predominant in CT. This, in turn, resulted in high surface roughness. However, due to the ‘ironing effect’, suppression of material side flow was observed during the UAT process. This decreased the peak-to-valley height, improving surface finish.

Figure 16 shows a clear visual difference between two processes, i.e. CT and UAT. It can be seen that CT process resulted in a glossy finish, whereas matte finish was observed during UAT process. This is in line with the research work of Patil et al. (2014). To further improve the surface finish, UAT with self-lubricating cutting insert was performed.

This technique was found to improve overall surface finish due to a formation of thin layer of lubrication at the tool–chip interface. Essentially, during UAT the penetration of lubricant to inaccessible areas of tool–chip contact is increased with the use of self-lubricating tools, thus reducing the frictional forces (Zou et al. 2015).

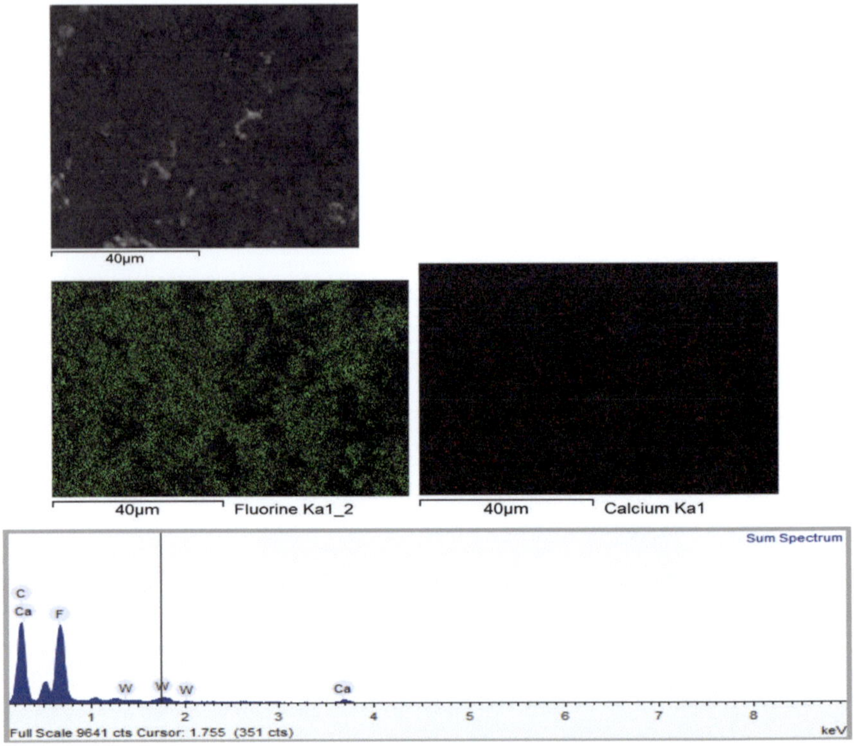

Fig. 17 EDX mapping of textured rake face after 3 min of machining (scale: 40 µm)

Figure 17 shows the EDX map of the top surface of the textured rake face after performing 3 min of cutting under UAT conditions. It can be clearly inferred that Ca and F particles which were earlier embedded into the groove depth hahad now risen above and formed a thin layer of lubrication on rake face. This, in turn, helped to lower the frictional force between tool and chip back surface. This lubrication of the self-lubricating cutting insert helped to attain better surface finish.

In order to study the effect of increase in percentage intensity of ultrasonic power (i.e. amplitude) on surface roughness, experiments were carried out at constant cutting parameters while varying ultrasonic power. Figure 18 shows the variation in surface roughness values (R_a) for UAT process.

An increase in ultrasonic power leads to a decrease in surface roughness. This is due to the fact that with an increase in the amplitude, the non-contact time of tool with the workpiece decreases, which in turn decreases the TWCR and the cutting forces (Gao et al. 2002).

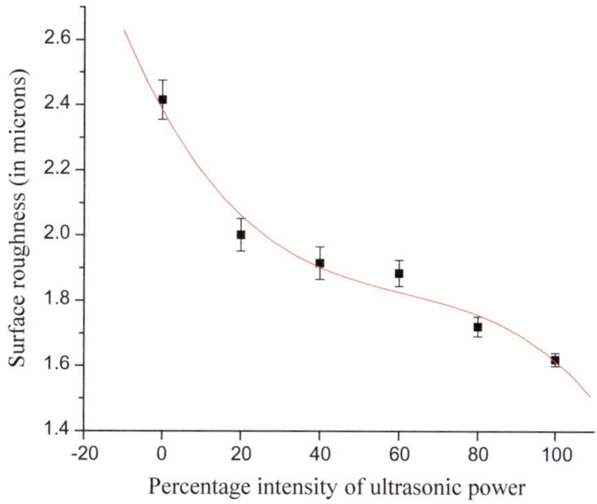

Fig. 18 Surface roughness variation with percentage intensity of ultrasonic power for PUAT process

5.2 *Residual Stress*

Figure 19 shows the 2θ plots as measured by Panalytical X'pert Pro stress software for un-machined specimen (M1), conventionally turned specimen (M2), machined specimen with UAT (M3) and machined specimen with self-lubricating cutting insert during UAT process (M4). A gradual shift in the intensity peaks is observed towards the right from M1 to M4. This clearly signifies that with an increase in 2θ angle there occurs a decrease in interatomic plane distance. This, in turn, helps to generate more compressive residual stress. Under same cutting and vibration conditions, UAT with self-lubricating cutting inserts results in more compressive residual stresses as compared to UAT performed with plain tool.

5.3 *FEM Modelling*

In order to study the surface residual stress, a 2D finite element simulation was carried out. Arbitrary Lagrangian–Eulerian method was used to model the process in a commercially available FE package ABAQUS. The thermo-mechanical properties of material undergoing plastic deformation is assumed to be adequately represented by the Johnson–Cook material model. The constitutive relation of 4340 hardened steel is defined as follows:

$$\bar{\sigma} = \left(A + B\varepsilon_p^n\right)\left[1 + C\ln\left(\frac{\varepsilon}{\varepsilon_0}\right)\right]\left[1 - \left(\frac{T - T_{\text{room}}}{T_{\text{room}} - T_{\text{melt}}}\right)^m\right], \quad (10)$$

where $\bar{\sigma}$ is equivalent stress, *A* denotes initial yield stress, *B* represents hardening modulus, ε_p is plastic strain, ε is reference strain rate, ε_0 denotes operating

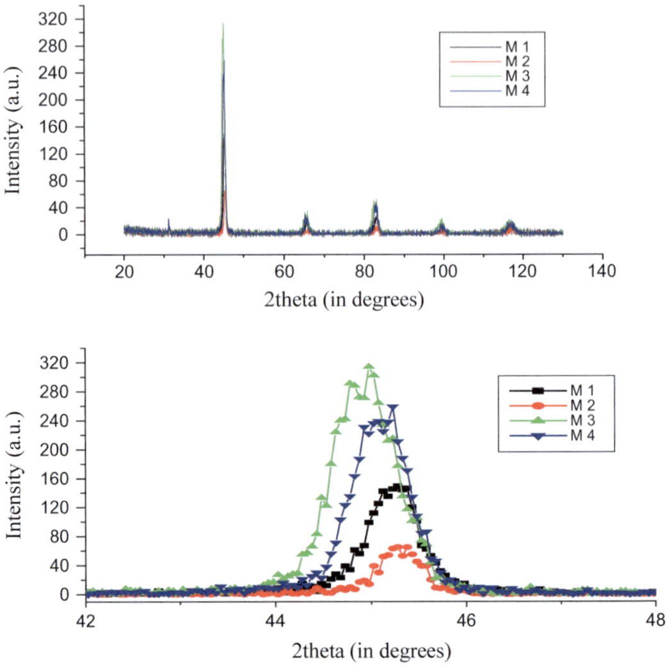

Fig. 19 2θ variation for M1: un-machined, M2: CT, M3: PUAT and M4: TUAT as measured by XRD analysis

temperature, T_{room} is room temperature and T_{melt} denotes melting temperature, C is strain rate coefficient and m is thermal softening coefficient. The Coulomb's law of friction has been used to define the coefficient of friction between tool–chip interface. Four-noded bilinear plane strain quadrilateral (CFE4R) elements with hourglass control and reduced integration were used for meshing the tool and the workpiece. The workpiece was meshed with 12,000 elements with an approximate global size of 0.005 mm, whereas 250 elements with a global size of 0.033 mm were created in the deformable tool. The properties of tool and workpiece used during the simulation process can be found elsewhere (Sharma 2017).

From Fig. 20, it can be inferred that the surface residual stress becomes more compressive in nature during UAT process (c) as compared to CT process (a). Pockets on rake face in a textured tool helps to remove the heat at a faster rate, which results in more compressive residual stress generation in TCT (631.4 MPa) as compared to PCT (899.9 MPa). Moreover, in order to further enhance the fatigue life of the component the superimposition of ultrasonic vibration to textured cutting insert was found to increase compressive residual stress generation in machined specimen (587.1 MPa). The same has been experimentally validated from the shifting of 2θ peaks shown in Fig. 19.

Table 1 presents the experimental conditions for the residual stress measurement during PUAT and TUAT process. It could be inferred that the extent of

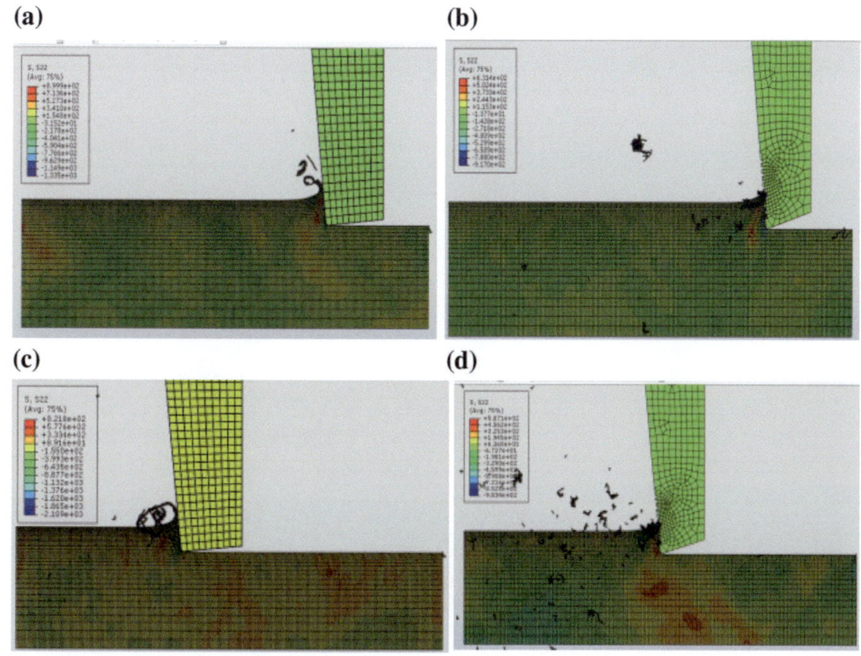

Fig. 20 Residual stress (in MPa) variation in **a** PCT, **b** TCT, **c** PUAT and **d** TUAT

Table 1 Experimental conditions for residual stress measurement

Exp. No.	Depth of cut (mm)	Feed rate (mm/rev)	Cutting speed (m/min)	% intensity of ultrasonic power	Residual stress (MPa)	
					TUAT	PUAT
1.	0.2	0.15	70	40	−123.6	−78.09
2.	0.3	0.20	90	60	256.6	343.00
3.	0.5	0.20	50	60	−239.5	−197.00
4.	0.4	0.25	70	80	−61.7	20.31

compressive residual stress generation was more during TUAT process as compared to PUAT process. The mechanical loading due to UAT combined with cooling effect due to lubrication was used to explain the generation of compressive residual stress in the process.

5.4 *Microhardness*

Microhardness testing was performed to demarcate the machine affected zone (MAZ) in machined specimen during CT and UAT process. LEICA VM HT AUTO

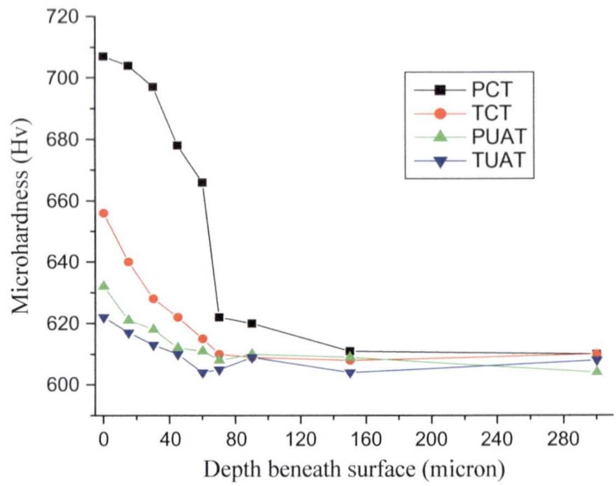

Fig. 21 Microhardness variation with depth of cut (indentation load = 0.1 kg)

was used for measuring the microhardness variation along the material cross section at 0.1 kg of indentation load. The sample preparation was carried out as discussed in Sect. 2. Figure 21 depicts the microhardness variation along the depth from the machined surface.

From Fig. 21, we observe that the material microhardness reaches the hardness of the bulk specimen after a particular depth of the machined specimen. For PCT, the MAZ is about 60 m from the top surface. However, the MAZ decreases to 40 μm for TCT, PUAT and TUAT turned specimen. This clearly signifies that UAT process is more delicate to the machined specimen as reported by Babitsky et al. (2004). However, the average hardness at the top surface for TUAT process (622 Hv) is 12.86% less than the hardness of PCT turned specimen (702 Hv). This is in line with research work of Babitsky et al. (2004) and Silberschmidt et al. (2014) who proposed less deformed grains in UAT specimen as compared to CT process.

6 Conclusions

A comparative study of conventional and ultrasonic assisted turning with plain and textured tools was carried out. The following conclusions are drawn from the study.

- Ultrasonic assisted turning process generates better surface finish when compared to CT process. An increase in the penetrability of lubricant to tool–chip contact area due to vibration results in better surface finish during ultrasonic assisted turning with textured tool. An improvement in surface finish of the order of 47.5% has been found in PUAT process in comparison to PCT process.

- Ultrasonic assisted turning reduces the tensile residual stresses on the surface, which is helpful for enhancing the fatigue life of the machined component. Sometimes, it may produce compressive residual stresses as well.
- The average hardness of the topmost surface of the machined specimen is less in ultrasonic assisted turning than in CT. It has been found that the average hardness in TUAT process is 12.86% less than PCT process.
- Ultrasonic assisted turning with textured tool results in less MAZ than in CT. One can optimize the cutting and vibration parameters to attain minimum cutting forces accompanied with better surface roughness with ultrasonic assisted turning using self-lubricating cutting inserts.
- The present research has successfully attempted the use of textured cutting inserts during UAT process. This has paved the future research to be intended in the direction of sustainable manufacturing by combining different sustainable techniques like minimum quantity lubrication, cryogenic cooling, etc. in conjunction with UAT process.

Acknowledgements Funding from the engineering and physical sciences research council (UK) through grant EP/K028316/1 and department of science and technology (India) through grant DST/RC-UK/14-AM/2012, project modelling of advanced materials for simulation of transformative manufacturing process (MAST), is gratefully acknowledged. The authors would also acknowledge the help provided by Mr. Shantanab Dinda for his help while performing the experimentation.

References

Ahmed, N., A.V. Mitrofanov, V.I. Babitsky, and V.V. Silberschmidt. 2006. Analysis of material response to ultrasonic vibration loading in turning Inconel 718. *Materials Science and Engineering A* 424 (1–2): 318–325.

Babitsky, V.I., A.V. Mitrofanov, and V.V. Silberschmidt. 2004. Ultrasonically assisted turning of aviation materials: Simulations and experimental study. *Ultrasonics* 42: 81–86.

Brehl, D.E., and T.A. Dow. 2008. Review of vibration-assisted machining. *Precision Engineering* 32 (3): 153–172.

Brinksmeier, E., and R. Gläbe. 1999. Elliptical vibration cutting of steel with diamond tools experimental setup. In *LFM laboratory for precision machining*. Germany: University of Bremen.

Dixit, U.S., D.K. Sarma, and J. Paulo Davim. 2012. *Environmentally friendly machining*. New York: Springer.

Gao, G.F., B. Zhao, F. Jiao, and C.S. Liu. 2002. Research on the influence of the cutting conditions on the surface microstructure of ultra-thin wall parts in ultrasonic vibration cutting. *Journal of Materials Processing Technology* 129: 66–70.

Lin, G., and Z.W. Zhong. 2006. Ultrasonic assisted turning of an aluminium-based metal matrix composite. *International Journal of Advanced Manufacturing Technology* 27: 1077–1081.

Maurotto, A., R. Muhammad, A. Roy, V.I. Babitsky, and V.V. Silberschmidt. 2012. Comparing machinability of Ti-15-3-3-3 and Ni-625 alloys in UAT. *Procedia CIRP* 1: 330–335.

Moriwaki, T., and E. Shamoto. 1995. Ultrasonic elliptical vibration cutting. *CIRP Annals—Manufacturing Technology* 44 (1): 31–34.

Muhammad, R., M.S. Hussain, A. Maurotto, C. Siemers, A. Roy, and V.V. Silberschmidt. 2014. Analysis of a free machining titanium alloy using conventional and ultrasonically assisted turning. *Journal of Materials Processing Technology* 214: 906–915.

Nath, C., M. Rahman, and S.S.K. Andrew. 2007. A study on ultrasonic vibration cutting of low alloy steel. *Journal of Materials Processing Technology* 192–193: 159–165.

Nath, C., M. Rahman, and K.S. Neo. 2009. A study on the effect of tool nose radius in ultrasonic elliptical vibration cutting of tungsten carbide. *Journal of Materials Processing Technology* 209 (17): 5830–5836.

Nestler, A., and A. Schubert. 2014. Surface properties in ultrasonic vibration assisted turning of particle reinforced aluminium matrix composites. *Procedia CIRP* 13: 125–130.

Patil, P., S. Joshi, A. Tewar, and S.S. Joshi. 2014. Modelling and simulation of effect of ultrasonic vibrations on machining of Ti6Al4V. *Ultrasonics* 54 (2): 694–705.

Siddiq, A., and T.E. Sayed. 2011. Acoustic softening in metals during ultrasonic assisted deformation via CP-FEM. *Materials Letters* 65: 356–359.

Silberschmidt, V.V., S.M.A. Mahdy, M.A. Gouda, A. Naseer, A. Maurotto, and A. Roy. 2014. Surface-roughness improvement in ultrasonically assisted turning. *Procedia CIRP* 13: 49–54.

Sharma, V., and P.M. Pandey. 2016a. Comparative study of turning of 4340 hardened steel with hybrid textured self-lubricating cutting inserts. *Materials and Manufacturing Processes* 31 (14): 1904–1916. https://doi.org/10.1080/10426914.2015.1127951.

Sharma, V., and P.M. Pandey. 2016b. Geometrical design optimization of hybrid textured self-lubricating cutting inserts for turning 4340 hardened steel. *International Journal of Advanced Manufacturing Technology*. https://doi.org/10.1007/s00170-016-9163-6.

Sharma, V., and P.M. Pandey. 2016c. Study of ultrasonic assisted turning of 4340 steel with plane and self-lubricating cutting inserts. In *Proceedings of manufacturing science and engineering conference ASME MSEC2016 June 27–July 1, 2016, Virginia Tech, Blacksburg, USA*. https://doi.org/10.1115/msec2016-8565.

Sharma, V., and P.M. Pandey. 2016d. Optimization of machining and vibration parameters for residual stresses minimization in ultrasonic assisted turning of 4340 hardened steel. *Ultrasonics* 70: 172–182.

Sharma, V., and P.M. Pandey. 2016e. Recent advances in turning with textured cutting tools: A review. *Journal of Cleaner Production* 137: 701–715.

Sharma, V., P.M. Pandey, A. Roy, U.S. Dixit. 2016. Study of surface integrity in conventional and ultrasonic assisted turning with self-lubricating cutting inserts. In *6th international & 27th all india manufacturing technology, design and research conference (AIMTDR-2016),* December 16–18, 2016 at College of Engineering, Pune, Maharashtra, India.

Sharma, V. 2017. *Experimental investigations and modelling of ultrasonic assisted turning with self-lubricating textured cutting inserts*. Ph.D. thesis, IIT Delhi.

Skelton, R.C. 1969. Effect of ultrasonic vibration on the turning process. *International Journal of Machine Tool Design and Research* 9: 363–374.

Vivekananda, K., G.N. Arka, and S.K. Sahoo. 2014. Design and analysis of ultrasonic vibratory tool (UVT) using FEM, and experimental study on ultrasonic vibration-assisted turning (UAT). *Procedia Engineering* 97: 1178–1186.

Wang, L.J., and J. Zhao. 1987. Influence of surface roughness in turning with ultrasonic vibration tool. *International Journal of Machine Tools Manufacturing* 27: 181–190.

Zou, P., Y. Xu, Y. He, M. Chen, and H. Wu. 2015. Experimental investigation of ultrasonic-vibration assisted turning of 304 austenitic stainless steel. *Shock and Vibration*. https://doi.org/10.1155/2015/817598.

Pulsed Nd:YAG Laser Surface Texturing of Pure Titanium Material

G. Kibria, A. Sen, H. M. Tariq Aziz, B. Doloi and B. Bhattacharyya

Abstract Laser surface texturing is one of the key technologies in micromachining domain for generating defined surface features in micro-components in recent times. In this process, highly focused laser beam is irradiated on the material surface and laser scanning is carried out at different scanning patterns to alter surface conditions for improving the tribological properties of that surface. In this research study, an attempt has been made to carry out laser surface texturing on pure titanium material having a thickness of 1 mm with the aid of a pulsed Nd:YAG laser system. The various process parameters considered for the present research works are laser beam average power, pulse frequency, laser beam scanning speed, and transverse feed. Utilizing response surface methodology, experimentation has been planned. The responses measured are surface roughness, Ra and Rz in lateral and transverse directions and contact angle (θ). The chapter also highlights experimental results for validation of the developed empirical models. The test results have been analyzed using various surface plots. Multi-performance optimization has been conducted to obtain minimum values of surface roughness and contact angle. Using optical microscopic images, influence of process parameters on responses have also been discussed.

Keywords Laser surface texturing · Pulse Nd:YAG laser · Titanium
Surface roughness · RSM · Optimization · Contact angle

G. Kibria (✉) · H. M. Tariq Aziz
Mechanical Engineering Department, Aliah University, Kolkata 700156, India
e-mail: prince_me16@rediffmail.com

H. M. Tariq Aziz
e-mail: hazaritariq@gmail.com

A. Sen · B. Doloi · B. Bhattacharyya
Production Engineering Department, Jadavpur University, Kolkata 700032, India
e-mail: abhishek.sen1986@gmail.com

B. Doloi
e-mail: bdoloionline@rediffmail.com

B. Bhattacharyya
e-mail: bb13@rediffmail.com

S. S. Pande and U. S. Dixit (eds.), *Precision Product-Process Design and Optimization*, Lecture Notes on Multidisciplinary Industrial Engineering,
https://doi.org/10.1007/978-981-10-8767-7_14

1 Introduction

The present era belongs to the utmost demand of minimization of the products with the requirement of stringent quality control. The margin of era for fabricating miniaturized products almost tends to zero. The demand of micro-products in the area of automotive, biomedical, and aerospace engineering is ever increasing. Therefore, a number of micromachining processes have been developed successfully in order to meet those demands (Gentili 2005). Furthermore, several micromachining strategies have also been implemented successfully to manufacture micro-parts with high accuracy, reliability, and durability. With the progress of material science research, various high strength and temperature resisting (HSTR) materials are being developed for successfully implementing them in product manufacturing (Jackson 2015; Ali 2017). Laser beam micromachining processes are being utilized for micromachining of superalloys to advanced newly developed materials to meet the utmost demand from the industries like micro-reactors, electronics, biomedical, chemical and sensors–actuators (Meijer 2004; Chang et al. 2012). The laser micromachining approaches toward generating precise micro-features on the difficult to machine materials have a great potential to generate complicated featured micro-products with several unique characteristics such as high potentiality and effectiveness, exclusion of finishing stages, better quality products, higher material utilization and coverage, material processing regardless of its electrical conductivity, etc. (Mishra and Yadava 2015; Chavoshi and Luo 2015). In laser micromachining processes, highly intensive laser beam is transmitted on desired location of surface of material to simultaneously melt and vaporize the material. By utilizing assisted air or gas pressure, the melted and/or vaporized materials are removed from the focused zone creating crater on the surface (Kibria 2015).

Surface texturing using high intense laser beam is a promising advanced micromachining technology for the generation of defined surface features in engineering materials to improve the tribological performance of metallic and ceramic materials (Cunha 2016; Dunn et al. 2015; Toyserkani 2014). This process is a viable technology for surface engineering that can produce significant improvement of engineering components in various aspects such as wear resistance, wetting characteristics, friction coefficients, load capacities, and part lubrication. In laser texturing process, laser beam is focused and exposed on the workpiece surface with defined patterns by relatively moving the workpiece in X and Y directions. Thus, laser beam scanning was applied on the surface with overlaps between two consecutive laser scan tracks. The amount of this overlap is controlled by the value of transverse feed. In this way, laser surface texturing is carried out on workpiece surface to obtain square or rectangular type scan area. With the change in the number of scan passes during laser surface texturing process, a specific depth as well as surface features can be achieved. In laser surface texturing process, the controllable process variables such as laser beam power, pulse frequency or pulse repetition rate, scanning speed, etc., have great effect to obtain quality and defined surface features. These process parameters are dynamic and most of the performance criteria are expressively connected with instantaneous distinction of parameters during processing of advanced engineering materials.

Titanium is an attractive material for numerous industries such as aerospace, biomedical, sports, architecture, etc., for its diversified properties such as extremely tough, lightweight, high strength to weight ratio, corrosion resistance, biological compatibility, low density, low coefficient of thermal expansion, stability at elevated temperature, etc. Pure titanium is one of the biologically inert biomaterials as it exhibits property of high biocompatibility with tissue and blood. Due to this, this material is largely used for fabricating biomedical implants. In the literature, limited research work has been performed by various researchers and academia in the area of laser surface texturing on different engineering materials using various types of laser. Cunha (2016) carried out research on surface texturing process by means of direct laser writing using ultrashort laser beam on titanium and Ti–6Al–4V alloy. A wide range of created textures like laser-induced periodic surface structures (LIPSS), nanorange pillars and columns and composite type textures have been studied by the authors. The researchers also found that these specific generated textures have an accumulation amount of ability of governing the wettability performance for various biological fluids and other fluids on the material surface. Kovalchenko et al. (2004) carried out tribological experiments on pin-on-disk assessment device at a sliding speed of 0.15–0.75 m/s and minimal interaction pressure of 0.16–1.6 MPa. The authors found that the surface texturing process using laser prolonged the range of speed-load factors for hydrodynamic lubrication. Li et al. (2016) established the manufacture of steady super-hydrophilic and super-hydrophobic textured titanium surface using femtosecond pulsed laser. In scanning strategy by laser in two perpendicular orders enclosed by nanometer-sized ripples, are directed for the development of spikes of micrometer size. The results show that textured hydrophilic or super-hydrophilic Ti surfaces became high hydrophobicity or super-hydrophobicity after salinization. In Fig. 1, the laser-textured surface of Ti material machined at laser fluence of 1.5 J/cm^2 is shown at two different magnifications (500× an 50000×). Soveja et al. (2008) investigated the effect of working factors on laser texturing by means of two investigational design approach (Taguchi and RSM methodology) on Ti–6Al–4V alloy. The authors found that pulse frequency and pulse energy are the most significant factors. The results also reveal that MRR is directly proportional to laser pulse energy and pulse frequency. Moreover, it was seen that these values of surface roughness were inversely relational to these process parameters. Tripathi et al. (2015) studied the influence of laser surface texturing process on friction and wear behavior of graphite cast iron. In this work, authors successfully produced dimples of dimension of diameter 60 μm and depth 30 μm with a dimple pitch of 80–200 μm.

Velasquez (2013) investigated the effect of surface texturing on the surgical blade by developing micro-dimples of diameter of 110 μm and depth of 30 μm on the surface of cutting edge of blades (surgical) by a picosecond pulse Nd:YVO4 laser. A larger friction reduction is observed by production of an arrangement of dimples with combined rims on a side of blade cutting edge in comparison with

Fig. 1 SEM images of laser texturing Ti surface at **a** 500× and **b** 50000× magnifications machined at laser fluence of 1.5 J/cm^2 (Li et al. 2016)

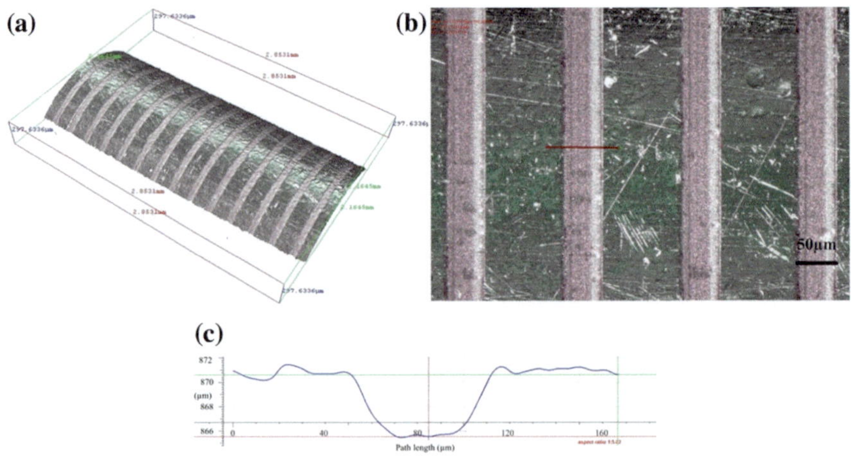

Fig. 2 3D and 2D profiles of machined micro-channels at axial overlap of 0.80, circumferential overlap of 0.88 and overscan number of 2 (Wang et al. 2015)

production of these dimples at both sides. Wang et al. (2015) carried out investigational analysis to study the effect of overlap and overscan number and found that both the factors have significant effects on results of depth achieved and surface roughness values. It was found that machined depth has a linear relationship with the axial overlap and circumferential overlap. The results also reveal that the overscan number is mostly effective factor for machined depth compared to the overlap factors. The 3D and 2D scan profiles of the micro-channel machined surface are shown in Fig. 2 at axial overlap of 0.80, circumferential overlap of 0.88 and overscan number of 2. Yang et al. (2016) studied the modification of wettability on titanium material. The results show that accumulation of carbon on the surface is the key factor to improve surface hydrophobicity. The SEM images of three type of

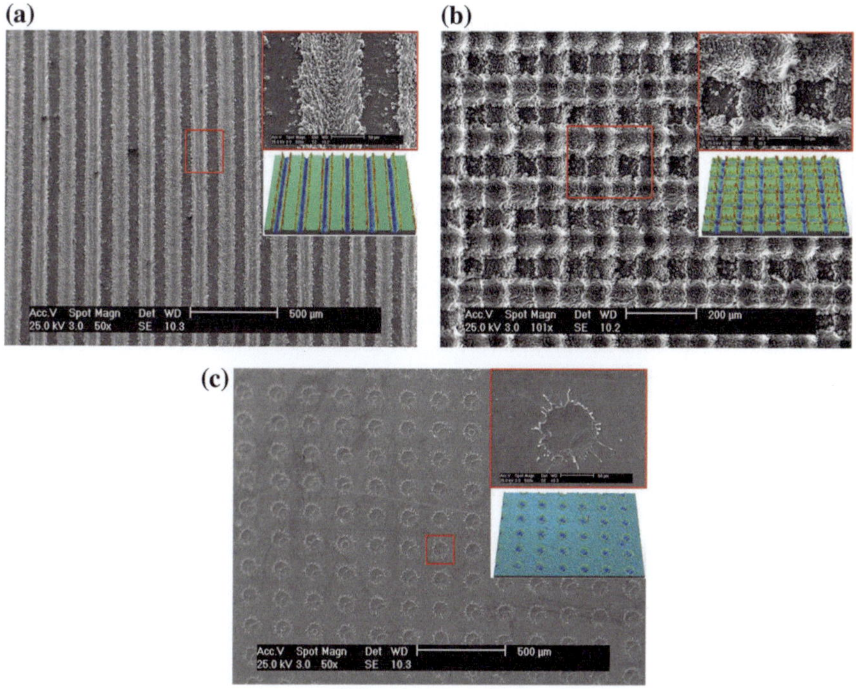

Fig. 3 SEM images of standard microstructures **a** line pattern **b** grid pattern and **c** spot pattern under the irradiation of laser pulses (Yang et al. 2016)

patterns such as line, grid, and spot patterns are shown in Fig. 3a–c. From the literature review, it was found that very few experimental researches have been carried out in order to find out the influencing process variables on surface characteristics and the effect of overlap factors on surface conditions during laser surface texturing process. Furthermore, as the process involves a number of process parameters, the selection of suitable process parameter setting is a very difficult task for achieving the desired surface characteristics. Therefore, to meet the micro-manufacturing demands and also to explore the influences of different process parameters on surface characteristics, in the present chapter, attempts have been made to conduct experiments in laser surface texturing process on titanium material to analyze the effect of laser average power, laser beam scanning speed, pulse frequency, and transverse feed on the response criteria. The research work also attempts to develop empirical models for responses through response surface methodology to analyze the effect of aforesaid process inputs. Finally, optimization has been carried out to develop optimal parametric combinations to achieve desired surface characteristics.

2 Laser Surface Texturing

One of the key technologies for manufacturing of defined and patterned surface microstructures is laser surface texturing (LST). This process renders tribological characteristics on the surface of work material to improve load capacity, wear rate as well as to reduce coefficient of friction of the surface. In this process, relative movements between the laser beam and work surface is given at different patterns to create patterned microstructures such as crossed grooves, linear grooves, and dimple-shaped impressions. In laser texturing process, lateral overlap and transverse overlap are the major criteria for achieving qualitative and defined surface. Therefore, the basic understanding of lateral overlap and transverse overlap of laser beam scanning is very much needed. The schematic representation of laser beam scan pattern with two overlaps is viewed in Fig. 4. The schematic representation also shows the Gaussian energy distributions of laser spots as well as the craters generated by these laser spots. The amount of lateral overlap and transverse overlap should be chosen very carefully as these overlaps depend upon the various parametric settings of process parameters. The mathematical relationship between various factors and the abovementioned overlaps can be calculated as follows (Vora and Dahotre 2015):

$$\text{Lateral overlap } (O_{\mathrm{L}}) = \frac{D - D_{\mathrm{L}}}{D} \times 100\% \tag{1}$$

$$\text{Transverse overlap } (O_{\mathrm{T}}) = \frac{D - D_{\mathrm{T}}}{D} \times 100\% \tag{2}$$

Here, D is the laser beam spot diameter at focused zone in mm, D_{L} ($D_{\mathrm{L}} = V_{\mathrm{in}}/f$) is lateral overlap distance in mm, V_{in} is the laser beam scanning speed in mm/s, f corresponds to pulse frequency in Hz, and D_{T} is the transverse overlap distance in mm. According to Eq. (1), it is very clear that by increasing the pulse frequency or

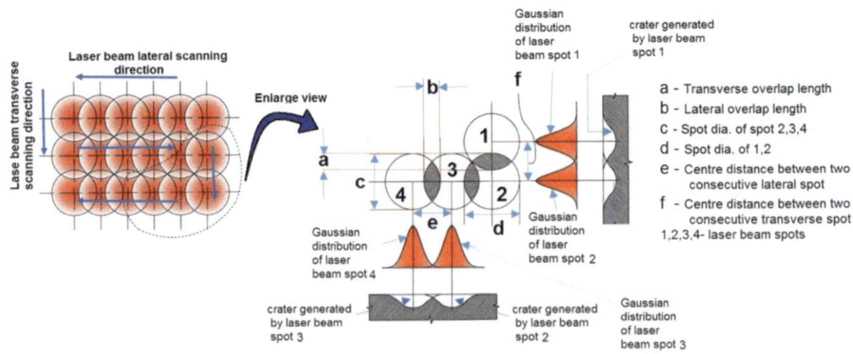

Fig. 4 Schematic view of laser scanning strategy showing lateral and transverse overlap and Gaussian energy distribution pattern for laser spots

by reducing the laser beam scanning speed, the lateral overlap area can be incremented. Equation (2) depicts the relation of transverse overlap with transverse feed. It is theoretically very clear than the transverse overlap area, i.e., overlap of two consecutive laser spot in transverse direction, can be increased by decreasing the transverse feed or reducing the center distance between two consecutive spots.

3 Materials, Methods, and Machining Conditions

All the experimentation were carried out in CNC controlled pulse Nd:YAG laser beam micromachining system of average power 50 W (make: M/S Sahajanand Laser Technology Limited, Pune, India). In the present experimental investigation, the process parameters, which have been selected for conducting laser surface texturing process of pure titanium, are average power of laser, pulse frequency, laser beam scanning speed, and transverse feed. The list of process parameters with actual and coded values is shown in Table 1. Finally, a uniform rotatable central composite experimental design based on response surface methodology (RSM) for modeling of laser surface texturing process has been carried out and listed in Table 2. The details of machine setup specifications and other parameters which are kept constant during the present experimentation are shown in Table 3. Each of the four process parameters having four levels requires a total of 31 numbers of experiments. All the experiments have been performed randomly, i.e., without following the order mentioned in the first column. Considering the ranges of process parameters taken in this research study, the values of lateral overlap and transverse overlap have been calculated using Eqs. (1) and (2). In Fig. 5, the variation of lateral overlap with laser scanning speed at various pulse frequency values is shown. From this figure, it is revealed that the values of lateral overlap are more than 95%, which ensures that quality surface can be achieved by laser surface texturing process. The variation of transverse overlap with transverse feed is shown in Fig. 6. The calculated values of transverse overlap are in the range of 50–90%. This wide range of overlap has been considered to study and analyze the effect of these overlaps on surface characteristics to be obtained. Flat titanium workpiece of

Table 1 Process parameters and various levels of laser surface texturing experimentation

Parameters	Symbol	Unit	Levels				
			−2	−1	0	+1	+2
Average power	X1	W	4	5.5	7	8.5	10
Pulse frequency	X2	Hz	600	800	1000	1200	1400
Scanning speed	X3	mm/s	1	2	3	4	5
Transverse feed	X4	mm	0.01	0.02	0.03	0.04	0.05

Table 2 Experimental plan with various values of process parameters

Expt. No.	Average power (W)	Pulse frequency (Hz)	Scanning speed (mm/s)	Transverse feed (mm)
1	7.0	1000	3	0.03
2	8.5	1200	4	0.04
3	8.5	800	2	0.04
4	5.5	800	4	0.02
5	8.5	800	4	0.04
6	5.5	1200	2	0.02
7	8.5	800	2	0.02
8	8.5	1200	2	0.02
9	7.0	1000	3	0.03
10	5.5	1200	4	0.04
11	5.5	800	2	0.04
12	7.0	1000	1	0.03
13	10.0	1000	3	0.03
14	7.0	1000	3	0.05
15	8.5	1200	2	0.04
16	7.0	1000	3	0.01
17	7.0	1000	3	0.03
18	7.0	1000	3	0.03
19	5.5	1200	2	0.04
20	5.5	1200	4	0.02
21	7.0	1000	5	0.03
22	7.0	600	3	0.03
23	7.0	1000	3	0.03
24	4.0	1000	3	0.03
25	7.0	1000	3	0.03
26	8.5	800	4	0.02
27	7.0	1000	3	0.03
28	5.5	800	4	0.04
29	7.0	1400	3	0.03
30	8.5	1200	4	0.02
31	5.5	800	2	0.02

dimension 50 mm × 50 mm × 1 mm is considered for carrying out the experimentation. After conducting experiments, response criteria surface roughness (Ra and Rz) both in lateral and transverse directions and wettability (in terms of contact angle) were measured. Surface roughness values were measured using roughness measuring instruments, Mitutoyo Surftest SJ-210. During surface roughness evaluation, the cutoff length was taken as 0.8 mm and total length of measurement was 5 mm. The values of surface roughness along the laser scan line (lateral direction) are termed as surface roughness (Ra and Rz) in lateral direction whereas the values

Table 3 Experimental conditions for laser surface texturing

Condition	Description
Laser type	Nd:YAG laser (pulsed)
Wavelength	1064 nm
Polarization of beam	Random
Type of Q-switch	Q-switched (pulsed)
Mode of operation	Fundamental mode (TEM_{00})
Mode of laser beam	Acousto optic
Pulse width, % of duty cycle	Variable
Mirror reflectivity	Rear mirror 100% and front mirror 80%
Air pressure (kgf/cm^2)	1.3
Z feed rate (mm/s)	0.01

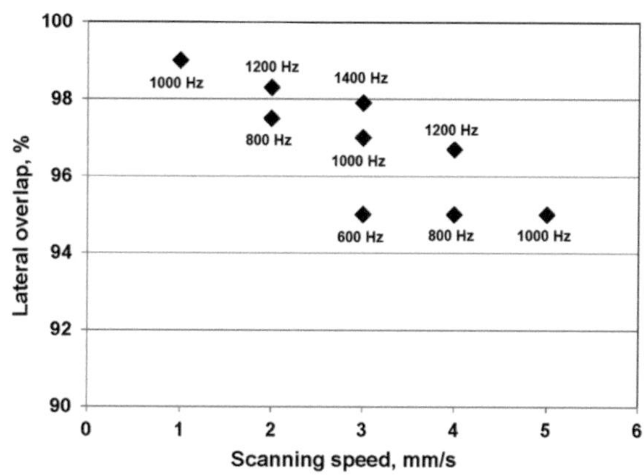

Fig. 5 Variation of lateral overlap with laser scanning speed at various pulse frequency values

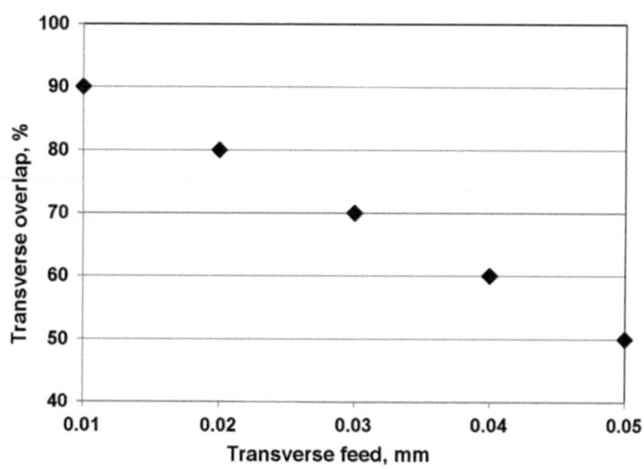

Fig. 6 Variation of transverse overlap with transverse feed

of surface roughness obtained along the perpendicular of lateral direction (i.e., transverse direction) are termed as surface roughness (Ra and Rz) in transverse direction. Wettability is another response criterion that was considered in the present experimentation. Wettability is the effect of variation of contact angle on the surface that decides whether the liquid drop over the surface is able to wet the surface or not. As this is a direct function of contact angle (θ), therefore, wettability is calculated by the angle of contact of the liquid resting on the surface. If two different fluids make contact with the surface of a solid, the configuration of balance of these two different fluid segments (say air and water) relies on the comparative value of the surface tension among every pair of these three phases. Each surface tension executes toward its respective interface, and defines the angle (θ) at which the liquid contacts the surface. This is known as contact angle. The contact angle is generally measured in the liquid side. When the contact angles are >90°, the wettability characteristics of the liquid on the surface are very low, i.e., hydrophobic. At the same time, when contact angles are <90°, the wettability characteristics are very high on the surface, i.e., hydrophilic. Even a contact angle of more than 150° shows super-hydrophobicity. Wettability of the machined surface was observed by measuring contact angle of pre-quantified distilled water drop. The schematic view of measuring the contact angle using water droplet is represented in Fig. 7. The wettability of the textured surface in terms of contact angle, angle measuring setup is indigenously developed and the schematic representation of this setup is viewed in Fig. 8. For each water drop on the machined surface, contact angle is measured twice at both ends of water drop and for analysis; the average value of these two angles is taken. MINITAB statistical software of version 16.1 is used to analyze the responses.

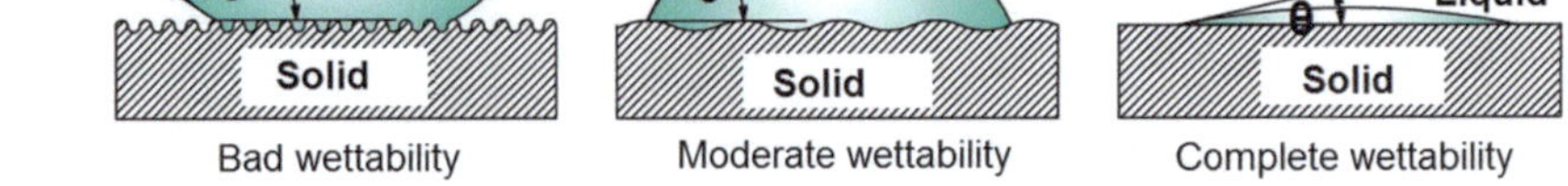

Fig. 7 Schematic of contact angle of liquid drops in contact with solid surfaces (hydrophobic and hydrophilic)

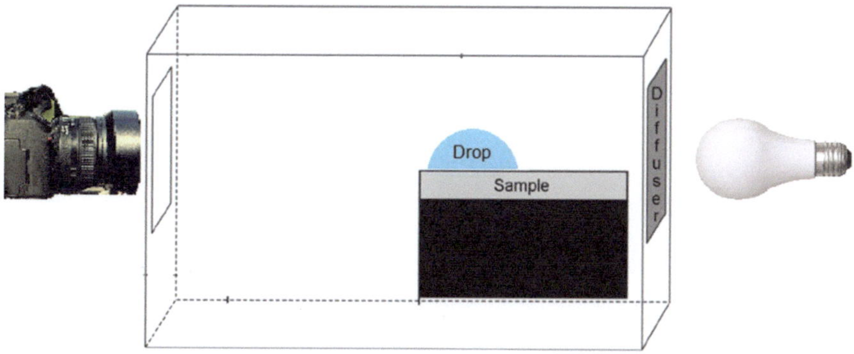

Fig. 8 Schematic view of indigenously developed contact angle measuring arrangement

4 Results and Discussion

Based on the results of laser surface texturing on pure titanium material, various statistical analyses were conducted to validate the developed mathematical models of responses. Moreover, various surface plots were analyzed to examine the parametric influences of process parameters on performance criteria.

4.1 Development of Predictive Models and ANOVA Test

According to the results of response criteria, i.e., surface roughness (Ra and Rz) and contact angle achieved as shown in Table 4, mathematical models of responses have been established to develop the relationship between various process parameters and performance criteria. The details of developed mathematical models of responses are shown below.

$$\begin{aligned} Y_{\mathrm{RaL}} = & -59.44 + 10.35 \times X_1 + 0.05 \times X_2 + 5.83 \times X_3 + 262.35 \times X_4 \\ & - 0.55 \times X_1^2 - 1.25 \times X_3^2 - 7203.72 \times X_4^2 + 0.08 \times X_1 X_3 \\ & - 28.54 \times X_1 X_4 + 0.20 \times X_2 X_4 + 29.94 \times X_3 X_4 \end{aligned} \quad (3)$$

$$\begin{aligned} Y_{\mathrm{RaT}} = & -24.08 + 6.79 \times X_1 + 0.01 \times X_2 + 4.13 \times X_3 \\ & - 47.01 \times X_4 - 0.36 \times X_1^2 - 0.71 \times X_3^2 - 3416.52 \times X_4^2 \\ & - 0.23 \times X_1 X_3 + 5.96 \times X_1 X_4 + 0.10 \times X_2 X_4 + 34.31 \times X_3 X_4 \end{aligned} \quad (4)$$

$$\begin{aligned} Y_{\mathrm{RzL}} = & -313.5 + 50.4 \times X_1 + 0.3 \times X_2 + 34.4 \times X_3 \\ & + 1604.7 \times X_4 - 2.3 \times X_1^2 - 6.1 \times X_3^2 - 38642.1 \times X_4^2 \\ & - 0.1 \times X_1 X_3 - 143.0 \times X_1 X_4 + 1.2 \times X_2 X_4 + 28.1 \times X_3 X_4 \end{aligned} \quad (5)$$

Table 4 Experimental results of a total 31 experiments

Expt. No.	Responses				
	Surface roughness (lateral), Ra (μm)	Surface roughness (transverse), Ra (μm)	Surface roughness (lateral), Rz (μm)	Surface roughness (transverse), Rz (μm)	Contact angle, θ (degree)
1	11.75	6.91	61.06	37.49	52.60
2	5.02	4.99	29.86	25.60	115.10
3	6.98	7.11	42.46	35.99	76.55
4	8.68	6.32	47.05	33.03	25.00
5	11.50	8.25	59.66	41.66	102.00
6	3.68	3.62	21.54	20.11	15.10
7	13.66	9.91	72.54	48.15	67.30
8	7.32	4.56	40.50	23.55	30.30
9	12.70	7.73	65.42	38.95	53.30
10	4.91	4.55	31.62	23.80	132.45
11	7.53	5.22	40.15	25.63	82.40
12	6.07	4.86	36.60	29.20	38.90
13	12.06	7.35	72.18	37.43	26.70
14	9.84	7.41	53.13	34.30	75.15
15	5.13	5.32	33.60	31.14	31.40
16	10.23	7.57	53.91	35.80	25.90
17	11.10	7.62	55.45	39.12	71.55
18	10.25	8.62	60.59	40.54	51.25
19	3.57	3.58	24.16	18.48	95.95
20	6.25	5.32	40.50	27.22	73.20
21	9.73	7.18	52.22	39.17	117.85
22	10.64	9.53	57.04	45.85	48.65
23	11.93	8.18	68.73	42.35	55.10
24	3.82	3.95	24.27	20.94	80.50
25	13.86	9.65	74.94	46.47	22.10
26	12.88	7.07	72.90	39.37	26.60
27	14.64	9.62	75.54	48.81	31.10
28	6.97	7.08	38.36	35.41	82.70
29	5.70	5.82	32.43	28.82	91.90
30	7.50	4.70	42.53	25.50	32.20
31	10.25	7.17	53.76	34.08	21.30

$$\begin{aligned} Y_{\mathrm{RzT}} = & -135.2 + 32.5 \times X_1 + 0.1 \times X_2 + 20.6 \times X_3 \\ & + 166.8 \times X_4 - 1.7 \times X_1^2 - 2.5 \times X_3^2 - 22934.8 \times X_4^2 \\ & - 1.2 \times X_1 X_3 + 37.2 \times X_1 X_4 + 0.6 \times X_2 X_4 + 100.0 \times X_3 X_4 \end{aligned} \quad (6)$$

$$
\begin{aligned}
Y_\theta = {} & 103.38 + 37.30 \times X_1 - 0.20 \times X_2 - 106.27 \times X_3 \\
& + 1525.91 \times X_4 + 0.60 \times X_1^2 + 7.54 \times X_3^2 + 5793.90 \times X_4^2 \\
& - 0.04 \times X_1 X_2 - 1.18 \times X_1 X_3 - 376.04 \times X_1 X_4 \\
& + 0.6 \times X_2 X_3 + 0.65 \times X_2 X_4 + 768.44 \times X_3 X_4
\end{aligned} \tag{7}
$$

Here, X_1, X_2, X_3, and X_4 are the actual values of laser beam average power, scanning speed, pulse frequency, and transverse feed, correspondingly. The analysis of variance (ANOVA) test for these four empirical models was performed and the calculated F-values are shown in Tables 5, 6, 7, 8 and 9 for surface roughness (lateral and transverse), Ra and Rz as well as contact angle (θ). The typical F-value of lack-of-fit is 4.06 for confidence level of 95% as taken from F distribution table. On the other hand, the evaluated F-values for surface roughness (lateral), surface roughness (transverse), and contact angle are 1.43, 0.97, 1.51, 1.11, and 1.66, respectively, at 95% confidence level. These calculated F-values are obviously lower than the typical F-value. This entails that the RSM-based empirical models are satisfactory at 95% confidence level and valid in the ranges of process parameters. The p-values of lack-of-fit for these five response criteria are 0.344, 0.542, 0.317, 0.471, and 0.276, which indicate the models adequately fit. Based on the developed empirical models of responses, further a total of eight validation

Table 5 Results of analysis of variance (ANOVA) for surface roughness (lateral), Ra

Source	DF	Seq SS	Adj SS	Adj MS	*F*	*P*
Regression	14	277.304	277.304	19.8074	6.63	0.000
Linear	4	156.766	47.670	11.9175	3.99	0.020
Square	4	111.445	111.445	27.8613	9.33	0.000
Interaction	6	9.093	9.093	1.5154	0.51	0.794
Residual error	16	47.800	47.800	2.9875		
Lack-of-fit	10	33.651	33.651	3.3651	1.43	0.344
Pure error	6	14.149	14.149	2.3581		
Total	30	325.104				

Table 6 Results of analysis of variance (ANOVA) for surface roughness (transverse), Ra

Source	DF	Seq SS	Adj SS	Adj MS	*F*	*P*
Regression	14	84.828	84.8277	6.0591	5.80	0.001
Linear	4	47.363	20.9190	5.2298	5.01	0.008
Square	4	31.280	31.2797	7.8199	7.49	0.001
Interaction	6	6.185	6.1852	1.0309	0.99	0.466
Residual error	16	16.714	16.7143	1.0446		
Lack-of-fit	10	10.321	10.3207	1.0321	0.97	0.542
Pure error	6	3.394	6.3935	1.0656		
Total	30	101.542				

Table 7 Results of analysis of variance (ANOVA) for surface roughness (lateral), Rz

Source	DF	Seq SS	Adj SS	Adj MS	*F*	*P*
Regression	14	6796.22	6796.22	485.44	6.42	0.000
Linear	4	3953.21	1207.27	301.82	3.99	0.020
Square	4	2555.70	2555.70	638.93	8.45	0.001
Interaction	6	287.31	287.31	47.88	0.63	0.702
Residual error	16	1209.48	1209.48	75.59		
Lack-of-fit	10	865.79	865.79	86.58	1.51	0.317
Pure error	6	343.69	343.69	57.28		
Total	30	8005.70				

Table 8 Results of analysis of variance (ANOVA) for surface roughness (transverse), Rz

Source	DF	Seq SS	Adj SS	Adj MS	*F*	*P*
Regression	14	1858.63	1858.63	132.759	7.01	0.000
Linear	4	1095.62	457.70	114.424	6.04	0.004
Square	4	644.14	644.14	161.036	8.50	0.001
Interaction	6	118.87	118.87	19.812	1.05	0.433
Residual error	16	302.97	302.97	18.936		
Lack-of-fit	10	196.43	196.43	19.643	1.11	0.471
Pure error	6	106.54	106.54	17.756		
Total	30	2161.60				

Table 9 Results of analysis of variance (ANOVA) for contact angle, θ

Source	DF	Seq SS	Adj SS	Adj MS	*F*	*P*
Regression	14	25543.9	25543.9	1824.56	4.78	0.002
Linear	4	17659.4	3886.7	971.69	2.55	0.080
Square	4	2283.4	2283.4	570.86	1.50	0.250
Interaction	6	5601.0	5601.0	933.50	2.45	0.072
Residual error	16	6106.2	6106.2	381.63		
Lack-of-fit	10	4485.1	4485.1	448.51	1.66	0.276
Pure error	6	1621.1	1621.1	270.18		
Total	30	31650.0				

experiments are carried out within the range of the process parameters considered in the present experimental work.

In Table 10, the process parametric settings considered for verification experiments are enlisted along with experimental result of responses. The comparative results between the experimental results and the RSM predicted results are shown in Table 11. The prediction errors have been calculated according to the following equation:

Table 10 Results of responses of verification experiments

Expt. No.	Process parameters				Responses obtained from experiments				
	Average power (W)	Pulse frequency (Hz)	Scanning speed (mm/s)	Transverse feed (mm)	Surface roughness (μm)				Contact angle (θ)
					Ra (L)	Ra (T)	Rz (L)	Rz (T)	
1	7	800	2	0.02	6.12	7.25	38.26	21.28	109.12
2	5.5	900	4	0.04	5.23	6.86	42.15	38.52	76.29
3	8.5	1100	2.5	0.01	5.75	9.56	53.46	42.15	54.28
4	5	1300	3.5	0.05	9.86	8.62	68.54	25.14	113.81
5	8	1200	2.5	0.02	4.62	7.91	72.13	37.29	80.18
6	9	950	3	0.03	8.25	5.26	35.16	37.21	95.25
7	6	700	4	0.01	8.46	4.85	47.91	47.29	76.15
8	8	1050	4.5	0.02	5.92	6.34	52.38	36.21	82.65

$$\text{Prediction error } (\%) = \frac{\text{Experimental results} - \text{Predicted results}}{\text{Experimental results}} \times 100\% \quad (8)$$

As shown in the Table 11, the predicted average percentage errors of surface roughness (lateral), surface roughness (transverse), and contact angle are 3.34, 3.19, 3.08, 3.10, and 3.19%, respectively. The calculated overall percentage of prediction error considering the aforesaid responses is 3.18%. As the prediction errors are below 5%, the result can be considered within acceptable range. In addition to this, the predicted values of responses are closed to experimental results, which ensure that RSM models developed can predict response results moderately acceptable.

4.2 *Influence of Process Parameters on Surface Criteria*

The various response surfaces have been plotted between one process criteria and two process parameters at a time. The parametric analyses of surface roughness (Ra and Rz) and contact angle with respect to process variables such as pulse frequency, average power, scanning speed, and transverse feed have been discussed hereunder.

4.2.1 Parametric Effects on Surface Roughness (Ra)

The influence of surface roughness (Ra and Rz) in lateral and transverse directions has been analyzed with the aid of various response surface plots. In Fig. 9, the combined effect laser beam average power and pulse frequency on surface roughness (Ra) in lateral direction is depicted when laser beam scanning speed and transverse feed are kept as constant at 3 mm/s and 0.03 mm, respectively. The surface plot reveals that an increment in laser beam average power, the surface roughness (lateral) values are increasing at all pulse frequency settings. This is due

Table 11 Results of verification experiments between the experiments and RSM prediction

Expt. No.	Experimental results					RSM predicted results					Prediction error (%)				
	Ra (L)	Ra (T)	Rz (L)	Rz (T)	(θ)	Ra (L)	Ra (T)	Rz (L)	Rz (T)	(θ)	Ra (L)	Ra (T)	Rz (L)	Rz (T)	(θ)
	(in μm)				(in degree)	(in μm)				(in degree)					
1	6.12	7.25	38.26	21.28	109.12	5.88	7.03	37.05	20.77	106.51	3.87	2.93	3.16	2.35	2.39
2	5.23	6.86	42.15	38.52	76.29	5.09	6.63	40.63	36.97	73.741	2.65	3.25	3.59	4.01	3.34
3	5.75	9.56	53.46	42.15	54.28	5.56	9.21	52.13	40.50	52.564	3.14	3.62	2.48	3.91	3.16
4	9.86	8.62	68.54	25.14	113.81	9.53	8.39	66.16	24.45	110.88	3.26	2.65	3.47	2.73	2.57
5	4.62	7.91	72.13	37.29	80.18	4.46	7.63	70.09	36.33	77.381	3.29	3.45	2.82	2.56	3.49
6	8.25	5.26	35.16	37.21	95.25	7.94	5.11	34.04	35.91	90.716	3.69	2.69	3.16	3.47	4.76
7	8.46	4.85	47.91	47.29	76.15	8.19	4.65	46.54	46.28	73.728	3.15	4.01	2.84	2.13	3.18
8	5.92	6.34	52.38	36.21	82.65	5.70	6.15	50.72	34.87	80.426	3.69	2.96	3.16	3.69	2.69
Average prediction error											3.34	3.19	3.08	3.10	3.19
General prediction error											3.18				

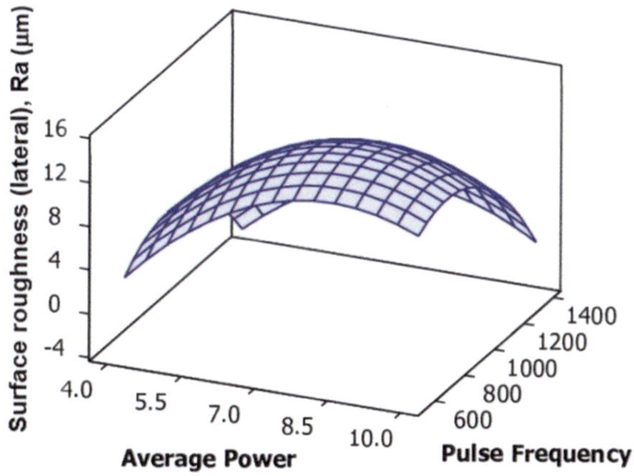

Fig. 9 3D surface plots of surface roughness (lateral), Ra against of average power and pulse frequency

to the fact that with increase of laser beam average power, the peak power of the beam is also increasing according to following equation (Kibria et al. 2012; Thawari et al. 2005):

$$\text{Peak power}\ (P_{\text{p}}) = \frac{\text{Average power}\ (P_{\text{A}})}{\text{Pulse frequency}\ (F_{\text{p}}) \times \text{Pulse duration}\ (\mu)} \tag{9}$$

Thus, the material from the laser irradiated zone gets adequate energy to melt and evaporate instantly and as a result, the machined surface has a lot of irregularities. According to Eq. (1), the increase of pulse frequency for any change in laser beam scanning speed results in increase of lateral overlap percentage. Thus, the altitude of the micro-peaks on the generated surface is less, which further decreases the surface roughness (lateral) value. It is also seen from this figure that increment in pulse frequency first leads to increasing of the surface roughness (lateral) values and later reversed back. It is because, when pulse frequency is increasing, the laser beam interaction time with material surface is also increasing subsequently and thus, the crater size is also increasing. Consequently, the roughness values are increasing.

Figure 10 illustrates the influence of laser scanning speed and transverse feed on surface roughness (lateral) values when other process variables remained constant as beam average power at 7 W and laser pulse frequency at 1000 Hz. It is observed from the figure that with increase of laser beam scanning speed, initially roughness values have a tendency to increase and then follows a gradual fall. With increasing values of scanning speed, as the spot overlap values decreases, therefore, the roughness is increasing. At high scanning speed values, the laser beam and material interaction time decreases. As a result of this, slight quantity of work material is removed by laser beam irradiation and the surface has less irregularities. The same plot also depicts that with the increase of transverse feed, the values of surface roughness (lateral) are also increasing. This phenomenon is observed by the increase of transverse feed, where the values of transverse overlap between two

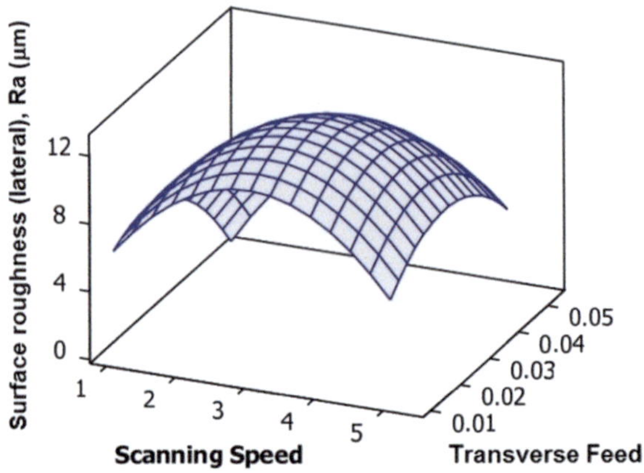

Fig. 10 3D surface plots of surface roughness (lateral), Ra against of scanning speed and transverse feed

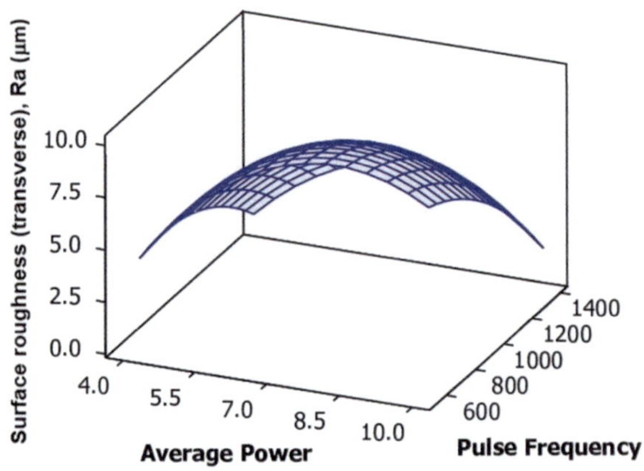

Fig. 11 3D surface plots of surface roughness (transverse), Ra against of average power and pulse frequency

laser scanning track decrease. Therefore, the machined surface has a lot of irregularities which ultimately increases the roughness.

Figure 11 shows the influence of laser beam average power and pulse frequency on surface roughness (transverse), Ra when other process parameters, i.e., scanning speed and transverse feed were kept constant at a value 3 mm/s and 0.03 mm, correspondingly. The figure reveals that surface roughness (transverse) shows an increasing trend with the variation of laser average power. This is due to the fact that when the average power is increasing, the peak power is also increasing according to Eq. (9). Thus, the material from the laser irradiated zone finds suitable thermal energy for melting and evaporation instantly and as a result, the machined surface has a lot of irregularities. The same plot also indicates that the surface roughness (transverse) values reduced with the variation of pulse frequency. This trend can be understood by Eq. (9) that shows with increasing pulse frequency, the peak power decreases, so the energy associated with the laser beam at the laser irradiated zone also decreases and less amount of metal is melted and vaporized

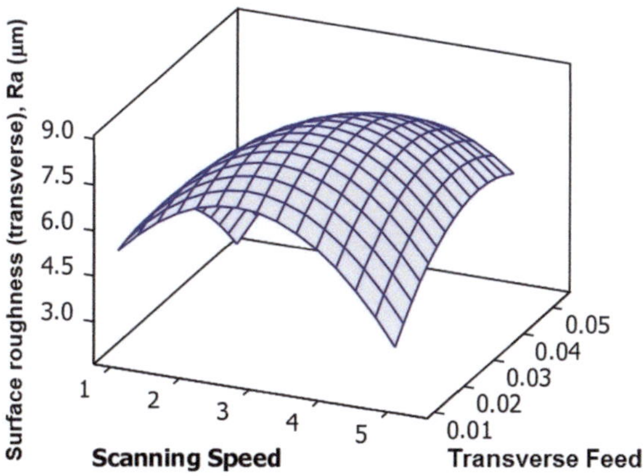

Fig. 12 3D surface plots of surface roughness (transverse), Ra against of scanning speed and transverse feed

from the zone. Therefore, irregularities with lower crater size are formed and the surface roughness value decreases.

Figure 12 shows the effects of surface roughness (transverse), Ra with the variation of scanning speed and transverse feed when the other two process parameters, i.e., laser beam average power and pulse frequency are kept constant at a value of 7 W and 1000 Hz. The figure depicts that surface roughness first gradually increases and then steep decreases with the increase of scanning speed parameter. This nature of the graph can be explained as when the scanning speed is increasing with constant average power and transverse feed, the overlap distance between the two consecutive lasers scanning track decreases. Therefore, the surface roughness also decreases. The same graph also depicts that the surface roughness (transverse) increases with the increasing of transverse feed. It is because according to Eq. (2), with increase of transverse feed, the values of transverse overlap between two laser scanning track decrease and thus, the machined surface has a lot of irregularities which ultimately increases the roughness.

4.2.2 Parametric Effects on Surface Roughness (Rz)

Figure 13 depicts the variation of surface roughness (lateral), Rz with laser average power and pulse frequency keeping other two process variables, i.e., scanning speed and transverse feed constant at a value of 3 mm/s and 0.03 mm, respectively. The graph indicates that the surface roughness values increase with laser beam average power. It is because of the increasing of peak power with the increase of laser beam average power according to Eq. (9). An increase value of peak power generates more energy in the laser zone and as a result of this, subsequent melting and vaporization takes place. Therefore, deeper layer of material is removed. This causes an increment in surface roughness. The same figure also depicts that the surface roughness (lateral) has a decreasing nature with the increment in pulse

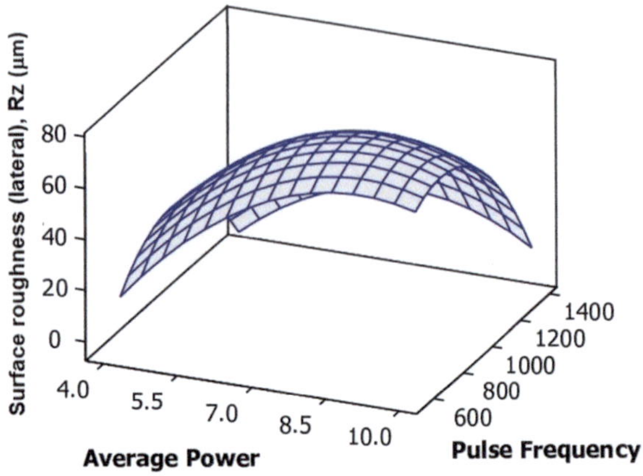

Fig. 13 3D surface plots of surface roughness (lateral), Rz against of average power and pulse frequency

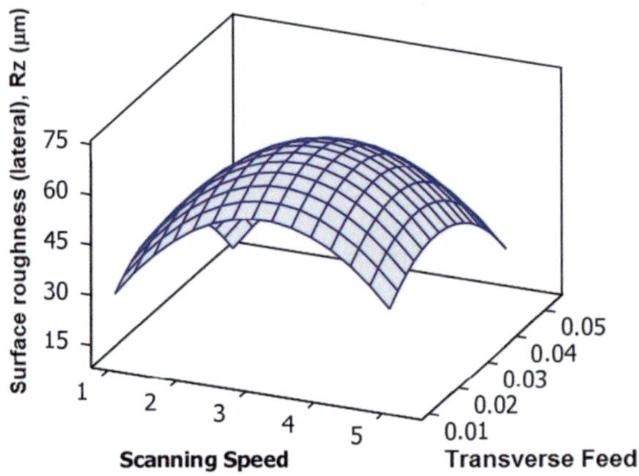

Fig. 14 3D surface plots of surface roughness (lateral), Rz against of scanning speed and transverse feed

frequency. Increased pulse frequency causes a decreasing peak power that further causes decrement in surface roughness.

Figure 14 shows the variation of surface roughness (lateral), Rz with scanning speed and transverse feed keeping the other two process parameters, i.e., laser average power and pulse frequency constant at 7 W and 1000 Hz, respectively. The figure shows a gradually increasing then followed by decreasing nature of the surface roughness curve with scanning speed. This is because, at lower value of scanning speed, the generated textured surface shows a decrease in overlap between the two spots and increased surface roughness. But, at high value of scanning speed, the duration of contact between laser beam and the material decreases significantly and it does not get enough time to melt and vaporize the material and as a result, the values of surface roughness decrease. The figure also shows that the surface roughness increases with increasing transverse feed. It is because with the increase of transverse feed, the values of transverse overlap between two laser

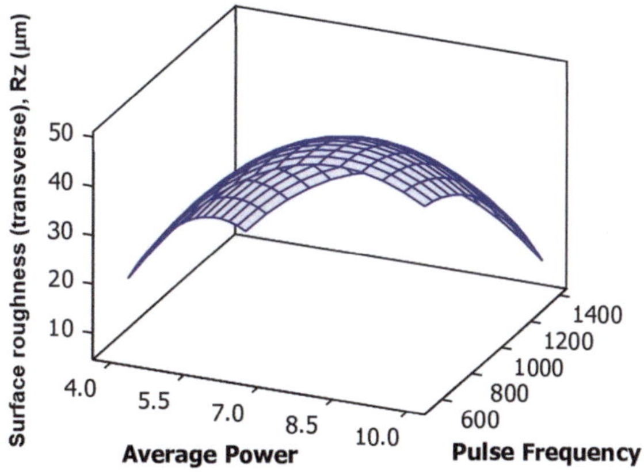

Fig. 15 3D surface plots of surface roughness (transverse), Rz against of average power and pulse frequency

scanning track decrease and thus, the machined surface has a lot of irregularities which ultimately increases the roughness.

Figure 15 illustrates the variation of surface roughness (transverse), Rz with beam average power and pulse frequency. The other two process parameters, i.e., scanning speed and transverse feed, are at constant at a value of 3 mm/s and 0.03 mm, respectively. The figure depicts that the surface roughness is increasing with beam average power. This is because the increased laser beam average power causes an increment in peak power and that generates more energy for melting and vaporization the work material. This causes the increasing of surface roughness. The same figure also shows that the surface roughness is decreasing when the pulse frequency goes on increasing as the peak power decreases with the increase in pulse frequency.

The variation of transverse surface roughness (Rz) with scanning speed and transverse feed is shown in Fig. 16, while laser beam average power and pulse frequency were kept constant at 7 W and 1000 Hz, correspondingly. The figure

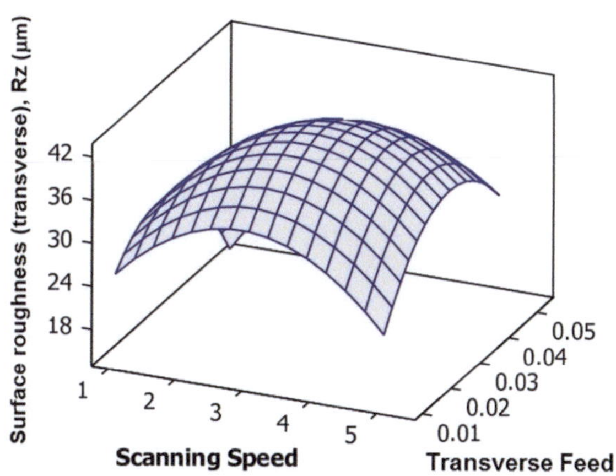

Fig. 16 3D surface plots of surface roughness (transverse), Rz against of scanning speed and transverse feed

shows a gradually increasing then decreasing nature of the surface roughness curve with scanning speed. This is because, at lower value of laser scanning speed, the generated textured surface shows decrease in overlap between the two spots and therefore, the surface roughness increases. But, at higher value of scanning speed, the interaction time between the laser beam and the work material decreases significantly without having enough time to melt and vaporize the material. As a result of this, the surface roughness decreases. The very same figure also shows that the surface roughness increases with increasing transverse feed. It is because with increase of transverse feed, the values of transverse overlap between two laser scanning track decrease and thus, the machined surface has lot of irregularities which ultimately increases the roughness.

4.2.3 Parametric Effects on Contact Angle (θ)

Figure 17 illustrates the variation of contact angle (θ) with beam average power and pulse frequency keeping other two process parameters, i.e., scanning speed and transverse feed, constant at 3 mm/s and 0.03 mm correspondingly. It is obvious from the surface plot that with increment of average power of laser, it has small effect on contact angle when pulse frequency value is less. However, at high value of pulse frequency settings, the contact angle sharply decreases. Consequently, the wettability between water drop and machined surface increases. It is also observed that at less value of laser beam average power, with increase of pulse frequency, contact angle increases sharply and consequently the wettability of the surface with water drop decreases.

Figure 18 illustrates the variation of contact angle (θ) with pulse frequency and scanning speed when laser beam average power and transverse feed remained constant at 1000 Hz and 0.03 mm correspondingly. The surface plot reveals that with the combination of low scanning speed and high values of pulse frequency, the changes in contact angle are minimal. On the contrary, at high scanning speed with

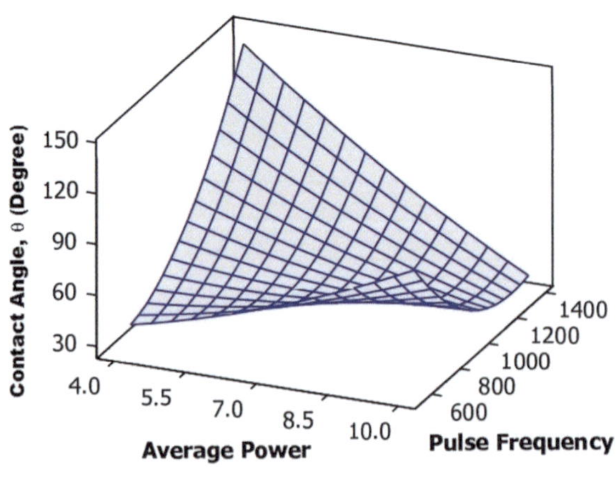

Fig. 17 3D surface plots of contact angle against of average power and pulse frequency

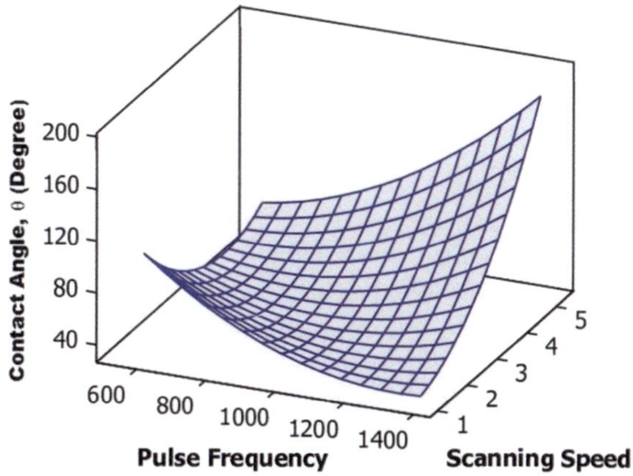

Fig. 18 3D surface plots of contact angle against of pulse frequency and scanning speed

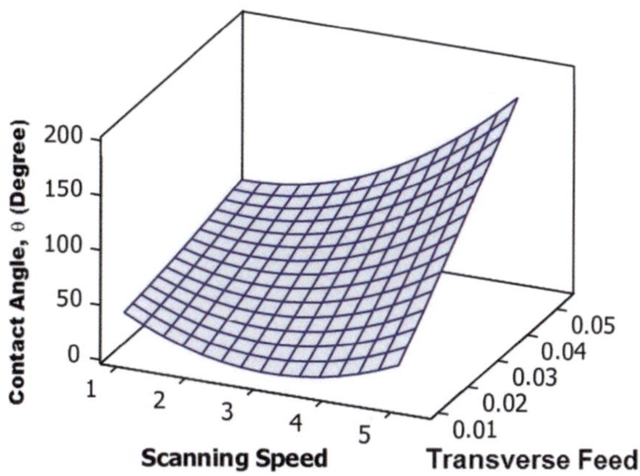

Fig. 19 3D surface plots of contact angle against of scanning speed and transverse feed

increasing pulse frequency contact angle increases and the wettability decreases. Simultaneously, at low pulse frequency, increasing of scanning speed has a very small effect on contact angle, but at high pulse frequency, increment in scanning speed causes the contact angle to increase sharply. From the plot, it is optimized that the contact angle has minimum value at higher value of pulse frequency and lower value of scanning speed. On the other hand, the contact angle has a maximum value at higher value of pulse frequency as well as scanning speed.

Figure 19 shows the variation of contact angle (θ) with scanning speed and transverse feed while remaining other process parameters such as laser beam average power and pulse frequency constant at 7 W and 1000 Hz, correspondingly. The surface plot of the contact angle against these two parameters shows that with the increase of scanning speed, the contact angle increases sharply at all value of transverse feed. This makes the surface hydrophobic in nature. At the same time, with increase of transverse feed, the value of contact angle has an increasing nature at all values of scanning speed. Thus, wettability of the surface against water drop decreases.

5 Multi-objective Optimization and Validation Experiments

For obtaining minimum values of surface roughness (Ra and Rz) and contact angle (θ), it is very essential to consider optimization statistical tool to achieve multi-objective optimization parametric combination of considered process parameters. Figure 20 shows the result of the optimized combination of the process variables in order to achieve the optimized tribological characteristics. In the aforesaid figure, process parameters and their ranges are considered in columns, whereas the output responses are plotted in each row. Each cell in the figure also represents how the output response varies with one of the process parameters, while the other parameters are remained as constant. At an average power of 9.02 W, pulse frequency of 1340 Hz, laser beam scanning speed of 2.04 mm/s, and transverse feed of 0.0227 mm, the optimal settings of the process parameters are

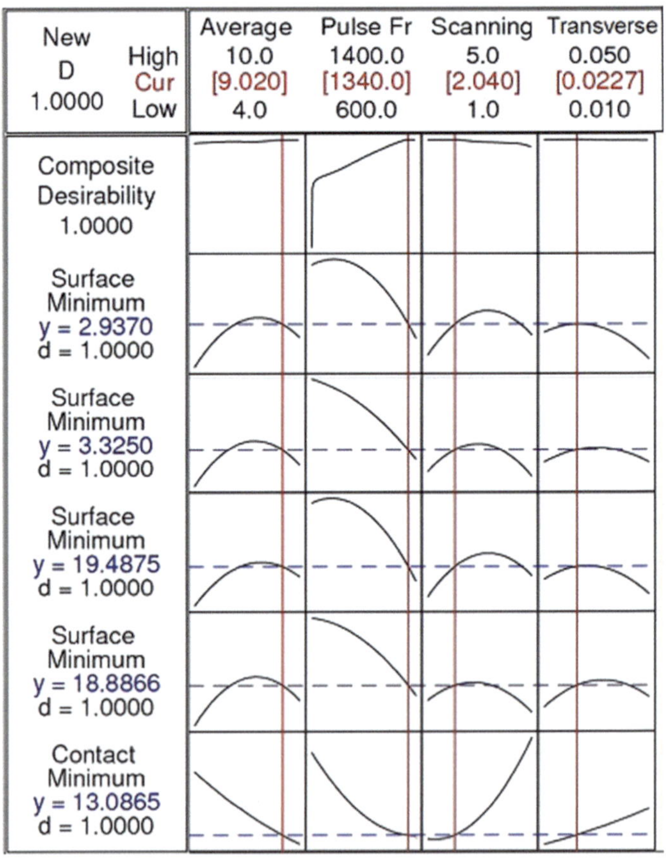

Fig. 20 Multi-objective optimization results of responses

achieved. The optimal surface roughness (lateral), Ra achieved was 2.93 μm, surface roughness (transverse), Ra 3.32 μm, surface roughness (lateral), Rz 19.48 μm, surface roughness (transverse), Rz 18.89 μm, contact angle (θ) 13.08°. With the composite desirability (D) value of 1, all the responses have been successfully optimized during multi-objective optimization.

Verification experiment of confirming the values of multi-objective optimization was conducted at the closer achievable parametric combination of multi-objective optimized setting. The RSM predicted and experimental results of responses with the percentage of prediction errors are shown in Table 12. It has been found that the optimum values of surface roughness (lateral and transverse) and contact angle (θ) were obtained as 3.01, 3.44 20.13, 19.45 μm, and 13.53° respectively. The calculated values of prediction errors were 2.71, 3.43, 3.18, 2.87, and 3.31% as obtained using Eq. (8). These values are within acceptable limit. Thus, the optimized parametric combination is valid within considered range of process parameters.

Table 12 Results of optimal parametric combinations and confirmation experiments

Optimization	Experimental parameters settings	Experimental results
Multi-objective optimization	Average power = 9.02 W Pulse frequency = 1340 Hz Scanning speed = 2.04 mm/s Transverse feed = 0.0227 mm	Surface roughness (lateral), Ra = 3.01 μm, Surface roughness (transverse), Ra = 3.44 μm Surface roughness (lateral), Rz = 20.13 μm Surface roughness (transverse), Rz = 19.45 μm Contact angle (θ) = 13.53° **RSM predicted results** Surface roughness (lateral), Ra = 2.93 μm, Surface roughness (transverse), Ra = 3.32 μm Surface roughness (lateral), Rz = 19.48 μm Surface roughness (transverse), Rz = 18.89 μm Contact angle (θ) = 13.08° **Percentage of error (%)** Surface roughness (lateral), Ra = 2.71 Surface roughness (transverse), Ra = 3.43 Surface roughness (lateral), Rz = 3.18 Surface roughness (transverse), Rz = 2.87 Contact angle (θ) = 3.31

6 Analysis of Optical Microscopic Images of Laser Textured Surfaces

Studies of optical microscopic images of textured surface machined at various parametric settings have been done in depth for investigating the influences of process parameters. Figure 21a–d show and compare the laser textured surfaces machined at parametric combinations of (a) and (b) 5.5 W/1200 Hz/2 mm/s/0.02 mm and 5.5 W/1200 Hz/4 mm/s/0.04 mm of average power, pulse frequency, scanning speed, and transverse feed. By closely observing these optical micrographs in Fig. 21a–d, it is seen that with the combination of higher value of laser scanning speed and transverse feed, the values of surface roughness (lateral and transverse) were increased from Ra = 3.68 μm and Ra = 3.62 μm to Ra = 4.91 μm and Ra = 4.55 μm. The corresponding values of surface roughness (lateral and transverse) were increased from Rz = 21.54 μm and Rz = 20.11 μm to Rz = 31.62 μm and Rz = 23.80 μm. The increase of surface roughness (both lateral and transverse) is due to decrease in lateral and transverse overlaps with laser beam scanning speed and transverse feed. Figure 22a–d shows and compare the laser

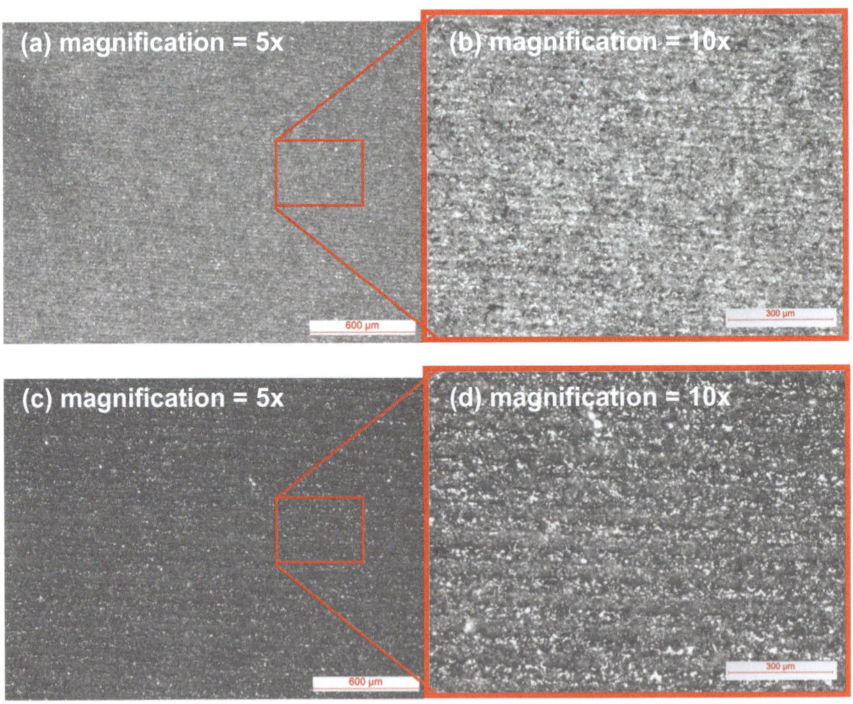

Fig. 21 Optical micrographs of laser textured surface machined at **a**, **b** 5.5 W/1200 Hz/2 mm/s/0.02 mm, **c**, **d** 5.5 W/1200 Hz/4 mm/s/0.04 mm of average power, pulse frequency, scanning speed and transverse feed

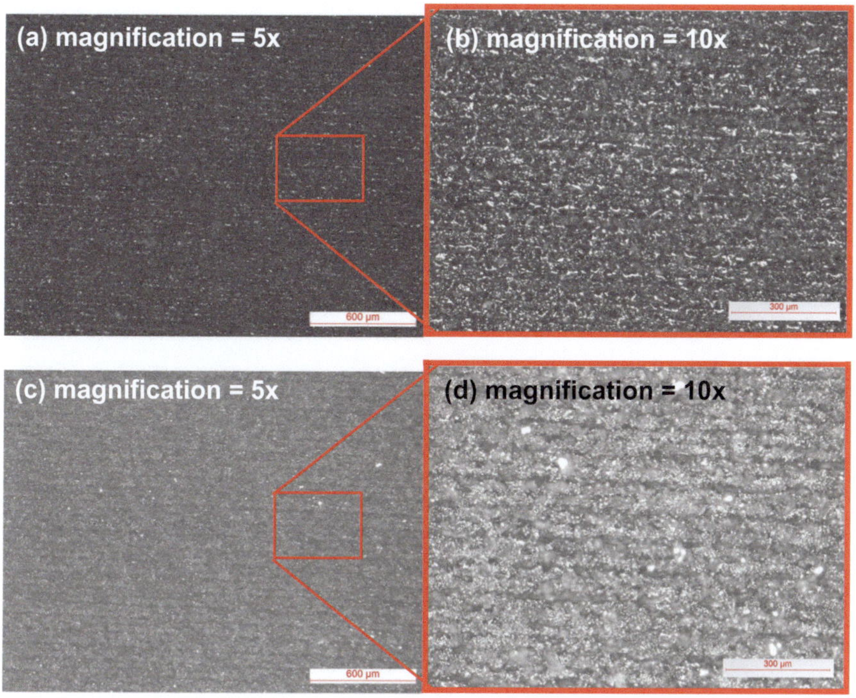

Fig. 22 Optical micrographs of laser textured surface machined at **a** and **b** 5.5 W/1200 Hz/2 mm/s/0.04 mm, **c**, **d** 8.5 W/800 Hz/2 mm/s/0.04 mm of average power, pulse frequency, scanning speed and transverse feed

textured surfaces machined at parametric combinations of (a) and (b) 5.5 W/1200 Hz/2 mm/s/0.04 mm and 8.5 W/800 Hz/2 mm/s/0.04 mm of average power, pulse frequency, scanning speed, and transverse feed. By closely observing these optical micrographs, it is seen that with the increase of laser beam average power and decrease of pulse frequency, the values of surface roughness (lateral and transverse) were decreased from Ra = 6.98 µm and Ra = 7.11 µm to Ra = 3.57 µm and Ra = 3.58 µm. The corresponding values of surface roughness (lateral and transverse) were decreased from Rz = 42.46 µm and Rz = 35.99 µm to Rz = 24.16 µm and Rz = 18.48 µm. The decrease of surface roughness (both in lateral and transverse) with an increase in average power and decrease in pulse frequency is due to increase in laser–material interaction time and eventually the formation of craters on the textured surfaces.

Figure 23 shows the photographic views of water droplets on laser textured surfaces machined at six different parametric combinations. These images have been taken utilizing the indigenously developed contact angle measuring arrangement as explained in Fig. 8. The figures also show the values of contact angle produced due to contact of solid titanium machined surface as well as due to surface tension for surrounding air. As the values of contact angle of water droplets shown

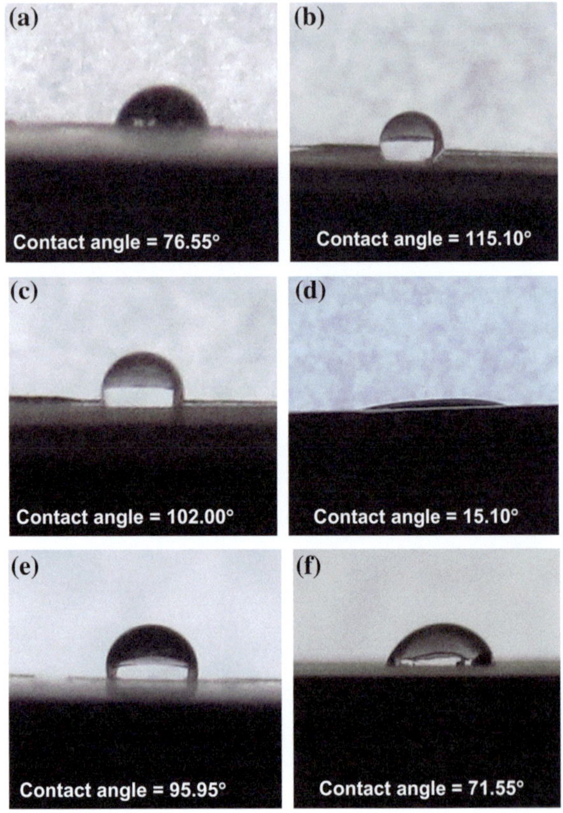

Fig. 23 Photographic views of water droplets on the laser textured surface machined at **a** 8.5 W/800 Hz/2 mm/s/0.04 mm, **b** 8.5 W/1200 Hz/4 mm/s/0.04 mm, **c** 8.5 W/800 Hz/4 mm/s/0.04 mm, **d** 5.5 W/1200 Hz/2 mm/s/0.02 mm, **e** 5.5 W/1200 Hz/2 mm/s/0.04 mm, and **f** 7 W/1000 Hz/3 mm/s/0.03 mm of average power, pulse frequency, scanning speed, and transverse feed

in Fig. 23b, c, e are 115.10°, 102.00°, and 95.95°, which are more than 90°, therefore, these machined surfaces can be termed as hydrophobic surface (wettability low). However, the machined surfaces shown in Fig. 23a, d, f create contact angles of water droplets as 76.55°, 15.10°, and 71.55°, which are less than 90°, therefore, these machined surfaces can be termed as hydrophilic surface (wettability high).

7 Conclusions

In the present chapter, the effect of various process parameters on laser surface texturing tribological characteristics of commercially pure titanium sheet (thickness of 1 mm) is investigated and analyzed subsequently. Further, development of empirical models was conducted to correlate the process parameters and various responses. Based on the experimental results obtained, the mathematical models of the responses have been developed. With analysis of variance (ANOVA) test, these models were validated statistically at 95% confidence level. Analysis of various surface plots reveals that all the parameters, i.e., laser beam average power, pulse frequency, scanning speed, and transverse feed have significant influences to achieve quality surface and defined wettability. Scanning speed and pulse frequency are the most significant and contributing parameters for surface roughness (Ra and Rz) in lateral direction, whereas scanning speed and traverse feed are the most relevant and contributing parameters for surface roughness (Ra and Rz) in transverse direction. Multi-objective optimization parametric combination was also obtained for achieving least surface roughness (Ra and Rz) in lateral and transverse directions. Optimized response values are achieved such as surface roughness (lateral), Ra of 2.93 μm, surface roughness (transverse), Ra of 3.32 μm, surface roughness (lateral), Rz of 19.48 μm, surface roughness (transverse), Rz of 18.89 μm, and contact angle (θ) of 13.08° at the parametric setting of average power at 9 W, pulse frequency at 1400 Hz, laser beam scanning speed at 2 mm/s, and transverse feed at 0.02 mm. The results of verification experiments show that the prediction errors are within acceptable limits. The experimental results achieved in this research will be useful in developing new strategies and guidelines for research scientists and manufacturing engineers for more exploration in this area of surface texturing process using laser beam on titanium material. Furthermore, this research will provide great impetus for effective and successful utilization of pulsed Nd:YAG laser during the tribological surface generation of specific surface properties for successful applications in the micro-parts manufacturing.

References

Ali, M.Y., and W.N.P. Hung. 2017. Micromachining. In *Comprehensive materials finishing*, ed. S. Hashmi, vol. 1: *Finish machining and net-shape forming*, 322–343. Elsevier.

Chang, C.W., C.Y. Chen, T.L. Chang, C.J. Ting, C.P. Wang, and C.P. Chou. 2012. Sapphire surface patterning using femtosecond laser micromachining. *Applied Physics A* 109 (2): 441–448.

Chavoshi, S.Z., and X. Luo. 2015. Hybrid micro-machining processes: A review. *Precision Engineering* 41: 1–23.

Cunha, A., V. Oliveira, and R. Vilar. 2016. Ultrafast laser surface texturing of titanium alloys. In *Laser surface modification of biomaterials: Techniques and applications*, 301–322. Woodhead Publishing.

Dunn, A., K.L. Wlodarczyk, J.V. Carstensen, E.B. Hansen, J. Gabzdyl, P.M. Harrison, J.D. Shephard, and D.P. Hand. 2015. Laser surface texturing for high friction contacts. *Applied Surface Science Part B* 357: 2313–2319.

Gentili, E., L. Tabaglio, and F. Aggogeri. 2005. Review on micromachining techniques. In *AMST'05 advanced manufacturing systems and technology*, vol. 486, 387–396. Vienna: CISM International Centre for Mechanical Sciences, Springer.

Jackson, M.J., M.D. Whitfield, G.M. Robinson, R.G. Handy, J.S. Morrell, W. Ahmed, and H. Sein. 2015. Micromachining from a materials perspective. In *Machining with nanomaterials*, 77–127. Switzerland: Springer International Publishing.

Kibria, G., B. Doloi, and B. Bhattacharyya. 2012. Optimisation of Nd:YAG laser micro-turning process using response surface methodology. *International Journal of Precision Technology* 3 (1): 14–36.

Kibria, G., B. Doloi, and B. Bhattacharyya. 2015. Pulsed Nd:YAG laser micro-turning process of alumina ceramics. In *Lasers based manufacturing*, ed. Shrikrishna N. Joshi, and Uday Shanker Dixit, 343–380. India: Springer.

Kovalchenko, A., O. Ajayi, A. Erdemir, G. Fenske, and I. Etsion. 2004. The effect of laser texturing of steel surfaces and speed-load parameters on the transition of lubrication regime from boundary to hydrodynamic. *Tribology Transactions* 47 (2): 299–307.

Li, B.J., H. Li, L.J. Huang, N.F. Ren, and X. Kong. 2016. Femtosecond pulsed laser textured titanium surfaces with stable superhydrophilicity and superhydrophobicity. *Applied Surface Science* 389: 585–593.

Meijer, J. 2004. Laser beam machining (LBM), state of the art and new opportunities. *Journal of Materials Processing Technology* 149: 2–17.

Mishra, S., and V. Yadava. 2015. Laser beam micro machining (LBMM)—A review. *Optics and Lasers in Engineering* 73: 89–122.

Soveja, A., E. Cicala, D. Grevey, and J.M. Jouvard. 2008. Optimisation of TA6 V alloy surface laser texturing using an experimental design approach. *Optics and Lasers in Engineering* 46 (9): 671–678.

Thawari, G., J.K.S. Sundar, G. Sundararajan, and S.V. Joshi. 2005. Influences of process parameters during pulsed Nd:YAG laser cutting of nickelbase superalloys. *Journal of Materials Processing Technology* 170 (1–2): 229–239.

Toyserkani, E., and N. Rasti. 2014. Ultrashort pulsed laser surface texturing. In *Laser surface engineering*, ed. D. Waugh, and J. Lawrence, 441–453. Woodhead Publishing.

Tripathi, K., B. Joshi, G. Gyawali, A. Amanov, and S.W. Lee. 2015. A study on the effect of laser surface texturing on friction and wear behavior of graphite cast iron. *Journal of Tribology* 138 (1): 011601–011610.

Velasquez, T., P. Han, J. Cao, and K.F. Ehmann. 2013. Feasibility of laser surface texturing for friction reduction in surgical blades. In *Proceedings of the ASME 2013 international manufacturing science and engineering conference*, V001T01A009. Madison, Wisconsin, USA.

Vora, H.D., and N.B. Dahotre. 2015. Surface topography in three-dimensional laser machining of structural alumina. *Journal of Manufacturing Processes* 19: 49–58.

Wang, X., Y. Xing, and M. Giovannini. 2015. Effect of overlap and overscan number in laser surface texturing of medical needles. *Applied Physics A* 120 (1): 229–238.

Yang, C.J., X.S. Mei, Y.L. Tian, D.W. Zhang, Y. Li, and X.P. Liu. 2016. Modification of wettability property of titanium by laser texturing. *International Journal of Advanced Manufacturing Technology* 87 (5): 1663–1670.

Sheet Thinning Prediction and Calculation in Incremental Sheet Forming

Harish K. Nirala and Anupam Agrawal

Abstract Incremental Sheet Forming (ISF) technique is an emerging process for die less forming. It has wide applications in many industries e.g., automobile and medical bone transplants. In ISF, forming of the sheet is done using Numerical Control (NC) single point forming tool, which incrementally deforms the sheet by highly localized plastic deformation. It gives better formability when compared with traditional forming processes, like deep drawing and spinning. ISF has few limitations out of which sheet thinning is one of the most critical limitations. In ISF, formability is generally measured by the limit of maximum formable wall angle and maximum permissible sheet thinning. Formability of the sheet during ISF or traditional forming processes can be presented by a Forming Limit Curve (FLC). The sheet thinning in ISF can be predicted through sine law. By assuming plastic incompressibility, sheet thinning can also be predicted by considering volume constancy concept. In this study, forming limit has been predicted for two wall profiles, viz., circular and elliptical wall. Further, a methodology has been presented as a way to predict and calculate sheet thinning during ISF. The developed methodology has been validated through numerical simulations followed by experimental investigations. An in-house Computer-Aided Manufacturing (CAM) module for incremental toolpath is developed for both simulations and experiments. The results are in correlation with considerable accuracy.

Keywords Incremental sheet forming (ISF) · FEA · Sheet thinning CAD/CAM · Sheet thinning models · FLC

H. K. Nirala · A. Agrawal (✉)
Department of Mechanical Engineering, Indian Institute of Technology Ropar, Rupnagar, Punjab 140001, India
e-mail: anupam@iitrpr.ac.in

H. K. Nirala
e-mail: harish.nirala@iitrpr.ac.in

S. S. Pande and U. S. Dixit (eds.), *Precision Product-Process Design and Optimization*, Lecture Notes on Multidisciplinary Industrial Engineering,
https://doi.org/10.1007/978-981-10-8767-7_15

Nomenclature

$\propto$, ø	Wall angle
x, y, and z	Coordinates
t_f	Final thickness
t_o	Initial thickness
f_t	Forming limit
v	Poisson's ratio

1 Introduction

Single Point Incremental Sheet Forming (SPISF) is a widely known alternative to traditional forming techniques. The sheet metal could be formed into desired symmetrical or asymmetrical shape without using any die. In this process, the sheet is clamped from its periphery on a suitable fixture mounted on a CNC machine and a hemispherical end shape forming tool is clamped to the machine spindle. The movement of forming tool over the sheet is controlled by the toolpath given as numerical code to the machine controller. The tool induces localized plastic deformation in the sheet (Young and Jeswiet 2004). These localized deformations are responsible for the forming of required geometry with higher formability as compared to conventional forming techniques. SPISF has drawn attention of manufacturing arena because it is very suitable and effective for small batches in production system (Jeswiet et al. 2005; Jackson and Allwood 2009). In 1967, a patent filed by Leszak initiated research in ISF (Leszak 1967). In 1978, work done by Mason at university of Nottingham gave a new direction to forming by ISF (Mason 1978). In 2001, Matsubara developed a double point ISF technique to improve its dimensional accuracy (Matsubara 2001). A comprehensive review of ISF is very effectively given by Jeswiet et al. (2005). Schematic representation of SPISF process is shown in Fig. 1.

Figure 1, briefly shows a SPISF process, a truncated cone geometry with a wall angle 'α' is being formed by an incremental spiral toolpath. The area of sheet metal which is in contact with the tool undergoes localized plastic deformation and results in sheet thinning. An intermediate geometry is also shown in the figure to give an idea of the deformation direction. A backing plate has been used in order to avoid undesired bending near the periphery of the geometry. Rotation of the tool is given in a clockwise direction with an angular velocity 'ω'.

This process has drawn the attention of many manufacturing research communities all around the world since last two decades. The reasons are greater process flexibility, improved formability and reduced forming forces when compared with conventional forming processes like deep-drawing, stretch forming and spinning. Ford Motor Company has recently released Ford Free Form Fabrication Technology (FFFFT) that uses Two-Point ISF (TPISF) for rapid prototyping of

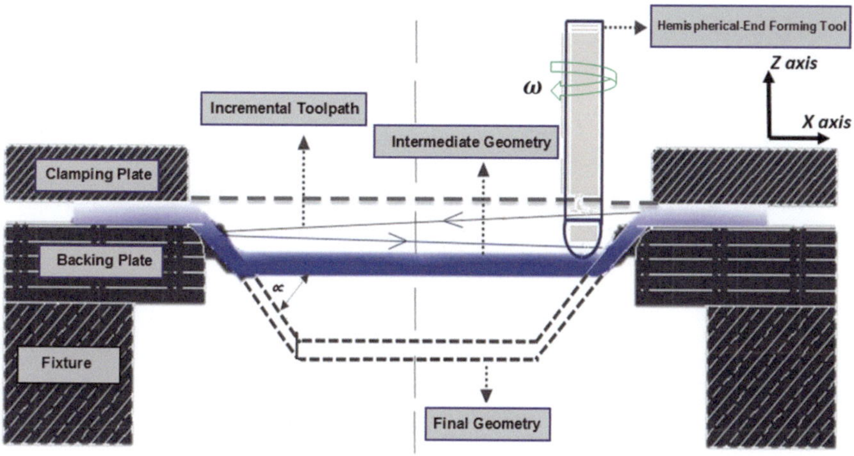

Fig. 1 Schematic representation of SPISF process

automobile parts.[1] When in SPISF any full or partial die is used then the process is termed as Two-Point Incremental Sheet Forming (TPISF). In SPISF there is only one contact point between the sheet and the tool while in TPISF there are two contact points. In TPISF, first contact point is in between forming tool and the inner surface of geometry and other contact point is between outer surface of geometry and die used during forming. ISF has potential application in the medical bone prosthesis. Bone implants with some cavity can be formed by using ISF process. Literature (Lu et al. 2015) shows its application in cranioplasty (Bagudanch et al. 2015), knee arthroplasty (Bhoyar and Borade 2015; Eksteen 2013; Eksteen and Van der Merwe 2012), denture plate (Milutinovića et al. 2014), maxillofacial implant (Araújo et al. 2013), heel bone replacement (Ambrogio et al. 2005) and many more. These implants need a customized fabrication, which can be possible through a flexible process like ISF. Figure 2, shows some application of ISF in different manufacturing sectors. The deformation mechanism of ISF includes, biaxial stretching at the start point, at the corners and at each increment in the negative *z* direction. The deformation, when the tool travels horizontally or spirally along the formed profile, is plain-strain stretching and shearing (Jackson and Allwood 2009). The deformation mechanism in ISF also includes pure bending of the sheet while forming (Jackson and Allwood 2009). Further, the deformation mechanism in conventional drawing process is purely biaxial, due to which the stress triaxial condition necessary for fracture is achieved far earlier in conventional forming than in incremental forming. Process parameters like feed rate, spindle speed, tool diameter, incremental depth, forming angle, sheet thickness and toolpath will have a

[1]http://corporate.ford.com/news-center/press-releases-detail/pr-ford-develops-advanced-technology-38244, 10 August, 2014, 8:30 pm IST.

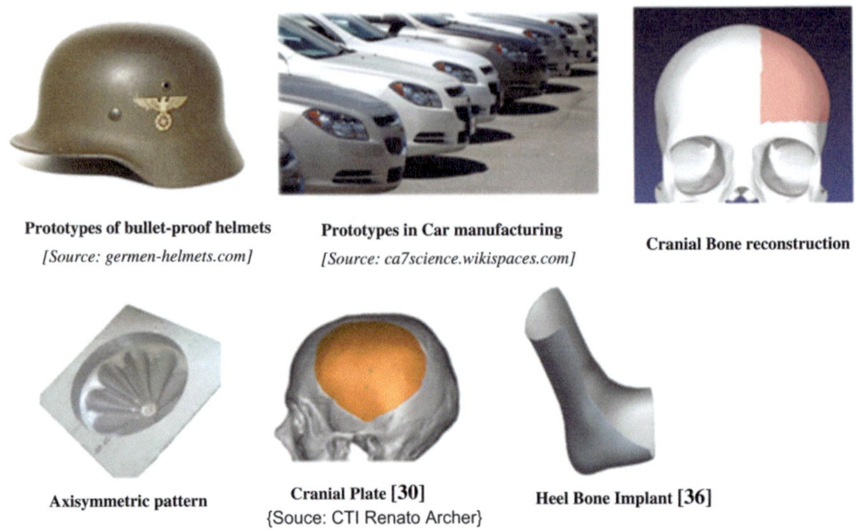

Fig. 2 Some applications of ISF Bagudanch et al. (2015), Ambrogio et al. (2005)

significant effect on the formability achieved (Kim and Park 2002; Gatea et al. 2016). In SPISF, formability of the sheet is generally presented by a Forming Limit Diagram (FLD). FLCs are generally plotted in a principal space of major and minor strain $(\varepsilon_1, \varepsilon_2)$. In Fig. 3, formability comparison between deep drawing and incremental sheet forming is presented (Silva et al. 2008a). Figure 3, clearly shows that ISF gives better formability than traditional forming techniques in a principal strain space (Silva et al. 2008b). There are few limitations of this process also like geometrical inaccuracy, sheet thinning, undesired bending and spring back. Out of these limitations, sheet thinning is the most critical limitation. Hence, it becomes very necessary to study sheet tinning in ISF. From literature, it has been found that sine law can be used to predict the sheet thickness variation along the formed profile (Duflou et al. 2008). The sine law can be utilized to predict the final part thickness T_{final} from the unprocessed sheet thickness T_{Initial} and the part wall angle α (Duflou et al. 2008). From Eq. (1), one can conclude that, as the wall angle (α) of the formed profile increases, the final profile thickness T_{final} decreases.

$$T_{\text{final}} = T_{\text{Initial}} \sin(90^\circ - \propto) \tag{1}$$

According to sine law, it is not possible to form vertical wall $(\alpha = 90^\circ)$, as it gives final sheet thickness as zero. However, researchers have successfully formed vertical walls using multistage ISF. When the sheet is being deformed strain hardening effect appears which reduces the deformation capability of the sheet. There will be more strain hardening, when deformation imposed on the sheet is large as in single-stage SPISF process. When few intermediate stages are introduced then the strain hardening effect will reduce and formability will improve

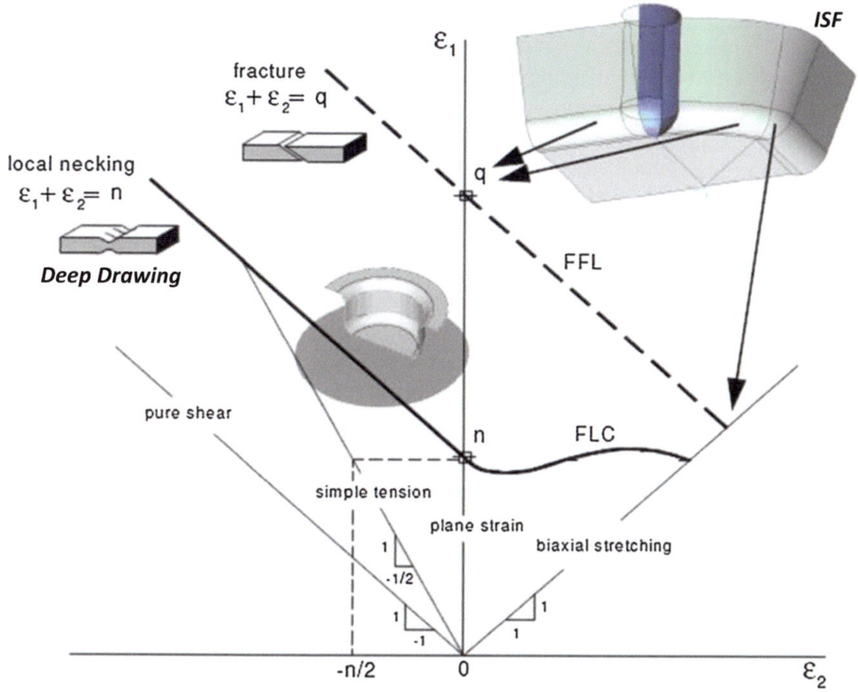

Fig. 3 Forming limit diagram for conventional forming and ISF (Silva et al. 2008a)

because the overall deformation will now be distributed among these intermediate stages (Nirala et al. 2017). Young and Jesweit (2004), investigated a double pass SPISF process. In double pass SPISF process, a preform geometry is formed before forming the final desired geometry. The formability of geometry formed by using double pass ISF process was higher than the single pass ISF process. The final thickness in double pass ISF can be given by:

$$T_f = T_p \frac{\sin(\propto)}{\sin(\phi)} \tag{2}$$

In Eq. (2), T_P is the thickness of preform obtained from sine law, ø is the wall angle of preform geometry, $\propto$ is the wall angle of final geometry, and T_f is the thickness of final geometry. Liu et al. (2013, 2014), developed some strategies for multistage ISF and conducted plane strain deformation analysis to analyze the strain distribution in these strategies. Filice (2006), developed some analytical relations to predict sheet thickness by the concept of volume constancy. Step features observed while forming with multistage SPISF strategy (Liu et al. 2014; Xu et al. 2012). Bambach (2010), given a more generalized sine law, which considers the evolution of surface from the sheet blank (Bambach 2010). Mirnia et al. (2014) have done modeling of SPISF using sequential limit analysis to predict the sheet thickness

(Mirnia et al. 2014). Cao et al. (2015) have developed an efficient method to predict sheet thickness in multistage incremental sheet forming. Lu et al. (2015) reinvestigate the deformation mechanism in ISF by the concept of stress triaxiality. Ambrogio et al. (2017) proposed a robust meta-modeling technique for the crucial problem of localized thinning. Skjodt et al. (2010), studied the formability of 1 mm AA1050-O sheets, their study includes evaluation of strain paths and fracture strains in single and multistage SPISF process. Silva et al. (2011), revisited the failure mechanism in ISF, they considered the influence of process parameters like tool diameter on fracture initiation and propagation. Silva et al. (2011), very effectively presented the state of stress and strain in plane strain and biaxial strain conditions, also the influence of tool diameter on localized deformation is presented. The strain evolution is then verified for truncated cone and truncated pyramid by experiments and shown using Forming Limit Curves (FLC's). Over the time, several techniques have been developed to improve the formability of ISF process like, laser assisted ISF (Jeswiet et al. 2005) and electrically assisted ISF (Nguyen-Tran et al. 2015). There are many online thickness measurement techniques available in the literature, like thickness measurement by using Digital Image Correlation (DIC) (Hild and Roux 2006; Chu et al. 1985), Dejardin et al. (2010), emphasized that for the industrialization and optimization of ISF, online monitoring of sheet thinning is very necessary. Paniti and Paroczi (2011), developed a noncontact type thickness measurement device based on Hall-effect. The methodology developed by these researchers is effective, but some are complex and time taking, when it comes to implementation. The present methodology can predict and calculate sheet thickness of formed profile in an effective, simpler and easy manner.

This research work presents a methodology to predict and calculate sheet thinning in single-stage ISF process. The sheet is assumed to be isotropic and incompressible, the deformation is only considered as plane strain, and the effect of process parameters like feed, tool diameter, etc., is neglected. To predict the sheet thinning in single stage ISF, a further improvement in the model proposed by Hussain and Gao (2007) is proposed for better flexibility and implementation. These analytical models are discussed in detail in further sections of this study. Anon-contact-based technique is presented to calculate sheet thinning in single stage ISF. This technique uses the strain distribution data after experiments and fits that data into a theoretical relation (Sect. 3). The effectiveness of the abovementioned models is then verified through simulations followed by real experiments.

2 Thickness Prediction

In this section, analytical model to predict sheet thinning along forming depth in ISF is discussed. In this model, the effect of elastic behavior of the sheet is neglected and the sheet is considered as isotropic. The deformation is considered as plane strain deformation. The thickness prediction results are then compared with experimental and FEA results, in results and discussion section, which describes the effectiveness of the proposed model. Circular and elliptical wall profiles are formed using ISF. In circular wall profile, wall angle varies from 30° to 75° and for elliptical wall profile, wall angle varies from 20° to 85°. For simplicity, the elastic behavior of the sheet is neglected, even it effects the sheet when unclamped from the fixture. The overspinning of the sheet during forming is neglected, overspinning refers to sheet thinning by compressive stresses applied by the forming tool. The material behavior and friction between the sheet and tool are neglected. On the whole, the problem is plane strain type. There are certain limitations of the model proposed by Hussain and Gao (2007), it specifically calculates the thinning limit of the formed geometry. It requires coordinate data points of the toolpath at every incremental depth, hence, use of commercial CAM software is not advisable to implement the concept. The quality of the formed part is very much dependent on the toolpath strategy. At present, many existing CAM (Computer Aided Manufacturing) software packages can be used to generate the toolpath for SPISF process like CATIA, Siemens NX 8.0 and Delcam PowerMILL 10. To overcome the abovementioned limitations an in-house CAM module is developed which is capable of giving constant z toolpaths for any axisymmetric component. The CAM module is subdivided into two sections. The first program calculates the wall angle at every increment for a generalized equation driven curve by calculating the slope at every incremental depth (Eq. 7). The second program, generates a 3D incremental toolpath, for any choice of incremental depth, for the required geometry. In order to calculate the wall angle of any profile, one needs to know the equation of that curve first. Most of the equation driven curves like parabola, hyperbola, and spline have their predefined equations. For free-form curves (asymmetric), rational Bezier curves can be used for describing the 2D cross-section of the profiles to be formed by ISF. A rational Bezier curve of *n*th degree can be defined as (Cao et al. 2015; Piegl and Tiller 1997):

$$c(u) = \frac{\sum_{i=0}^{n} B_{i,n}(u)\omega_i P_i}{\sum_{i=0}^{n} B_{i,n}(u)\omega_i} 0 \leq u \leq 1 \tag{3}$$

where, $P_i = (x_i, y_i, z_i)$ are the control points, and ω_i is the scalar for each point. In Eq. (3), the basis function $B_{i,n}(u)$ is the Bernstein polynomial with *n*th degree. However, in this research work, curvilinear geometric profiles with predefined equations (circular and elliptical) for their curves have been used. A pseudo code for incremental toolpath generation for any profile in *MATLAB*® is given in the Appendix. Also, procedure to conduct numerical simulations and experiments from this CAM module is shown in Fig. 4.

In Fig. 5, a circular generatrix profile is shown, *A* and *C* are the extreme ends of the profile, while *B* is the point where crack has been occurred, *h* is the forming depth of the profile. Thickness (t_f) and amount of thinning (f_t) at failure point '*B*' are given by following equations:

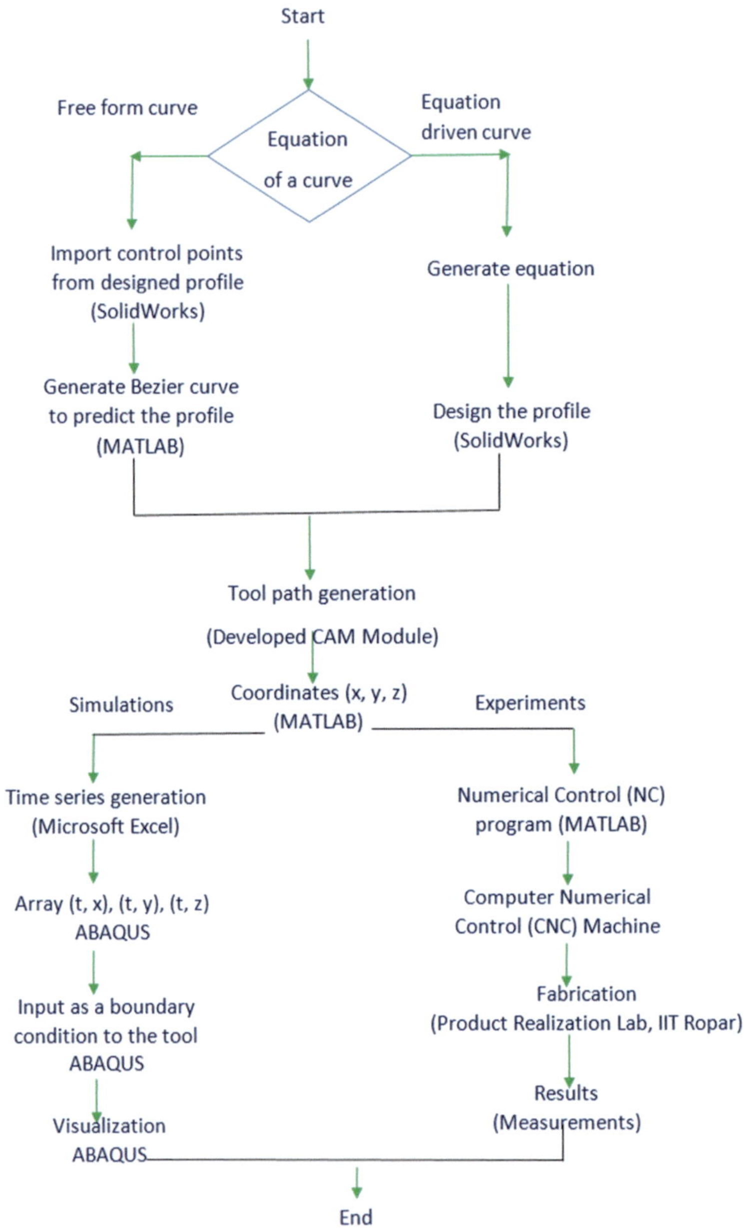

Fig. 4 Flowchart for toolpath generation, simulations and experiments

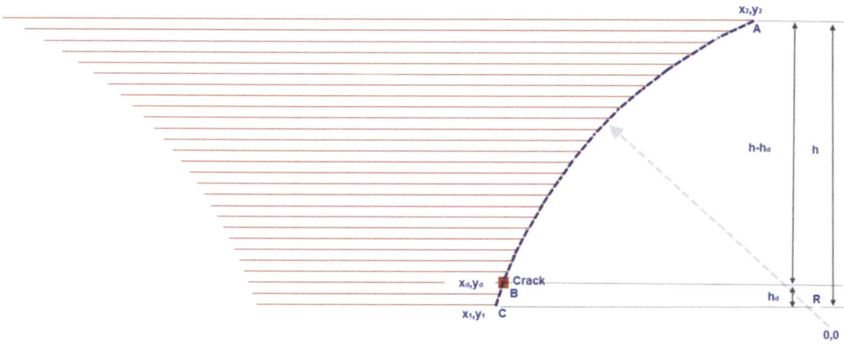

Fig. 5 Illustration of a generatrix with MATLAB toolpath

$$t_f = \frac{t_0}{R}(y_2 - \{h - h_d\}) \tag{4}$$

$$f_t = 100\left[1 - \frac{1}{R}(y_2 - \{h - h_d\})\right] \tag{5}$$

where, t_f is the final thickness, t_0 is the initial thickness and f_t is the amount of thinning. One can predict the thickness for overall profile, based on the modifications (Eqs. 6 and 7) in conjunction with abovementioned equations (Eqs. 4 and 5). Based on the below modifications (Eqs. 6 and 7) in conjunction with abovementioned equations. Equations (6) and (7), give a modified version to the sine law, which is capable of predicting sheet thickness of asymmetric curvilinear wall profiles.

$$\frac{\{y_2 - \{h - h_d\}\}}{R} = \sin(90 - \propto) \tag{6}$$

where,

$$\propto = \tan^{-1}\frac{(\mathrm{d}y/\mathrm{d}z)}{(\mathrm{d}z/\mathrm{d}x)} \tag{7}$$

The major benefits of this model over other models are as follows:

- It is capable of predicting thickness of any geometric profile.
- It is basically an improved version of sine law for curvilinear profiles.
- It is capable of predicting thickness at any forming depth.
- Continuous thickness prediction throughout the profile.
- Predict thickness even when there is variation in step depth of incremental toolpath.

3 Thickness Calculation

The most common and effective methods of measuring strain or plastic deformation are grid marking, extensometers, strain gauges and ultrasound thickness measurements. Grid deformation analysis is a most common and easy method of measuring the deformation of a formed sheet (Kim and Lee 1996). Grid marking consists of marking square or circular grid pattern over the sheet metal blank (Diameter of circle ranges between 1 and 8 mm, which depends on the size of the component) (Hariharan and Balaji 2009).[2] The incremental sheet forming process usually considered as plane strain deformation process because the thickness of the sheet is generally very less as compared to other dimensions of the sheet. Due to the plane strain nature of deformation, surface strains measurement becomes very important. Forming limit diagram can also be plotted using the calculated surface strains. In this study, a simplified technique (Eqs. 8–11) is presented to measure thinning of the sheet by using strains in meridional direction (ε_{11}), circumferential direction (ε_{22}) and through thickness direction (ε_{33}). In Eq. (8), E is the modulus of elasticity and v is the Poisson's ratio. In Eq. (11), t_0 is the initial sheet thickness and t_f is the final sheet thickness. This method of thickness prediction is basically employed by ABAQUS,[3] it computes the lateral and longitudinal strain for every shell element undergoing plane strain deformation. In this study, the present method has been used in conjunction with the strain distribution data collected from the grid deformation analysis.

$$\frac{Ev}{(1-2v)(1+v)}\varepsilon_{11}+\frac{Ev}{(1-2v)(1+v)}\varepsilon_{22}+\frac{Ev}{(1-2v)(1+v)}\varepsilon_{33}=0 \tag{8}$$

$$\varepsilon_{33}=-\left(\frac{v}{1-v}\right)(\varepsilon_{11}+\varepsilon_{22}) \tag{9}$$

$$\ln\frac{t_f}{t_0}=\varepsilon_{33} \tag{10}$$

$$t_f=t_0e^{\varepsilon_{33}} \tag{11}$$

The final sheet thickness (t_f) of curvilinear profiles, calculated from the calculation model is then compared with FEA and experimental results. A comparison of the effectiveness of this method is discussed further in this study.

[2]http://lectroetch.com/product-category/grid-marking-equipment/, 31st July, 2017, 1:45 am. IST.
[3]ABAQUS/Explicit Theory Manual Version 6.3, vol. 1–2, 2002.

4 Finite Element Analysis

Finite Element Analysis (FEA) is a tool to approximately predict the results of a process. The accuracy of results in ABAQUS depends on the quality of technical data available as an input to the FEA model. Few parameters like E, v material density, strain hardening exponent, friction coefficient and the material constant need to be accurately known for any sheet metal forming simulation. Meshing effects the accuracy of results and computational time. Generally, adaptive meshing is preferred to optimize computational time with the accuracy of results. A 3D elastic-plastic FEM (Finite Element Model) for ISF is established in ABAQUS software suite. An isotropic sheet blank with 'S4R' shell elements is considered for this study. The forming tool is considered as a rigid body. The elastic and plastic properties of the sheet blank (Aluminium 6106-T6) are considered during simulations. The thicknesses of the sheet and tool diameter are taken as 1 and 8 mm respectively for the simulations and experiments. The interaction between the sheet and forming tool is implemented by a pure master-slave contact algorithm. Additionally, the coefficient of friction between blank and the tool is taken as 0.1 by using Coulomb's law of friction (Li et al. 2012). The forming tool moves in a 3D path according to the boundary conditions assigned to it in x, y, and z coordinates. The time series generation for the amplitude of these boundary conditions is based on Eqs. (12) and (13); where x, y, and z are the coordinates of the toolpath are extracted from MATLAB (Nirala et al. 2015).

$$\text{Time step } (t) = \frac{D(\text{Step distance})}{\text{Velocity of tool } (v)} \tag{12}$$

$$\text{Step distance } (D) = \sqrt{(x_2 - x_1)^2 + (y_2 - y_1)^2 + (z_2 - z_1)^2} \tag{13}$$

The time step and step distance used during simulations are derived from Eqs. (12) and (13). An array of time steps and x, y, z coordinates are established while assigning boundary condition to the forming tool. As a result, the toolpath in the real process is same as that in simulations. The velocity of forming tool in simulations is generally taken more than the real process but below a critical value in order to reduce computational time in simulations (Hirt et al. 2005; Hadoush and Van den Boogaard 2008). A simple setup for carrying out simulations is shown in Fig. 6. The toolpath designed in MATLAB and FEA results of thickness distribution for circular and elliptical wall profiles are shown in Fig. 7. In Fig. 7, the output variable (sheet thickness) values for shell elements are averaged within 75% in correlation with each other (nodal averaging scheme in Abaqus).

The circular and elliptical wall profiles are simulated up to a designed forming depth of 30 mm. From the results of the simulations, it is quite clear that the

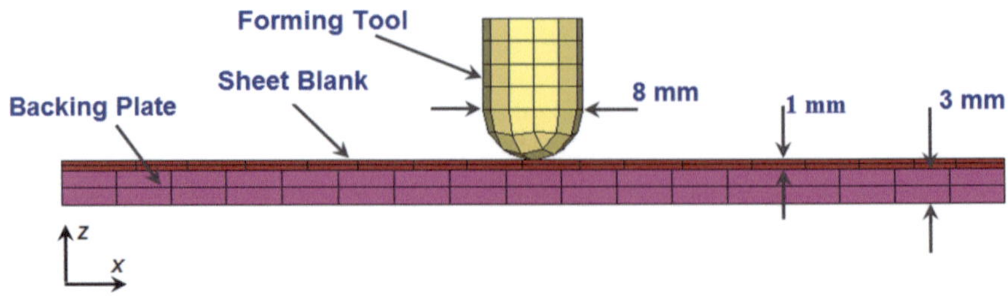

Fig. 6 Numerical simulation setup for ISF

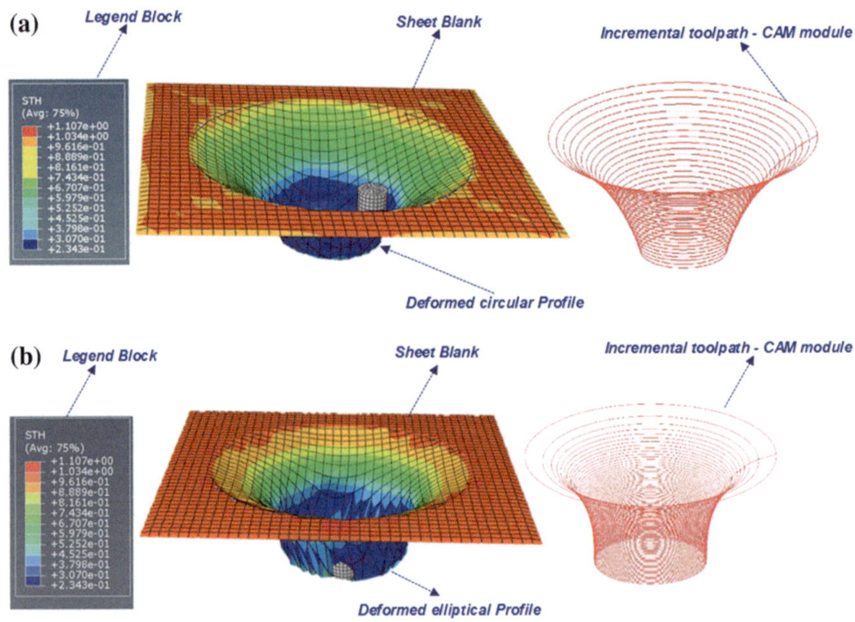

Fig. 7 FEA results of shell thickness for **a** circular wall, **b** elliptical wall profiles

deformation in SPISF is plane deformation i.e., stretching in the radial direction and compressing in the normal direction. The results of the simulations are well in accordance with the previous studies (Nirala et al. 2015; Han and Mo 2008). In previous studies, the authors have conducted similar simulations and found similar decreasing trends of sheet thickness along forming depth. In all three cases, the minimum thickness is observed near the base of the formed profile. From the stress-strain relationship, it can be concluded and also observed in simulations that over the formed profiles, where the strain is higher, stress will also be reported as higher (Fig. 7). Hence, in simulated profiles, maximum stress will be observed in the region of maximum thinning (Fig. 7). The advantage of using numerical analysis is to predict the sheet thinning (Nirala et al. 2017) in ISF before doing real experiments. Fine meshing in simulations gives accurate results (percentage error <5%) but fine meshing takes very long computational time. It takes few days or even weeks to do the analysis of a geometry when very fine meshing is used (shell element size <0.2 mm, for considered geometries).

5 Experimental Exploration

One cannot solely depend on simulation results because every solver of a numerical analysis software gives different results for the same problem. This is primarily due to the various assumptions considered while designing a numerical analysis setup. Hence, the results from the simulations need to be verified by the experiments for confirmation. The *x*, *y* and *z* coordinate extracted from the toolpath program can be utilized to make an NC part program for a 3 axis vertical milling machine (here, BFW, Model: VF 30 CNC VS). A fixture to conduct the experiments is designed and fabricated as shown in Fig. 8. A backing plate is generally used to avoid undesired bending near the periphery of the geometry (Duflou et al. 2008). For each of the experiments, a tungsten carbide made hemispherical end shape forming tool with 8 mm diameter is chosen. The sheet blanks used in the present study are 1-mm-thick aluminum sheets (70 × 70, Al 6101-T6). Al 6101-T6 is a commonly used alloy in aerospace, aviation, and marine industry. The mechanical properties of Al 6101-T6 are shown in Table 1. Feed of 500 mm/min is provided to the tool which is rotating at 500 RPM.

The sheet metal blanks of aluminum alloy are prepared by trimming operation and then clamped over the fabricated fixture. Serigraphy method is used to print the

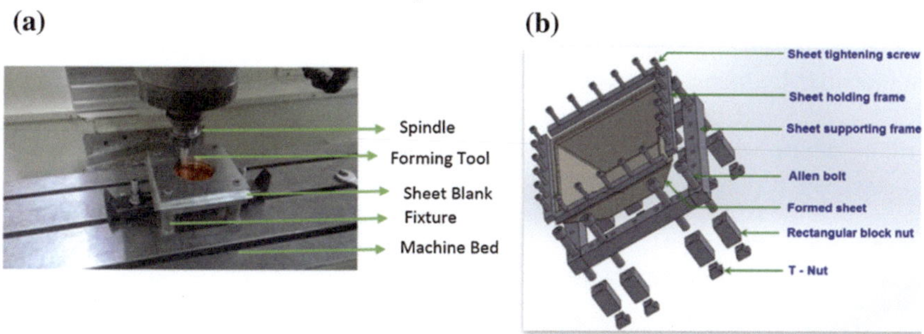

Fig. 8 Fixture for ISF. **a** Experimental setup and **b** CAD model

Table 1 Mechanical properties of Aluminium 6101-T6

Properties	Magnitude
Density	2.7 g/cc
Melting point	600 °C
Poisson's ratio	0.33
Tensile strength	97 MPa
Modulus of elasticity	70 GPa
Yield strength	76 GPa

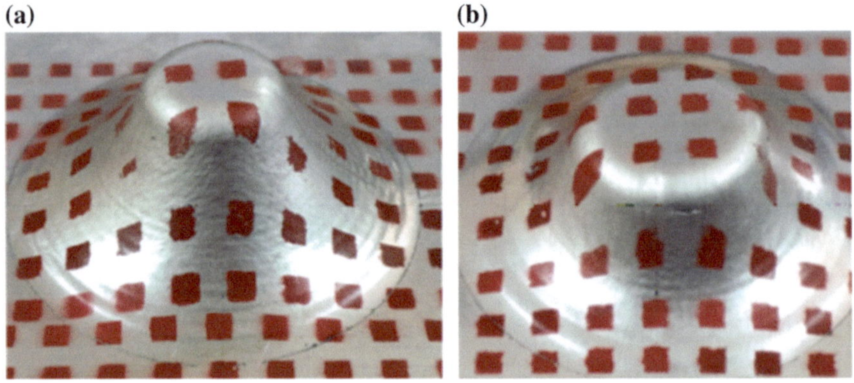

Fig. 9 Formed profiles. **a** Circular wall and **b** elliptical wall

grid pattern over the sheet. This method is one of the easiest and cost-effective methods of grid marking (Hariharan and Balaji 2009). This is done in order to measure the surface strains after deformation. The profiles are chosen so that crack can be observed, which the most critical limitation of this process. Multiple experiments were carried out for the different range of profiles, so that the crack initiation could be observed at shallow forming depth. Formed components, circular and elliptical wall profiles, by SPISF are shown in Fig. 9. To obtain the thickness profiles, the formed geometries are mounted from the fixture and their cut sections are prepared by using Electrical Discharge Machining (EDM). The thickness profiles are prepared by measuring the sheet thickness of formed geometries along forming depth using a micrometer with least count ± 1 μm.

6 Results and Discussion

In this section, results of thickness and strain distribution with forming depth are presented and discussed. Also, a comparison based study of the formed components (circular and elliptical wall profiles) is presented using Fracture Forming Limit

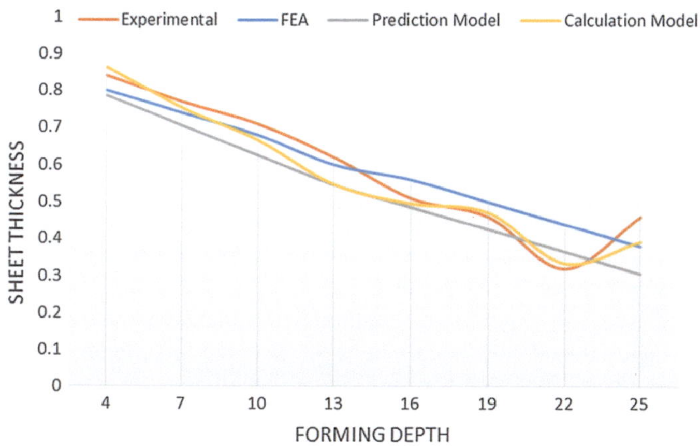

Fig. 10 Thickness distribution comparison for circular wall profile (Nirala and Agrawal 2016)

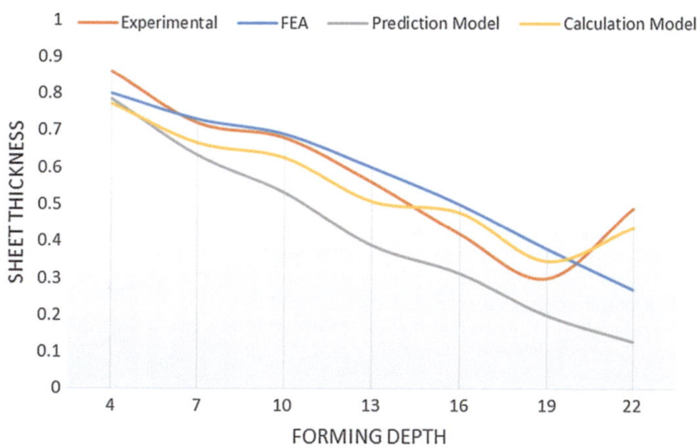

Fig. 11 Thickness distribution comparison for elliptical wall profile (Nirala and Agrawal 2016)

(FFL) diagram. From Figs. 10 and 11, it can be observed that the thickness of wall reduces with forming depth. The main reasons for thickness reduction are stretching, shearing and squeezing of the sheet in meridional, circumferential and through-thickness directions respectively (Jackson and Allwood 2009). It has been observed during experiments that the thickness near the base is slightly more than the region where the crack has occurred. The possible reason for this can be the accumulation of stretched material near the tool contact point. FEA and prediction model do not consider this factor which results in higher average percentage errors.

However, up to crack region, the values of percentage error remain below 10%. From Fig. 10, circular wall profile, the average percentage errors of FEA, prediction model and calculation model from experimental data are 11, 12, and 5% respectively. From Fig. 11, elliptical wall profile, the average percentage errors of FEA, Prediction model and calculation model on comparison with experimental data are 15, 29, and 10% respectively.

It took nearly 48 h to complete one FEA simulation of the single stage ISF. One can clearly see (Figs. 10 and 11) that the prediction model can be used over FEA to predict the approximate thickness distribution up to crack initiation. Calculation model can be used over experimental method to calculate thickness distribution. On comparing Figs. 10 and 11, it can be seen that the maximum forming depth achieved before crack initiation for the circular and elliptical wall is 25 and 22 mm respectively. Repeatability tests were also conducted to come to any conclusion. Hence, it can be concluded from the available results that, the prediction and calculation models fit better for the curvilinear profiles, where the wall angle variation from top to bottom of the wall is less (circular wall profile). For curvilinear profiles, with the high variation of wall angle from top to bottom of the wall (elliptical profile), the proposed models show high variance from actual experimental results.

Figure 12, shows the Fracture Forming Limit (FFL) diagram containing the Forming Limit Curve (FLC). The FLC is plotted from the data available in the

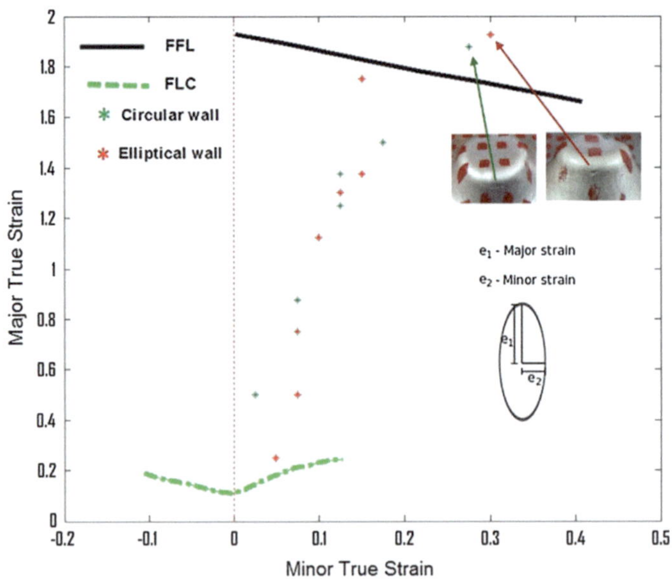

Fig. 12 FFL comparison for circular wall and elliptical wall (Nirala and Agrawal 2016)

literature, which was collected by conducting a combination of simple tensile test and Dome test (Silva et al. 2011). The reason of showing FLC in this study is to highlight the higher formability in ISF. Sheet cracked in the case of circular and elliptical wall profiles, whereas conical geometry successfully formed without any fracture. The points where the crack is observed are shown above the FFL line. In this study, analytical model available in the literature (Duflou et al. 2008; Hussain and Gao 2007), for thickness prediction is modified. Also, the thickness calculation model available in the literature (Filice 2006) is further derived and implemented using grid deformation analysis. However, the developed methodology by which this study has been conducted has not been found in the available literature.

7 Summary

In this work, a methodology is presented to predict and calculate the sheet thickness for the components fabricated by ISF process. The prediction model is developed to predict the sheet thickness before real experiments, while calculation model is developed to calculate the sheet thickness after real experiments.

- Thickness prediction model can predict sheet thickness under acceptable limits (approximate average % error—15%) up to crack initiation depth.
- Thickness calculation model can calculate the thickness within acceptable limits up to full forming depth.
- For a curvilinear profile, when the range of angle between two extreme ends is less and not very steep then the analytical models give better results. The range of angle between two extreme ends of the circular profile is less than the elliptical profile. Hence, results in the circular profile are better than the elliptical profile.
- The results from analytical models and calculation models are in good agreement with FEA and experimental results.
- For future study, material modeling using advanced material models like GTN, to predict damage initiation and propagation in single stage can be implemented in conjunction with developed prediction and calculation models.

Appendix

Pseudo Code

```
clc; clear all; clear;

Decide a constant value of stretching / Scallop Height; %For variable increment in negative z direction

Create empty arrays for x, y and z values;

Create an array for calculating wall angle at every increment;

Define initial constant step depth;

Define major base radius=r;

Program of wall angle (theta) calculation                % Based on equation 7

Forming Depth = 0;

While (angle >initial wall angle && angle<final wall angle)

    Call program of wall angle calculation;

        Generation of profile contour with forming depth

    Forming depth = Forming depth + step depth;

    If (depth>required depth)

        Break;

    End

End
```

References

Ambrogio, G., L. De Napoli, L. Filice, F. Gagliardi, and M. Muzzupappa. 2005. Application of incremental forming process for high customised medical product manufacturing. *Journal of Materials Processing Technology* 162: 156–162.

Ambrogio, G., C. Ciancio, L. Filice, and F. Gagliardi. 2017. Innovative metamodelling-based process design for manufacturing: an application to Incremental Sheet Forming. *International Journal of Material Forming* 10 (3): 279–286.

Araújo, R., P. Teixeira, M.B. Silva, A. Reis, and P. Martins. 2013. Single point incremental forming of a medical implant. *Key Engineering Materials* 554: 1388–1393. Trans Tech Publications.

Bagudanch, I., L.M. Lozano-Sánchez, L. Puigpinós, M. Sabater, L.E. Elizalde, A. Elías-Zúñiga, and M.L. Garcia-Romeu. 2015. Manufacturing of polymeric biocompatible cranial geometry by single point incremental forming. *Procedia Engineering* 132: 267–273.

Bambach, M. 2010. A geometrical model of the kinematics of incremental sheet forming for the prediction of membrane strains and sheet thickness. *Journal of Materials Processing Technology* 210 (12): 1562–1573.

Bhoyar, P.K., and A.B. Borade. 2015. The use of single point incremental forming for customized implants of unicondylar knee arthroplasty: a review. *Research on Biomedical Engineering* 31 (4): 352–357.

Cao, T., B. Lu, D. Xu, H. Zhang, J. Chen, H. Long, and J. Cao. 2015. An efficient method for thickness prediction in multi-pass incremental sheet forming. *The International Journal of Advanced Manufacturing Technology* 77 (1–4): 469–483.

Chu, T.C., W.F. Ranson, and M.A. Sutton. 1985. Applications of digital-image-correlation techniques to experimental mechanics. *Experimental Mechanics* 25 (3): 232–244.

Dejardin, S., Gelin, J.-C., Thibaud, S: On-line thickness measurement in incremental sheet forming process. In *Proceedings of the 13th International Conference on Metal Forming, Toyohashi*, 19–22 Sept 2010, 938–941.

Duflou, J.R., J. Verbert, B. Belkassem, J. Gu, H. Sol, C. Henrard, and A.M. Habraken. 2008. Process window enhancement for single point incremental forming through multi-step toolpaths. *CIRP Annals-Manufacturing Technology* 57 (1): 253–256.

Eksteen, P. D. W. 2013. *Development of incrementally formed patient-specific titanium knee prosthesis*. Doctoral dissertation, Stellenbosch: Stellenbosch University).

Eksteen, P.D.W., and A. Van der Merwe. 2012. Incremental sheet forming (ISF) in the manufacturing of titanium based plate implants in the bio-medical sector. *Computers and Industrial Engineering* 42: 15–18.

Filice, L. 2006. A phenomenology-based approach for modelling material thinning and formability in incremental forming of cylindrical parts. *Proceedings of the Institution of Mechanical Engineers, Part B: Journal of Engineering Manufacture* 220 (9): 1449–1455.

Gatea, S., H. Ou, and G. McCartney. 2016. Review on the influence of process parameters in incremental sheet forming. *International Journal of Advanced Manufacturing Technology* 87 (1): 479–499.

Hadoush, A., and A.H. Van den Boogaard. 2008. Time reduction in implicit single point incremental sheet forming simulation by refinement-derefinement. *International Journal of Material Forming* 1: 1167–1170.

Han, F., and J.H. Mo. 2008. Numerical simulation and experimental investigation of incremental sheet forming process. *Journal of Central South University of Technology* 15 (5): 581–587.

Hariharan, K., and C. Balaji. 2009. Material optimization: A case study using sheet metal-forming analysis. *Journal of Materials Processing Technology* 209 (1): 324–331.

Hild, F., and S. Roux. 2006. Digital image correlation: from displacement measurement to identification of elastic properties—A review. *Strain* 42 (2): 69–80.

Hirt, G., J. Ames, and M. Bambach. 2005. A new forming strategy to realize parts designed for deep-drawing by incremental CNC sheet forming. *Steel Research International* 76 (2–3): 160–166.

Hussain, G., and L. Gao. 2007. A novel method to test the thinning limits of sheet metals in negative incremental forming. *International Journal of Machine Tools and Manufacture* 47 (3): 419–435.

Jackson, K., and J. Allwood. 2009. The mechanics of incremental sheet forming. *Journal of Materials Processing Technology* 209 (3): 1158–1174.

Jeswiet, J., F. Micari, G. Hirt, A. Bramley, J. Duflou, and J. Allwood. 2005. Asymmetric single point incremental forming of sheet metal. *CIRP Annals-Manufacturing Technology* 54 (2): 88–114.

Kim H. J., and Lee D. 1996. Further development of experimental methods to verify computer simulations. In *Proceedings of NUMISHEET, Dearbom, Michigan, USA*, 316–323.

Kim, Y.H., and J.J. Park. 2002. Effect of process parameters on formability in incremental forming of sheet metal. *Journal of Materials Processing Technology* 130: 42–46.

Leszak, E. 1967. U.S. Patent No. 3,342,051. Washington, DC: U.S. Patent and Trademark Office.

Li, J.C., L.I. Chong, and T.G. Zhou. 2012. Thickness distribution and mechanical property of sheet metal incremental forming based on numerical simulation. *Transactions of Nonferrous Metals Society of China* 22: s54–s60.

Liu, Z., Y. Li, and P.A. Meehan. 2013. Vertical wall formation and material flow control for incremental sheet forming by revisiting multi-stage deformation path strategies. *Materials and Manufacturing Processes* 28 (5): 562–571.

Liu, Z., W.J. Daniel, Y. Li, S. Liu, and P.A. Meehan. 2014. Multi-pass deformation design for incremental sheet forming: Analytical modeling, finite element analysis and experimental validation. *Journal of Materials Processing Technology* 214 (3): 620–634.

Lu, B., Y. Fang, D.K. Xu, J. Chen, S. Ai, H. Long, and J. Cao. 2015. Investigation of material deformation mechanism in double side incremental sheet forming. *International Journal of Machine Tools and Manufacture* 93: 37–48.

Mason, B. 1978. *Sheet metal forming small batches*. Bachelor's thesis, University of Nottingham.

Matsubara, S. 2001. A computer numerically controlled dieless incremental forming of a sheet metal. *Proceedings of the Institution of Mechanical Engineers, Part B: Journal of Engineering Manufacture* 215 (7): 959–966.

Milutinovića, M., R.L.M. Potranb, D. Vilotića, P. Skakuna, and M. Plančaka. 2014. Application of single point incremental forming for manufacturing of denture base. *Journal for Technology of Plasticity* 39 (2): 15–23.

Mirnia, M.J., B.M. Dariani, H. Vanhove, and J.R. Duflou. 2014. Thickness improvement in single point incremental forming deduced by sequential limit analysis. *The International Journal of Advanced Manufacturing Technology* 70 (9–12): 2029–2041.

Nguyen-Tran, H.D., H.S. Oh, S.T. Hong, H.N. Han, J. Cao, S.H. Ahn, and D.M. Chun. 2015. A review of electrically-assisted manufacturing. *International Journal of Precision Engineering and Manufacturing-Green Technology* 2 (4): 365–376.

Nirala, H. K, Paul, A., Singh, A., Agrawal, A. 2015. Adaptive step depth for uniform stretching in single point incremental forming. In *International Conference on Precision, Meso, Micro and Nano Engineering (COPEN 2015)*, December 10–12, IIT Bombay, India.

Nirala, Harish Kumar, and Anupam Agrawal. 2016. Approach for prediction and calculation of sheet thinning in incremental sheet forming (AIMTDR-2016), December 16–18, 2016 at College of Engineering, Pune, Maharashtra, INDIA.

Nirala, H.K., P.K. Jain, J.J. Roy, M.K. Samal, and P. Tandon. 2017. An approach to eliminate stepped features in multi-stage incremental sheet forming process: Experimental and FEA analysis. *Journal of Mechanical Science and Technology* 31 (2): 599–604.

Paniti, I., & Paroczi, A. 2011. Design and modeling of integrated hall-effect sensor based on-line thickness measurement device for incremental sheet forming processes. In *2011 IEEE/ASME International Conference on Advanced Intelligent Mechatronics (AIM)*, 297–302. IEEE.

Piegl, L., and W. Tiller. 1997. *The NURBS book (Monographs in visual communication)*, 2nd ed. New York: Springer.

Silva, M.B., M. Skjodt, A.G. Atkins, N. Bay, and P.A.F. Martins. 2008a. Single-point incremental forming and formability—Failure diagrams. *The Journal of Strain Analysis for Engineering Design* 43 (1): 15–35.

Silva, M.B., M. Skjodt, P.A. Martins, and N. Bay. 2008b. Revisiting the fundamentals of single point incremental forming by means of membrane analysis. *International Journal of Machine Tools and Manufacture* 48 (1): 73–83.

Silva, M.B., P.S. Nielsen, N. Bay, and P.A. Martins. 2011. Failure mechanisms in single-point incremental forming of metals. *The International Journal of Advanced Manufacturing Technology* 56 (9–12): 893–903.

Skjodt, M., M.B. Silva, P.A.F. Martins, and N. Bay. 2010. Strategies and limits in multi-stage single-point incremental forming. *The Journal of Strain Analysis for Engineering Design* 45 (1): 33–44.

Xu, D., R. Malhotra, N.V. Reddy, J. Chen, and J. Cao. 2012. Analytical prediction of stepped feature generation in multi-pass single point incremental forming. *Journal of Manufacturing Processes* 14 (4): 487–494.

Young, D., and J. Jeswiet. 2004. Wall thickness variations in single-point incremental forming. *Proceedings of the Institution of Mechanical Engineers, Part B: Journal of Engineering Manufacture* 218 (11): 1453–1459.

Optimum Machines Allocation in a Serial Production Line Using NSGA-II and TOPSIS

Masood Ashraf, Faisal Hasan and Qasim Murtaza

Abstract In the present scenario, manufacturing industries are witnessing large fluctuation in the product demands. The need of the hour is to have a responsive manufacturing system, which can cope up with these kinds of stochastic events. Reconfigurable Manufacturing System (RMS) is considered as modern manufacturing paradigm that offers customised functionality and capacity as and when required. The key enablers for this customisation in functionality and capacity are the Reconfigurable Machine Tools (RMTs) or simply Reconfigurable Machines (RMs). Selecting these multi-functionality and capacity machines along stations/stages of a product flow line is an important aspect for operating RMSs as it has direct implications on the operational performance of such system. In this paper, optimal configurations are selected under conflicting objectives based on cost, operation capability and reliability of the machines. The objective is to assign those machine configurations that minimises the cost while maximising the operation and reliability of the production line. The problem is framed as a multi-objective optimisation problem and is solved using NSGA-II. The results so obtained are discussed in terms of the performance of the reconfigurable system.

Keywords RMS · NSGA-II · Optimality

M. Ashraf (✉) · F. Hasan · Q. Murtaza
Department of Mechanical Engineering, Z.H. College of Engineering and Technology, Aligarh Muslim University, Aligarh 202002, Uttar Pradesh, India
e-mail: masoodashraf18@gmail.com

F. Hasan
e-mail: f.hasan.me@amu.ac.in

Q. Murtaza
e-mail: qasimmurtaza@gmail.com

S. S. Pande and U. S. Dixit (eds.), *Precision Product-Process Design and Optimization*, Lecture Notes on Multidisciplinary Industrial Engineering,
https://doi.org/10.1007/978-981-10-8767-7_16

1 Introduction

Until 1970s, minimising cost was the prime objective of manufacturing enterprises. In 1970s, the Japanese industry confirmed that good quality products do not require additional cost to produce. Thus, producing good quality products at inexpensive price became a new aim of the manufacturing firms. However, in the early 1990s, it was predicted that globalisation would turn into a most important factor for industrial competition, with characteristics of short openings of market opportunity, recurrent product introductions and swift alterations in product demand (Elmaraghy 2011; Koren 2013).

Over previous some years, this dynamic economic sector has met substantial changes due to globalisation which includes lack of specialised and trained workforce, its apprehension about the environs and necessity for high-class custom-made products at competitively economical rates (Elmaraghy 2011). Nowadays, the manufacturing enterprises have moved into a new era in which customisation, on time delivery and adjustment for demands fluctuations are to be looked upon. Amplified consumers' purchasing power has led to increased worldwide competition resulted in regular introduction of customised products which causes large variations in product demands (Dashchenko 2006; Ashraf and Hasan 2015). Besides this, short product life cycles, volatile demands, high demand of custom-made products and fast moving process technology have compelled the industries to acclimatise with the changeable manufacturing according to their requisite (Ashraf and Hasan 2015; Al-Zaher and ElMaraghy 2013; Haddou-Benderbal 2016). The progressively difficult market dynamics due to shorter life cycles and unexpected forecasts requires higher flexibility in the supply chain and in particular with the manufacturing systems (Matt and Rauch 2013). Therefore, developing manufacturing systems with the capability to rapidly respond to market variations has turn out to be a determined goal for the modern industry (Elmaraghy 2011; Koren 2013).

At the early start of twentieth century, enterprises started implementing the conventional manufacturing systems. These manufacturing systems have progressed from job shops productions, which are characterised by general-purpose machines, high variety, low capacity and substantial human participation, to high capacity, low-variety Dedicated Manufacturing Lines (DMLs) motivated by the economy of scale (Koren and Shpitalni 2010). Specifically, in the 1900s, the way of production had dramatically changed due to the implementation of mass production and dedicated manufacturing lines (Gola 2012). DML is one of the conventional manufacturing approaches that involve the production of identical items at a time, though it is cost-effective for bulk production but with no variety at all (Koren and Shpitalni 2010; El Maraghy 2006). Hence, the DMLs failed to respond to changes as and when required and that ultimately led to losses to the organisation.

1.1 Dedicated Manufacturing Lines

It is well known that the introduction of the dynamic assembly line in the year 1913 established the commencement of the mass production model. Yet, it is rarely known that mass production was made conceivable only by the invention of Dedicated Manufacturing Lines (DMLs) which are known for having high production volumes for producing any single part type. Since, owing to large volume, unit cost of production is less, thus providing large profits for the products having high demand rates. Till 1990s, these DMLs were proved to be the most cost-effective systems for making large quantities of identical products. These dedicated lines tend to be cost-effective for the period when demand accords to the supply. However, there may be situations in which DMLs may not operate at its full capacity, thereby reducing profits or eventually loss occurs. Producing a range of products in terms of variety is impossible with a DML, and therefore their role in today's manufacturing environment is decaying (Koren 2006).

1.2 Group Technology and Cellular Manufacturing Systems

In 1960s, the concept of Group Technology (GT) was introduced and consequently, the idea of Cellular Manufacturing Systems (CMSs) was presented. The first article was written on GT by Mitrofanov (1966) and the first journal article with the notion of CMS was published in the year 1969 (Opitz et al. 1969). A statistical review paper on GT and CMS over the period of 1965–1995 was also published (Reisman et al. 1997). CMS is one of the utmost vital implications of GT principles to manufacturing which emphasises on grouping of similar parts into families and interrelated machines into clusters so that single or multiple part families can be manufactured within one machine group. This process of defining machine groups and part families is called as the Cell Formation (CF) problem (Yin and Yasuda 2006). The problem involved in validation of CM systems has received much attention. Most of the works focused on the comparative performance measure between cellular layout and functional layout (Yin and Yasuda 2006), though the CMS fails in providing flexibility outside the part family domain. When machine cells are designed for part families, it is hard to physically rearrange the facilities of the cell as per new manufacturing requirements. This inflexibility inhibits CMSs in dealing with contemporary challenges, e.g. dynamic part mix and demand fluctuation, lead-time reduction which are the primary requisite of agile manufacturing (Pattanaik et al. 2007).

1.3 Flexible Manufacturing

With the invention of Numerically Controlled (NC) machines and later introduction of computerised numerically controlled machines (Koren 1983) laid down the foundation of flexible manufacturing. The concept of flexible manufacturing to some extent answers the need of customisation and has some degree of responsiveness to variations in products, production technology and market trends (El Maraghy 2006). In second half of the twentieth century, FMS was introduced and in 1981, for the first time a mathematical model and its corresponding solution were suggested for FMS (Stecke and Solberg 1981). These systems were introduced to respond moderate volume-variety production requirements. It anticipated the differences in parts and built-in flexibility on priority basis; thus, they are more robust but have high initial capital cost and the flexibility features are occasionally underutilised which lead to the loss to the organisation (El Maraghy 2006; Hasan and Jain 2015).

All these conventional manufacturing systems failed to fulfil the modern manufacturing objectives of flexibility and responsiveness (Hasan 2014), specifically when dealing with the stochastic demands, high customisation and shorter product life cycles. These volatile circumstances forced the industrialists to handle the manufacturing systems more efficiently (Galan et al. 2007). Summarily, it can be concluded that these traditional manufacturing systems cannot respond to these new market challenges, and even the FMSs failed to cater these new requirements in a profitable manner.

1.4 Reconfigurable Manufacturing Systems

In order to cope up these present-day challenges, the concept of RMS was proposed (Koren 1997). RMS aims at joining the high throughput characteristic of DML with the flexibility attribute of FMSs, conserving the capability for dealing with diversity of products and volumes in a profitable manner. This is attained through rapid modification in its structure, both in its hardware and software modules, for complying rapid adjustment of the exact capacity and functionality as and when needed (El Maraghy 2006). In a nutshell, it can be said that RMS is a new class of manufacturing system which are characterised by adjustable capacity and functionality in order to cope up with product variety and production volume (El Maraghy 2006). Thus, RMSs have custom-built flexibility to handle product variety and is open-ended so that it can be modified and reconfigured, rather than substituted in future (Mehrabi et al. 2000). Moreover, one of the distinguishing features of these systems is that the focus is both 'on part' as in DMLs and 'on machines' as in the case of FMSs. Thus, RMS can be viewed as systems which provides customised capacity and functionality as and when needed and its configuration can be either dedicated or flexible, or in between and can be modified as and when required as depicted in Fig. 1.

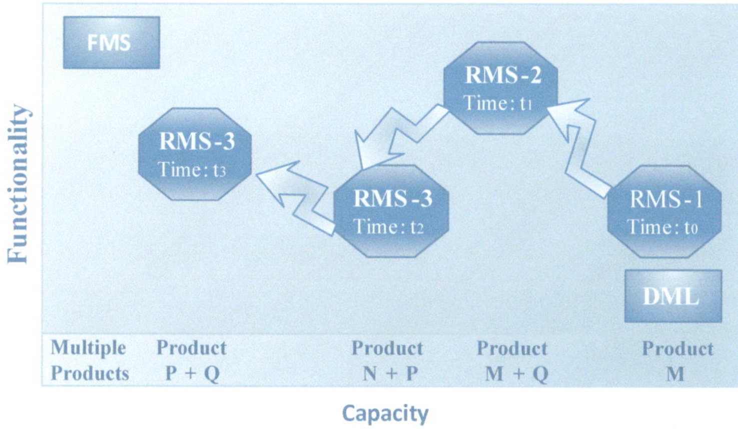

Fig. 1 Manufacturing systems: DML and FMS are non-dynamic systems, while an RMS is a dynamic system

The essential and distinctive ingredients of RMSs are the reconfigurable machines, which facilitate changes in the organisation structure to house new product types and their required volume. The conceptualisation, design and development of reconfigurable machines were proposed by Koren (1999), and later contributions towards the development of modular machines were suggested by Landers (2001) as well as Moon and Kota (2002) also. The different modules of these machines are generally classified as basic and auxiliary modules (Goyal et al. 2012).

The structural elements of the machines like base, slideways, columns, bed, etc. are termed as basic or essential modules while the auxiliary modules are taking care of dynamic action of the machines; this comprises components like tool changers, spindle heads, adapter plates, indexing units, angle structure, etc. These machines are referred as Reconfigurable Machine Tools (RMTs) which involve basic and auxiliary module library as shown in Fig. 2.

In a simple serial reconfigurable product flow line, the jobs are processed on RMTs from one stage or station to subsequent stages or stations. The type of operation required on stages is determined by the operation precedence. The configuration of selected machines and other subsystems within the production system may either contribute or impede the system's receptiveness in terms of throughput, responsiveness, convertibility and scalability (Koren 2013). For a serial production flow line, various applied configurations can be re-adjusted relying upon (1) the manner in which the machines are settled in the stages, and (2) which machine configuration is used for the ideal operation on the stages. The selection of machine configurations is crucial for the production in terms of throughput or production rate (Hasan and Jain 2015). Thus, optimal selection of machine configurations plays a vital role in deciding the performance of these systems. In the following section, some relevant literature on RMSs are presented.

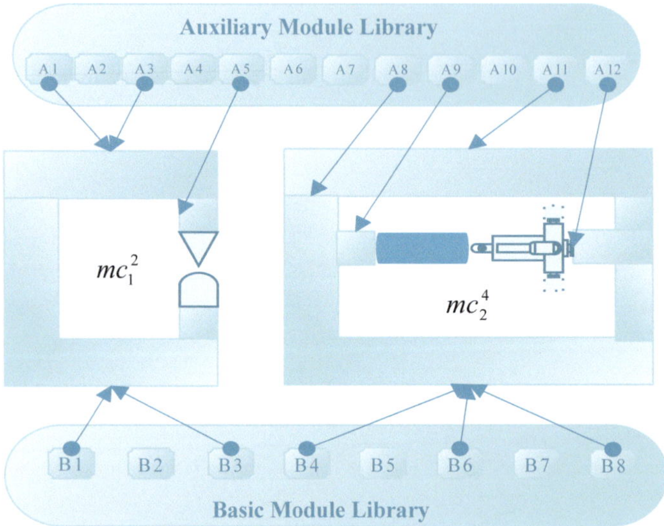

Fig. 2 RMT basic and auxiliary module library

In formulating an optimal configuration selection problem, (Xiaobo et al. 2000) devised two algorithms and evaluated the efficiency of those algorithms. A work (Goyal et al. 2012) based on the module interfaces and machine capability measured the machine reconfigurability, cost and operational capability of an RMT. Similarly, another work (Ashraf 2016) selected optimal configurations on the basis of cost maximisation, minimisation of operation capability and reliability of the selected machine configurations which were taken as multiple objectives for the optimal machine assignment for an SPFL allowing paralleling of alike machines. These problems (Goyal et al. 2012; Ashraf 2016) were solved using NSGA-II. Likewise, (Hasan and Jain 2015) proposed an optimal machine assignment model based on a weighted objective function comprising reliability, cost, production time and operational capability as performance parameters. The role of performance parameters in selecting optimal machine configurations for different stages across any SPFL was also demonstrated. A GA-based methodology (Hasan et al. 2014) for optimal configuration selection in an SPFL was proposed, and it provided feasible solution towards optimal configuration selection in RMS, based on throughput maximisation. Other works, i.e. (Youssef and ElMaraghy 2007; Youssef 2006), proposed the selection of the near-optimal alternative configurations for each feasible demand scenario over the given configuration phases. Bensmaine (Abderrahmane et al. 2012) proposed an approach of candidate selection of reconfigurable machines amongst the available sets for the minimisation of the total cost and the total completion time. Likewise, another problem of an SPFL (Maniraj 2015) used ant colony optimisation approach for minimising the capital cost of the RMSs. Furthermore, a model (Hasan et al. 2014) was proposed for determining the optimal configuration needed for multiple part family orders and their corresponding optimum sequence on the basis of maximum revenue earned for a particular system configuration.

In the present work, the optimal configuration selection problem based on three parameters namely cost, operational capability and reliability is formulated and solved. These chosen parameters are linked to economic benefits, degree of reconfigurability and failure rate respectively. For the optimal selection of machine configurations, it is desired to select those machine configurations which would maximise the operational capability of RMT as well as of system, reliability and productivity of the system along with minimisation of the system cost. The problem is solved by employing NSGA-II and solution are ranked using TOPSIS technique.

2 Problem Formulation and Notations

During the production phase, RMS behaves as DML and can readily be reconfigured as per capacity and functionality needs (Koren 1997). The simplest layout for any production system is a serial flow line layout, which is being considered here too. Further, in literature, most of the RMS modelling approaches (El Maraghy 2006; Hasan and Jain 2015; Goyal et al. 2012; Dou et al. 2009) used serial flow line for modelling and analysis, which is also one of the reasons it has been used for the present problem. The following notations are used in the present work:

mc_p^q	Machine p in its qth configuration $1 < p \leq P$ and $1 < q \leq Q$
n_p^q	Number of machines p in qth configuration required to satisfy the demand Δ
Δ	Demand rate (parts per hour)
$\mathbb{F}_o$	A set of feasible alternative machine configurations to perform the oth $(1 < o < O)$ operation, $\{(p_1, q_1), (p_2, q_2), \ldots, (p_f, q_f), \ldots, (p_{F_o}, q_{F_o})\}$
f	Feasible alternative $(1 < f < \mathbb{F}_o)$ is defined as (p_f, q_f), where p_f specifies the feasible machine and q_f specifies the feasible machine configuration
C_p^q	Cost of pth machine in qth configuration
$pr_{p,o}^q$	Production rate of machine p with qth configuration for performing oth operation
$\phi_{p,o}^q$	$\begin{cases} 1 & \text{if operation o is performed} \\ 0 & \text{otherwise} \end{cases}$
$\mathrm{TC}_{p,q}$	Total cost of a pth feasible alternative machine in its qth configuration to perform oth operation to satisfy a specific demand Δ
$\mathrm{OC}_{p,q}$	Operational capability of a pth feasible alternative machine in its qth configuration
s	Number of stages in flow line $1 < s \leq S$
$R_{p,q}$	Reliability of a pth feasible alternative machine in its qth configuration
R_{system}	Reliability of the overall system

The three performance parameters that are considered for the choosing machine configurations on various stages/station of the product line are cost, operational capability and reliability. The details of these performance measures are as under.

2.1 Cost

Cost or capital cost is the price incurred while using a particular machine configuration for a desired operation on a job. Since cost is an economical parameter, this non-beneficial attribute is to be minimised (Goyal et al. 2012). Cost depends on the nature of operation to be executed on any feasible machine configuration, i.e. the cost changes, if for the same operation any other feasible configuration is used. The total cost ($TC_{p,q}$) of a feasible alternative machine configuration for carrying out the operation at a specified demand rate is determined using Eq. (1):

$$\begin{aligned} \mathrm{TC}_{p,q} &= n_p^q \times C_p^q \\ \text{where, } n_p^q &= \left\lceil \frac{\Delta}{pr_{p,o}^q} \right\rceil \end{aligned} \tag{1}$$

The fraction $\Delta / pr_{p,o}^q$ is approximated to higher integer, as the number of machines cannot be a fraction.

2.2 Operational Capability

The operational capability of a machine configuration depends upon the range of operations that can be readily executed by an RMT on a particular configuration. The operational capability ($\mathrm{OC}_{p,q}$) of machine configurations is computed using the following relationship (Goyal et al. 2012):

$$\mathrm{OC}_{p,q} = \left[\left(\sum_{o=1}^{O} \phi_{p,o}^q \right) - 1 \right]^{\alpha} \tag{2}$$

It is to be noted that a machine must perform at least two operations in a particular configuration to have at least an operational capability of one.

2.3 Reliability

Reliability describes the system uptime. Reliability simply estimates the probability that a given system is producing parts under given environment for a stated time interval, i.e. working without failure. Higher probability results in less system failure and consequently less downtime and loss of production. Reliability is a beneficial attribute and is to be maximised. Reliability of serial configuration is evaluated using Eq. (3)

$$\mathrm{R_{system}} = \prod_{s} (\text{Reliability of } sth \text{ station group}) \tag{3}$$

3 Objective Functions

The objective functions for the present problem are presented in Eq. (4). The overall optimisation problem is solved as a minimisation problem; therefore, the objective functions F_2 and F_3 are multiplied by -1 to convert them into a minimisation problem.

$$\left.\begin{aligned} \text{Minimize}, F_1 &= \sum_{S=1}^{S} \mathrm{TC}_{p,q} \\ \text{Maximize}, F_2 &= \sum_{s=1}^{S} \mathrm{OC}_{p,q} \\ \text{Maximize}, F_3 &= \prod_{s=1}^{S} R_{\text{system}} \end{aligned}\right\} \tag{4}$$

3.1 Multi-objective Optimisation Problem

Various day-to-day problems encompass various performance parameters or objectives that are needed to be optimised simultaneously. These parameters are called as objective functions and such kind of optimisation problems which entail multiple objective functions are known as Multi-Objective Optimisation Problems (MOOPs). The multiple objective functions in a MOOP may be optimised independently from each other; however, suitable solutions for the overall problem can seldom be found in this way (Fonseca and Fleming 1995). Hence, a MOOP must involve concurrent optimisation of a number of objectives functions which are to be either maximised or minimised. It may enfold multiple constraints where any viable solution (i.e. all optimal solutions) needs to satisfy (Deb 2011). Since the formulated problem is a MOOP with conflicting objectives, cost is to be minimised and other two objectives are to be maximised. Hence, the MOOP for the present problem is represented by Eq. (4).

For a MOOP, it is not feasible to have solely one solution which at once optimises all objectives. Thus, NSGA-II is used in the present MOOP problem. Most of the multi-objective problems use the idea of domination where two solutions are picked and are compared on the measure that one dominates the other solution or not. For a minimisation problem, a solution is assumed to dominate the other solution if the following both conditions are followed:

i. The solution $x^{(1)}$ is not inferior than $x^{(2)}$ in all objectives.
ii. The solution $x^{(1)}$ is strictly superior than $x^{(2)}$ in at least one objective.

4 Revised Non-dominated Sorting Genetic Algorithm

Due to some drawbacks of evolutionary algorithms (Coello 2007), a revised Non-dominated Sorting Genetic Algorithm (NSGA-II) was introduced by Deb et al. (2002), which is now established as an efficient algorithm for solving MOOPs. Deb et al. (2002) noticed that in most of the problems, NSGA-II has ability to discover considerably better spread of solutions and improved convergence that can be achieved close to the true Pareto-optimal front in contrast to the other evolution strategies. One of the significant features of NSGA-II is that the individuals having higher finesses are chosen from a combined pool of parent and offspring solutions (created after crossover and mutation of the parent population), and these solutions become the parents for the succeeding generation (Deb et al. 2002). NSGA-II preserves elitism which shrinks the diversity of the gene pool, i.e. the solutions having higher fitness remain preserved over the generations along with the other solution while the solutions with poor fitness get sweep out with growing number of generations. Also, it is comparatively faster with reduced computational complexity and gives better convergence of the solutions (Deb 2000). NSGA-II implements the concept of selection of non-dominated solutions and then forms the Pareto-fronts, the solutions obtained in the first front are the non-dominated solutions. Flowchart of NSGA-II has been presented in Fig. 3 and the following steps are performed for implementation of NSGA-II:

(1) A random population of size N is initialised.
(2) The initialised population is non-dominated sorted into different fronts.
(3) Individual chromosomes of each front are ranked based on front in which they belong to. In addition to ranking, crowding distance is calculated for each individual.
(4) Parents are chosen from the population by performing binary tournament selection based on the rank and crowding distance. The selected population produces offspring from crossover and mutation operators.
(5) The parent population is combined with offspring population to give a size of $2N$ and is sorted again based on non-domination criteria and ranked.
(6) The highest ranked fronts are used to create new parent population until N individuals are reached. The last front is selected based on rank and the crowding distance.
(7) Copy the new population to parent population and go to step 4 until stopping criteria met.
(8) The first non-dominated front of the newly selected population is solution of problem.

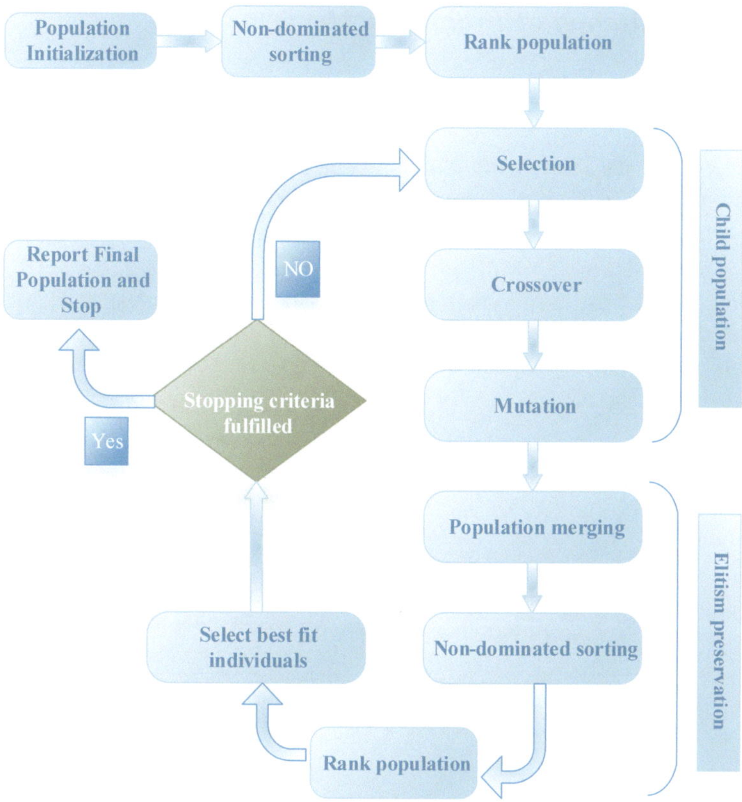

Fig. 3 Flowchart of NSGA-II

5 Technique for Order of Preference by Similarity to Ideal Solution (TOPSIS)

NSGA-II provides a large set of solutions that contains enormous number of non-dominated solutions, so it becomes difficult to choose the best-ranked solution along with good precision. Hence, an inclusive technique has been applied to rank the non-dominated solutions using TOPSIS method.

TOPSIS is a practical and convenient method for ranking and choosing a number of feasible alternatives (Hwang and Yoon 2011). The basic concept of TOPSIS is the chosen alternative that should have simultaneously shortest Euclidean distance from the positive ideal solution and the farthest Euclidean distance from the negative ideal solution. The positive ideal solution is obtained from a hypothetical alternative that has the best values for all considered criteria, whereas the negative ideal solution is identified with a hypothetical alternative that has the worst criteria values. Further, the Shannon Entropy Weight Theory (SEWT) is used for calculating the weight of each objective under consideration.

5.1 Normalisation of Evaluation Matrix

The normalisation process makes over different scales and units among various attributes into collective measurable units to allow the comparison of different parameters. The decision matrix $\widehat{X}$ comprising normalised values is obtained from non-dominated solutions X_{ij}, which is further treated to assign the rank to the non-dominated solutions. The element $\widehat{X}_{ij}$ of the decision matrix $\widehat{X}$ represents the value of the jth attribute for the ith solution, where $i = 1, 2...m$ and $j = 1, 2...n$. Equations 5 and 6 are used to perform normalisation of the evaluation matrix, such that for all the beneficial and non-beneficial attributes having higher values are desirable. Thus, the normalised evaluation matrix can be formed which is represented by Eq. (7):

$$\widehat{X}_{ij} = \left\{ \frac{\max_i\{X_{ij}\} - X_{ij}}{\max_i\{X_{ij}\} - \min_i\{X_{ij}\}}, \right. \quad \text{for non-beneficial attributes} \tag{5}$$

$$\widehat{X}_{ij} = \left\{ \frac{X_{ij} - \max_i\{X_{ij}\}}{\max_i\{X_{ij}\} - \min_i\{X_{ij}\}}, \right. \quad \text{for beneficial attributes} \tag{6}$$

$$\widehat{X} = \left\{ \widehat{X}_{ij} \right\}_{m \times n} \tag{7}$$

5.2 Parametric Weight Calculation by Shannon Entropy Theory

Shannon (1948) introduced the model of entropy into information theory which is called as Shannon entropy theory. Further, the concept of determining the objective weights based on information entropy which can measure the amount of useful information in the data was proposed by Zeleny and Cochrane (Zeleny 1973). In order to evaluate the objective weights by the entropy measure, the decision matrix elements in Eq. (7) need to be normalised for each attribute to obtain the projection value $\left(\widehat{P}_{ij}\right)$ of each attribute $\{A_j | j = 1, 2...n\}$ for all the alternative solutions $\{S_j | j = 1, 2...m\}$ as given in Eq. (8):

$$\widehat{P}_{ij} = \frac{1 + X_{ij}}{\sum_{i=1}^{m} \left(1 + X_{ij}\right)} \quad | \forall i \in \{1, 2...m\}; \forall j \in \{1, 2, ...n\} \tag{8}$$

Further, the normalised values from the decision matrix are used to calculate the entropy using Eq. (9):

$$E_j = -\frac{1}{\ln(m)}\sum_{i=1}^{m} P_{ij} \times \ln(P_{ij}) \quad |\forall i \in \{1, 2...m\}; \forall j \in \{1, 2, ...n\}\Big\} \tag{9}$$

5.3 Calculating Normalised Weight by Information Entropy Theory (IET)

The normalised weight is calculated by determining the respective degree of divergence (D_j) and objective weight $(\widehat{\omega}_j)$ of the attribute A_j. It characterises the essential contrast intensity of any attribute A_j. The larger will be the value of D_j, the more will be the significance of that attribute A_j in the problem. The expression of D_j for each attribute, the objective weight $(\widehat{\omega}_j)$ and weighted normalised values $\left(\widehat{V}_{ij}\right)$ are represented by Eqs. (10)–(12):

$$D_j = 1 - E_j \quad |\forall j \in \{1, 2, ...n\} \tag{10}$$

$$\widehat{\omega}_j = \frac{D_j}{\sum_{j=1}^{n} D_j} \quad |\forall j \in \{1, 2, ...n\} \tag{11}$$

$$\widehat{V}_{ij} = \omega_j \times \widehat{P}_{ij} \quad |\forall i \in \{1, 2, ...m\}; \forall j \in \{1, 2, ...n\}\Big\} \tag{12}$$

5.4 Determining the Separation Measure

The distance of an alternative from the best and worst alternatives is computed by determining the Positive Ideal Solution (PIS) and Negative Ideal Solutions (NIS). The PIS $(\widehat{V}_+)$ is the best desired alternative, while NIS $(\widehat{V}_-)$ is the least desired alternative for the particular criterion. Therefore, the values of PIS and NIS are evaluated for all the attributes $\{A_j | j = 1, 2...n\}$ as defined in Eqs. (13)–(14):

$$\widehat{X}_+ = \{\max(X_{i1}), \max(X_{i2}), ...\max(X_{in})\} \quad |\forall i \in \{1, 2, ...m\}; \forall j \in \{1, 2, ...n\}\Big\} \tag{13}$$

$$\widehat{X}_- = \{\min(X_{i1}), \min(X_{i2}), ...\min(X_{in})\} \quad |\forall i \in \{1, 2, ...m\}; \forall j \in \{1, 2, ...n\}\Big\} \tag{14}$$

The overall performance measure of an alternative is acquired by calculating its separation measure, i.e. the distance of each alternative $\{A_j | j = 1, 2 \ldots n\}$ from PIS and NIS which is given by Euclidean distance as defined in Eqs. (15)–(16):

$$\widehat{D}_{j_+} = \sqrt{\left(\sum_{j=1}^{j=n} \widehat{\omega}_j \left(\widehat{X}_{ij} - X_{j_+}\right)\right)^2} \quad \left| \forall i \in \{1, 2, \ldots m\}; \forall j \in \{1, 2, \ldots n\}; 0 < \widehat{D}_{j_+} < 1 \right\} \tag{15}$$

$$\widehat{D}_{j_-} = \sqrt{\left(\sum_{j=1}^{j=n} \widehat{\omega}_j \left(\widehat{X}_{ij} - X_{j_-}\right)\right)^2} \quad \left| \forall i \in \{1, 2, \ldots m\}; \forall j \in \{1, 2, \ldots n\}; 0 < \widehat{D}_{j_-} < 1 \right\} \tag{16}$$

5.5 *Calculating the Relative Closeness and Ranking the Priority*

The relative closeness (C_j) of the *j*th particular alternative to ideal solution is defined by Eq. (17) as follows:

$$C_j = \frac{D_{j_-}}{D_{j_+} + D_{j_-}} \quad | \forall i \in \{1, 2 \ldots m\}; \forall j \in \{1, 2, \ldots n\} \tag{17}$$

After evaluating the relative closeness using Eq. (17), if *j*th alternative is the PIS, then $C_j = 1$; however, if the *j*th alternative is the NIS, then $C_j = 0$. In other words, if the value of C_j is closer to 1 or 0, the *j*th alternative will be closer to PIS or NIS, respectively. The set of alternatives $\{A_j | j = 1, 2 \ldots n\}$ can be ranked now according to the descending order of values of C_j. Thus, how organisation may select the top-ranked solution according to need of industry.

6 Optimal Machine Assignment for Serial Production Line

Three performance indices considered for present study are applied to a serial production line (SPL). SPL allows paralleling of identical machines. An SPL with an operation sequence $1 \rightarrow 11 \rightarrow 20 \rightarrow 4 \rightarrow 13$ is considered. Each operation is consigned to a stage as per the precedence constraint of operations. After each stage is assigned an operation, the machine type and its configuration are selected suitable for that operation as per Table 1. The optimal machine assignment is tackled by

Table 1 RMT production rate in parts/h for performing various operations

mc_p^q	*Operation (o)*																			
	1	2	3	4	5	6	7	8	9	10	11	12	13	14	15	16	17	18	19	20
mc_1^1	–	–	–	12	–	–	–	13	–	–	–	30	–	–	–	22	–	–	–	–
mc_1^2	–	–	–	–	18	–	–	–	15	–	–	–	–	–	–	–	–	18	–	–
mc_1^3	–	–	25	–	–	–	17	–	–	–	–	–	–	–	–	16	–	–	–	–
mc_1^4	–	–	–	–	–	–	–	–	–	14	–	–	–	–	–	–	–	–	18	–
mc_2^1	24	–	–	–	–	15	–	–	–	–	–	12	–	–	–	–	–	–	–	20
mc_2^2	–	22	–	–	–	–	–	–	–	–	–	–	20	–	23	–	–	–	–	–
mc_2^3	–	–	26	–	–	–	–	25	–	–	20	–	–	–	–	–	17	–	–	–
mc_2^4	–	24	–	–	14	–	15	–	–	–	–	–	–	28	–	–	–	–	–	–
mc_2^5	–	–	–	19	–	–	–	–	–	–	–	–	23	–	–	–	–	19	–	17
mc_3^1	–	19	–	–	–	–	–	–	18	–	–	25	–	–	–	–	14	–	–	–
mc_3^2	26	–	–	30	–	–	–	24	–	–	30	–	–	–	24	–	37	–	21	–
mc_4^1	–	–	–	–	–	22	–	–	–	32	–	–	–	–	–	–	–	31	–	–
mc_4^2	17	–	–	–	–	–	–	–	–	–	–	19	–	–	–	–	24	–	–	22
mc_4^3	–	25	–	27	–	–	–	17	–	–	–	–	25	–	–	29	–	–	30	–
mc_5^1	23	–	–	–	–	–	17	–	–	–	19	–	–	15	–	–	–	31	–	–
mc_5^2	–	–	19	–	23	–	–	–	–	28	–	–	–	–	–	–	15	–	–	24
mc_5^3	–	–	–	35	–	–	–	–	32	–	–	–	–	–	18	–	–	–	–	–
mc_5^4	17	–	–	–	–	27	24	–	–	–	–	–	–	16	–	23	–	–	28	–

NSGA-II, taking cost, operational capability and reliability as the considered objectives. To apply NSGA-II in present study, the set of feasible alternative machine configuration ($\mathbb{F}_o$) for all the operations is recorded. Each member of set of feasible machine configuration ($\mathbb{F}_o$) is characterised by two parameters, one is the machine type and the other is its configuration. To perform an operation o $(1 < o < O)$, total number of alternative machine is $\mathbb{F}_o$. The feasible alternative machine configurations for all the operations are recorded for the constraints handling by implementing the real encoded chromosomes.

6.1 Chromosome Encoding and Solution Mapping

The chromosome length is equal to the number of stages; an operation has to be performed on each stage according to the prescribed operation sequence. A set of feasible alternative machine configurations $\mathbb{F}_o$ for each operation is prior recorded to limit the number of feasible machine alternatives on each stage/station of the line. Now, at each stage, a feasible machine configuration is to be assigned which is mapped in the present study through the coded chromosomes schema with parameters shown in Table 2. The first production stage (Stage-1) gene value is taken as 0.3401, and operation 1 is to be performed (Table 3). The set of feasible alternative machine configurations for operation-1 at stage-1 ($\mathbb{F}_1$) can be read as $\{mc_2^1, mc_3^2, mc_4^2, mc_5^1$ and $mc_5^4\}$, as presented in Table 1.

Table 3 shows the gene value that has to be multiplied to the number of feasible alternative machine configurations ($\mathbb{F}_o$) which are available to accomplish the operation, as here in this case it is 5. The final value is rounded off to the higher integer. This value represents the order of the selected machine configuration alternative from the already recorded feasible alternatives. This technique guarantees that for an arbitrary gene value, a feasible machine configuration is allocated to each production stage at each time. The crossover and mutation will not alter the feasibility of the solution due to the mapping scheme carried out in the present work. The non-dominated solutions obtained after applying NSGA-II are quite large in number, so the further ranking of the non-dominated solutions is performed by applying the TOPSIS approach.

Table 2 NSGA-II parameters

Demand rate	Population size	P_c	P_m	η	η_m
80	80	0.8	0.2	10	20

Table 3 Non-dominated results after 120 generations

Chromosomes					Machine assigned					Fitness values		
S-1	S-2	S-3	S-4	S-5	S-1	S-2	S-3	S-4	S-5	Cost (F1)	Operational capability (F2)	Reliability (F3)
0.3819	0.4516	0.7973	0.4059	0.6772	2	2	4	3	3	28550	149	0.9308
0.6381	0.6796	0.7534	0.8319	0.2022	4	3	4	5	1	20130	61	0.9958
0.7278	0.3620	0.9951	0.6175	0.6886	4	2	4	4	3	23575	118	0.8914
0.8556	1.0000	0.8990	0.3778	0.3274	5	3	4	2	1	23965	70	0.9961
0.9572	0.3683	0.8155	0.5702	0.6869	5	2	4	3	3	27125	138	0.9335
0.7959	0.9018	0.9548	0.4050	0.5170	4	3	4	3	2	22475	93	0.9756
0.3401	0.3349	0.7784	0.8872	0.4374	2	2	4	5	2	25365	106	0.9725
0.6026	0.6525	0.7761	0.7026	0.3404	4	2	4	4	2	22175	102	0.9136
0.8794	0.9365	0.9913	0.1952	0.3947	5	3	4	1	2	24525	75	0.9951
0.2951	0.3086	0.8319	0.9495	0.3328	2	1	4	5	1	23890	74	0.9894
0.2634	0.6451	1.0000	0.4301	0.6212	2	2	4	3	2	27150	133	0.9540
0.9771	0.3987	0.8271	0.5877	0.5620	5	2	4	3	2	25725	122	0.9568
0.6367	0.6530	0.8082	0.8658	0.1925	4	2	4	5	1	21105	81	0.9767
0.3992	0.4368	0.9479	0.3276	0.6336	2	2	4	2	2	26925	106	0.9730
0.8134	0.6210	0.8677	1.0000	0.3519	5	2	4	5	2	23940	95	0.9753
0.9864	0.4004	0.7968	0.6022	0.5254	5	2	4	4	2	24450	111	0.9136
0.7032	0.6846	0.8150	0.8693	0.3548	4	3	4	5	2	20690	66	0.9945
0.2533	0.6816	0.9547	0.5018	0.5484	2	3	4	3	2	26175	113	0.9727
0.3561	0.7005	0.9376	0.1577	0.2764	2	3	4	1	1	25390	81	0.9935
0.7938	0.3917	0.2964	0.0265	0.3297	4	2	2	1	1	23215	74	0.9778
0.2728	0.5403	0.7816	0.7843	0.2702	2	2	4	4	1	25315	117	0.9121
0.6201	0.6477	0.7791	0.5418	0.2853	4	2	4	3	1	22890	108	0.9581

(continued)

Table 3 (continued)

Chromosomes					Machine assigned					Fitness values		
S-1	S-2	S-3	S-4	S-5	S-1	S-2	S-3	S-4	S-5	Cost (F1)	Operational capability (F2)	Reliability (F3)
0.6395	0.6395	0.8020	0.7109	0.2314	4	2	4	4	1	21615	97	0.9148
0.6938	0.3350	0.8699	0.5129	0.7597	4	2	4	3	3	24850	129	0.9336
0.3596	1.0000	0.8606	0.0866	0.4455	2	3	4	1	2	25950	86	0.9922
0.8100	0.6019	0.8708	0.9850	0.3312	5	2	4	5	1	23380	90	0.9766
0.6222	0.7211	0.8040	0.4084	0.2004	4	3	4	3	1	21915	88	0.9769
0.9479	0.3728	0.4718	0.5576	0.5121	5	2	2	3	2	26275	115	0.9572
0.2269	0.5136	0.9995	0.4395	0.2078	2	2	4	3	1	26590	128	0.9553
0.3530	0.3175	0.8308	0.9571	0.3345	2	1	4	5	2	24450	79	0.9881
0.8175	0.7733	0.8370	0.9808	0.2546	5	3	4	5	1	22405	70	0.9957
0.8025	0.3994	0.2720	0.0155	0.3028	5	2	2	1	1	25490	83	0.9777
0.6082	0.7448	0.8170	0.3751	0.2801	4	3	4	2	1	21690	61	0.9962
0.6115	0.7265	0.8418	0.7432	0.3700	4	3	4	4	2	21200	82	0.9315
0.3654	0.3338	0.8216	0.9256	0.3174	2	2	4	5	1	24805	101	0.9738
0.3377	0.6335	0.9640	0.1630	0.1704	2	2	4	1	1	26365	101	0.9745
0.6079	0.6673	0.8370	0.6369	0.2609	4	3	4	4	1	20640	77	0.9327

6.2 Optimisation

The initial random population is generated as a set of five genes; the value of each gene is random value which lies between 0 and 1. After initial population is created, real-coded chromosomes are decoded for objective function calculation. Each gene value of chromosome represents stages of flow line. At each stage, a set of feasible alternative machine configuration is assigned. Number of alternative machine configurations assigned for performing the operation at each stage are multiplied by their respective gene value and rounded off to higher integer, which represent the respective member from the set of feasible alternative machine configuration. After feasible machines are selected, performance parameters are computed. Non-dominated solutions are sorted based on the performance parameters, ranked in fronts, and tournament selection operator is applied. Mutation and crossover operators are applied which creates child population. After child population is created, it is again decoded, mapped to feasible solution. Now both populations, i.e. parent population and child population, are combined and non-dominated sorted and ranked to preserve best solution from both populations. The procedure is repeated until the stopping criteria are satisfied.

In the present study, the stopping criteria assumed are the number of generations until nearly identical solutions are achieved. In our case, it is observed that 120 generations are enough to meet these criteria in each of the cases considered.

Briefly, the step-by-step procedure adapted in the problem is described under the following points (Fig. 4):

a. The feasible solutions have been identified.
b. The non-dominated solutions have been determined by applying NSGA-II to the feasible set of solutions.
c. Non-dominated solutions are ranked using TOPSIS.
d. Weights for the beneficial and non-beneficial objective functions are obtained using SEWT.
e. The best optimal solutions are further chosen from the set of ranked non-dominated solutions.

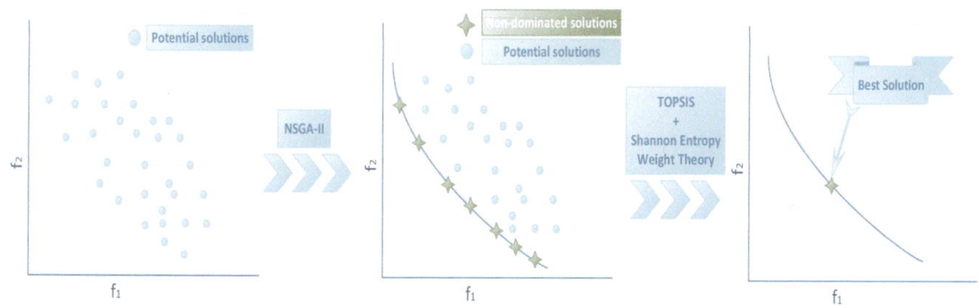

Fig. 4 NSGA II, TOPSIS, and Shannon entropy weight applied to feasible solutions

7 Results and Discussion

The proposed methodology has been implemented on a single operation sequence for optimal machine assignment in an SPL. One level of demand is considered, i.e. 80 parts per hour with variable mutation probability. The example of a part family processed on operation sequence $1 \rightarrow 11 \rightarrow 20 \rightarrow 4 \rightarrow 13$ having demand rate 80 parts per hour is considered for the illustration of developed approach. The first step is to identify and record set of feasible alternative machine configuration for each operation. After each stage is assigned feasible alternative machine configuration, NSGA-II is applied along with the decoding scheme as illustrated above. Non-dominated solutions obtained from NSGA-II are too large to be decided subjectively, i.e. which solution is better to which; therefore, further ranking is done by applying TOPSIS ranking and Shannon entropy theory to avoid any imprecision. The results obtained are shown in Tables 3 and 4.

Results show that the non-dominated solutions obtained are 37 in numbers (Table 3). Since these solutions are large in numbers, it becomes difficult to choose any one from among these solutions as all these solutions may have optimal value of any one of the objective functions. Hence, the non-dominated solutions are ranked using TOPSIS and the top five solutions are recorded which are presented in Table 4. The cost of machine which is a non-beneficial attribute is found to be minimum, while the operation capability which is a beneficial attribute is found to be minimum in the set-5 solutions. The operation capability is found to be maximum for the set-4 solutions while the reliability of the system is found to be minimum in the same set.

A similar problem (Goyal et al. 2012) considered performance measures, i.e. cost, operation capability and machine reconfigurability for the optimal machine assignment in a SPFL. The number of solutions found using NSGA-II in all the cases was observed varying from 22 to 48 (Goyal et al. 2012), while in another work (Ashraf 2016) it was ranged between 30 and 48 when cost, operation capability and reliability of machines were taken as objective functions.

The solutions obtained in other works [i.e. in (Goyal et al. 2012) and in (Ashraf 2016)] were ranked using TOPSIS technique. Other problems (i.e. Abderrahmane et al. 2012 and Bensmaine 2014) related to the candidate selection of machines

Table 4 Best ranked non-dominated results after 120 generations

Set No.	Solution (machine assigned)					Fitness value $[x_{ij}]$		
	S-1	S-2	S-3	S-4	S-5	Cost (F1)	Operational capability (F2)	Reliability (F3)
1	4	2	4	3	1	22890	108	0.9581
2	4	3	4	3	1	21915	88	0.9769
3	2	2	4	5	2	25365	106	0.9725
4	4	2	4	3	2	23450	113	0.9569
5	4	2	2	5	1	21655	74	0.9771

involving minimisation of total cost were solved by applying an adapted version of NSGA-II which revealed a list of selected machines for each non-dominated solution with their corresponding total costs. A similar kind of problem (Youssef and ElMaraghy 2007) performed the selection of near-optimal configurations under several demand scenarios for the continuous optimisation of capital cost, reported the least amount of reconfiguration effort without sacrificing the system performance objectives.

8 Conclusion

Based on the formulation of the problem and its subsequent analysis, the following conclusions are drawn:

(1) The performance indices selected in this work are from the perspective of economy (cost), system performance (reliability) and responsiveness (operational capability) which are directly related to throughput (productivity).
(2) The performance indices used in this study would help the management to enhance the decision-making diversity while dealing with reconfigurable manufacturing system by providing an apparent trade-off between economy, responsiveness and performance.
(3) Since conflicting objectives are involved in the problem, therefore, NSGA-II was first applied to identify Pareto frontiers generating large number of alternatives available for management. The real-coded chromosomes are used for dealing with the sparse population of viable solutions along with enabling the crossover and mutation.
(4) The non-dominated results obtained by NSGA-II are in the range of 30–48. It is quite difficult for the selector to choose best alternative. Hence, non-dominated results were further analysed using TOPSIS to rank the solutions based on their distance from the best solution and worst solution.
(5) Shannon entropy concept is used for computing the weights for the attributes, which is based on the information theory. Considering the ranking of the solutions, the decision-maker may select a proper candidate amongst the top-ranking solutions for corroborating the objectives set by the organisation along with the current market state. Therefore, one can decide whether to select a rapid but relatively expensive production system, or the opposite, i.e. a low-cost system with a normal production output.
(6) Rather providing solely one solution to the decision-maker, this work provides a solution set that gives more flexibility for selection.
(7) It is very much apparent that the application of reliability and cost parameters along with operational capability will lead to a reduction in reconfiguration efforts as well as selection of machines that will have lesser failure probability and higher productivity while planning multiple period configurations in the RMS.

Acknowledgements The contribution from the Council of scientific and Industrial research (CSIR), Human resource development group (HRDG), India under file no. 09/112 (0552)2K17 EMP-I in support of this research is greatly acknowledged.

References

Abderrahmane, B., M. Dahane, and L. Benyoucef. 2012. A non-dominated sorting genetic algorithm based approach for optimal machines selection in reconfigurable manufacturing environment. *Computer and Industrial Engineering* 66: 519–524.

Al-Zaher, A., and W.H. ElMaraghy. 2013. CIRP annals—manufacturing technology design of reconfigurable automotive framing system. *CIRP Annals-Manufacturing Technology* 62: 491–494.

Ashraf, M., M.A. Khan, F. Hasan, M.I. Siddiqui, Q. Murtaza, and S.N. Ali. 2016. Optimum allocation of machines with multiple objectives using NSGA-II. In: *6th international & 27th all india manufacturing technology, design and research conference* (*AIMTDR-2016*), 1723–1728. Pune.

Ashraf, M., and F. Hasan. 2015. Product family formation based on multiple product similarities for a reconfigurable manufacturing system. *International Journal of Modelling in Operations Management* 5: 247–265.

Bensmaine, A., M. Dahane, and L. Benyoucef. 2014. Design of reconfigurable manufacturing systems: Optimal machines selection using nondominated sorting genetic algorithm (NSGA-II). In: *41st international conference on computers & industrial engineering, 2014,* 110–115.

Coello, C.A.C., G.B. Lamont, D.A., Van Veldhuizen. 2007. *Evolutionary algorithms for solving multi-objective problems*. Springer.

Dashchenko, A.I. 2006. *Reconfigurable manufacturing systems and transformable factories*. Moscow: Springer.

Deb, K., S. Agrawal, A. Pratap, T. Meyarivan. 2000. A fast elitist non-dominated sorting genetic algorithm for multi-objective optimization: NSGA-II. Kanpur.

Deb, K. 2011. Multi-objective optimization using evolutionary algorithms: An introduction. In *Multi-objective evolutionary optimisation for product design and manufacturing*, ed. L. Wang, A. Ng, and K. Deb, 1–24. Kanpur: Springer.

Deb, K., A. Pratap, S. Agarwal, and T. Meyarivan. 2002. A fast and elitist multiobjective genetic algorithm: NSGA-II. *IEEE Transactions on Evolutionary Computation* 6: 182–197.

Dou, J., X. Dai, and Z. Meng. 2009. Graph theory-based approach to optimize single-product flow-line configurations of RMS. *International Journal of Advanced Manufacturing Technology* 41: 916–931.

El Maraghy, H.A. 2006. Flexible and reconfigurable manufacturing systems paradigms. *International Journal of Flexible Manufacturing Systems* 17: 261–276.

Elmaraghy, H.A. 2011. Enabling manufacturing competitiveness and economic sustainability. In *4th international conference on changeable, agile, reconfigurable and virtual production*, ed. H.A. Elmaraghy, 2–9. Montreal: Springer.

Fonseca, C.M., and P.J. Fleming. 1995. An overview of evolutionary algorithms in multiobjective optimization. *Evolutionary Computation* 3: 1–16.

Galan, R., J. Racero, I. Eguia, and J.M. Garcia. 2007. A systematic approach for product families formation in reconfigurable manufacturing systems. *Robotics and Computer-Integrated Manufacturing* 23: 489–502.

Gola, A., and A. Swic. 2012. Directions of manufacturing systems evolution from the flexibility level point of view. In: *Innovations in management and production engineering*, 226–238. Oficyna Wydawnicza, Opole.

Goyal, K.K., P.K. Jain, and M. Jain. 2012. Optimal configuration selection for reconfigurable manufacturing system using NSGA II and TOPSIS. *International Journal of Production Research* 50: 4175–4191.

Haddou-Benderbal, H., M. Dahane, and L. Benyoucef. 2016. Hybrid Heuristic to minimize machine's unavailability impact on reconfigurable manufacturing system using reconfigurable process plan. In *8th IFAC conference on manufacturing Modelling, management and control MIM 2016*—Troyes, France, 28–30 June 2016. 1626–1631. Elsevier B.V.

Hasan, F., P.K. Jain, and D. Kumar. 2014a. Performance issues in reconfigurable manufacturing system. In: *DAAAM international scientific book*, 295–310.

Hasan, F., and P.K. Jain. 2015. Genetic Modelling for selecting optimal machine configurations in reconfigurable manufacturing system. *Applied Mechanics and Materials* 789–790: 1229–1239.

Hasan, F., S. Jain, and S.N. Ali. 2014b. Configuration selection for optimal throughput from a reconfigurable product line using genetic algorithm. *Som* 2014: 669–674.

Hasan, F., P.K. Jain, and D. Kumar. 2014c. Optimum configuration selection in reconfigurable manufacturing system involving multiple part families. *Opsearch.* 51: 297–311.

Hwang, C.-L., and K. Yoon. 2011. *Multiple attribute decision making*. London: CRC Press.

Koren, Y. 1983. *Computer control of manufacturing systems*. New york: McGraw-Hill Inc.

Koren, Y., and A.G. Ulsoy. 1997. *Reconfigurable manufacturing systems.* Ann Arbor (1997).

Koren, Y., and S. Kota. 1997. *Reconfigurable machine tool.* https://www.google.com/patents/US5943750.

Koren, Y. 1999. *Reconfigurable machining systems vision with examples*. Ann Arbor (1999).

Koren, Y. 2006. General RMS characteristics. Comparison with dedicated and flexible systems. In *Reconfigurable manufacturing systems and transformable factories,* ed. Dashchenko, A.I., 27–45. Berlin, Heidelberg: Springer.

Koren, Y. 2013. The rapid responsiveness of RMS. *International Journal of Production Research* 51: 6817–6827.

Koren, Y., and M. Shpitalni. 2010. Design of reconfigurable manufacturing systems. *Journal of Manufacturing Systems* 29: 130–141.

Landers, R.G., B.K. Min, and Y. Koren. 2001. Reconfigurable machine tools. *CIRP Annals-Manufacturing Technology* 50: 269–274.

Maniraj, M., V. Pakkirisamy, and R. Jeyapaul. 2015. An ant colony optimization-based approach for a single-product flow-line reconfigurable manufacturing systems. *Proceedings of the Institution of Mechanical Engineers, Part B: Journal of Engineering Manufacture,* 1–8.

Matt, D.T., and E. Rauch. 2013. Design of a network of scalable modular manufacturing systems to support geographically distributed production of mass customized goods. *Procedia CIRP* 12: 438–443.

Mehrabi, M.G., A.G. Ulsoy, and Y. Koren. 2000. Reconfigurable manufacturing systems: Key to future manufacturing. *Journal of Intelligent Manufacturing* 11: 403–419.

Mitrofanov, S.P. 1966. Scientific Principles of Group Technology.

Moon, Y.-M., and S. Kota. 2002. Design of reconfigurable machine tools. *Journal of Manufacturing Science and Engineering* 124: 480–483.

Opitz, H., W. Eversheim, and H.P. Wiendahl. 1969. Workpiece classification and its industrial application. *International Journal of Machine Tool Design and Research* 9: 39–50.

Pattanaik, L.N., P.K. Jain, and N.K. Mehta. 2007. Cell formation in the presence of reconfigurable machines. *The International Journal of Advanced Manufacturing Technology* 34: 335–345.

Reisman, A., A. Kumar, J. Motwani, and C.H. Cheng. 1997. Cellular manufacturing: A statistical review of the literature (1965–1995). *Operations Research* 45: 508–520.

Shannon, C.E. 1948. A mathematical theory of communication. *Bell System Technical Journal* 27: 379–423.

Stecke, K.E., and J.J. Solberg. 1981. Loading and control policies for a flexible manufacturing system. *The International Journal of Production Research* 19: 481–490.

Xiaobo, Z., J. Wang, and Z. Luo. 2000. A stochastic model of a reconfigurable manufacturing system Part 2: Optimal configurations. *International Journal of Production Research* 38: 2829–2842.

Yin, Y., and K. Yasuda. 2006. Similarity coefficient methods applied to the cell formation problem: A taxonomy and review. *International Journal of Production Economics* 101: 329–352.

Youssef, A.M.A. 2006. *Optimal configuration selection for reconfigurable manufacturing systems*. http://link.springer.com/10.1007/s10696-007-9020-x%5Cn. https://www.infona.pl//resource/bwmeta1.element.springer-92540f5e-f764-3369-b248-7dac20aab687

Youssef, A.M.A., and H.A. ElMaraghy. 2007. Optimal configuration selection for reconfigurable manufacturing systems. *International Journal of Flexible Manufacturing Systems* 19: 67–106.

Zeleny, M., and J.L. Cochrane. 1973. *Multiple criteria decision making*. University of South Carolina Press.

MIX
Papier aus verantwortungsvollen Quellen
Paper from responsible sources
FSC® C105338

If you have any concerns about our products,
you can contact us on
ProductSafety@springernature.com

In case Publisher is established outside the EU,
the EU authorized representative is:
Springer Nature Customer Service Center GmbH
Europaplatz 3, 69115 Heidelberg, Germany

Printed by Libri Plureos GmbH
in Hamburg, Germany